Symbol	Meaning
j	Sometimes used to represent subscripts in a double summation: $\sum_{j=1}^{m}\sum_{i=1}^{n} X_{ij} = (X_{11}+X_{21}+\cdots+X_{n1}) + (X_{12}+X_{22}+\cdots+X_{n2}) + (X_{1m}+X_{2m}+\cdots+X_{nm})$ p.465
K	Test statistic for Kruskal-Wallis one-factor analysis-of-variance test p.536 (14-8)
M	Population median p.43
MSA	Mean square (between columns); used to calculate the F statistic in a two-factor analysis of variance when A represents one of the factors about which inferences are to be made p.485
MSB	Mean square (between rows); used to calculate the F statistic in a two-factor analysis of variance when: (1) B represents the blocking variable p.483 (13-18) (2) B represents one of the factors about which inferences are to be made p.485
$MSCOL$	Mean square (between columns); used to calculate the F statistic for a latin-square design in a three-factor analysis of variance p.492
MSE	Error mean square; used to calculate the F statistic: (1) (within columns) for a one-factor analysis of variance p.468 (13-8) (2) (residual) for a two-factor or a three-factor analysis of variance p.483 (13-19)
$MSROW$	Mean square (between rows); used to calculate the F statistic for a latin-square design in a three-factor analysis of variance p.492
MST	Mean square; used to calculate the F statistic: (1) (between columns) for the treatments in a one-factor analysis of variance p.468 (13-7) (2) (between letters) in a three-factor analysis of variance using a latin-square design p.492
m	(1) Sample median p.43 (2) Number of variables used in multiple regression and correlation analysis p.393
μ (mu)	Arithmetic mean of a population p.40
μ_0	Value of the population mean assumed under the null hypothesis p.300
$\mu_1, \mu_2, \mu_3, \ldots$	Means of populations; used in analysis of variance p.463
μ_A, μ_B	Means of populations A and B, which are compared using two samples p.404
$\mu_{Y \cdot X}$	Conditional mean for values of the dependent variable Y for a specified value of the dependent variable X; used in regression analysis, where $\mu_{Y \cdot X}$ represents the height of the true regression line p.351 (9-14)
$\mu_{Y \cdot 12}$	Conditional mean for values of the dependent variable Y for a specified value of the independent variables X_1 and X_2; used in multiple regression analysis, where $\mu_{Y \cdot 12}$ represents the height of the true regression plane p.389 (10-6)
N	Number of observations in a population (that is, the population size) p.54
n	(1) Number of observations in a sample (that is, the sample size) p.40 (2) Number of trials in a Bernoulli process p.171
$n!$	n factorial, or the product $n \times (n-1) \times (n-2) \times \cdots \times 2 \times 1$ p.135 (4-11)
n_A, n_A	Sample sizes for two-sample tests comparing the parameters of two populations A and B p.404
n_a, n_b	Number of observations of type a and type b obtained in a sample; applicable to the number-of-runs test p.525
n_j	Size of the jth sample group; used in one-factor analysis of variance with Kruskal-Wallis test p.536
P	(1) Proportion of sample observations having a particular characteristic p.60 (2-11) (2) Proportion of successes from a Bernoulli process p.174
$P[A]$	Probability of event A p.98
$P[A\|B]$	Conditional probability of event A given event B p.115 (4-7)
$P[A \text{ and } B]$	Probability of the intersection of events A and B; event occurs only when both A and B occur (also called the joint probability of A and B) p.107 (4-8)
$P[A \text{ or } B]$	Probability of the union of A and B; occurs when either A or B occurs, or when both A and B occur p.107 (4-3)
$P[X=x]$	Probability that the random variable X assumes one of several particular values p.157
$P[X \leq x]$	Cumulative probability that X assumes any value less than or equal to a particular value x p.177
P^*	Critical value for P, used in one-sided hypothesis-testing procedure p.295 (8-1)
P_A, P_B	Proportions of observations having a particular characteristic for samples from populations A and B p.420
P_C	Combined sample proportion; used in two-sample inferences for comparing proportions of two populations p.421 (11-21)
P_r^n	Number of permutations of r items taken from a collection of size n when different orders of selection are counted p.136 (4-12)
π (pi)	(1) Proportion of a population having a particular characteristic p.60 (2) Probability of a trial success in a Bernoulli process p.171
π_0, π_1	Values of the population proportion assumed under the null and the alternative hypotheses p.288

STATISTICS Meaning and Method

Statistics

Meaning and Method

LAWRENCE L. LAPIN

San José State University

HARCOURT BRACE JOVANOVICH, INC.

New York *Chicago* *San Francisco* *Atlanta*

TO MY PARENTS, NAOMI AND LEO LAPIN

ISBN: 0-15-583772-9

Library of Congress Catalog Card Number: 74-24951

Printed in the United States of America

Illustrations by Vantage Art, Inc.

Photo credits appear on page 570.

Preface

In writing this introductory textbook, my principal aim was to design a vivid presentation of statistics—to make it more interesting, more concrete, and easier to grasp. Students today often approach statistics armed with a prejudice that the subject is dull, that it does not touch their lives, that it is more complex than it ought to be.

To show that in fact the reverse is true—that statistics *is* a lively and useful study—I have tried to arrive at generalizations by way of concrete examples which hold interest beyond their value as illustrations of technical mathematical concepts. Thus, I suggest the importance of frequency distributions by applying them to cryptanalysis; I demonstrate hypothesis testing by considering the theory that explains why most people are right-handed; to emphasize the fact that sampling is only one potential source of error, I cite some government blunders in taking the Decennial Census. I have derived many of the exercises and examples from such contemporary issues as conservation, health, and the environment. In an effort to suggest the wide and useful application of statistical concepts, I consider a variety of areas: education; the social, physical, biological, and health sciences; the professions of medicine, law, engineering, and administration. To achieve this goal of tying statistics to the everyday world we know, I have devoted far more space than is customary to such examples.

A course in high-school algebra is the only background required. And although this book should prove more accessible than most, the relevant nature of its presentation has not been achieved at the cost of avoiding reputedly difficult material. More attention has been given to such topics as probability and hypothesis testing. These and other more difficult topics receive generous explanation, often from several viewpoints. The reader is encouraged to rely more on intuition than on rote memory; less than the customary emphasis is placed upon the mechanical and computational aspects of statistics. The computer's role in statistical analysis and in managing the more onerous calculations is highlighted throughout the book.

All the material presented here has been thoroughly tested and thereby improved in many ways. The exercise problems have been solved by a large cross section of students and have been modified and revised to make them more effective. There is a rich variety of problems, increasing in difficulty within each group of exercises. Most of these problems appear at the end of sections within chapters, permitting the student to easily relate the questions to the concepts just covered. Each chapter ends with an additional set of problems for purposes of review, so that the student may gain experience in determining which procedures and concepts to apply. This allows the instructor considerable flexibility in making assignments and in emphasizing special areas of application.

The chapters have been arranged to make it easier for the instructor to design a course to fit individual needs. The core material is contained in the first seven chapters. The remaining chapters may be used in any number of ways. A short course, for example, might cover Chapter 8 (hypothesis testing) and Chapter 9 (regression and correlation analysis), followed perhaps by Chapter 10 (multiple regression and correlation); or, two-sample procedures (Chapter 11) might be considered. Topics of special interest include chi-square applications (Chapter 12) and analysis of variance (Chapter 13). Nonparametric statistics is introduced in Chapter 14. In a longer course, the instructor might include most of the later chapters, skipping one or two, or perhaps might cover the entire book. Since problem material is readily identified with separate sections, time devoted to the various chapters can be easily shortened by deleting certain areas. For example, inferences regarding variances (discussed in Chapter 12 and Chapter 13) may be deleted with no loss of continuity; for this reason, the chi-square and the F distributions are introduced first in their more important applications: testing for independence and the analysis of variance. The modular arrangement also makes it convenient to emphasize certain concepts or to play them down. It is possible, for example, to give a lesser or a greater role to small-sample statistics, since the normal and the Student t distributions are employed in different sections within the later chapters. Similarly, two-sample inferences can be covered in terms of confidence intervals only by skipping the portions of Chapter 11 concerned with hypothesis testing. Or, coverage of probability can be expanded, perhaps by including the discussion of Bayes' Theorem (an optional section in Chapter 4).

A glossary of statistical symbols is provided on the endpapers for easy reference. Abbreviated answers to all even-numbered exercises are included at the back of the book. Complete solutions to all exercises are available in the Instructor's Manual, which also contains 100 problems suitable for test use and 325 examination questions and their solutions.

I am greatly indebted to the many people who have assisted me in preparing this book. Special thanks go to my colleagues whose comments were invaluable in setting the tone: C. Randall Byers, University of Idaho; Kenneth Eberhard, Chabot College; Kenneth Masters, Pennsylvania State University; Sue Solomon, University of Maryland. I am deeply grateful to my students who over many semesters of debugging the manuscript helped identify the problems a reader might face. There is not enough space to list everyone, but special mention goes to Les Morgan, who checked the manuscript for errors; to Kirby Root, who wrote the computer programs for curve plotting and the square root table; and to Ed Reed, who wrote the computer program for the table of cumulative probabilities for the binomial distribution.

Lawrence L. Lapin

Contents

THE STATISTICAL SAMPLING STUDY 66

PROBABILITY 92

PROBABILITY DISTRIBUTIONS AND EXPECTED VALUE 152

THE NORMAL DISTRIBUTION 202

STATISTICAL ESTIMATION 244

HYPOTHESIS TESTING 284

REGRESSION AND CORRELATION ANALYSIS 328

MULTIPLE REGRESSION AND CORRELATION 380

INFERENCES USING TWO SAMPLES 402

CHI-SQUARE APPLICATIONS 434

ANALYSIS OF VARIANCE AND RELATED TOPICS 460

NONPARAMETRIC STATISTICS 500

STATISTICS **Meaning and Method**

1 Introduction

The true foundation of theology is to ascertain the character of God. It is by the aid of Statistics that law in the social sphere can be ascertained and codified, and certain aspects of the character of God thereby revealed. The study of statistics is thus a religious service.

Florence Nightingale

There is perhaps no other subject quite like statistics. It is used by everybody, from some of the remotest people, such as the Bushmen of Africa's Kalahari Desert, to space scientists planning grand tours of the planets. Surprisingly, the use to which statistics is universally applied is not what most people envision. The word statistics commonly brings to mind masses of numbers, graphs, and tables. These do play a role in statistics, but a limited one. The Kalahari Bushmen cannot count very high, know little mathematics, draw no graphs, and assemble no tables, yet they do use statistics.

1-1 THE MEANING OF STATISTICS

One common thread linking scientist and Bushman is that both *must make inferences in the face of uncertainty*. Such inferences are generalizations based upon incomplete or imperfect information that lead to a particular decision, estimate, or prediction. For example, a successful Bushman locates water in a manner puzzling to some geologists. He sucks through a reed stuck in the sand in what his experience tells him is a likely spot, as if he were drinking a thick milk-shake from a straw. Sometimes no water is found, but often enough he does find a "drink hole." His very survival depends on his skill at locating drink holes—on using his experience to cope with uncertainty.

Scientists also depend on experience for determining an optimal space-vehicle configuration—one that can withstand the rigors of a far more hostile environment than that of earth and its atmosphere and survive enough years to fly past Jupiter and eventually to even more distant planets. The communications system must not fail, or the mission itself will be a failure. The system must be super-reliable, designed so that its chances for survival are high. Millions of man-hours of experience can be drawn upon to facilitate design decisions. Even so, there are no guarantees that the chosen design will work as intended. Again, inferences must be made in the face of uncertainty.

Both the Bushman's and the space scientist's mechanisms for making choices utilize limited experience, and in this respect they are not substantially different. The remaining element making them statistical is that *both use numerical evidence*. To evaluate communications equipment reliability, components must be tested under stress to determine the number of hours they can function before failing. The scientist's evidence is obviously numerical. But what about the Bushman's? He knows that one set of conditions favors the presence of water more than another. Every wet hole reinforces his learning of factors positive to water, and each dry hole strengthens his awareness of the negative factors. Thus the Bushman's numerical evidence is the *frequency* of successful drink holes found under various prevailing conditions. Even though presence of water is a qualitative factor, frequencies of occurrence are themselves numerical.

The goal of this book is to describe some of the more useful methods and procedures that can be applied to numerical evidence to facilitate decision-making in the face of uncertainty. So that the reader may achieve a greater appreciation and understanding of how and why these work, certain basic principles will be explained within a theoretical framework.

A concise statement summarizing the subject of this book is the

DEFINITION ***Statistics*** is a body of methods and theory applied to numerical evidence in making inferences in the face of uncertainty.

This definition treats statistics as a separate area of academic endeavor, just like physics, history, or botany. As a discipline, statistics has advanced rapidly during the twentieth century to become recognized as a branch of mathematics. Partly due to the relative newness of statistics as a subject of study, the word *statistics* has acquired several meanings.

In its earlier and still most common usage, statistics means a collection of numerical facts or data. In sports, for example, batting averages, rushing yardage, and games won or lost are all statistics. Closing prices from the New York Stock Exchange, figures showing sources of income and expenditures for the federal budget, and distances between major cities are also statistics. Your reported income on last year's tax return is a statistic. Examples of this usage are the *vital statistics* for figures on births, deaths, marriages, and divorces, and the titles of publications listing population and economic data, such as the *Statistical Abstract of the United States*. The main distinction between our definition and this usage of the word is its form, plural versus singular. Statistics (singular) *is* a subject of study, while statistics (plural) *are* numerical facts.

1-2 THE ROLE OF STATISTICS

Our definition of statistics indicates that it can be a powerful tool for analyzing numerical data. This is what makes it so crucial in the sciences, engineering, education, and business. In the physical sciences, statistics can determine whether or not experimental results should be incorporated into the general body of knowledge. In the biological and medical sciences, it guides researchers in ferreting out significant experimental findings worthy of further attention. Thus, the physician uses it to assess the effectiveness of a particular treatment; and statistics also helps the pharmacologist to evaluate a proposed drug. In some fields, such as genetics, statistics is thoroughly integrated.

The role of statistics in the social sciences—especially psychology, sociology, and economics—is a critical one. Here the behavior of individuals and organizations must often be monitored through numerical data in order to lend support and credence to models and theories which cannot stand on rationale alone. The professional fields of engineering, education, and business all employ statistics in planning, in establishing policies, and in setting standards. The school superintendent may use it to mold his curriculum; the civil engineer can use statistics to determine the properties of various materials; the company president may employ it to forecast sales, design products, and produce goods more efficiently.

1-3 DESCRIPTIVE AND INFERENTIAL STATISTICS

The emphasis upon the decision-making aspects of statistics is a recent one. In its early years, the study of statistics largely consisted of methodology for summarizing or describing numerical data. Any aspects facilitating choice were secondary in importance to the then essentially reportorial nature of the subject. This area of study has become known as *descriptive statistics* because it is concerned largely with summary calculations and graphical displays. These methods are in contrast with the modern approach, where generalizations are made about the whole, called the *population*, by investigating a portion, referred to as the *sample*. Thus, the average income of *all* familes in the United States can be estimated from figures obtained from a *few hundred* families. Such a prediction or estimate is an example of an inference. The study of how inferences are made from numerical data is thus called *inferential statistics*.

Our formal definition for the field of statistics applies to both the descriptive and inferential forms. This book emphasizes inferential statistics because of the dynamic role it can play in decision-making. But descriptive statistics remains an important element, and our focus is placed here in Chapters 2 and 3, where we begin by describing the population. Later, we will study how generalizations can be made of the population by means of the sample.

Inferential statistics acknowledges the potential for error that exists in making generalizations from a sample. The seasonal controversy over decisions made by television networks to drop certain programs illustrates some of the principles and problems involved. The information used in these decisions is obtained from a sample of a few hundred viewing families believed to be a representative microcosm of the viewing public at large. Only those shows indicated by the sample to be most popular are allowed to continue. Although articulate segments of the public complain that their tastes are not represented, the agency responsible for the collection of sample data has demonstrated that its audience is selected in accordance with accepted scientific sampling procedures.

The sampling agency knows that its errors are likely to be insignificant because it has selected its sample *randomly*. Using probability theory, the likelihood of large sampling errors is known to be small. Probability theory measures the chances that an untypical sample will be selected from a population whose characteristcs are known. Inferential statistics is based on probability theory, extending its concepts to the measurement of the chance of erroneous generalizations, even when the characteristics of the population are unknown or uncertain.

Because of the important role it plays in statistical inference, we shall discuss probability in Chapters 4 and 5, immediately following descriptive statistics for populations having *known* properties, in order to set the stage for developing statistical inferences about populations whose characteristics are *unknown*.

1-4 DEDUCTIVE AND INDUCTIVE STATISTICS

There is another dichotomy encountered in statistics. Deduction ascribes properties to the specific, starting with the general. For example, probability tells us that if a person is chosen by lottery from a group containing nine men and one woman, then the odds against picking the woman are 9 to 1. We deduce that in about 90 percent of such samples the person will be male. The use of probability to determine the chances of getting a particular kind of sample result is known as *deductive statistics*.

Chapters 4 through 6 present deductive techniques where we know everything in advance regarding the population, and there we are concerned with studying the characteristics of the possible samples that may arise from known populations. In much of the remainder of the book, we reverse directions and use *inductive statistics*. Induction involves drawing general conclusions from the specific. In statistics this means that inferences about populations are drawn from samples. The sample is all that is known, and uncertain characteristics of the population must be gleaned from the incomplete information available.

Inductive and deductive statistics are completely complementary. We must study how samples are generated before we can turn the tables to generalize from the sample. Deductive statistics is not wholly satisfactory in helping us perform much statistical analysis. The pragmatic aspects of statistics are largely inductive. But we can only understand inductive statistics by first studying deductive statistics.

1-5 STATISTICAL ERROR

Statistics is characterized as partly art and partly science. It is an art because it relies so heavily upon experience and judgment in choosing from the vast panoply of procedures available for analyzing sample information. But it embodies, to various degrees, all elements of the scientific method. The most noteworthy of these is its focus on *error*. Because statistics concerns uncertainty, there is always a chance of making erroneous inferences. Statistical procedures are available for both controlling and measuring the risks of erroneous conclusions. To illustrate this point, we may again consider television viewer sampling.

Those who disagree with the network choices also base their arguments on a sample—the opinions of their own friends and acquaintances. The problem here is that such a sample is biased in favor of persons with similar tastes, education, and social experience. The validity of the sample agency's claims rests upon random selection from the public at large, which allows everyone an equal chance of representation. As we shall see in Chapter 7, the agency chooses a sample large enough to keep the chance of error small.

Otherwise the survey agency cannot be legitimately criticized, unless it can be shown that its samples were not random and were therefore biased in favor of specific groups. Even with large samples, bias could be introduced by statistical errors due to improper procedures. Some of the more serious sources of bias will be discussed in Chapter 3.

2 Frequency Distributions and Summary Measures

Measurement does not necessarily mean progress. Failing the possibility of measuring that which you desire, the lust for measurement may, for example, merely result in your measuring something else—and perhaps forgetting the difference—or in ignoring some things because they cannot be measured. . . .

George Udny Yule

We have said that descriptive statistics involves the arrangement and display of observed data, which are then summarized and analyzed by means of inferential statistics in order to reach some decision. The manner in which the data are described and the procedures followed for their analysis depend upon the decision-making goal and the nature of the data.

2-1 THE POPULATION AND THE SAMPLE

An observation may be a physical measurement (weight or height), an answer to a question (yes or no), or a classification (defective or nondefective). Observations relevant to a particular decision constitute a population. Stated more formally, we have the following

DEFINITION A statistical ***population*** is the collection of all possible observations of a specified characteristic of interest.

Note that a statistical population exists whether all, some, or none of the possible observations are actually made. It may be real (such as the height of

30-year-old males in Schenectady in June 1963) or hypothetical (such as the longevity of laboratory rats fed a special diet that is not yet widely used). Because it consists of all possible observations, a population is often referred to synonymously as a *universe*.

Populations and Samples

In contrast to the statistical population, which consists of all possible observations, the *sample* contains only some of the observations. Whether a collection of observations is a population or just a sample depends upon the purpose of the analysis. For example, a medical association may wish to compare physicians' fees to the costs of other kinds of hospital services in order to absolve doctors from total blame for the skyrocketing hospitalization costs experienced in recent years. Here, one population (physicians' fees per hospital stay of all persons treated) would be compared to another population (the costs per patient for ancillary medical and housekeeping services). Depending upon the objectives of the medical association, the populations would comprise fees and costs for the entire nation, for a state, or for a county.

If the population of ancillary costs covered all hospitalized patients released during a particular calendar year, then the charges for all patients released on *a single day* would be a sample. The costs for patients at *a single hospital* during the entire year would also be a sample of the same population.

Again, if the medical association is concerned only with fees and charges for patients discharged on March 1, those data would constitute populations rather than samples. Similarly, if fees at different hospitals were to be compared, then the costs at one hospital would be a separate population.

Elementary Units

A statistical population consists of *observations* of some characteristic of interest associated with the individuals concerned, *not* the individual items or persons themselves. A company may need to know the ages of its employees in order to analyze proposed changes in its retirement program. To find these, the employee records would be searched and the ages determined from dates of birth. Each number determined constitutes an observation of the age characteristic. The entire collection of numbers so obtained is the population; the employees themselves will be referred to as the *elementary units of the population*.

A limitless variety of different populations may be obtained from the same elementary units, depending upon the characteristic of interest. This principle is illustrated in Figure 2-1. Thus an employer could have different populations for political affiliation, sex, marital status, years of education, height, weight, eye color, job classification, and years on present job, where each population is made up of observed characteristics of the same elementary units—the employees.

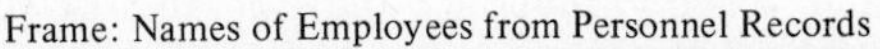

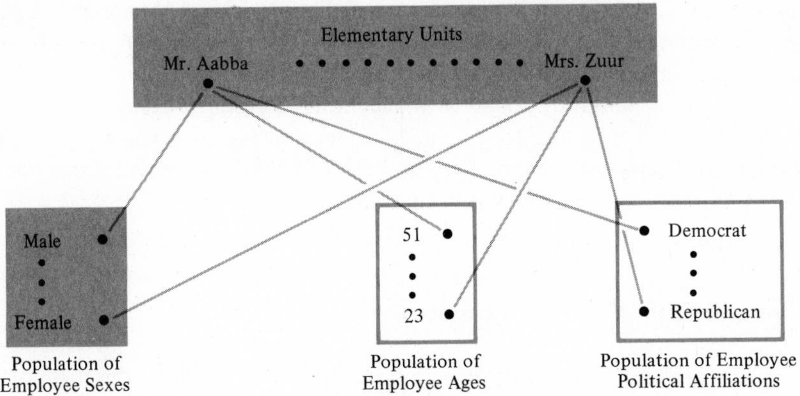

FIGURE 2-1 Illustration of how several different populations may be obtained from the same elementary units.

A population may be *multidimensional*, in which case two or more characteristics are observed for each elementary unit. For example, if we wished to compare the effect of education upon income, observations of both factors would be required for each elementary unit of the population.

The Frame

How do we define the elementary units in a population? The employer can define his employees as those persons on the payroll, so that the names of all employees may be obtained from pay records. Such a listing of the elementary units is called a *frame*.

The frame is very important in statistical studies because it serves partly to define the population. Suppose, for instance, that a sample is taken of voter preference toward candidates in order to predict the outcome of an upcoming election. The sample should represent the population of votes cast for a specific office. The population of interest is referred to as a *target population*. In selecting the sample of voters, the only frame available to the opinion surveyor is the roll of registered voters, and the elementary units listed in this frame may differ from those of the target population, because the names of some people who will not vote are included. The votes to be cast by these persons constitute the *working population*, because its frame is the only one currently available. The sample must be drawn from this population.

The potential exists for erroneous sample results, because the candidate preferences may be obtained from elementary units not in the target population. In such a case, it may be wise to include in the sample only those persons who are likely to vote. This illustration highlights one difficulty in using a sample to make an inference about a target population. Great care must be exercised in selecting a frame that closely matches that of the target population. The telephone book serves as an inadequate frame for a working population of voter attitudes, because it does not allow certain persons to be presented in the statistical study. Some persons are not listed, because they have recently moved into the community or because they do not own telephones. A danger exists that a sample of persons chosen from the telephone book will be biased toward a particular opinion not reflective of the target population. In Chapter 3 we will discuss many of the other pitfalls associated with sampling.

Qualitative and Quantitative Populations

There are two basic kinds of populations, distinguished by the form of the characteristic of interest. When the characteristic can be expressed numerically, as with height, weight, cost, or income, then the population is *quantitative*. When the characteristic is non-numerical, such as sex, marital status, occupation, or college major, the population is *qualitative*. As we proceed through our presentation of statistics, these two kinds of populations will be discussed separately

TABLE 2-1 Illustration of Possible Observations for Various Quantitative and Qualitative Populations

Quantitative Population Observations

Elementary Unit	*Characteristic of Interest*	*Unit of Measurement*	*Possible Variate*
person	age	years	21.3 yrs
microcircuit	defective solder joints	number	5
tire	remaining tread	inches	$\frac{1}{2}$ in.
account balance	amount	dollars	$5,233.46
employer	female employees	percent	17%
common stock	earnings per share	dollars	$3.49
keypuncher	errors	proportion	.02
light bulb	lifetime	hours	581 hrs
can of food	weight of contents	ounces	15.3 oz

Qualitative Population Observations

Elementary Unit	*Characteristic of Interest*	*Possible Attributes*
person	sex	male or female
security	type	bond, common stock, or preferred stock
building	exterior materials	brick, wood, aluminum
employee	experience	applicable or not applicable
television	quality	defective or nondefective
firm	legal status	corporation, partnership, or proprietorship
patient	recovery status	ill, well
student	residence	on-campus, off-campus

but in parallel. Different methods will be introduced for describing and summarizing each type. The difference in techniques stems from the fact that arithmetic operations can be performed on numbers, so that, for example, we can calculate the average height for a collection of men, but other procedures must be employed to summarize a qualitative population.

We refer to a particular observation of a qualitative characteristic as an *attribute*. Thus, for the characteristic of marital status, any one of the following attributes might be observed: single, married, divorced, or widowed. An observation of a quantitative characteristic such as income, will result in a particular numerical value or *variate*, such as: $11,928.23, $21,234.15, or $1,095.61. Table 2-1 shows examples of observations from various qualitative and quantitative populations.

EXERCISES

2-1 Consider the data for the number of days employees lost due to illness during August 1975. Give an example of a goal where we may consider the data to be a population; a sample.

2-2 For each of the following problem situations, (1) provide an example of an elementary unit; (2) provide an example of a characteristic of interest; and (3) state whether the population would be qualitative or quantitative.
(a) An aircraft manufacturer's investment plans will be affected by the fate of a congressional appropriation bill containing funds for a new fighter plane. The company wishes to conduct a survey to facilitate planning.
(b) A politician plans to study the voter response to pending legislation to determine his platform.
(c) The Civil Aeronautics Board is seeking operating cost data that will help it establish new rate levels for the New York–Miami airline routes.

2-3 For each of the following situations, discuss whether the suggested frame would be suitable.
(a) An insurance company is using home theft claims processed in the past year as its frame to determine the number of thefts from its policyholders of items valued below $50.
(b) A public health official uses doctors' records of persons diagnosed as suffering from flu as his frame to study the effects of a recent flu epidemic.
(c) A stock exchange official studying the investment attitudes of owners of listed securities uses the active accounts of member brokerage firms as his frame.

2-2 THE FREQUENCY DISTRIBUTION

Finding a Meaningful Pattern for the Data

The ages of a sample of 100 statistics students is shown in Table 2-2. If we wish to describe this sample, how should we proceed? Because the values in Table 2-2 have not been summarized or rearranged in a meaningful manner, we

TABLE 2-2 Ages of a Sample of 100 Statistics Students

Age in Years									
20.9	33.4	18.7	24.2	22.1	18.9	21.9	20.5	21.9	37.3
57.2	25.3	24.6	29.0	26.3	19.1	48.7	23.5	23.1	28.6
21.3	22.4	22.3	20.0	30.3	31.7	34.3	28.5	36.1	32.6
33.7	19.6	18.7	24.3	27.1	20.7	22.2	19.2	26.5	27.4
22.8	51.3	44.4	22.9	20.6	32.8	27.3	23.5	23.8	22.4
18.1	23.9	20.8	41.5	20.4	21.3	19.3	24.2	22.3	23.1
22.9	21.3	29.7	25.6	33.7	24.2	24.5	21.2	21.5	25.8
21.5	21.5	27.0	19.9	29.2	25.3	26.4	22.7	27.9	22.0
23.3	28.1	24.8	19.6	23.7	26.3	30.1	29.7	24.8	24.7
23.5	22.9	26.0	25.2	23.6	21.0	30.9	21.7	28.3	22.1

refer to them as *raw data.* We might begin by grouping the ages in a meaningful fashion.

A convenient way to accomplish this is to group them into categories that are two calendar years long, beginning with age 18. Each age will then fall into one of the categories 18.0–under 20.0, 20.0–under 22.0, and so on, listed in the first column of Table 2-3. We refer to such groupings of values as *class intervals.*

TABLE 2-3 Sample Frequency Distribution of Student Ages

Age	*Tally*	*Number of Persons*
18.0–under 20.0	𝍸 𝍸	10
20.0–under 22.0	𝍸 𝍸 𝍸 ///	18
22.0–under 24.0	𝍸 𝍸 𝍸 𝍸 ///	23
24.0–under 26.0	𝍸 𝍸 ////	14
26.0–under 28.0	𝍸 𝍸	10
28.0–under 30.0	𝍸 ///	8
30.0–under 32.0	////	4
32.0–under 34.0	𝍸	5
34.0–under 36.0	/	1
36.0–under 38.0	//	2
38.0–under 58.0	𝍸	5
	Total	100

Now suppose that we simply count the number of ages falling into each class interval. By summarizing the raw data in this manner, we ought to be able to identify some properties of the sample.

First we list the class intervals in increasing sequence. Then, going down the list of raw data for each successive age, we place a tally mark beside the corresponding class interval and a check mark beside the number on the original list so that it will not be counted twice. When the tally is complete, we find the total number of observations falling into each class interval by counting the marks. The final result shows the frequency with which ages occur in each class interval and is therefore called a *sample frequency distribution*. We find that 10 of the 100 statistics students fall into the class interval 18.0–under 20.0. The number 10 may be referred to as the *class frequency* for the first class interval.

Each class interval has two limits. For the first interval, the *lower class limit* is 18.0 and the *upper class limit* is "under 20.0" (we use the word under to distinguish this limit from 20.0 exactly, which serves as the lower limit of the second class interval). No matter how precisely the raw data are measured, this designation provides no ambiguity in classifying an observation to a particular interval. For example, the age 19.99 years is represented by the first class, while 20.01 years would fall into the second class (20–under 22.0). The *width* of each class interval is found by subtracting the lower limit from the upper limit. For example, the width of the first class interval is $20.0 - 18.0 = 2.0$. All classes but the last one have the same width; as we shall see, beginning or ending class intervals must sometimes be treated differently.

Population Data

The frequency distribution illustrated above applies to sample data. When they are available, the raw data for an entire population could be similarly arranged. In this case, a table of frequencies would constitute a *population frequency distribution*. Ordinarily, only sample results are available, but the present discussion applies to raw data from either a sample or population.

Graphical Displays

The Histogram

A visual display can be a very useful starting point in describing a frequency distribution. Figure 2-2 graphically portrays the same information that is provided in Table 2-3. In this figure, age is represented on the horizontal axis, which is divided into class intervals two years wide. The classes are represented by bars of varying heights, corresponding to the class frequency—the number of observations falling into each interval. Thus the vertical axis represents frequency. Such a graphical portrayal of a frequency distribution is called a *histogram*.

Since there are 23 ages 22.0 or above to just below 24.0, the corresponding bar is of height 23. Only one person is between 34.0 and below 36.0 years old, so the height of that bar is 1. The bars shown for these two class intervals cover the values from 22.0 to just below 24.0 and from 34.0 to 36.0, respectively. All of the bars touch, emphasizing that age varies continuously on a time scale.

Note that the highest class interval is 20 years wide—ten times as wide as the standard intervals. With only five ages falling between 38.0 and 58.0, there is, on the average, one-half an observation every two years in this interval. This

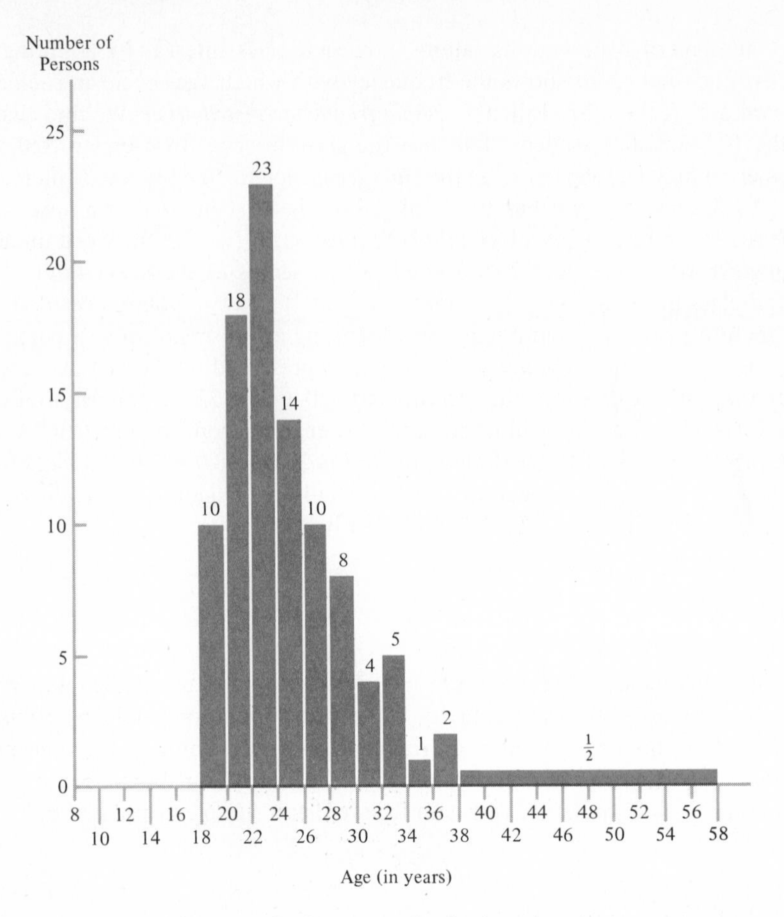

FIGURE 2-2 Histogram for frequency distribution of student ages.

bar is therefore drawn with height $\frac{1}{2}$, so that the total area under the bar is $\frac{1}{2}$ times the number of standard two-year class intervals, or $\frac{1}{2}(10) = 5$, the number of observations in this category. Thus, treating the two-year width as a standard unit, *the area under any bar* (its height times its width, in standard units) *equals the number of observations in the corresponding class interval.*

The Frequency Polygon

An alternative graphical portrayal of a frequency distribution is the *frequency polygon*, shown in Figure 2-3 for the same population of ages in Figure 2-2 and using the same axes. In Figure 2-3, each class interval is represented by a dot positioned above its midpoint at a height equal to the class frequency. The midpoints are defined by the average of the interval's class limits. Thus, the midpoint of the first interval is $(18.0+20.0)/2 = 19.0$; the second midpoint is 21.0, and so on. The value used for the last interval is $(38.0+58.0)/2 = 48.0$. The dots are connected by line segments to facilitate reading the graph.

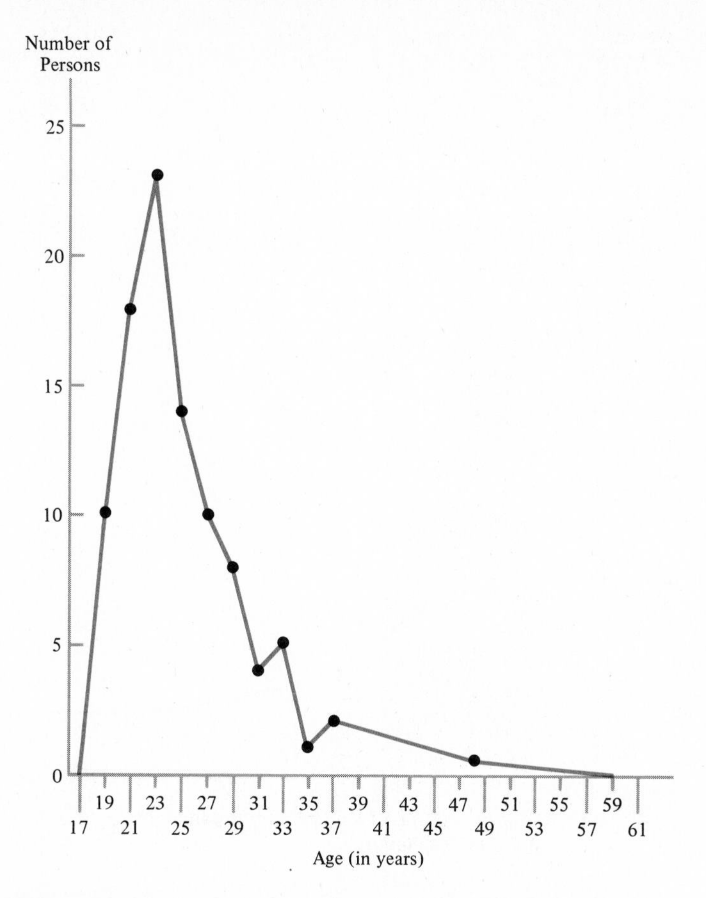

FIGURE 2-3 Frequency polygon for sample of student ages.

Geometrically, a polygon is a closed figure having many sides. To complete the frequency polygon, line segments are drawn from the first and last dots to the horizontal axis at points one-half the width of a class interval below the lowest and above the highest class intervals, in this case touching the axis at 17 and 59. The frequency polygon provides a meaningful frequency only for the observation values corresponding to the midpoints beneath the dots; these are interpreted as the typical values for all observations in the respective class intervals.

The Frequency Curve

When the number of observations is large, the polygon for a sample frequency distribution presents a shape similar to the shape which might be obtained for raw data consisting of the entire population. But since data for the entire population are not ordinarily available, the population frequency distribution is usually portrayed in terms of a *frequency curve* like the one shown

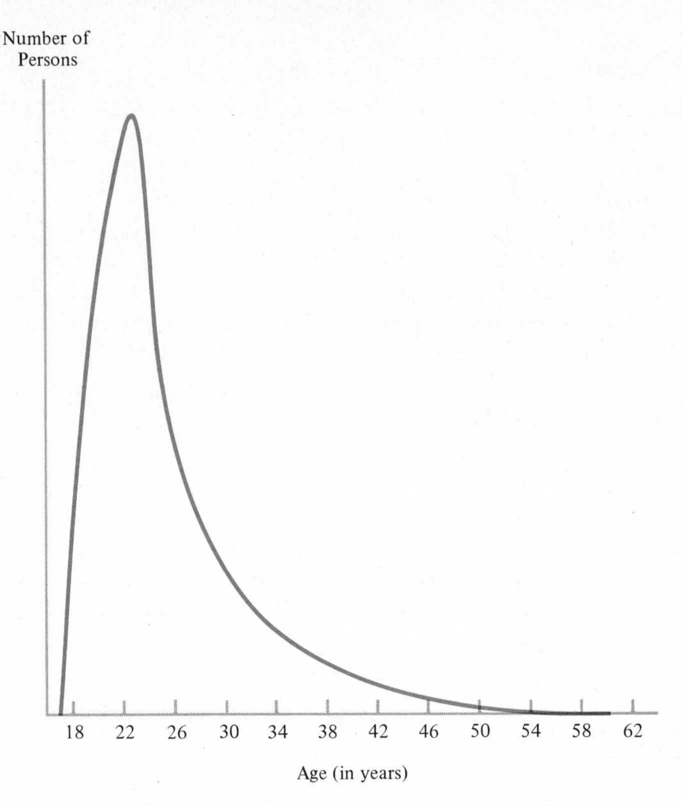

FIGURE 2-4 Suggested shape of smoothed frequency curve for the entire population.

in Figure 2-4. The basic shape of a population's frequency curve is usually suggested by the histogram or frequency polygon originally found for sample data. Because populations are usually quite large in relation to the number of observations in a sample, the frequency curve would resemble a frequency polygon or a histogram with many class intervals of tiny widths. If these intervals were laboriously plotted, the entire collection of population data would present a less jagged graph than that obtainable from sample data alone, and this graph would be almost totally smooth (like the curve in Figure 2-4). Later in this chapter we will discuss some of the shapes commonly encountered for populations.

Descriptive Analysis

We have managed to translate the confusion of the raw data into a pattern based upon frequencies. The frequency distribution tells us two things: it shows how the observations cluster around a central value, and it illustrates the degree of dispersion or difference between observations.

We see that none of the students is younger than 18 and that ages below 28 are most typical. Of these, the most common age is somewhere between 22 and 24, which (from general information obtained from the registrar's office) we know to be higher than usual for the student who enters college right after high school and graduates at about age 22. The students in the sample are generally older; but we can rule out the possibility that they might be graduate students because of the substantial number of younger persons. One guess is that the population could be made up of night students, with the older persons working on their degrees on a part-time basis while holding full-time jobs.

The appropriateness of this conclusion may be substantiated by the incidence of persons over 30, who most likely have financial burdens and cannot afford to be full-time students. Predominantly young, the sample is peppered with persons approaching or exceeding the age of 40. Five of these are relatively so much older that they have been put into a special group—the "over 38s." Their ages (41.5, 44.4, 48.7, 51.3, and 57.2) are spread so thinly along the age scale that it was more convenient to lump these more extreme ages into a single special category.

The foregoing descriptive analysis provides us with an image of this sample that is not immediately available from the raw data. The entire description is based upon frequencies of occurrence—the heart of all statistical analysis. As we shall see in Chapter 4, frequency is the basis for probability theory and, as such, has a fundamental role in all procedures of statistical inference.

A vivid illustration of how frequency of occurrence can be used to reduce the confusion of raw data is provided by *cryptanalysis*. This also serves to explain frequency distributions for qualitative populations.

The Frequency Distribution of a Qualitative Population

Role of Frequency Distribution in Cryptanalysis

Cryptanalysis is the breaking of codes or ciphers used to keep secret those communications made in the "never-never land" of espionage, foreign intrigue, diplomacy, or matters military. In its simplest form, a *cipher* is a message where another alphabet has been substituted for the ordinary one. Consider the example in Table 2-4. To decipher this message, we must find the counterparts to the letters in the original message.

TABLE 2-4 Cipher Message

AOYNS	YIRXJ	AJRRS	OOYIR	YGDYP
MQQCY	CYMOQ	JAPQM	QDPQD	RMGMI
MGXPD	PDQMG	GHVPS	PQJRO	YMQYJ
OKYOA	AJHRJ	IASPD	JIHYM	IDIBA
SGGXM	OOMIB	YKISH	UYOPR	MIQYG
GSPMP	QJOXM	IKCYG	LSPRC	JJPYH
YQCJK	PMIKL	OJRYK	SOYPA	JOBYI
YOMGD	ZDIBM	UJSQL	JLSGM	QDJIP

The first step in cryptanalysis is to determine the frequency distribution for the letters in the cipher message, as in Table 2-5. The ciphertext letters are the elementary units from a qualitative population where the possible attributes are the symbols of the alphabet. The frequency distribution for qualitative data provides a count of the number of occurrences for each attribute, so that there is a category for each number that is analogous to the class interval for quantitative data.

TABLE 2-5 Frequency Distribution of Letters in Cipher Message

Letter	*Tally*	*Number of Letters*	*Letter*	*Tally*	*Number of Letters*
A	𝍸 ///	8	N	/	1
B	////	4	O	𝍸 𝍸 𝍸	15
C	𝍸	5	P	𝍸 𝍸 𝍸 /	16
D	𝍸 𝍸	10	Q	𝍸 𝍸 ////	14
E		0	R	𝍸 𝍸	10
F		0	S	𝍸 𝍸 /	11
G	𝍸 𝍸 //	12	T		0
H	𝍸	5	U	//	2
I	𝍸 𝍸 𝍸	15	V	/	1
J	𝍸 𝍸 𝍸 //	17	W		0
K	𝍸 /	6	X	////	4
L	////	4	Y	𝍸 𝍸 𝍸 𝍸 /	21
M	𝍸 𝍸 𝍸 ///	18	Z	/	1
				Total	200

The cryptanalyst knows the frequency with which each letter occurs in the ordinary English language. This is provided in Table 2-6, where the letters are shown in lower case to distinguish them from the ciphertext alphabet.

TABLE 2-6 Frequency Distribution of Letters in 200 Characters of Ordinary English Language Text

Letter	*Frequency*	*Letter*	*Frequency*	*Letter*	*Frequency*
a	16	j	1	s	12
b	3	k	1	t	18
c	6	l	7	u	6
d	8	m	6	v	2
e	26	n	14	w	3
f	4	o	16	x	1
g	3	p	4	y	4
h	12	q	$\frac{1}{2}$	z	$\frac{1}{2}$
i	13	r	13	Total	200

SOURCE: David Kahn, *The Codebreakers: The Story of Secret Writing* (New York: Macmillan, 1967), p. 100. Copyright © 1967 by David Kahn.

Graphical Portrayal

The frequency distribution of the letters is graphed as a histogram in Figure 2-5. When the observations are attributes rather than variates, the bars do not touch each other, reflecting the fact that each category is discrete. Sometimes vertical spikes are used instead of bars to emphasize that the observations do not range along a continuous scale.

For samples from qualitative populations, where the characteristic of interest takes the form of various attributes (such as male or female, defective or nondefective, or professional categories such as doctor, lawyer, engineer, or educator), the graphical portrayal of the frequency distribution would be similarly represented.

Sometimes a quantitative frequency distribution is handled in the same way. For example, the number of children in a collection of families would be so represented. Here the possible variates are whole numbers or *integers*, such as 0, 1, 2, 3, or 4. The number of children cannot be fractional, such as 2.67, so

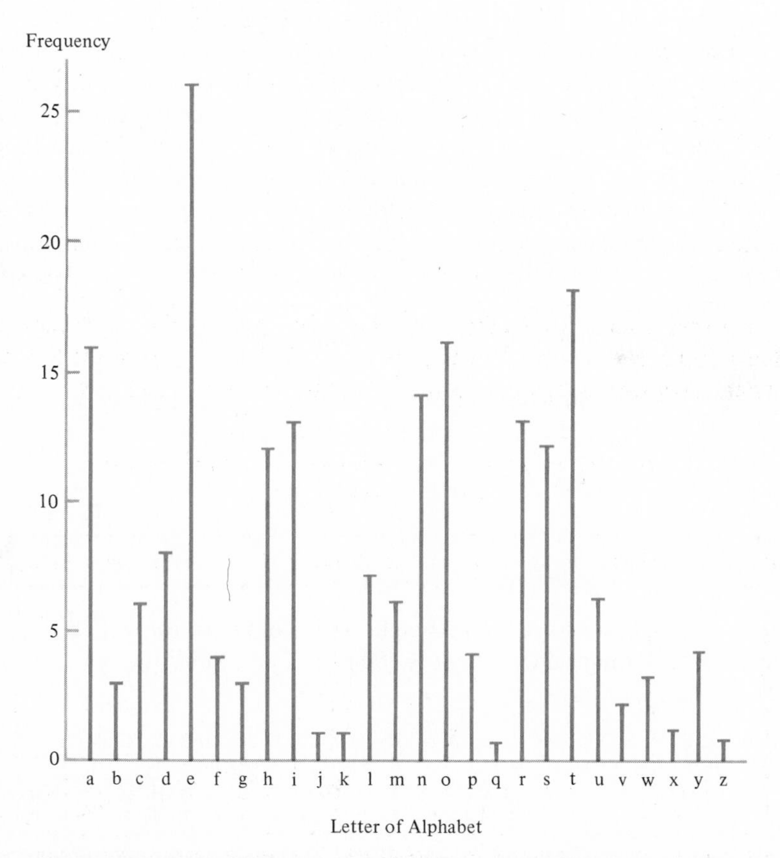

FIGURE 2-5 Qualitative frequency distribution for 200 letters of ordinary English text.

they should not be represented on a continuous scale; the possible values are *discrete*. When quantitative observations are discrete, they are graphically portrayed in the same manner as the attributes of a qualitative population.

Frequency Analysis

The letters occurring in ordinary English may be arranged in sequence of decreasing frequency:

English letters	e	t	a	o	n	i	r	s	h	d	l	u	c	m	p	f	y	w	g	b	v	j	k	x	q	z
Frequency	26	18	16	16	14	13	13	12	12	8	7	6	6	6	4	4	4	3	3	3	2	1	1	1	½	½
					High											*Medium*								*Low*		

A similar arrangement of the cipher text letters may be made:

Cipher text	Y	M	J	P	O	I	Q	G	S	D	R	A	K	C	H	B	L	X	U	N	V	Z	E	F	T	W
Frequency	21	18	17	16	15	15	14	12	11	10	10	8	6	5	5	4	4	4	2	1	1	1	0	0	0	0

There is no reason why we should be so fortunate that this particular message will provide an identical match—letter for letter—strictly on the basis of frequency. But the frequencies can be used to find likely possibilities. For instance, they tell us that the letter e is most frequently used, so that this probably corresponds to either the Y, M, J, P, O, or I of the cipher. Similarly, we can identify English text letters likely to correspond to the other letters. Notice the rather dramatic drops in frequency between h and d and again between v and j, which provide clues to considerably narrow our search for a solution to the ciphertext.

A more refined analysis is usually the next step in cryptanalysis. It is based upon how frequently each of the higher count ciphertext letters combines with each of the other letters of the alphabet. From experience the cryptanalyst knows that certain two-letter combinations occur more often than others. By a process of elimination, the more common vowels and consonants can be matched to the ciphertext.

Using the first 50 letters in our message, assuming that we have found Y = e, Q = t, M = a, J = o, D = i, I = n, P = s, and O = r, we obtain the partial translation:

A	O	Y	N	S	Y	I	R	X	J	A	J	R	R	S	O	O	Y	I	R	Y	G	D	Y	P
	r	e			e	n			o		o				r	r	e	n		e		i	e	s

M	Q	Q	C	Y	C	Y	M	O	Q	J	A	P	Q	M	Q	D	P	Q	D	R	M	G	M	I
a	t	t		e		e	a	r	t	o		s	t	a	t	i	s	t	i		a		a	n

In the second row, we have the first eight letters of *statistical*, with the c and l missing. This is strong evidence that R = c and G = l. Substituting these values into the message, we have

A	O	Y	N	S	Y	I	R	X	J	A	J	R	R	S	O	O	Y	I	R	Y	G	D	Y	P
	r	e			e	n	c		o		o	c	c		r	r	e	n	c	e	l	i	e	s

M	Q	Q	C	Y	C	Y	M	O	Q	J	A	P	Q	M	Q	D	P	Q	D	R	M	G	M	I
a	t	t		e		e	a	r	t	o		s	t	a	t	i	s	t	i	c	a	l	a	n

A good guess is that S = u and C = h. Filling in the corresponding blanks leads us to further conclude that A = f and N = q.

As further original text letters become known, the remaining letters are increasingly easier to identify, thus filling in the blank spaces below the ciphertext. Finally the cipher is broken; the complete key is

Cipher text:	M	U	R	K	Y	A	B	C	D	E	F	G	H	I	J	L	N	O	P	Q	S	T	V	W	X	Z
Original text:	a	b	c	d	e	f	g	h	i	j	k	l	m	n	o	p	q	r	s	t	u	v	w	x	y	z

The cipher is constructed by using the word MURKY for the first five original text alphabet letters and listing the remaining cipher letters in alphabetical sequence. The text of the full message reads:

> **Frequency of occurrence lies at the heart of statistical analysis. It allows us to create order from confusion. Meaningfully arranged, numbers can tell us a story and help us choose methods and procedures for generalizing about populations.**

Guidelines for Constructing a Frequency Distribution

The purpose of a frequency distribution is to convert raw data into a meaningful pattern for statistical analysis. In doing this, detail must be sacrificed for insight. Because it lumps individual variates into class intervals, the frequency distribution cannot tell us how the observations within a certain category differ from each other. Only the raw data can provide this information. Thus some less important information is lost in order that more useful information can be gained.

How can the raw data best be condensed into class intervals? This question leads to two others: What range of values should be included in a single category? How many intervals should be used?

Width of Class Intervals

It is desirable that the class intervals be of the same width. This facilitates comparisons between classes and provides simpler calculations of the even more concise summary values to be discussed later in this chapter. As we noted with the distribution of the age data in Table 2-3, it is sometimes better to lump the extreme values in a single category. The need for this becomes quite evident during construction of a frequency distribution for individual annual earnings. Most persons have incomes less than $25,000. A few earn between $25,000 and $100,000. But the rare extremely well-to-do have incomes substantially higher. A frequency distribution graph with equal intervals of income width $5,000 would be hard to draw and would not be very readable. The resulting histogram would be somewhat analogous to a few mountain peaks (representing frequency for the lower income levels) sitting to the left of very many successively smaller anthills (representing the number of persons in each progressively rarer, higher income category). This difficulty may be avoided by grouping the higher incomes into a single class.

Thus, higher incomes might be lumped into a category such as "$100,000 or more." Usually such a class interval has no stated upper limit. This is partly due to lack of available data. Another reason is that such an upper limit may dwarf the other intervals to the point that all the categories appear "out of line." Often there is no theoretical upper limit, as might be the case with a frequency distribution for times taken by electronic components to fail. The histogram of a frequency distribution with no upper limit would not usually show a bar for the extreme values, although these should be mentioned with a note.

We cannot really know how wide the class intervals ought to be without knowing how many intervals are to be used.

Number of Class Intervals

There are no rules indicating how many class intervals to use. It is generally recommended that there be between five and 20, but the number is really a matter of personal judgment. The main consideration should be that the interval is not so narrow or so wide that useful information cannot be obtained. Figure 2-6 shows two extreme cases. In diagram (a), the class intervals are so tiny that

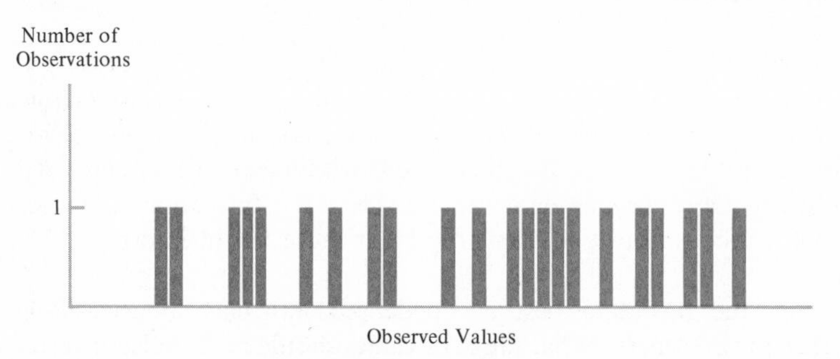

(a) A separate class for each observation

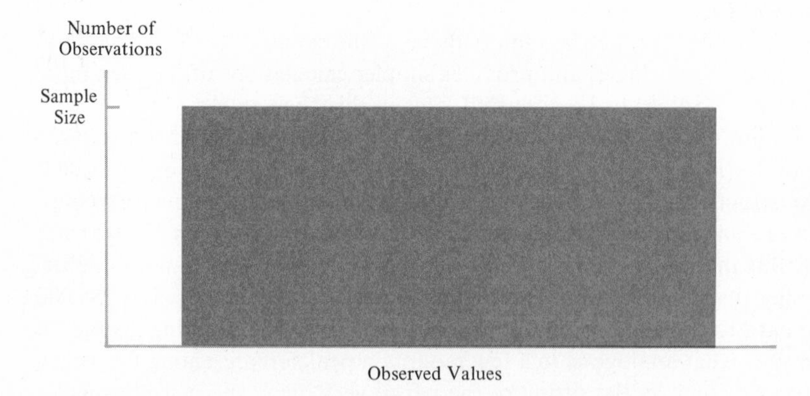

(b) A single class for all observations

FIGURE 2-6 Illustration of two extreme cases for number of frequency distribution classes.

each has height 1 and the frequency distribution is no better than the raw data themselves. On the other hand, a single class interval like the one in diagram (b) lumps all data into one category, which can only indicate the total amount of dispersion in the observations. We desire something in between these two.

When intervals are to be the same size width, the following rule may be applied to find the required ***class interval width***.

$$\text{Class interval width} = \frac{\text{Largest value} - \text{Smallest value}}{\text{Number of class intervals}} \tag{2-1}$$

This is used when the extreme values are not lumped into a special category.

We illustrate this for the 100 observations of fuel consumption in miles per gallon for a fleet of cars listed in Table 2-7. Suppose we desire 5 class intervals

TABLE 2-7 Miles per Gallon Achieved with 100 Medium-Sized Cars (raw data)

19.0	20.8	22.0	22.7	20.0	18.9	16.6	16.8	20.8	14.7
15.1	21.8	21.1	21.5	21.1	15.5	19.3	15.1	20.6	16.8
18.2	20.5	15.3	16.2	16.3	22.8	22.7	21.9	22.5	17.1
19.1	21.6	19.0	18.3	18.6	22.1	17.5	22.9	21.7	18.7
21.9	20.2	14.5	14.1	22.9	20.2	17.3	22.6	19.3	21.7
21.5	22.6	18.7	19.2	22.8	21.6	21.7	20.5	22.7	20.4
18.8	15.1	16.5	20.5	19.1	17.4	19.7	19.2	16.4	21.9
14.3	19.2	19.7	17.1	21.4	21.9	21.7	19.2	23.9	19.6
20.9	18.5	20.2	18.2	20.2	22.4	20.4	21.6	21.3	22.4
20.5	18.1	20.7	21.3	16.9	20.3	23.9	18.8	21.1	21.9

of equal width. We find the required width by taking the difference between the largest and smallest value, $23.9 - 14.1 = 9.8$, and then dividing by 5, getting 1.96 miles per gallon. To ease our task, we can use 2.0 instead of 1.96 for the width and 14.0 instead of 14.1 for the lower limit of our first class interval. The frequency distribution is provided in Table 2-8 and graphed in Figure 2-7(a).

TABLE 2-8 Frequency Distribution for Miles per Gallon of 100 Medium-Sized Cars Using 5 Classes

Miles per Gallon	*Number of Cars*
14.0–under 16.0	9
16.0–under 18.0	13
18.0–under 20.0	24
20.0–under 22.0	38
22.0–under 24.0	16
Total	100

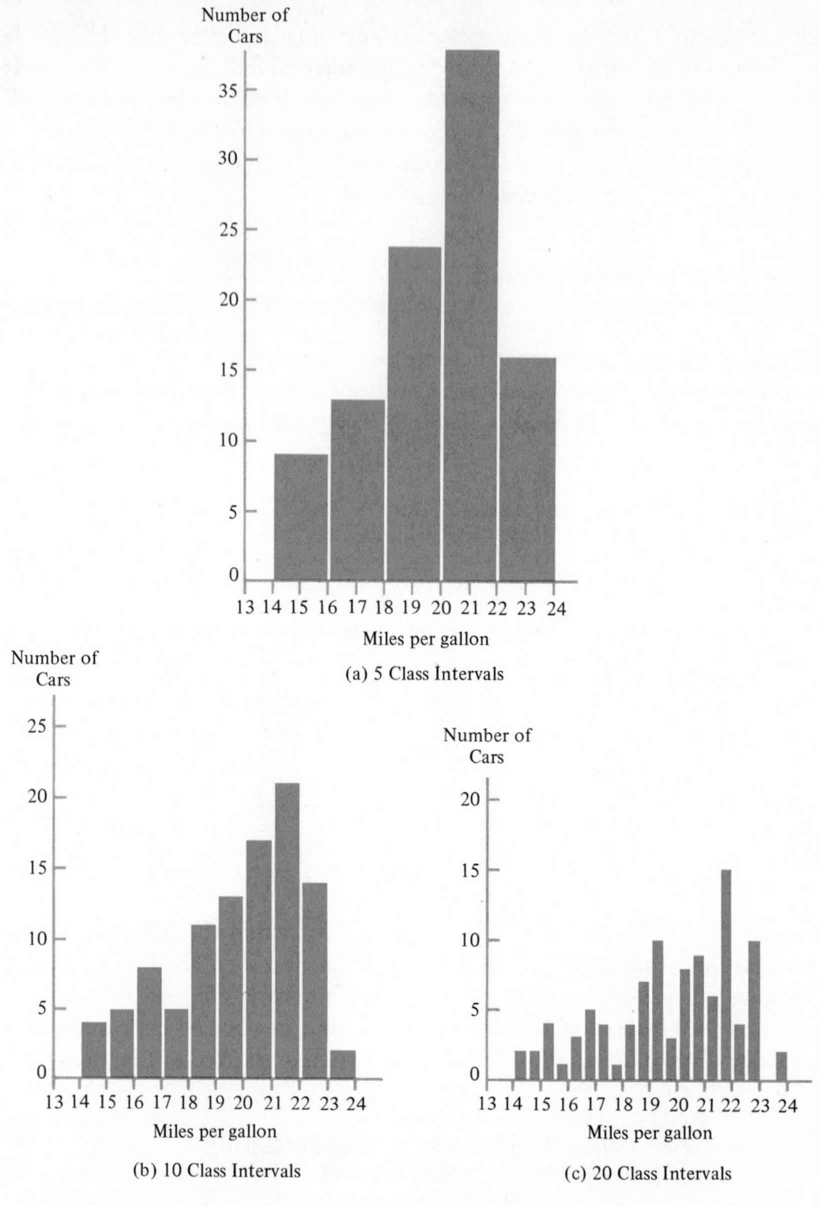

FIGURE 2-7 Histograms for gasoline consumption of 100 medium-sized cars, using three different sets of class intervals.

Figure 2-7 also contains histograms for the same gasoline consumption data using 10 class intervals (b) and 20 class intervals (c). Notice that the histograms become progressively "lumpier" as the number of classes increases. Both the five-interval and ten-interval graphs provide concise summaries of the data, but a pronounced "saw-tooth" effect occurs with 20 intervals. The poorness of the 20-interval summary reflects the large number of intervals in relation to the size of the sample. One way to decide how many class intervals to use would be to try several—plotting a histogram for each and selecting the one that provides the most logical explanation of the underlying population pattern. A large number of class intervals will provide more data, but too many will show meaningless oscillations.

EXERCISES

2-4 Using the income data below:

$ 3,145	$15,879	$ 6,914	$ 4,572	$11,374
12,764	9,061	8,245	10,563	8,164
6,395	8,758	17,270	10,755	10,465
7,415	9,637	9,361	11,606	7,836
13,517	7,645	9,757	9,537	23,957
8,020	8,346	12,848	8,438	6,347
21,333	9,280	7,538	7,414	11,707
9,144	7,424	25,639	10,274	4,683
5,089	6,904	9.182	12,193	12,472
8,494	6,032	16,012	9,282	3,331

(a) Construct a frequency distribution having six class intervals of equal width. Round the width to the nearest thousand dollars and use $3,000 as the lower class limit of the first interval.
(b) Draw a histogram showing the distribution from part (a).

2-5 Using the family income data in Exercise 2-4:
(a) Construct a frequency distribution having 12 class intervals of width $2,000, with $3,000 as the lower limit of the first class.
(b) Construct a frequency polygon using the frequency distribution from part (a).
(c) Do you think a better data summary could be gained by using fewer or more class intervals? Discuss the reasons for your choice.

2-6 A state university has two campuses. All entering freshmen are required to submit verbal Scholastic Aptitude Test (SAT) scores. The score frequency distributions for the current classes are provided below.

Old Campus		*New Campus*	
SAT Score	*Number of Students*	*SAT Score*	*Number of Students*
200–under 300	50	200–under 300	100
300–under 400	100	300–under 400	250
400–under 500	150	400–under 500	100
500–under 600	300	500–under 600	100
600–under 700	100	600–under 700	50
700–under 800	50	700–under 800	0

(a) Plot frequency polygons for each campus separately but on the same graph.
(b) Combine the old and the new campus data into a single, all-university frequency distribution.
(c) Plot a frequency poygon for the data obtained in part (b).
(d) Compare the graphs in part (a) to the one in part (c). Which presentation do you think provides more meaningful information? Explain.

2-7 Construct a frequency distribution, with a separate category for each alphabetic character, of the first 200 letters of Abraham Lincoln's Gettysburg address:

> Four-score and seven years ago our fathers brought forth on this continent, a new nation, conceived in Liberty, and dedicated to the proposition that all men are created equal.
> Now we are engaged in a great civil war, testing whether that nation, or . . ..

Treat upper and lower case letters the same. Comparing your distribution with that in Table 2-6, do you think that the English used by Lincoln is "ordinary?" Explain.

2-8 Comment on the appropriateness of each of the following class interval widths for a frequency distribution of rates of pay for 100 selected hourly workers in a metropolitan area: \$0.05, \$0.25, \$0.50, \$1.00, \$2.00.

2-9 The following data have been obtained for the monthly rentals for 45 apartments in a large metropolitan area:

\$100	\$130	\$130	\$305	\$175
155	150	95	295	210
80	270	135	130	335
230	235	75	90	285
65	345	110	135	185
300	70	250	125	180
150	305	170	95	90
145	90	160	130	80
490	235	75	60	425

(a) Construct a frequency distribution for the above data.
(b) How many classes did you use? Justify your choice.
(c) Did you use equal or unequal class interval widths? Why?
(d) Plot a histogram of your results.
(e) Write a brief summary of your findings.

2-10 Criticize each of the following designations for class intervals of monthly household electricity bills:

(a)	(b)	(c)
\$10–15	\$11–22	\$10–under 15
15–20	21–32	16–under 20
20–25	31–42	21–under 25
etc.	etc.	etc.

2-3 RELATIVE AND CUMULATIVE FREQUENCY DISTRIBUTIONS

Two useful extensions of the basic frequency distribution are the *relative* and the *cumulative* frequency distributions.

A *relative frequency* is the ratio of the number of observations falling into a particular category to the total, and may be determined for both quantitative and qualitative data. Relative frequency is a convenient basis for the comparison of similar groups of different size. For example, consider comparing the earnings of workers in sparsely populated Nevada to their counterparts in neighboring California, the most populous state. It would not be very meaningful to state that Nevada has only 6,000 persons achieving a particular level of income, whereas California has 250,000. But if the percentage of such workers is 1 percent in Nevada and only 0.8 percent in California, a realistic comparison between workers in the two states can be made. Analogously, to determine whether the men at a large university have more opportunities to date women on campus than men at a small college have, we must compare the proportion of male students—not the total frequencies.

A *cumulative frequency* is the sum of the frequencies for successively higher classes, and only applies when the observations are numerical. For example, we might find that 8,439 men in a population of 10,000 are less than 6 feet tall; stated differently, 84.39 percent of the population is shorter than 6 feet. Cumulative frequencies can be very useful descriptions of a population, especially when they are expressed relatively as percentages or proportions. We can judge our chances for being accepted by a graduate school, for example, partly according to the percentage of persons obtaining a lower score on the admissions screening examination. (Cumulative frequency is relevant to some probability concepts we will encounter in Chapter 5.)

Relative Frequency Distributions

The relative frequencies for a population are readily obtained for each category. Since a relative frequency is the proportion of all observations falling into a particular category, it is calculated by dividing the number of observations in this category by the total number of observations. Once the frequency distribution itself is obtained, the relative frequency computations are a matter of simple arithmetic. This is illustrated in Table 2-9 for each class interval of data on the age of department store accounts (accounts 180 days or older are written off as uncollectable). The relative frequency of accounts less than 30 days old (class interval 0–under 30) is 532/1,615 = .330, which says that 33 percent of all accounts are less than 30 days old. The other relative frequencies are calculated in the same manner, by dividing the class frequency in column (2) by the total

TABLE 2-9 Calculation of Relative Frequency Distribution of Age of Department Store Accounts

(1) *Age* *(days)*	(2) *Number of* *Accounts* *(frequency)*	(3) *Relative* *Frequency*
0–under 30	532	.330
30–under 60	317	.196
60–under 90	285	.176
90–under 120	176	.109
120–under 150	158	.098
150–under 180	147	.091
Totals	1,615	1.000

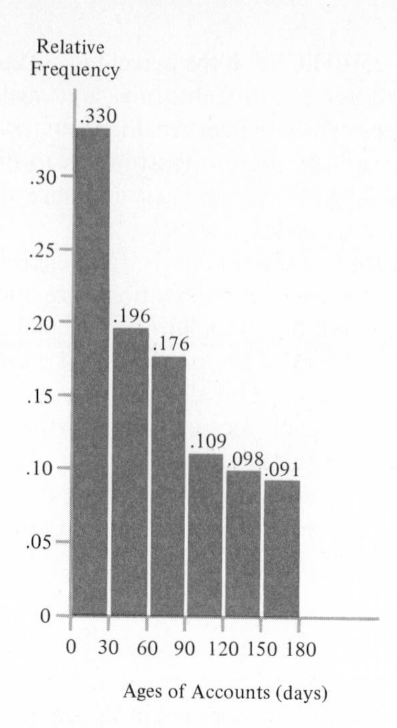

FIGURE 2-8 Relative frequency histogram for age of department store accounts.

number of accounts. The sum of all the relative frequencies must always be 1. Columns (1) and (3) constitute the *relative frequency distribution* for these data. The histogram for this distribution is shown in Figure 2-8. A histogram constructed using relative frequencies will have the same shape as one constructed using original tallies, or absolute frequencies. Only the scale on the vertical axis will change.

Analysis of relative frequencies can sometimes be useful when analysis of absolute frequencies is not. A department store's credit sales will fluctuate considerably over the seasons. If the credit worthiness of the store's customers remains about the same as it has in the past, then the relative frequency distributions for the age of accounts obtained during the same calendar month in two successive years should not be noticeably different when growth is steady. This is because the same seasonal pattern will generally persist, even though the total level of sales is higher in the second year. If there is a substantial difference in relative frequencies, it may be due to erratic growth or to a change in the quality of credit users or in collection procedures. Absolute frequencies cannot isolate the effect of sales growth from the effect of changes in credit procedures.

The relative frequency of men on the two campuses mentioned earlier shows how relative frequency may be used to compare two qualitative populations. Table 2-10 shows the frequency distribution by sex of students in the two schools. The number of men competing for dates in the large university is 6,000; the number in the small college is only 400. The relative frequency distributions are provided in Table 2-11. In terms of relative frequency, there is less competition in the university, where the proportion of men is smaller (.60 in the university versus .80 at the college).

TABLE 2-10 Frequency Distributions of Students by Sex

Large University		*Small College*	
Sex	*Frequency*	*Sex*	*Frequency*
Male	6,000	Male	400
Female	4,000	Female	100
Totals	10,000		500

TABLE 2-11 Relative Frequency Distributions of Students by Sex

Large University		*Small College*	
Sex	*Frequency*	*Sex*	*Frequency*
Male	.60	Male	.80
Female	.40	Female	.20
Totals	1.00		1.00

Cumulative Frequency Distributions

The cumulative frequencies may be determined by adding the frequency for each class interval to the frequencies for preceding intervals. Table 2-12 illustrates this for a sample of 100 verbal SAT (Scholastic Aptitude Test) scores achieved by the students in one high school. Here, we find it convenient to express the upper limit of each class interval as "less than" the lower limit of the succeeding interval. The cumulative frequency for scores less than 300 is

TABLE 2-12 Calculation of Cumulative Frequencies for Verbal SAT Scores

(1) *Score*	(2) *Number of Students (Frequency)*	(3) *Cumulative Frequency*
100–less than 200	0	0
200–less than 300	5	0+ 5= 5
300–less than 400	39	5+39= 44
400–less than 500	31	44+31= 75
500–less than 600	16	75+16= 91
600–less than 700	6	91+ 6= 97
700–less than 800	3	97+ 3=100

found by adding the frequency of scores at or above 100 (but less than 200) to the number of students having scores at 200 or above (but less than 300), or $0+5=5$. The cumulative frequency for a class interval represents the total number of scores falling in or below that interval. Columns (1) and (3) of Table 2-12 constitute the *cumulative frequency distribution* for the SAT score data.

The common graphical portrayals for cumulative frequency distributions are shown in Figure 2-9 for the verbal SAT data. The ordinate (vertical height) of a plotted point represents the cumulative frequency for values less than the abscissa (the horizontal scale value). Such a curve is called an *ogive.*

Sometimes it is desirable to calculate cumulative relative frequencies. These are found by adding successive relative frequencies. The largest possible cumulative relative frequency would be 1.0.

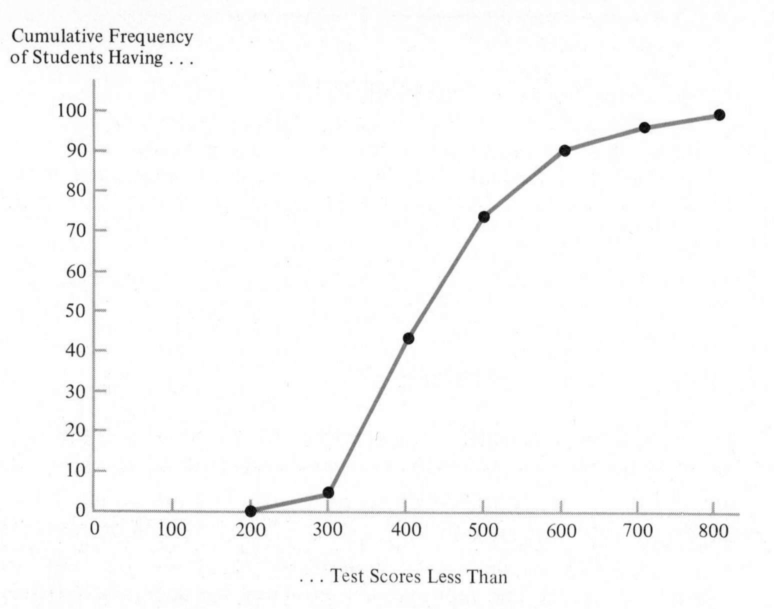

FIGURE 2-9 Ogive for cumulative frequency distribution of verbal SAT scores.

EXERCISES

2-11 The frequency distribution of times between arrivals of injured patients at a hospital emergency room is as follows:

Time (minutes)	*Number of Patients*
0.0–under 3.0	210
3.0–under 6.0	130
6.0–under 9.0	75
9.0–under 12.0	40
12.0–under 15.0	20
15.0–under 18.0	15
18.0–under 21.0	10

(a) Determine the relative frequency distribution. Then plot the resulting distribution as a histogram.
(b) Determine the cumulative frequency distribution. Then plot this distribution as a "less than" ogive.

2-12 The following is the frequency distribution of consecutive days absent by students in a surburban school district:

Consecutive Days Absent	*Number of Students*
0–4	3,100
5–9	810
10–14	510
15–19	320
20–24	120
25–29	30
30–59	110

(Students absent a total of 60 or more consecutive days are classified as inactive. Each student is counted just once—in the category corresponding to his longest absence.)
(a) Determine the relative frequency distribution.
(b) Determine the cumulative *relative* frequency distribution.
(c) Plot a "less than" ogive using the data obtained in part (b).

2-13 A school principal keeps records of student deportment in five classes. One factor considered is the frequency of tardy and absent students during a particular month:

Number of Students by Class

	A	B	C	D	E
Tardy	35	60	20	42	6
Absent	35	30	40	18	36
On time	630	490	590	540	508
Totals	700	580	650	600	550

(a) Determine the relative frequency distributions for each class.
(b) Using class A as a standard, which classes have an excessively high number of tardy students?
(c) Compared to class A, which classes have an excessively high number of absent students? Do you think that the teachers of these classes are "softer" than the teacher of class A? Explain.

2-14 The cumulative relative frequency distribution for the size of 1,000 small cities is provided below.

Number of People	*Cumulative Proportion of People*
0–under 5,000	.36
5,000–under 10,000	.63
10,000–under 15,000	.81
15,000–under 20,000	.90
20,000–under 25,000	.97
25,000–under 30,000	1.00

(a) Determine the relative frequency distribution.
(b) Using your answer to part (a), determine the original frequency distribution.
(c) Using your answer to part (b), determine the cumulative frequency distribution.

2-4 COMMON FORMS OF THE FREQUENCY DISTRIBUTION

Statistical methodology has been developed for analyzing samples taken from populations having various general forms. Thus, by placing populations into various categories, it is possible to increase our analytical powers considerably by using some of the techniques that apply to all populations belonging to the same *distribution family*. A population's membership in a particular family is determined largely by the shape of the histogram for the sample distribution. The underlying frequency distribution of each family may be approximated by various frequency curves that can be described by mathematical equations.

In Figures 2-10 to 2-14, examples of several of the more common general shapes of frequency curves are sketched alongside representative sample histograms. Figure 2-10 represents the relative frequency distribution of the diameters of a production batch of one-inch thick steel reinforcing rods. Note that the histogram is fairly symmetrical about the interval 0.995–1.005 inches, with frequency dropping for the next higher and lower intervals. On the left and right, the histogram bars become progressively shorter, tapering off at about 0.91 and 1.09 inches, respectively. The smoothed frequency curve beside the histogram is bell-shaped and belongs to a class of populations having the *normal distribution*. We will have more to say about normal curves in Chapter 6. A great many populations have frequency distributions in this category, including many physical measurements.

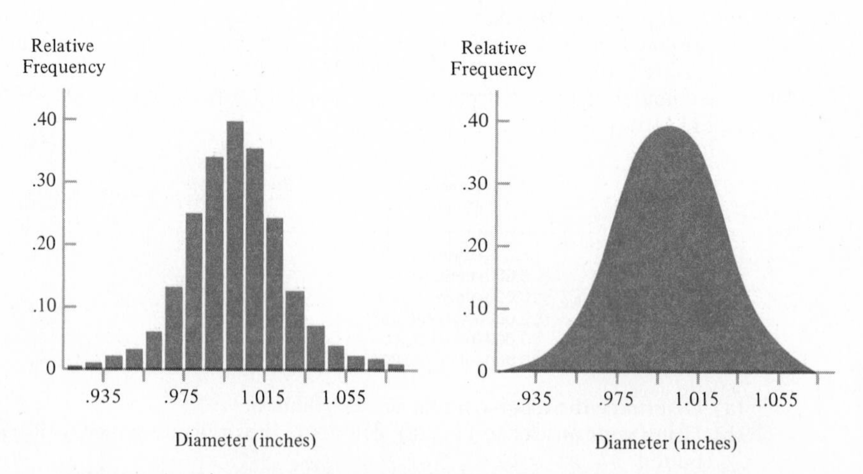

FIGURE 2-10 Sample histogram (left) and population frequency curve (right) for steel rod diameters.

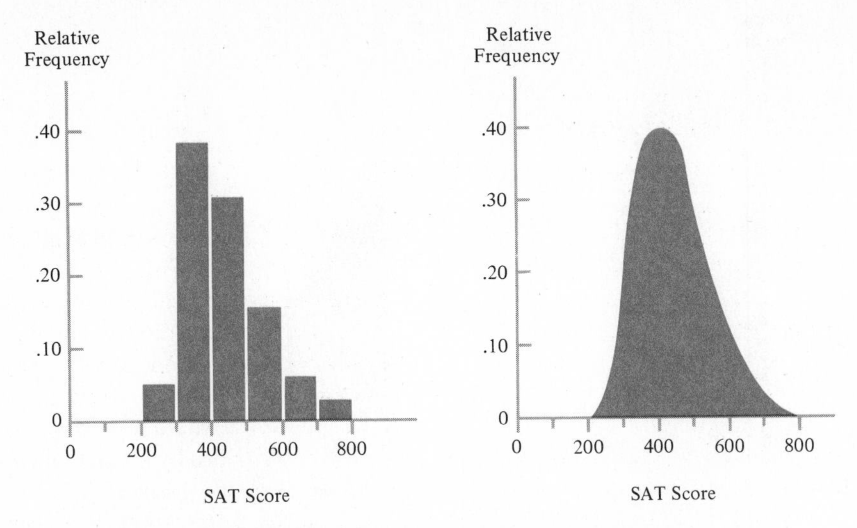

FIGURE 2-11 Sample histogram (left) and population frequency curve (right) for verbal SAT scores.

Figure 2-11 shows the histogram for the SAT data from Table 2-12. Here the data are skewed, because a few students scored quite highly. Beside the histogram is a frequency curve with the right "tail" longer than the left. This general shape corresponds to a class of skewed distributions where there are few small observations and proportionately more large observations fall over a wide range of values.

Another type of skewed distribution is provided in Figure 2-12, where the

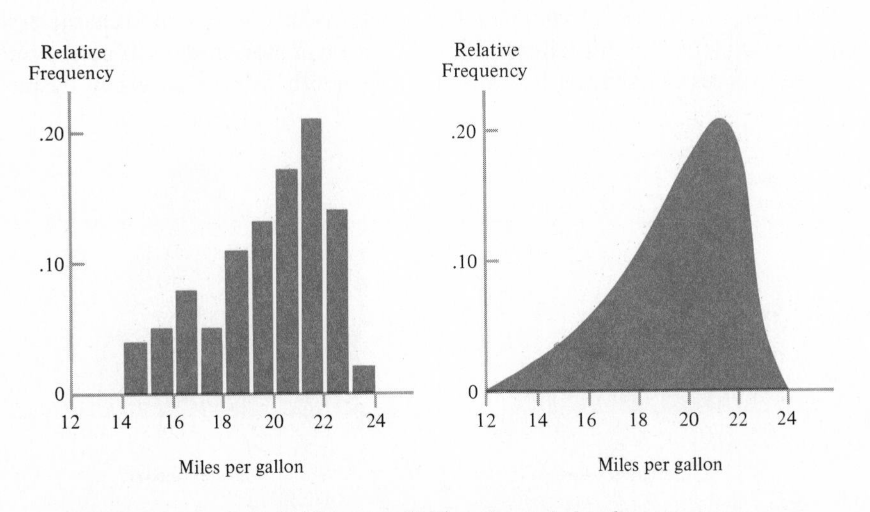

FIGURE 2-12 Sample histogram (left) and population frequency curve (right) for gasoline mileage.

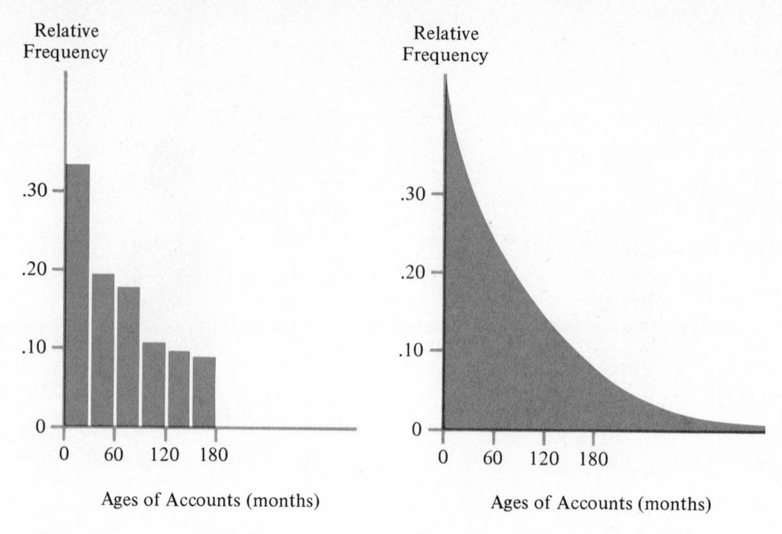

FIGURE 2-13 Sample histogram (left) and population frequency curve (right) for ages of department store accounts.

gasoline mileage frequency curve has a longer left tail. In this case, the smaller values are widely spread from 14 through 18 miles per gallon, while the rarer large values fall into a single class, 23.0–under 24.0.

The distribution for the ages of the department store accounts receivable given in Table 2-9 is plotted in Figure 2-13. Here the approximating smoothed curve has the general shape of a reversed letter J. This type of distribution is sometimes called the *exponential distribution*. Such a frequency curve approximates a great many populations where the observations involve items whose status changes over time. Exponential distributions have been used to characterize equipment lifetimes until failure. They are also common in describing the time between successive arrivals by cars at a toll booth or by emergency hospital

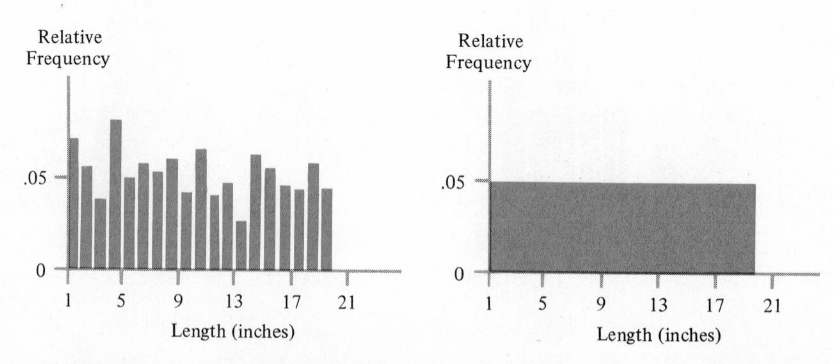

FIGURE 2-14 Sample histogram (left) and population frequency curve (right) for lengths of scrap pieces of lumber.

patients and are therefore very useful in the analysis of waiting-line or queueing situations.

The left-hand side of Figure 2-14 is the histogram for the lengths of pieces of roofing scrap lumber from a construction job. Any piece shorter than 20 inches (the width between studs) cannot be used. Because the lumber comes from the mill in varying lengths, a piece of scrap is likely to be any length between 1 and 20 inches (scrap shorter than 1 inch is not counted here). The histogram is therefore approximated by a rectangle. Because no particular width is favored, such a population is a member of the *uniform distribution* family.

Later we will encounter other, less common shapes and will discuss how one particular shape—the bimodal—can help to identify nonhomogeneous population influences. By classifying the various shapes of frequency distributions, we obtain better population descriptions and can be selective in choosing statistical techniques for analysis.

EXERCISES

2-15 For each of the following populations for the *ages* of persons, sketch an appropriate shape for the frequency distribution and explain the reasons for your choice.
(a) Science fiction novel readers.
(b) Persons not holding full-time jobs.
(c) Persons having false teeth.
(d) Children in a grammar school.

2-16 For each of the following populations, sketch an appropriate shape for the frequency distribution and explain the reasons for your choice.
(a) Times required by a pharmacist to fill new prescriptions.
(b) Last two digits in numbers assigned to telephones in a metropolitan area.
(c) Number of persons employed by manufacturing firms.
(d) Duration of stay by patients hospitalized with nonchronic diseases.

2-5 SUMMARY MEASURES

Now that we have seen how the frequency distribution arranges raw data into a meaningful pattern, we are ready to discuss summary measures. These are numbers that precisely measure various properties of the observations. For quantitative data, there are two major classes of summary numerical values. One of these measures *central tendency* or *location*, a value around which the observations tend to cluster and which typifies their magnitude. The *arithmetic mean* is one of the more commonly used measures of central tendency. Another broad category of numbers provides measures of *dispersion* or *variability* among the observation values. These indicate how the variates differ from each other. Conceptually, the simplest of these is the *range*, which expresses the difference between the largest and smallest observations.

Summary data measures may correspond either to populations or to samples. A summary measure based upon population data is referred to as a *population parameter* and summarizes a particular property of the entire collection of potential observations. Ordinarily, not all of the population will be observed, and usually only a sample is taken. A summary measure obtained from sample data is referred to as a *statistic* and is a number conveying a property of just those data actually observed. Corresponding to every population parameter is an analogous sample statistic that can be computed in a similar manner. For instance, the arithmetic mean for a population has a counterpart that is calculated from the sample. Usually the population's mean is of unknown value and may be estimated from an arithmetic mean calculated from sample data.

Population parameters serve a role in statistical analysis that is not provided by frequency distributions alone. For example, a grammar-school principal may wish to monitor the reading progress of his pupils by making periodic adjustments in curriculum and teaching methods. He may find that an average of each pupil's past reading test scores provides the best indicator of his future performance. A frequency distribution for the scores achieved by first graders would provide less specific information for his periodic assessments than the average score. Similarly, a hospital chief of staff can use recovery time data in evaluating the effectiveness of a particular treatment. A frequency distribution of recovery times may be less adequate for this purpose than an average time per patient.

When the data are qualitative, the parameter used is the *proportion* of observations having a particular attribute. Thus, a federal agency investigating sex discrimination in employment may be interested in the proportion of females employed in various job categories. If this proportion is judged too low in certain jobs, pressures may be applied to remedy the situation. In quality control, the proportion of defective items often serves as a guideline for accepting or rejecting a production batch or for initiating remedial action.

Populations may be easily compared using parameter values. For example, a young person choosing a career may consider the typical earnings levels achieved by members of the various professions. He may find that half of all the doctors in his state earn above $21,000, a *median* income level of this population, while the median income for lawyers is only $15,000. This information may influence him to pursue a medical career rather than a legal one.

When generalizing about a population from a sample, the most common inferences made are those regarding the value of a parameter. The median income of all physicians would be estimated from a limited sample of doctors. The average test score of all persons who will be taking an aptitude test can be projected from the scores of the initial few persons taking it. A decision based upon the proportion of defectives in a shipment is made using the results of detailed testing conducted on a sample of items. The procedures of statistical inference are largely concerned with estimating the value of a particular population parameter or with testing to determine whether an assumed value holds.

2-6 THE ARITHMETIC MEAN

The arithmetic mean is the most commonly encountered and best understood of the measures of central tendency. Consider the heights (in inches) of five men: 66, 73, 68, 69, 74. The mean height is calculated by adding these values and dividing the sum by the number of men (5):

$$\frac{66+73+68+69+74}{5} = 70 \text{ inches}$$

If the five heights represent but a sample from a population consisting of the students in a university, then 70 inches would be the value of the *sample mean*. If, instead, they are the entire set of observations for the population of the heights of men in a five-man office, then the above calculation yields a *population mean* equal to 70 inches.

Symbolic Expressions for Calculating the Arithmetic Mean

Since the arithmetic mean is calculated in the same fashion for any group of raw data, it will prove convenient to express it symbolically. Because the value of a particular observation may not yet be known, each observation can also be given a special symbol. The letter X is traditionally used to represent the arithmetic mean. To distinguish each observation, we will use the numbers 1, 2, 3, . . . as *subscripts*. Thus, X_1 represents the first observation value, X_2 the second, X_3 the third, and so on. The symbol X_1 is referred to as "X sub 1." Another advantage to expressing observation values by symbols is that we may use algebraic expressions to state precisely how each statistical measure is to be calculated, so that the same procedure is indicated for any set of data.

The sample mean is itself represented by a special symbol $\bar{X}$, which we call "X bar." In calculating the sample mean, we must divide the sum of the values by the number of observations made. This is the *sample size*, which is represented by the letter n. We may use the following formula to calculate the ***sample mean***

$$\bar{X} = \frac{X_1+X_2+\cdots+X_n}{n} \tag{2-2}$$

This expression applies for any sample result. In our previous illustration, the sample size was $n = 5$ and the observed values were $X_1 = 66$, $X_2 = 73$, $X_3 = 68$, $X_4 = 69$, and $X_5 = 74$.

A similar calculation applies for the *population mean*, which is traditionally referred to by the Greek letter *mu*, μ. (Several other population parameters are also represented by Greek letters; μ is the equivalent of "m," the first letter in mean.) When all the observation values from the population are available, then μ is found in the same way $\bar{X}$ is. Generally, we do not compute μ directly because the entire population is not usually observed. Ordinarily, the value of μ remains unknown; only $\bar{X}$ is calculated and this value then serves as an estimate of μ.

Sometimes it is convenient to use an even more concise expression than the above formula. Just as the plus sign + means that two quantities are added together, we may represent the sum of several values by a *summation sign*. For this purpose, mathematics employs the upper case Greek *sigma*, $\sum$. We may then express the sample mean by

$$\bar{X} = \frac{X_1 + X_2 + \cdots + X_n}{n} = \frac{\sum_{i=1}^{n} X_i}{n}$$

The term involving the summation sign summarizes the procedure: take all the X's, add them together, and then divide this result by n. In most statistical applications, where the successive subscripts are the increasing sequence of integral values, $1, 2, \cdots, n$, we may drop the subscripts entirely and use instead the further abbreviated formula:

$$\bar{X} = \frac{\sum X}{n} \tag{2-3}$$

Calculating the Mean from Grouped Data

When the number of observations is very large, calculation of the mean can be a tedious chore, fraught with potential error. If the raw data are contained in a file accessible by a digital computer, then experssion (2-3) indicates the best procedure for finding the sample mean. Otherwise, a shortcut method may sometimes be used.

Generally, the first step in analyzing or describing a collection of raw data is to construct a frequency distribution by arranging the raw data into groups or classes according to size. It is usually possible to obtain a good approximation to the sample mean using only typical values from each group and the corresponding class frequencies. Not only can such a shortcut procedure save time and effort, but it may be the only practicable way to obtain a mean, for sometimes data are published already grouped into classes and the raw data are unavailable.

Representing the class frequencies of successive groups by the letters $f_1, f_2, \cdots$, denoting the corresponding class interval midpoints by the letters $X_1, X_2, \cdots$, and letting n represent the sample size, we may apply the following

shortcut expression for calculating the sample mean from grouped data

$$\bar{X} = \frac{f_1 X_1 + f_2 X_2 + \cdots + f_C X_C}{n} = \frac{\sum fX}{n} \qquad (2\text{-}4)$$

where C is the number of classes.

We illustrate the calculation of the mean in Table 2-13 using the grouped data for the gasoline mileage frequency distribution given in the first two columns.

All values falling within a class interval are represented by the midpoint. The first of these, 15, is found by averaging the limits of the first class interval: $(14.0 + 16.0)/2 = 15$. The remaining class interval midpoints can then be easily found by adding one class interval width (here, 2 miles per gallon) to the preceding class midpoint.

The calculation in expression (2-4) is an example of a *weighted average*. Each class midpoint X_i is weighted by the respective frequency f_i. The sum of the weighted products is then divided by n, which equals the sum of the frequencies or weights. We will encounter weighted averages again in Chapter 5 when we discuss special topics in probability.

The value 19.78 found in Table 2-13 is only an *approximation* of the sample mean and can be expected to differ from the value calculated using expression (2-2). The reason for this is that all gasoline mileages falling within a particular class interval are represented by a single number. Furthermore, other groupings of gasoline mileages into classes would result in different midpoint values and frequencies, and the calculation using expression (2-4) might turn out to be either larger or smaller than 19.78. The raw gasoline mileage data are provided in Table 2-7 (p. 25). The arithmetic mean of these, calculated without grouping the data, is 19.718 miles per gallon. This differs only slightly from the value calculated for $\bar{X}$ using the shortcut method. For practical purposes, the approximation is usually close enough.

TABLE 2-13 Calculation of the Mean Gasoline Mileage Using Shortcut with Grouped Data

Gasoline Mileage (miles per gallon)	*Number of Cars* f	*Class Interval Midpoint* X	fX
14.0–under 16.0	9	15	135
16.0–under 18.0	13	17	221
18.0–under 20.0	24	19	456
20.0–under 22.0	38	21	798
22.0–under 24.0	16	23	368
Totals	100		1,978

$$\bar{X} = \frac{\sum fX}{n} = \frac{1{,}978}{100} = 19.78 \text{ miles per gallon}$$

Open-Ended Class Intervals

Special difficulties arise when the data have been grouped so that either the lowest or highest class interval is open-ended, for the midpoint of such an interval is undefined. For example, the frequency distribution of personal earnings may have a class interval of "$100,000 or more." In place of the appropriate fX value for this interval, the total earnings above $100,000 for all persons must be used. If the exact figure is unobtainable, then it must be estimated in some appropriate manner.

EXERCISES

2-17 The following numbers of new welfare cases were assigned in one county during each month of a certain year:

14, 6, 12, 19, 2, 35, 5, 4, 3, 7, 5, 8

Calculate the arithmetic mean cases per month.

2-18 A data processing manager has bought remote terminal processing time for running special jobs at two different computer "utilities." He wishes to sign a long-term contract with the firm whose computer causes, on the average, the least delay. The numbers of minutes of delayed processing encountered per week during trial periods with each firm are

CompuQuick	210, 15, 47, 93, 104
Dial-a-Pute	18, 341, 523, 25, 19, 293, 115, 203

Assuming that trial experience is representative of future performance, which firm should get the business? Substantiate your answer with appropriate calculations.

2-19 A medical researcher is testing the susceptability of five breeds of rabbits to a particular virus strain. In doing this he has exposed a sample of each rabbit population in ten different environments to the virus. The following percentages of rabbits became infected:

Percentage of Infected Rabbits

Environment	*Breeding Stock*				
	A	B	C	D	E
1	36	9	11	33	18
2	43	16	5	17	17
3	49	21	6	45	23
4	18	14	14	29	6
5	17	33	25	17	31
6	32	8	12	27	42
7	24	19	11	61	19
8	19	17	9	47	13
9	28	26	28	35	26
10	36	11	14	14	33

(a) Find the mean percentage of infection for each rabbit stock.
(b) Using your answers to part (a), determine the mean percentage of infected rabbits as a whole.
(c) Which breed do the sample data indicate suffers the least from the virus?

2-20 From the frequency distribution given below, calculate the sample mean length of logs arriving at a sawmill:

Length (feet)	*Number of Logs* f
10.0–under 15.0	107
15.0–under 20.0	253
20.0–under 25.0	1,213
25.0–under 30.0	2,412
30.0–under 35.0	506
35.0–under 40.0	77
40.0–under 45.0	3

2-21 Consider the percentages of infected rabbits in Exercise 2-19 as a single collection of raw sample data.

(a) Construct a sample frequency distribution, using six classes of equal width and beginning with 5.0–under 15.0.

(b) Calculate the sample mean using the grouped data from part (a).

(c) Calculate the sample mean using the original ungrouped data. By how much is this above or below your answer to part (b)?

2-7 THE MEDIAN AND THE MODE

The Median

After the mean, the most common measure of central tendency is the *median*. Like the mean, the median provides a numerical value that is typical of quantitative sample observations. The *sample median*, which we denote by the letter m, is the central observation when all the data are arranged in increasing sequence. For the heights 66, 68, 69, 73, and 74 inches, the median is $m = 69$ inches, the central value. If a person 70 inches tall were added to the initial group, then the median would be obtained by averaging the two central values from the following: 66, 68, 69, 70, 73, 74. Thus, the sample median would be $m = (69 - 70)/2 = 69.5$ inches.

In general, the ***median is the value above or below which lies an equal number of observations.*** When there is an even number of values, the median is the number splitting the data exactly into two halves. We find the median from the raw data by sequencing the observations from lowest to highest, and then selecting the central value if there is an odd number of values or averaging the two central values if there is an even number of observations.

When the data have already been grouped into a frequency distribution, the median will fall somewhere in the first class interval above which fewer than half the observations lie. (Its approximate position in this case can be found by using an interpolation procedure too complex for the scope of this book.)

The counterpart population parameter to m is the *population median*, denoted by a capital letter M. This median may be found in the same fashion from the entire collection of observed values. Ordinarily, we directly calculate a value for m only, and, as with $\bar{X}$ for μ, this value may then serve as the basis for estimating the value of M.

The Median Contrasted with the Mean

The arithmetic mean and the median are both averages, each in its own sense. The mean is the arithmetic average of variates, and the median is the average of position. When we use the term *average*, we mean the arithmetic mean.

A mean has nice mathematical properties. It can be algebraically manipulated; the combined mean of two populations may be calculated from the individual means. Due to its mathematical properties, many statistical techniques employing the mean have been developed. The median is not as well suited to mathematical operation. For example, the median of a combined population cannot be obtained from the separate component population medians. Because of the mathematical difficulties associated with the median, fewer statistical techniques employ it.

On the other hand, the mean is influenced by extreme values to a much greater degree than the median. Consider the net worth levels of your close friends. Suppose that you meet one of the world's few billionaires and include him in your circle of friends. The mean level of wealth would be so distorted by this person that, according to the mean, your friends would, "on the average," all be multimillionaires—hardly a meaningful summary. But the median would not be significantly influenced by the billionaire. The median is more democratic, each elementary unit having but one "vote" in establishing the central location.

Generally, the median provides a better measure of central tendency than the mean when there are some extremely large or small observations.

The Mode

A third measure that may be used to describe central tendency is the *mode*, the *most frequently occurring* value. Consider the simple illustration provided by the collection of five observations: 2, 3, 4, 4, and 7. The mode is 4, because 4 occurs most often.

The interpretation of the statical mode is analogous to that of the fashion mode. A person dressing in the current style is "in the mode." But a current fashion can be a poor description of what most persons are wearing, because of the variety of styles worn by the general public. In statistics, the mode only tells us which single value occurs most often; it may therefore represent a minority of the observations.

As a basis for decisions, the mode can be insidiously undemocratic. For example, most shoe stores stock only the most popular sizes, so the mode is their deciding parameter. One reason for this is that the turnover on unusual sizes is so low that they are unprofitable. When such sizes are stocked, there is ordinarily little choice of color or style (and what there is will usually be on the conservative side). A significant number of persons have great difficulty buying shoes and may have to resort to mail order purchasing. Not being of modal size, they are forced to be hopelessly out of mode in the sense of fashion as well.

When the data are grouped into classes, the mode is represented by the midpoint of the interval having the greatest class frequency. We refer to this group as the *modal class.* When the frequency distribution is portrayed as a smoothed curve like the one in Figure 2-15, the mode corresponds to the possible observation value lying beneath the highest point on the frequency curve—the location of maximum clustering. The mode can therefore serve as a basis for comparing the typicality of the other measures of central tendency.

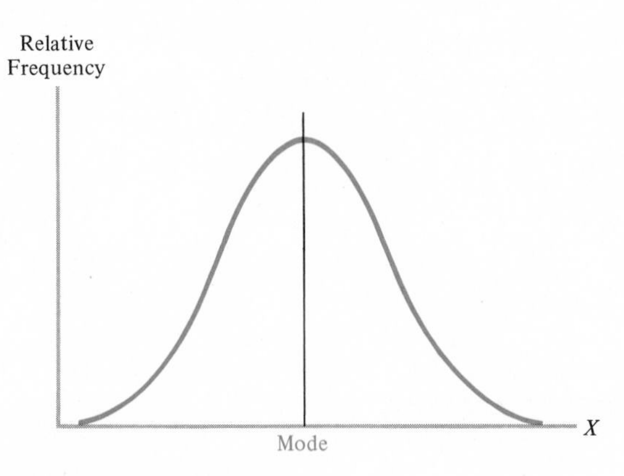

FIGURE 2-15 Illustration of location of mode.

Positional Comparison of Measures

When the population has a *symmetrical* frequency curve like the one in Figure 2-16(a), the mean, median, and mode coincide. When the population is not symmetrical, the mean and median will lie to the same side of the mode. The frequency curve in Figure 2-16(b) has a tail tapering off to the right. Such an asymmetrical frequency distribution is *skewed to the right*, so that the population values cluster around a relatively low value, although there are some extremely large observation values. In this case, the mean lies to the right of the mode, reflecting the influence of the larger values in raising the arithmetic average. The median, less sensitive to extremes, must lie to the left of the mean. A rightward skewed distribution will have the center of cluster below the point dividing the population into two groups of equal frequency; that is, the mode will be smaller than the median. The median will therefore lie somewhere between the mode and the mean.

Because a rightward skewed population will always have a median smaller than its mean, subtracting the median from the mean results in a positive difference. For this reason, such a frequency distribution is said to be *positively skewed.*

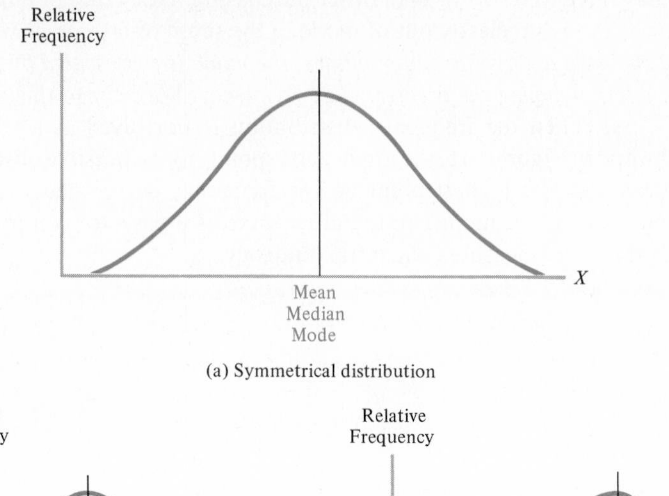

(a) Symmetrical distribution

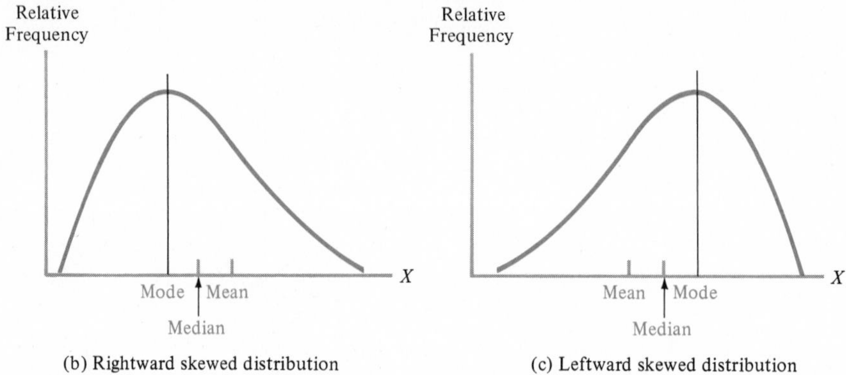

(b) Rightward skewed distribution

(c) Leftward skewed distribution

FIGURE 2-16 Positional comparison of center measures for symmetrical and skewed frequency distributions.

This expression also reflects the fact that the tail of the distribution tapers off in a positive direction from the center of the population.

Generally, a characteristic having a lower limit but no theoretical upper boundary will result in positively skewed populations. This would be the case with the annual wages earned by construction workers in the United States. A lower limit would be established by the legal minimum wage. Some blue-collar aristocrats, such as operating engineers (who drive heavy equipment), plumbers, electricians, and steeplejacks, earn more than many doctors (when their overtime premiums are considered). In total numbers, however, the high earners are relatively sparse, creating a long, thin, rightward tail in the frequency curve of this population. Much economic data have positively skewed frequency distributions.

Figure 2-16(c) shows a frequency curve *skewed to the left*—the direction toward which its tail points. Because the extremes are relatively small values, the mean lies below the mode. Again, the median will lie somewhere in between the mean and the mode. Subtracting the median from the mean (a smaller value) results in a negative difference. The frequency curve is therefore said to be *negatively skewed.*

Negatively skewed distributions result when the observed values have an upper limit and no significant lower boundary. For example, the ages of viewers watching a famous dance band leader's television program, where the musical taste runs to "old favorites," would be a negatively skewed population. Aimed at a preponderantly middle-aged and elderly audience, with commercials touting denture deodorants, laxatives, and home health care remedies, the show would not be too appealing to the younger set. Some young people might watch it, but not many; a few tots with late bedtimes might appreciate the hoopla and congeniality of the program. These young persons would bring the mean viewing age below what should be considered typical, so that here the *median* age would more truly represent a typical population value.

The median is the most realistic measure of location for data having skewed distributions.

Bimodal Distributions

The mode has been defined as the most frequently occurring population value. But suppose there are two or more values of equal or nearly equal occurrence. Then there are two or more modes. When a population has two modes it is said to be *bimodal.*

The presence of more than one mode has a special significance in statistical analysis: it indicates potential trouble. It is usually dangerous to compare bimodal populations or to draw conclusions about them, because they usually arise when there is some nonhomogeneous factor present in the population. Figure 2-17(a) presents the frequency curve of a population of height measurements for adult patients admitted to a large hospital during the past year. The curve is bimodal, with the two modes at 5′3″ and 5′10″. In this case, the nonhomogeneous factor is sex, for the two modes result from the fact that an equal number of male and female patients are measured. It would be more meaningful to separate the population into male and female populations and analyze the two more homogeneous populations separately.

To illustrate the potential difficulty, suppose we wish to compare patient heights in the year before the new maternity ward opened to those in the current year. The proportion of female patients would be higher in the second year, so the overall mean height would be less. The patients would not have become smaller as a cursory analysis might indicate. By separately comparing the two female populations, the mean heights would not noticeably change over the two-year period.

Figure 2-17(b) shows another population having a double-humped frequency curve. This curve represents the high-school grade-point average of students at a university. Humps in frequency occur at 2.8 and 3.5. Although the mode is 3.5, we would still classify this population as bimodal because there

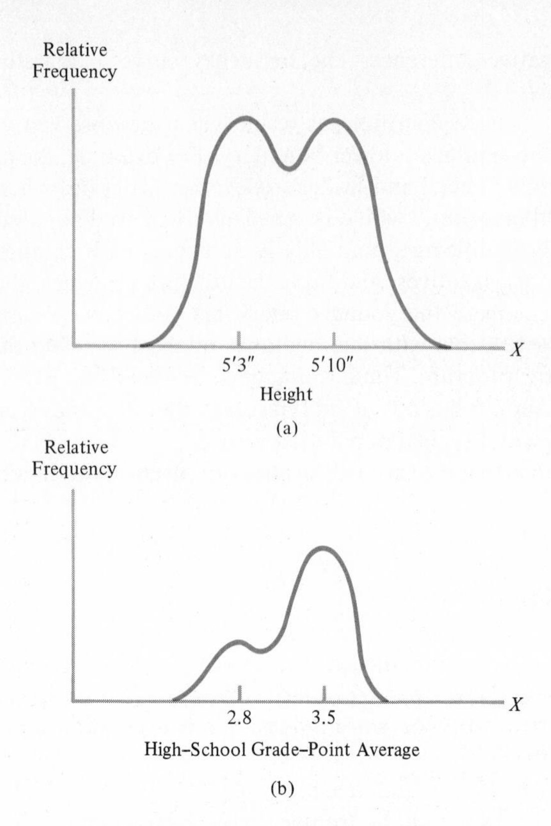

FIGURE 2-17 Illustrations of bimodal frequency distributions.

is a nonhomogeneous factor influencing the high-school grades of the students. In this case, the factor is due to a relatively small group of disadvantaged students who are to be given remedial work. Although comparatively ill prepared at present, it is contemplated that their *college* grade-point averages will not substantially differ from their fellow students.

EXERCISES

2-22 The numbers of major automobile accidents occurring each month during the past year in a certain city were

0, 1, 3, 4, 5, 2, 2, 6, 7, 2, 0, 1

Letting these data represent sample data for accidents in cities of similar size, calculate the sample mean, median, and mode.

2-23 The sample frequency distribution for the lifetimes of a particular stereo record playing cartidge is as follows:

Lifetime (hours)	*Number of Cartridges*
0.0–under 50.0	5
50.0–under 100.0	16
100.0–under 150.0	117
150.0–under 200.0	236
200.0–under 250.0	331
250.0–under 300.0	78
300.0–under 350.0	27
350.0–under 400.0	8

(a) Calculate the sample mean.
(b) Find the sample mode.
(c) Suppose that the sample median has value $m = 205$. Does the mean lie above or below the median? Does this indicate that the distribution is positively or negatively skewed?

2-24 Using the following family installment debt data, determine the sample median.

$2,032	$ 232	$ 493	$5,555
597	4,893	4,432	4,444
203	796	978	329
97	852	1,427	972
3,333	1,712	2,121	438
1,212	1,940	5,067	705
5,769	1,843	4,337	3,976
2,347	3,525	5,213	3,034
2,137	3,414	4,896	5,035
2,049	4,327	2,172	4,222

2-25 The instructor in a statistics course gives four exams of equal weight. From the scores on these, he will determine a single central score value. Suppose that he lets each student decide *in advance of the first test* whether his particular grade will be determined by his mean or by his median test score.
(a) Would you request the mean or the median? Why?
(b) For each of the following hypothetical sets of test scores, indicate whether the mean or the median would provide the greatest central value:

(1)	(2)	(3)	(4)
95	80	80	60
60	50	75	80
75	75	65	90
65	70	60	90

2-26 For each of the following situations, indicate a possible source of nonhomogeneity in the data and discuss why the population ought or ought not to be split to better serve the purposes of the statistical study.
(a) Six months ago, a new driver safety program was initiated by a state's highway patrol. For the past year, the governor has kept records of the number of weekly accidents. A study is being made to evaluate highway construction standards.
(b) An appliance manufacturer wishes to make a powerful home vacuum cleaner. To be most effective, the cleaner must be heavier than the normal weight for such an appliance. A representative group of employees, including men and women, is used to obtain data on how much weight each person can carry up a flight of stairs without excessive fatigue. These data will be used to set a maximum vacuum cleaner weight.
(c) An automobile manufacturer has obtained data on the total man-hours required to assemble each car made at two identically equipped plants of the same size. Each plant produces cars of makes and models as ordered by dealers in their respective geographical regions. These data will be used to establish production standards for each car model.

2-8 MEASURING VARIABILITY

Alone, any value of central tendency inadequately summarizes the observations. One value does not indicate how different from each other all the values are. Practically all samples and populations exhibit variability. Nowhere is this more true than in nature. Just as persons come in different shapes and sizes, so all natural phenomena vary in important characteristics. There are wet and dry years. Earthquakes occur with varying intensities. People vary in their talents. Economic growth is unsteady.

Importance of Variability

Measuring variability is just as important as finding central tendency. The role of variability is illustrated in Figure 2-18. Here, we compare the frequency curves of the earnings for dairies and ranches. These populations, reflecting both good and bad years, were determined in such a way that the median incomes were the same. In each case, the populations are positively skewed; losses occurred for a portion of both dairies and ranches. But the losses incurred by some ranchers were more frequent and were often larger than those of the worst performing dairy. On the other hand, many ranches were more profitable than the best dairies. Ranch earnings exhibit *greater variability* than dairy incomes. Although both forms of agriculture involve cattle and must meet comparable feed requirements, the cost of which has fluctuated wildly in recent years, the dairy farmer is blessed with stable milk prices (often due to regulation), while the rancher must sell his beef at the volatile market price. The greater variability in income makes ranching a riskier venture than owning a

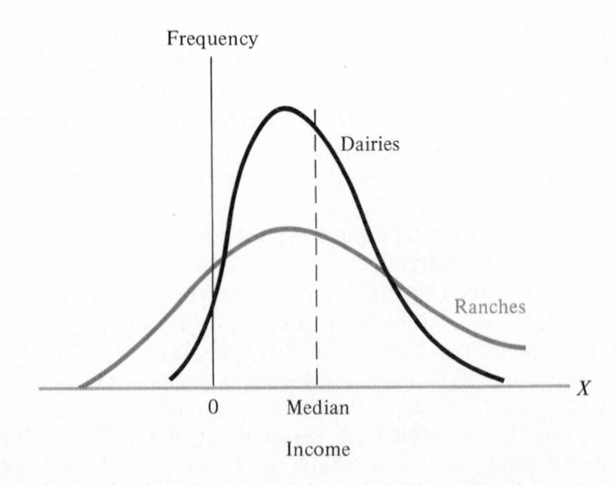

FIGURE 2-18 Frequency curves for the earnings of dairies and ranches.

dairy, even though both agricultural enterprises have had identical central tendency (equal medians). Here, statistical variability might help the new agricultural college graduate in choosing between ranching or operating a dairy.

Another illustration of the importance of variability in statistical analysis is provided by a decision to modify a chemical process. Figure 2-19 shows the frequency distributions of times required under two methods being pilot tested. Method A has a mean of 12 hours, whereas method B takes an average of 10 hours. Method B shows substantially greater variability than A. Assuming that each process costs the same on an hourly basis, B appears to be cheaper. But process A would be easier for planning purposes, because its completion times are less varied. If the final product is perishable or must be made to customer order, so that it cannot be stored in quantity for very long, then method A might turn out to be superior. *In a great many populations, reduction in variability is itself an improvement.*

Variability or dispersion may be measured in two basic ways: in terms of *distances* between particular observation values, or in terms of the *average deviations* of individual observations about the central value.

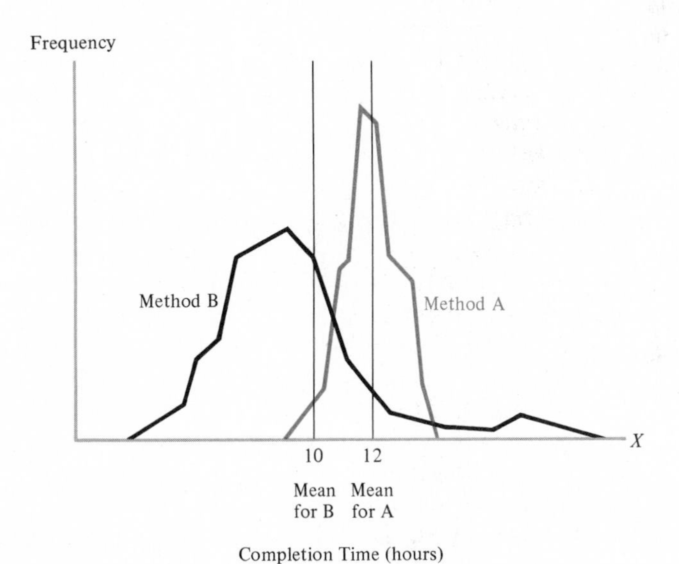

FIGURE 2-19 Frequency distributions for the completion times of two chemical processes.

Distance Measures of Dispersion

Distance measures of dispersion are popular when the only purpose is to describe a collection of data. The most common of these is the *range*, which is obtained by subtracting the smallest observation from the largest. For example, suppose that the five students in a college chemistry honors program have the following IQs: 111, 118, 126, 137, 148. The range of these values is 148 − 111 = 37. The figure 37 represents the total spread in these observations. The range provides a very concise summary of the total variation in a sample or a population. But its major disadvantage as a useful measure is that it ignores all but the two most extreme observations. These two numbers may be very far from typical values, even among the higher and lower observations.

Other distance measures ignore the most extreme observations. These are *interfractile ranges*, which express the differences between two values called fractiles. A *fractile* is a point below which some specified proportion of the values lies. The median is the .50 fractile because half of all observations lie below this value. When expressed as percentages, the analogous measure is called a *percentile*; the median is the 50th percentile. The most commonly used fractiles or percentiles are the *quartiles*. These divide the observations into four groups, each containing roughly an equal number of observations. The .25 fractile, or 25th percentile, is the *first quartile*. This is the value below which 25 percent of the observations lie. The median is the *second quartile*, while the 75th percentile is the *third quartile*.

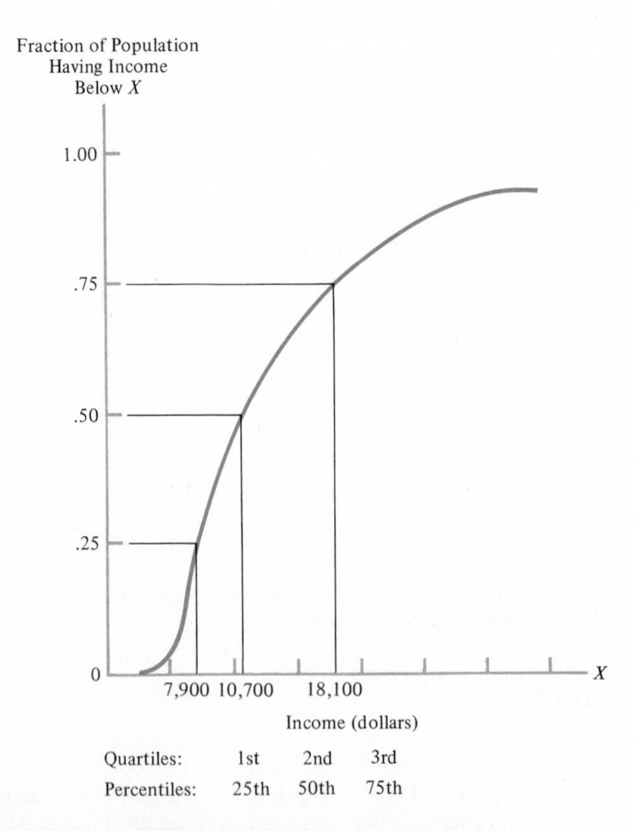

FIGURE 2-20 Cumulative relative frequency distributions for family income.

The most common dispersion measure based upon fractiles is the *interquartile range*. This is the difference between the third and first quartiles and thus *represents the middle 50 percent of the observation values.* Figure 2-20 shows an ogive for family income. The fractile values may be read from this graph. The first quartile is \$7,900 and the third quartile is \$18,100, so that the interquartile range is \$18,100 − \$7,900 = \$10,200.

Other interfractile ranges, such as those representing the middle 90 percent or 99 percent, may be used, but seldom are.

Measures of Average Deviation

The main disadvantage of the distance measures of dispersion is that they do not consider every observation. To include all observations, we can calculate how much each one deviates from the central value and then combine these deviations through averaging. The most common deviation considered is the difference between the observed value and the mean:

$$\begin{array}{ll} X - \overline{X} & \text{for sample data} \\ X - \mu & \text{for population data} \end{array}$$

A straightforward averaging of the deviation values will always result in zero. For example, consider the five values 1, 2, 3, 4, and 5. Their mean is 3. Subtracting 3 from each number, we obtain deviations −2, −1, 0, 1, and 2. These add up to zero, so the average deviation is zero. We might avoid this difficulty by ignoring the minus signs. That is, we could average the *absolute values* of each deviation (2, 1, 0, 1, and 2) and obtain the *mean deviation* 1.2, which reflects every observation. But because of mathematical difficulties in working with absolute values, two other measures of dispersion are more commonly used.

The Variance and Standard Deviation

The most important measure of variability is found by averaging the *squares* of the individual deviations; the resulting value is the *mean of the squared deviations.* This calculation may be performed on the IQs for the five chemistry students previously cited. The mean value is

$$\frac{111+118+126+137+148}{5} = 128$$

The deviations from the mean value are

$$\begin{aligned} 111 - 128 &= -17 \\ 118 - 128 &= -10 \\ 126 - 128 &= -\ \ 2 \\ 137 - 128 &= \ \ \ 9 \\ 148 - 128 &= \ \ 20 \end{aligned}$$

and the mean of the squared deviations is:

$$\frac{(-17)^2+(-10)^2+(-2)^2+(9)^2+(20)^2}{5}$$

$$=\frac{289+100+4+81+400}{5}=\frac{874}{5}=174.8$$

The measure of variability obtained in the above manner is referred to as the *variance*. It may be calculated using data from either the sample or the entire population. When all the population data are available, the above procedure provides the *population variance*, which is represented in abbreviated form by σ^2. (The symbol σ is the lower-case Greek *sigma*, and σ^2 is called "sigma squared.") The expression used for calculating the population variance is

$$\sigma^2=\frac{\Sigma(X-\mu)^2}{N} \tag{2-5}$$

Here the sum of the squared deviations from the population mean μ is found, and—because every possible observation is made—this sum is divided by the *population size*, denoted by the letter N. In the above illustration the *entire population* consisted of $N=5$ observations.

In practice, a value for σ^2 cannot be computed because most populations are so large that usually only a sample is taken. For sample data, the counterpart measure of variability is the *sample variance*, denoted symbolically by s^2. From the following analogous expression *using sample size n in place of N and sample mean $\bar{X}$ in place of μ*, we calculate the ***sample variance***

$$s^2=\frac{\Sigma(X-\bar{X})^2}{n-1} \tag{2-6}$$

Here the squared deviations from $\bar{X}$ are averaged using only the *n sample observations*. Notice that the expression for s^2 has a slightly different form than that for σ^2, because $n-1$ is the divisor instead of the complete sample size n. For reasons which will be explained in Chapter 7, doing this makes the resulting s^2 value a better estimate of the usually unknown value for σ^2.

There are two practical difficulties associated with using the variance. One is that the variance is usually a very large number compared to the observations themselves. Thus, if the observations are largely in the thousands, the variance will often be in the millions. Second, the units in which the variance is expressed are not the same as the observations. In the previous example, the variance is 174.8 *squared* IQ points. This is because the deviations, measured in IQ points, have all been squared. The variance of heights, originally measured in feet, will thus be expressed in square feet. The variance for gasoline mileages would be in units of squared miles per gallon.

In spite of these difficulties, the variance has some very nice mathematical properties that make it extremely important in statistical theory. Furthermore, the difficulties may be simply overcome by working with the square root of the variance, called the *standard deviation*. [As an aid, Appendix Table A (page 542) provides square roots for whole numbers between zero and 1,000.]

The Standard Deviation

Taking the positive square root of the variance in IQ levels, we have

$$\sqrt{174.8} = 13.22$$

which yields the standard deviation in IQ level. The standard deviation is expressed in the same units as the observations themselves; the value 13.22 is a point on the same numerical scale.

The *sample standard deviation*, represented by the letter s without its exponent 2. The following expression may be used to calculate the ***sample standard deviation***

$$s = \sqrt{\frac{\sum (X - \bar{X})^2}{n - 1}} \tag{2-7}$$

Notice that on both sides of the equation we simply have the square root of the expression for the sample variance. Likewise, for the population parameters, the *population standard deviation* is denoted by σ, and σ is equal to $\sqrt{\sigma^2}$. Both the variance and the standard deviation provide the same information; one can always be obtained from the other. The standard deviation is a practical descriptive measure of dispersion, whereas the variance is generally used in developing statistical theory.

The sample standard deviation is computed in Table 2-14 for the heights of $n = 5$ persons cited earlier. (These five observations are but a sample from a very large population.)

TABLE 2-14 Calculation of Sample Standard Deviation for Heights

	Height (inches) X	*Deviation* $X-\bar{X}$	*Squared Deviation* $(X-\bar{X})^2$
	66	66 − 70 = −4	16
	73	73 − 70 = 3	9
	68	68 − 70 = −2	4
	69	69 − 70 = −1	1
	74	74 − 70 = 4	16
Totals	350	0	46

$$\bar{X} = \sum X/n = 350/5 = 70 \text{ inches}$$

$$s = \sqrt{\frac{\sum (X-\bar{X})^2}{n-1}} = \sqrt{\frac{46}{5-1}}$$

$$= \sqrt{46/4} = \sqrt{11.5} = 3.391 \text{ inches}$$

Shortcut Calculations of Variance and Standard Deviation

The calculation of the variance and standard deviation can be quite a chore when the number of observations is large. Two shortcut procedures can ease this task. One of these reduces the amount of arithmetic involved. The other technique groups data into classes; s may then be calculated from the frequency distribution, rather than from the original observations, which may be unavailable.

The sample standard deviation may be calculated from individual values with an equation that is mathematically equivalent to expression (2-7) but usually simpler to use. ***The shortcut procedure for calculating the sample variance and standard deviation using ungrouped data is***

$$s^2 = \frac{\sum X^2 - n\bar{X}^2}{n-1} \qquad s = \sqrt{\frac{\sum X^2 - n\bar{X}^2}{n-1}} \tag{2-8}$$

We illustrate the use of the formula for the standard deviation in Table 2-15 for the above five heights. This procedure involves only one subtraction in place of the n needed to compute the deviations under the original procedures.

The shortcut procedure is recommended whenever s is computed by hand or with a calculator. (To ease this task, Appendix Table A provides the square roots for whole numbers up to 1,000.)

Just as the mean was calculated from the frequency distribution in Section 2-6, we can use grouped data to calculate the sample variance and standard deviation:

$$s^2 = \frac{\sum f(X-\bar{X})^2}{n-1} \qquad s = \sqrt{\frac{\sum f(X-\bar{X})^2}{n-1}} \tag{2-9}$$

TABLE 2-15 Calculation of Sample Standard Deviation Using Shortcut Procedure with Ungrouped Height Data

	Height (inches) X	X^2
	66	4,356
	73	5,329
	68	4,624
	69	4,761
	74	5,476
Totals	350	24,546

$$\bar{X} = \sum X/n = 350/5 = 70 \text{ inches}$$

$$s = \sqrt{\frac{\sum X^2 - n\bar{X}^2}{n-1}} = \sqrt{\frac{24{,}546 - 5\,(70)^2}{5-1}}$$

$$= \sqrt{11.5} = 3.391 \text{ inches}$$

However, we would ordinarily use the following ***shortcut procedure for calculating the sample variance and standard deviation from grouped data***

$$s^2 = \frac{\sum fX^2 - n\overline{X}^2}{n-1} \qquad s = \sqrt{\frac{\sum fX^2 - n\overline{X}^2}{n-1}} \tag{2-10}$$

These equations are similar to the shortcut expressions for ungrouped data (2-8).

The computations for the standard deviation are illustrated in Table 2-16, again using the gasoline mileages from Table 2-13. The first four columns of Table 2-16 are a reproduction of Table 2-13 for the calculation of $\overline{X}$. Two additional columns, for X^2 and fX^2, are needed to calculate the standard deviation. We find that $s = 2.34$ miles per gallon.

TABLE 2-16 Calculation of Sample Standard Deviation Using Shortcut Procedure with Grouped Gasoline Mileage Data

(1) *Gasoline Mileage (miles per gallon)*	(2) *Number of Cars* f	(3) *Class Interval Midpoint* X	(4) fX	(5) X^2	(6) fX^2
14.0–under 16.0	9	15	135	225	2,025
16.0–under 18.0	13	17	221	289	3,757
18.0–under 20.0	24	19	456	361	8,664
20.0–under 22.0	38	21	798	441	16,758
22.0–under 24.0	16	23	368	529	8,464
Totals	100		1,978		39,668

$\overline{X} = \sum fX/n = 1{,}978/100 = 19.78$ miles per gallon

$$s = \sqrt{\frac{\sum fX^2 - n\overline{X}^2}{n-1}} = \sqrt{\frac{39{,}668 - 100\,(19.78)^2}{100-1}}$$

$$= \sqrt{5.4865} = 2.34 \text{ miles per gallon}$$

As in calculating the mean from grouped data, only an *approximate* value for the standard deviation may be obtained with this procedure. For most purposes, this is close enough to the true value that would be calculated from either expression (2-7) or (2-8).

Meaning of Standard Deviation

The standard deviation is a parameter that, when combined with statistical techniques, provides a great deal of information. When the population has a special frequency distribution called the normal curve (to be discussed in detail in Chapter 6), we can find the percentage of observations falling within distances of one, two, or three standard deviations from the mean. When the frequency curve of the population has the form of a normal distribution, then about 68 percent of all observations lie within the region $\mu \pm 1\sigma$. For example, suppose a group of men has a mean height of $\mu = 5'9''$ and a standard deviation of $\sigma = 3''$. If these constitute a normal distribution, then 68 percent of all men will be between $\mu - \sigma = 5'6''$ and $\mu + \sigma = 6'0''$ tall. Furthermore, about 95.5 percent of the population will lie within $\mu \pm 2\sigma$, and 99.7 percent will fall within $\mu \pm 3\sigma$.

The normal curve is described mathematically in terms of just two parameters, μ and σ. Thus, for populations characterized by this curve, we can construct a close representation of the entire frequency distribution knowing only these two values, the mean and standard deviation. We will discuss this point more thoroughly in Chapter 6.

A theoretical result referred to as *Chebyshev's Theorem*, after the mathematician who proposed it, indicates that the standard deviation plays a key role in any population.

CHEBYSHEV'S THEOREM The proportion of observations falling within k standard deviations of the mean is at least $1-1/k^2$.

This says that regardless of the characteristics of the population, the following holds:

Standard Deviations k	*Minimum Proportion of Observations Within* $\mu \pm k\sigma$
1	$1-1/1^2 = 0$
2	$1-1/2^2 = .75$
3	$1-1/3^2 = .89$

Since the above principle applies for any population, it is too general to be of much practical use. Usually a more precise determination of how many observations lie within k standard deviations of the mean can be found when the form of a population's frequency distribution is known, as we have shown for the normal curve. The practical significance of Chebyshev's Theorem is that it shows the importance of information imparted by the population standard deviation.

EXERCISES

2-27 The following sample data for the low November temperatures achieved in a midwestern state have been recorded:

35, 47, 57, 16, 12, 33, 38

(a) Calculate the sample range.
(b) Calculate the sample variance and the sample standard deviation.

2-28 The following data have been obtained to represent the ages of patients in a certain hospital:

25, 16, 50, 19, 42, 37

Calculate the sample variance and standard deviation.

2-29 For each of the following decision situations involving populations, discuss why a central population value may not be wholly adequate by itself.

(a) The population of temperatures actually achieved in each room of a building with standard heating and cooling system control settings is being used to determine whether the system ought to be modified.
(b) Balancing an automobile assembly line requires that sufficient personnel and equipment be positioned at each work station so that no one station will be excessively idle. The population of task completion times at each station is used to help determine how this may be done.
(c) In planning for facilities expansion, a hospital administrator requires data on the convalescent time of surgical patients.

2-30 Using the following frequency distribution for the times to failure of a certain kind of fuse, calculate the mean and standard deviation.

Time to Failure (hundreds of hours)	*Number of Fuses*
0–under 20	500
20–under 40	250
40–under 60	125
60–under 80	61
80–under 100	15
100–under 120	6

2-31 The Educational Testing Service administers the verbal Scholastic Aptitude Test. The following cumulative relative frequency distribution for the base period is

Test Score	*Cumulative Relative Frequency*
200–below 250	.01
250–below 300	.03
300–below 350	.09
350–below 400	.19
400–below 450	.36
450–below 500	.55
500–below 550	.74
550–below 600	.88
600–below 650	.96
650–below 700	.99
700–below 750	1.00

(a) On graph paper, plot the ogive for the above data.
(b) Assuming that intermediate values may be read to a good approximation from your graph, determine the .25, .50, and .75 fractiles for SAT scores.
(c) Using your answers to part (b), calculate the interquartile range for SAT scores.

2-32 The following frequency distribution has been obtained for a sample of $n = 100$ university-student grade-point averages (GPA).

GPA	*Number of Students*
0–under 1.0	1
1.0–under 2.0	7
2.0–under 3.0	58
3.0–under 4.0	34

Using grouped approximation methods, determine the sample mean and standard deviation.

2-33 In describing the extent of quality control in his company, the president of a rubber company states that the mean weight of a particular tire is 40 pounds, with a standard deviation of 1 pound. He adds that around 68 percent of all tires weigh between 39 and 41 pounds, whereas nearly all tires weigh between 37 and 43 pounds. State the assumptions upon which these statements rest.

2-9 THE PROPORTION

In describing a qualitative population, there is one key measure of interest—the *proportion* of observations falling into a particular category. Like those measures already discussed, the population parameter is a separate entity from the sample statistic. The population parameter is referred to as the *population proportion* and is denoted by the lower-case Greek letter *pi*, π.* The counterpart sample statistic is the *sample proportion*, which is represented by P. The following ratio is used to calculate the ***sample proportion***:

$$P = \frac{\text{Number of observations in category}}{\text{Sample size}} \tag{2-11}$$

A proportion may assume various values between 0 and 1, according to the relative frequency at which the particular attribute occurs. Like the other population parameters, π is ordinarily of unknown value and may be estimated from the sample results by P. Statistical procedures for doing this are analogous to those used in estimating a mean from a sample. But there are substantial differences which require parallel statistical development throughout this book.

The proportion is important in many kinds of statistical analysis. For example, it is often used as the basis for taking remedial action. When a keypuncher makes a higher than usual proportion of errors, this will be singled out as a managerial problem. A machine producing a large proportion of oversized or undersized items should be adjusted or repaired. The level of impurities in a drug might be expressed as a proportion; if this figure is too high, the drug cannot be used. In may elections for public office in the United States, the winner is the candidate who receives a plurality, that is, the person who gets the highest proportion of votes.

If a sample of $n = 500$ persons contains 200 men and 300 women, then the sample proportion of women is $P = 300/500 = .60$. If a machine has produced a sample of $n = 100$ parts, 5 of which are defective, then the sample proportion of defective parts would be $P = 5/100 = .05$; this same machine might actually produce defective items at a consistently higher rate than experienced in the sample, so that the population proportion of defective parts could be a different value, such as $\pi = .06$. The proportion may also be expressed as a percentage. Thus, 60 percent of the first sample are women, while 5 percent of the items in the second sample are defective.

* Statistics has its own special notation, and here π is not the 3.1416 used in geometry to express the ratio of the circumference to the diameter of a circle. Just as μ and σ, the Greek equivalents of m and s, are the first letters in the words mean and standard deviation, so π, the Greek p, the first letter in the word proportion.

The proportion is the only measure available for qualitative data. It indicates relatively how many observations fall into a particular category. (When the observations are attributes, such as male or female, central tendency or variability have no meaning.) The type of question answered by P or π is *how many* rather than *how much*.

Sometimes we may wish to express the proportion of observations having a collection of attributes. For example, in Table 2-6 (p. 20), we have the frequency distribution of the letters in a representative sample of 200 characters of ordinary English text. We may wish to find the proportion of vowels. We find that "a" occurs 16 times, while "e" occurs 26 times, and the frequencies for "i," "o," and "u" are 13, 16, and 6, respectively. Thus, in 200 letters, vowels occur $16+26+13+16+6 = 77$ times. The sample proportion of vowels is therefore $P = 77/200 = .385$. These five letters, taken together, occur 38.5 percent of the time in the sample of ordinary English text.

The proportion is not limited to qualitative data. We may also use it to represent the relative frequency of a quantitative category. For example, the median is the .50 fractile. Thus, the proportion of observations falling below the sample median is $P = .50$. Likewise, the sample proportion of items falling below the third quartile is $P = .75$.

EXERCISES

2-34 A government investigator seeks companies exhibiting pronounced discrimination against employees based upon sex. The data on the number of men and women in management positions in five firms within the same industry are provided below:

Firm	A	B	C	D	E
Men	2,342	532	849	1,137	975
Women	156	115	57	145	139

Calculate the proportion of women managers in each firm. Assuming the investigator will more thoroughly study the firm with the lowest proportion of women managers, which firm would be selected?

2-35 Referring to the personal debt data in Exercise 2-24 of Section 2-7 (page 49), determine the sample proportion of families whose indebtedness lies below (a) \$1,000; (b) \$1,500; (c) \$3,000; (d) \$5,000.

2-36 A quality control inspector must reject incoming shipments if the proportion of inspected items found to be defective exceeds .05. For each of the following shipments: (a) determine the sample proportion of defectives found, and (b) state whether the inspector would accept or reject the shipment.

	(1)	(2)	(3)	(4)
Items inspected	100	500	600	1,000
Items defective	7	25	10	39

2-37 For each of the following situations, discuss whether a mean, a proportion, or both would be an appropriate parameter upon which to base decisions.

(a) A garment manufacturer wishes to ship dresses of the highest quality. Many things can cause a dress to be defective: incorrect sizing, improper seams, creases, missing stitches and so on.

(b) The Federal Trade Commission requires that the weight of ingredients in packaged goods be indicated on the label. A soap manufacturer wishes to comply. Underweight production batches (populations) are reprocessed rather than shipped. For the sake of efficiency, a large number of packages (elementary units) must be weighed simultaneously, with the weight of the packaging material subtracted to obtain the weight of the ingredients.

(c) A drug maker is testing a new food supplement which he believes will reduce levels of anemia. The supplement is not expected to work on all patients, but when it does, the extent to which it will reduce anemia by increasing the red corpuscle count in the blood will be measurable.

REVIEW EXERCISES

2-38 A sample of students has been selected at a university, and their ages have the frequency distribution provided below. Determine the relative and the cumulative frequency distributions.

Age (years)	Number of Employees
20–under 25	18
25–under 30	23
30–under 35	15
35–under 40	13
40–under 45	7
45–under 50	6
50–under 55	5
55–under 60	5
60–under 65	8
Total	100

2-39 The following data represent the marital status of the heads of households for a sample of welfare recipients:

File No.	Status	File No.	Status	File No.	Status	File No.	Status
00357	M	06315	D	10586	S	16627	M
01112	M	07448	D	10794	S	17051	M
01267	D	07496	S	10813	M	17834	M
01448	S	08523	S	10915	D	18215	D
02536	D	09117	M	17317	W	19006	M
02699	S	09856	W	12225	D	19853	D
03419	M	10013	M	12806	M	20017	W
03212	W	10074	M	13115	W	20556	W
04896	W	10176	M	14083	S	20562	S
05517	M	10324	D	14118	D	21010	M

where S = single, M = married, D = divorced, and W = widowed.

(a) Determine the frequency distribution for this qualitative population.

(b) Construct a histogram for the results obtained in part (a).

2-40 A statistics instructor needs to analyze the grade-point data for the grades of a random sample of ten statistics students. With 4 points for "A," the possible score range downward to 0 points for "F." The following data have been obtained from departmental records:

3	2	4	1	0
3	2	2	2	1

(a) Determine the sample range in grade points.

(b) Find the sample mean.

(c) Determine the sample variance for grade points.

(d) Determine the sample standard deviation in grade points.

2-41 The following gasoline mileage data for a taxicab fleet are incomplete:

Miles per Gallon	*Number of Cars*	*Relative Frequency*	*Cumulative Frequency*
6–under 8	—	—	—
8–under 10	23	—	29
10–under 12	—	.34	—
12–under 14	17	.17	—
14–under 16	—	—	92
16–under 18	—	—	—
Totals	100	1.00	

(a) Find the missing values.
(b) Plot the cumulative frequency ogive for this population.

2-42 The proportions of weed seeds found in boxes of lawn seed mixtures prepared by a grower have the following cumulative relative frequency distribution:

Proportion of Weed Seed	*Cumulative Relative Frequency*
.00–under .01	.05
.01–under .02	.12
.02–under .03	.25
.03–under .04	.37
.04–under .05	.50
.05–under .06	.67
.06–under .07	.75
.07–under .08	.84
.08–under .09	.89
.09–under .10	.90
.10–under .11	.95
.11–under .12	.98
.12–under .13	1.00

For the proportion of weed seed, determine the following:
(a) The .90 fractile.
(b) The 37th percentile.
(c) The first quartile.
(d) The third quartile.
(e) The interquartile range.
(f) The median.

2-43 The mean lifetimes for cartons of 100-watt, long-life lightbulbs have been established to be normally distributed, with mean 1,500 hours and standard deviation 100 hours. Find the upper and lower bounds for the following limits, and indicate the percentage of all cartons where the mean lifetimes fall within these limits:
(a) $\mu \pm 1\sigma$. (b) $\mu \pm 2\sigma$. (c) $\mu \pm 3\sigma$.

2-44 A group of 350 persons has been categorized by sex, marital status, and occupation. The number of persons in each category combination is provided below.

Sex	*Marital Status*	*Occupation*	*Number of Persons*
Male	married	blue collar	75
Male	married	white collar	37
Male	married	professional	12
Male	single	blue collar	55
Male	single	white collar	38
Male	single	professional	3
Female	married	blue collar	13
Female	married	white collar	32
Female	married	professional	2
Female	single	blue collar	12
Female	single	white collar	66
Female	single	professional	5
		Total	350

Construct the frequency distributions for the populations characterized by (a) sex; (b) marital status; and (c) occupation.

2-45 At a particular college, all entering freshmen must take the SATs, a series of aptitude tests. The following percentiles have been determined:

Score	*Percentile*
450	25th
573	50th
615	75th
729	90th
738	95th
752	99th

(a) Find the first, second, and third quartiles.
(b) What is the median score?
(c) Find the interquartile range.

2-46 The following frequency distribution has been obtained for sample data relating to the viable shelf life of a certain brand of bread:

Hours to Deterioration	*Number of Loaves*
90–under 110	23
110–under 130	37
130–under 150	26
150–under 170	14

Calculate the sample mean and variance using the grouped data.

2-47 The IQ scores achieved by students in a particular school have the cumulative relative frequency distribution provided below:

IQ Score	*Cumulative Proportion of Students*
70–less than 80	.02
80–less than 90	.12
90–less than 100	.24
100–less than 110	.46
110–less than 120	.65
120–less than 130	.79
130–less than 140	.88
140–less than 150	.97
150–less than 160	1.00

(a) Plot the ogive for the above data.
(b) From your graph, read the following values. (The figures obtained will be only estimates of the actual values.)
    (1) The 50th percentile.
    (2) The .75 fractile.
    (3) The first quartile.
    (4) The median.
    (5) The 90th percentile.
(c) Using your answers from part (b), compute the interquartile range.

2-48 (a) Calculate the sample mean and variance for the following ungrouped examination score sample data:

84	77	67	94	90
81	56	89	77	88
74	76	28	80	58
66	77	89	81	78
77	72	94	83	79
93				

(b) The above data have been grouped into the frequency distribution provided below. Calculate the sample mean and variance using the grouped data.

Class Interval	*f*
25–under 35	1
35–under 45	0
45–under 55	0
55–under 65	2
65–under 75	4
75–under 85	11
85–under 95	8
Total	26

(c) The calculation of population parameters using grouped data provides only approximations to their counterparts obtained directly from the ungrouped raw data. Find the amount of error in this approximation procedure for the sample mean by subtracting your answer in part (b) from the value you obtained in (a).

3 The Statistical Sampling Study

By a small sample, we may judge the whole piece.

Miguel de Cervantes

Statistical inference enables us to draw conclusions about a population without having to observe it in its entirety. This is accomplished by means of a *sample*, which is only part of the population. A sample is in contrast to a *census*, when observations are made of the entire population—a procedure often referred to as *complete enumeration*. Since a sample comprises only a portion of the whole, it is not always a very reliable indication of what the population is like, and inferences drawn from it may be erroneous. One question we may ask, then, is if samples may be inaccurate and unreliable, then why do we use them?

3-1 THE NEED FOR SAMPLES

There would be no need for statistical theory if samples were not used to obtain information about the population. Indeed, if the census were employed for this purpose, there would be little need for most of the material contained in this book. Unfortunately, for various reasons, a census may not be practicable and is almost never economical, so that sample information must be relied upon for most applications. There are six main reasons for sampling in lieu of the census: economy, timeliness, the large size of many populations, inaccessibility of the entire population, destructiveness of the observation, and accuracy.

Economic Advantages of Using a Sample

It should be obvious that the taking of a sample, which involves direct observation of only a portion of the population, would require fewer resources than would a census. Consider, for example, a consumer survey attempting to query all the owners of a popular automobile as to their opinions about several proposed colors for next year's model. A questionnaire is to be printed and mailed to all of the owners. Imagine the clerical chore of simply addressing the envelopes; the cost of the postage alone could easily run into five figures. Consider the bookkeeping problems in tabulating the replies.

Perhaps the greatest difficulty would be to insure a 100 percent response. A very high proportion of the questionnaires will simply be ignored along with the plethora of other "junk" mail. Most persons can be reached by telephone calls, telegrams, and personal visits; there will invariably be an irascible few who will slam their doors in the face of the company president coming to call, and some who will have to be bribed with a new car in order to respond. A great many car owners will be very hard to locate, having moved to parts unknown, so that the services of a detective agency may be required to track them down. Some of the owners will never be able to give their replies, having died before the survey is completed. The company would have more trouble completely enumerating its owner population than the FBI has in arresting its ten most wanted men.

As ludicrous as this example is, it vividly points out the very high cost of a census, and it is not an exaggeration. For the type of information desired, the above company would do well to interview a tiny portion of the car owners. Rare indeed is the circumstance that requires a census of the population, and even rarer is one that justifies its expense.

The Time Factor

A sample may provide an investigator with needed information quickly. Often speed is of paramount importance, as in political polling, where the goal is to determine voter preferences toward candidates for public office from sample evidence. The voting public as a whole is extremely fickle, fluctuating in preference, as polls have indicated, right up to the time of the election. If a poll is to be used to gauge public opinion, by necessity it must be very current; the opinions must be obtained, tabulated, and published within a very short period of time. It would not be physically possible to use a census in such an opinion survey; a census would require many months, negating the validity of the results because of significant shifts in public opinion over this time.

The Very Large Population

Many populations about which inferences must be made are quite large. Consider, for example, the population of high-school seniors in the United States—a group numbering about 3,000,000. The plans of these soon-to-be graduates will directly affect the status of universities and colleges, the military, and prospective employers—all of whom desire specific knowledge of the students' plans in order to make compatible decisions to absorb them during the coming year. But no matter how important this information might be to the organizations most affected, a census of the population of post-graduate plans is likely to be physically impossible due to its sheer size. Sample evidence may be the only means available for obtaining information feedback from high-school seniors.

Some populations appear to be infinite in size. Of course, a truly infinite population of physical objects does not exist. (Scientists have even estimated a finite value for the number of elementary particles in the universe.) Consider the theoretical problems that one would encounter in attempting to census a hypothetical infinite population. No matter what resources were used or how much time was expended, it would be impossible to observe all of the elementary units. At any point in time, the census would be only partially completed—the procedure would be eternal. *Information bearing upon the infinite population can only be obtained from a sample.*

The infinite population is a very convenient one to discuss in theory, because it has many properties that are ideally suited to mathematical techniques. Therefore, it is often desirable to label a whose class of populations "infinite" when they really are not. Most notable are populations whose elementary units result from a continuing process. Since it would be impossible to observe every elementary unit over a reasonable period of time because new ones are continually being created, it would be as difficult to take a census of such a population as it would be to take a census of an infinite population. Hereafter, when we refer to a population as being *infinite*, we mean it is *one for which census in a reasonable time period is impossible.*

Partly Inaccessible Populations

Some populations contain elementary units that are so difficult to observe that they are in a sense inaccessible. One example is the population of crashed aircraft; planes that crash and sink in the deep ocean are inaccessible and cannot be directly studied to determine the physical causes of their failure. Similar conditions hold in determining consumer attitudes. Not all of the users of a product can be queried. Some may be incarcerated in prisons or mental hospitals and have limited contact with the outside world. Some individuals may be insulated from harassment by survey takers for more noble reasons; consider the limited accessibility to the President of the United States or the Pope.

Whenever some elementary units are inaccessible, sampling must be used to provide the desired information about the population. The above illustrations demonstrate physical inaccessibility. As a practical matter, an elementary unit's accessibility may be limited for economic reasons alone: those observations which are deemed too costly to make are, in a real sense, also inaccessible.

Destructive Nature of the Observation

Sometimes the very act of observing the desired characteristic of the elementary unit destroys it for the use intended. Classical examples of this occur in quality control. For instance, in order to test a fuse to determine whether it is defective, it must be destroyed. In order to obtain a census of the quality of a lot of fuses, all of them must be destroyed. This negates any purpose served by the quality control testing, whose goal is to determine how good the entire lot is. Clearly, a *sample* of fuses must be used in order to assess the quality of the lot.

Accuracy and Sampling

A sample may be more accurate than a census. A census, if sloppily conducted, may provide less reliable information than a sample carefully obtained. Indeed, the 1950 U.S. Census provides us with an excellent example.

Example 3-1 Two statisticians noticed a very unusual anomaly in the change of two categories since 1940.* They found that an abnormally large number of 14-year-old widows were reported in the 1950 figures: there were 20 times as many as in 1940. Indian divorcees, who had also been extremely rare, had increased by a like magnitude. Unable to justify the changes by any sociological trends, they analyzed the data gathering and processing procedures of the census. After some investigation, it was concluded that the figures could have resulted from the erroneous reversal of the entries on a few of the punched cards used to process the mass of data gathered. In this instance, a census seemed to provide far from reliable or credible results.

Example 3-2 A large utility company had problems of a somewhat different nature with punched cards. The procedure for determining how to allocate the internal costs associated with the keypunching services of the data processing center had been of a census nature. At the end of each job, no matter what its size, each operator was required to make a detailed entry in a log, specifying how many cards were punched, for whom the job was done, and how much time it took. At the end of each accounting period, the supervisor was required to prepare a report for accounting based upon the logs. The data processing personnel complained about the bother involved in meticulously keeping these records, claiming that it detracted from overall productivity. Furthermore, the accounting people

* Ansley J. Coale and Frederick F. Stephan, "The Case of the Indians and Teen-age Widows," *Journal of the American Statistical Association*, LVII (June 1962), 338–437.

had been receiving complaints from the departments using the keypunching service that they were being inaccurately charged.

A company satistician was called in to analyze the situation and to see if improvements could be made. He found that the root cause of department complaints was attributable to the rather casual manner in which the keypunch operators kept their logs; it proved to be a demanding task, requiring skills that the average operator had not acquired in training. Furthermore, he found that the operators spent about 10 percent of their productive time keeping the records. He proposed to management that a *work sampling* technique be adopted, whereby, at random time intervals (three times daily), a bell would be rung; at this time all operators would stop work and fill out a simplified form pertaining only to the current job. Each form contained only the name of the department for which the job was being done. These records provided a basis for allocating keypunching costs to the various departments according to the percentage of forms returned for each department; over a period of a month, the heavier users of keypunching would have more forms indicating that they requested the service. Since each department used keypunching extensively, the proration of cost from the forms was a much more accurate representation of reality than the previous slipshod procedure. The productivity of the keypunchers increased as a result, and all parties involved were quite satisfied.

EXERCISES

3-1 For each of the following situations, indicate whether a sample or a census would be more appropriate. Explain your choice.
- (a) A manufacturer of automobile spark plugs knows that due to process variations a certain percentage of the plugs will be defective. A thorough test of a spark plug ruins it, yet the production output must be tested in order to determine whether remedial action is necessary.
- (b) NASA must determine the quality of the components of a manned space vehicle.
- (c) A hospital administrator attempting to decide how to improve patient services seeks the attitudes of persons treated.
- (d) A medical researcher wants to determine the possible harmful side effects from a chemical that eases the pain of arthritis.

3-2 For each of the following reasons, give an example of a situation for which a census would be less desirable than a sample. In each case, explain why this is so.

(a) economic	(b) timeliness	(c) size of population
(d) inaccessibility	(e) accuracy	(f) destructive observations

3-2 DESIGNING AND CONDUCTING A SAMPLING STUDY

Much of this book is concerned with how to collect samples in such a way that meaningful conclusions may be drawn about the entire population. Figure 3-1 shows the major stages of a sampling study. The first of these is *planning*, an area that requires careful attention so that the sampling results

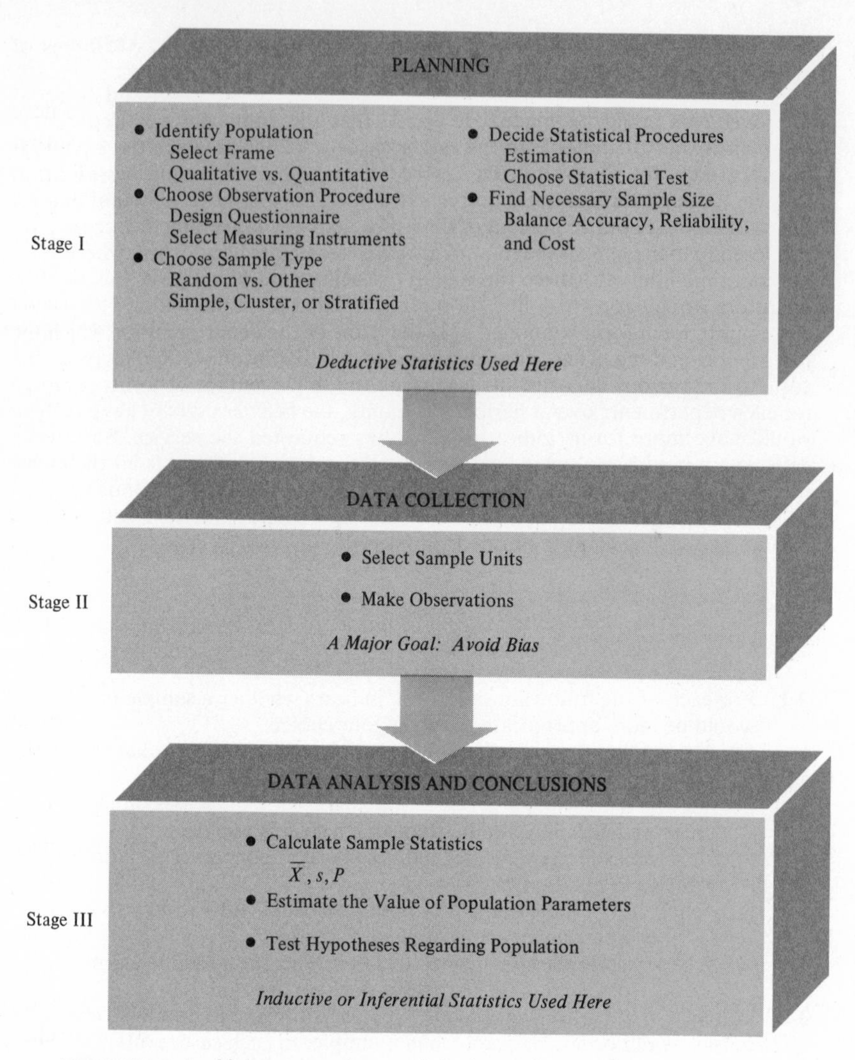

FIGURE 3-1 Major stages in a statistical sampling study.

can achieve the best impact. The second stage is *data collection*, where the plans are followed; here, a major goal is to insure that observations are free from bias. The last phase is *data analysis and conclusions*, an area to which a major portion of this book will be devoted.

The Importance of Planning

Planning a statistical study is the most important step. Inadequate planning can lead to needless expense when actual data collection begins. Poor planning may result in the eventual invalidation of the entire study. The outcome of the study may even prove strongly counterproductive. Several examples of extreme cases of poor planning will be given in this chapter.

Planning begins with the identification of a population which will achieve the study's goals. This is followed by selection of an observation procedure that might include the design of a questionnaire or the choice of an observation measuring instrument—a broad category that includes: mechanical devices (to record physical data, such as dimensions and weight); various tests (for intelligence, aptitude, personality, or knowledge); sources of data (for example, family income could be obtained from tax records, from employers, or by directly questioning a family member); and survey techniques (written reply, telephone response, or personal interview).

Of equal importance in planning is the choice of the sample type itself. Alternative kinds of samples range from the most "scientific" random samples, like that used in the major public opinion polls, downward to the convenience sample that campus straw votes represent, where practically no heed is paid to how representative the sample might be. The random sample is the most important one, because statistical theory applies to it alone. As we will see, there is a variety of ways to select a sample.

In order to draw proper and valid conclusions from sample data, it is very important that planning include a presampling choice of the statistical procedure to be used later in analyzing the results. Picking an analytical technique *after* sifting through the results may lead to the most insidious kind of bias: innocent or intentional selection of a tool that strongly supports the desired conclusion.

A very important planning consideration in any sampling study is how many observations to make. In general, the larger its size, the more closely a random sample is expected to represent its population. But where do we draw the line? Isn't a sample ten times as large ten times as good? The first question will be answered in Chapter 7. To the second, the answer is *no*. As we shall see, selection of sample size is largely an economic evaluation, where scarce resources (funds available for gathering data) must be traded off against the accuracy and reliability of the result. A point of "diminishing returns" is soon reached where an extra dollar buys very little in increased sample quality.

Choosing a statistical procedure and selecting an appropriate sample size each involve a heavy dose of deductive statistics, since both of these planning aspects require us to look at how samples are generated from "textbook" populations whose characteristics are fully known.

Data Collection

The second phase of the statistical sampling study—obtaining the raw data—is the one that usually requires the greatest amounts of time and effort. Adequate controls must be provided to prevent observations from becoming biased.

Potential problem areas associated with collecting and managing the data are described in detail in subsequent sections of this chapter. Such problems include observational biases that result when the sample units are measured, queried, or investigated. A poorly asked question and an improperly operated device both lead to incorrect responses. Another source of bias arises because some parts of the population are inaccessible or difficult to observe. Further difficulties may appear after the data are collected: dealing with the inevitable ambiguities and contradictions and processing the results so that statistical analyses can be made.

Data Analysis and Conclusions

This last stage of a statistical sampling study is extremely important. Here is where many of the descriptive tools we discussed in Chapter 2 are used to communicate the results. Our emphasis, however, is placed on drawing conclusions from the necessarily incomplete information available from any sample. This is when statistical inferences are made. The lion's share of statistical theory is directly related to or in support of making generalizations or *inferences* about populations from samples. Inferences generally fall into three broad categories: estimation, hypothesis testing, and association. The choice of which of these to use and the selection from a variety of techniques will depend upon the goals of the study and the resources available.

3-3 BIAS AND ERROR IN SAMPLING

Since a sample is a collection of observations made of a portion of the population, the manner in which the observed, units are chosen significantly affects the adequacy of the sample. There may be other factors influencing the representativeness of a sample due to the manner in which the observations are obtained. Here we will indicate some of the pitfalls commonly encountered in sampling and show how many of them may be avoided. Although essentially nonmathematical, avoidance of certain difficulties is a problem of paramount importance and will be necessary for the statistician to achieve his primary goal: selecting a sample at the lowest cost which will provide the best possible representation consistent with the objective of the study being made.

A sample is expected to mirror the population from which it comes. In choosing a sample, however, it must be acknowledged that there are no guarantees that any sample will be precisely representative of the population, as chance may dictate that a disproportionate number of untypical observations will be made. In practice, it is never known when a sample is unrepresentative and should be discarded, because some population characteristics are unknown (or else no sample would be needed in the first place). Steps must be taken during the planning of the study to minimize the chance that the sample will be untypical.

Sampling Error

What are some of the things that can make a sample deviate from its population? One of the most frequently mentioned is *sampling error*, for which we have the following.

> **DEFINITION** A ***sampling error*** comprises the differences between the sample and the population that are due solely to the particular elementary units that happen to have been selected.

To illustrate sampling error, suppose that a sample of 100 American men chosen is measured and that they are all taller than seven feet. An obvious conclusion would be that most American males are taller than seven feet. Of course, this is absurd, for most persons do not know a man that tall nor would they even be sure that one existed if some of the most noted basketball players were not that tall. And yet it is possible for the statistician to be unfortunate enough to obtain such a highly unrepresentative sample. However, this is highly unlikely; nature has distributed the rare seven-footers widely among the population.

More dangerous is the sampling error that is not so obvious and for which nature offers little protection. It is not difficult to envision a sample in which the average height is overstated not by a foot, but rather by an inch or two. It is the insidious, unobvious error which is our major concern.

There are two basic causes for sampling error. One is chance; bad luck may result in untypical choices. Since unusual elementary units exist, there is always a possibility that an abnormally large number of them will be chosen. The main protection against this type of error is to use a large enough sample.

Sampling Bias

Another cause of sampling error, *sampling bias*, is not as easy to remedy. Indeed, the size of the sample has little to do with the effects of bias. In the present context, we provide the

> **DEFINITION** ***Sampling bias*** is a tendency to favor the selection of elementary units having particular characteristics.

The bias may be present due to an intentional predilection, but usually it results from a poor sampling plan. A classical example of bias occurred in a poll conducted by the now defunct *Literary Digest* in 1936, when Democrat Franklin D. Roosevelt was running against Republican Alfred M. Landon for the presidency. The poll was a sample of voter preferences, and several million responses were obtained. Its conclusion that Landon would win by a record margin was the exact opposite of what actually happened: Roosevelt received one of the most lopsided victories in American history. The poll's erroneous results have since been attributed to sampling bias. The *Digest* had selected its sample from magazine subscription rolls and telephone directories; both sources, in the time of the Great Depression, contained a disproportionate number of prosperous persons who favored the laissez-faire Republican platform, while the disgruntled majority, strongly in favor of Roosevelt, largely did without telephones and magazines and hence did not receive adequate representation in the sample.

The avoidance of bias in sampling is a major concern of statisticians. Pains must be taken in designing a means of selecting the elementary units so that the more obvious forms of bias will be absent. In practice, it is very difficult to completely eliminate all forms of bias. Most notable is the *bias of nonresponse*, which results when—for whatever reason—some elementary units have no chance of appearing in the sample. The cost considerations involved in making sample observations almost guarantee that this particular form of bias will be present to some degree in any sample.

An excellent example is provided by consumer surveys involving food products. Because the housewife is usually easily accessible in her home, it is relatively cheap to obtain her opinion about a company's product. However, a high proportion of women work during the day, and unless the surveyors work in the evenings or on weekends, the opinions of these women—who also make food-buying decisions—will not be represented. An obvious danger is that a company attempting to determine whether to expand its line of convenience foods may obtain survey results indicating that such products are not extremely desirable; obviously, being pressured for time, the working woman, whose opinion was not obtained, would most likely prefer greater availability of convenience foods. The net result would be an erroneous marketing decision by the company, putting it at a competitive disadvantage.

Nonsampling Error

The other main cause of unreflective samples is *nonsampling error*. This type of error may occur whether or not a census is taken.

DEFINITION A ***nonsampling error*** is an error that is due solely to the manner in which the observation is made.

The simplest example of nonsampling error is inaccurate physical measurement. In weighing people, for instance, this error would exist when a malfunctioning scale is used. A more subtle example is provided by the measurement of metallic objects at different temperatures, for metals expand with increasing temperatures.

Another type of nonsampling error exists in personal interviews due to the manner in which a response is elicited; no two interviewers are alike, and the same person may provide different answers to different interviewers. The manner in which a question is formulated can also result in inaccurate responses. A good example of this concerns a person's age. Some people will lie about their age; asking an age in years can be dangerous, as people ordinarily use their chronological age at their previous birthday, even if it is more than 11 months past. Both problems can be alleviated by asking the date of birth, since the liar would have to do some quick arithmetic to fake the date, and a date of birth is much more accurate than a person's age at his last birthday.

Nonsampling error may occur whether a census or a sample is being used. As with sampling error, it may be either willfully produced by participants in the statistical study or it may be an innocent byproduct of the sampling plans and procedures.

The inaccurate measuring instrument—be it a physical device or a human being—is a very common source of nonsampling error. Biased observations due to inaccurate measurement can be very innocent but are obviously devastating. For example, a French astronomer once proposed a new theory based upon some spectroscopic measurements of light emitted from a star; believing his proposition to be incorrect, his colleagues discovered that the measuring equipment had been contaminated by cigarette smoke, and the unusual findings were attributed to that factor. In surveys of personnel characteristics, unintended errors may crop up because of (1) the manner in which the response is elicited, (2) the purpose of the study, (3) the idiosyncracies of the persons queried, or (4) the personal biases of the interviewer or question writer.

The observation of human weights is another example. It is necessary that all persons be weighed under controlled conditions and that they not provide their own weights. People will have weighed themselves on different scales in various states of poor calibration, so that no two responses are of equal reliability. A person's weight fluctuates diurnally by several pounds, so that the time of day they weighed themselves will affect the figure. A person's weight will also depend upon his state of undress when he steps onto the scales. Responses will not be of comparable validity unless all persons are weighed under the same circumstances.

Knowledge of the purpose for which the study is being conducted may tend to create incorrect responses. A classical example of this is the answer to the question: What is your income? If a government agency is asking, then a different figure may be provided than would be given on an application for a home mortgage. A teacher's union seeking to justify a wage increase may get a different response from a member than would his brother-in-law; the union would not likely be given (nor would it really want) the full income (incorporating amounts derived from moonlighting and summer employment), whereas the brother-in-law would probably get a more favorable picture, not only being informed about the income derived from these sources but also hearing about successful investments. One way to protect against such bias is to camouflage the study's goals, perhaps even to employ an independent opinion survey firm in order to keep secret the identity of the course of the investigation. Another remedy is to make the questions very specific, allowing very little room for personal interpretation. For example: Where are you employed?, followed by: What is your salary? Then: Do you have any extra jobs? A sequence of such questions may lead to a more accurate answer to the basic question.

Individuals tend to provide false answers to particular questions. A significant source of such errors may be psychological. For example, the answer to a question could be untrue because the respondee wishes to impress the interviewer. This is especially prevalent with regard to knowledge of current events and to intellectual accomplishments. Numerous examples exist of persons "knowing" about fictitious famous names or stating that they have read books that don't exist. This type of error is the most difficult to prevent, for it involves outright deceit on the part of the respondee. It is important to acknowledge, however, that certain psychological factors induce incorrect responses, and great care must be taken to design the study in such a way as to minimize their effect.

Finally, it should be noted that the personal prejudices of either the study designer or the data collector may tend to induce bias. It is not good practice for a person to design a questionnaire relating to a subject about which he has strong feelings. Questions may be slanted in such a way that a particular response will be obtained even though it is inaccurate. An example of this is a preference survey sponsored by the makers of brand A, prefaced by the question: Of all the following factors making brand A a superior quality product, which do you think is most important? This could then be followed by a question asking for a preference ranking of a list of products, including brand A (most blatantly, brand A may be the first name on the list). The preliminary question starts the respondee thinking about the good features of brand A, creating an atmosphere that will make the later comparison unfair. Another instance of induced error may occur in an experiment for which a technical specialist selects the elementary units to be treated with a new product. For example, when a doctor favoring a new drug decides which patients are going to receive it, he may select (consciously or not) those he feels have the best chance of benefiting from its administration. An agronomist may choose certain key plots upon which to apply a new fertilizer, knowing full well that they will provide more favorable yields than others. To protect against induced bias, an individual trained in statistics should have a measure of control over the design and implementation of the entire statistical study; at the very least, someone who is aware of the pitfalls should serve in an auditing capacity.

EXERCISES

3-3 A government official has delegated the responsibility for determining his employees' attitudes toward a training program run by the personnel department. What dangers are there in asking the personnel manager to dispose of this matter? Suggest a procedure that would overcome these difficulties.

3-4 Those persons who filled out the "long" form for the 1970 U.S. Census were asked to answer questions about their employment. For example, one question asked: How may hours did you work last week? Succeeding questions on the matter all related to the same week, requesting information on type of employer, duties, and so on. The public was asked to provide the data as of April 1, 1970, which fell on the Wednesday right after Easter, so that most of the nation's educators had been on a holiday for the entire week in question. Comment on the possible errors which may result from this set of questions. How might the questionnaire have been improved in order to avoid them?

3-5 A drug store has recently experienced a significant drop in sales due to the presence in the city of some new discount stores. A questionnaire, filled out by customers who visited the store during one week, asked for comparisons with respect to prices, convenience, and level of service. What do you think of the manner in which the information was obtained? What procedure would you recommend?

3-6 A questionnaire prepared for a study on consumer buying habits contains the question: What is the value of your present automobile(s)? What errors may result from this question? Suggest a better way to obtain the desired response.

3-7 Many personality profiles require that an individual provide answers to multiple-choice questions. His answers are then compiled, and various trait ratings are found. Comment on some of the pitfalls of this procedure. Suggest how some of them may be alleviated.

3-8 The testing of a new drug therapy sometimes involves two patient groups. The patients in one group receive the medication being investigated. The others may receive no medication at all and are often given an innocuous substance, called a *placebo* (often just a sugar pill), that has an appearance identical to the drug being tested. Why is the placebo used?

3-9 A well-known television rating agency provides the networks and advertisers with estimates of the audience size for all nationally televised programs. These estimates are based upon detailed logs maintained by persons in the households surveyed, indicating which stations were watched during each period of the day. The agency has a difficult time recruiting households for its sample, as considerable bother and inconvenience is involved in keeping the logs. It has been estimated by independent observers that an average of 50 families must be approached before one agrees to join the sample. Comment upon the pitfalls which might be encountered when the sample results are taken as representative of the nation's television viewing habits. Can you suggest any ways to minimize these pitfalls? Do you think that the sample observations obtained are in danger of being untypical? Explain.

3-4 SELECTING THE SAMPLE

In the preceding section, we encountered the most common problems associated with statistical studies. The desirability of a sampling procedure depends on both its vulnerability to error and its cost. However, economy and reliability are competing ends, because to reduce error often requires an increased expenditure of resources.

Of the two types of statistical error, only sampling error can be controlled by exercising care in determining the method for choosing the sample.

Sampling error may be due to either bias or chance, as we have established. The chance component exists no matter how careful the selection procedures are, and the only way to minimize chance sampling errors is to select a sufficiently large sample. Sampling bias, on the other hand, may be minimized by judicious choice of procedure.

Types of Samples

There are three primary kinds of samples: the convenience sample, the judgment sample, and the random sample. They differ in the manner in which their elementary units are chosen. Each sample will be described in turn, and its desirable and undesirable features will be discussed.

The Convenience Sample

A convenience sample results when the more convenient elementary units are chosen from a population for observation. Dependence on a convenience sample is epitomized by the tendency people have to draw conclusions about the nature of things by relying upon their personal experience as a source of observations. This often results in considerable consternation when more reliable information refutes their inferences.

Convenience samples are the cheapest to obtain. Their major drawback is the extent to which they may be permeated with sampling bias, which tends to make them highly unreliable. An important decision made upon the basis of a convenience sample is in great danger of being wrong. The letters received by a congressman constitute a convenience sample of constituent attitudes. The legislator need do nothing in order to receive his mail. But the opinions voiced there will rarely be indicative of the attitudes of the entire constituency. Persons with special interests to protect are far more likely to write than the typical voter.

The bias of nonresponse, when present, usually results in a convenience sample. This type of bias can be expected when the telephone—a very convenient device—is used to obtain responses, for the incidence of nonresponse may be relatively high: many calls will be unanswered, there are unlisted numbers, and some persons have no telephone.

In spite of the obvious disadvantages of convenience samples, they are sometimes suitable, depending upon the purpose of the study. If only approximate information is needed about a population, then a convenience sample may be adequate. Funds for the study may be so limited that only a convenience sample can be afforded. But in such cases the obvious drawbacks must be acknowledged. A convenience sample may lead to only a very insignificant source of bias. For example, when there is no basis for believing that nonresponders would provide responses that would, on the whole, be different from those readily available, then there may be little danger in using a telephone survey; but one can never be sure.

The Judgment Sample

A judgment sample is one that is obtained according to the discretion of someone familiar with the relevant characteristics of the population. Judgmental selection of the elementary units is made when the population is very heterogeneous, when the sample is to be very small, or when special skill is required to insure a representative collection of observations.

Example 3-3 The Consumer Price Index of the U.S. Department of Labor is partly a judgment sample. Those items to be included in establishing the index are chosen on the basis of judgment in order to measure a dollar's purchasing power. Thus, only items or commodities that are used by most people are included; for example, television sets are included, but tape recorders are not.* Over time, certain items become obsolete and must be replaced by others; for instance, refrigerators have supplanted iceboxes. Judgment must be employed in deciding what will be replaced, what will be substituted, and when the index will be spliced. Even within a particular category, there may be several specification categories, and judgment must be exercised by field workers to see that, for example, an item has a particular level of quality or has been changed in terms of quantity. This was the case with the nickel candy bar for 50 years. To maintain price, the quantity of candy in a bar had been declining steadily over the years—each change being tantamount to a price increase—until there was so little candy left that the price was finally raised instead. Certain foods are seasonal, and substitute items must be included in each periodic index determination. Judgment is also used in establishing weights for each item; toothpaste should only be weighted in proportion to its position in a typical family's total expenditure—the typical expenditure is itself a matter of judgment. Finally, judgment is employed in selecting cities to be represented in the surveys; these, too, must be assigned weights.

The judgment sample is obviously prone to bias. Its adequacy is limited by the discernibility of the individual who selects the sample. All of the dangers mentioned in connection with induced bias apply to the judgment sample. Clearly, there are instances in which this type of sample is preferred. Indeed, the Consumer Price Index would be unworkable were it not partly based upon judgment samples.

The Random Sample

Perhaps the most important sample type is the random sample. *A random sample is one that allows for the equal probability that each elementary unit will be chosen.* For this reason, it is sometimes referred to as a *probability sample*.

In its simplest form, a random sample is selected in the manner of a raffle. For example, a random sample of ten symphony conductors out of a population of 100 could be obtained by lottery. The name of each symphony conductor would be written on a slip of paper and placed into a capsule. All 100 capsules would then be placed in a box and thoroughly mixed. An impartial party would select the sample by drawing ten capsules from the box.

* Rothwell, Doris P. *The Consumer Price Index Pricing and Calculation Procedures* (Washington: U.S. Bureau of Labor Statistics: March 17, 1964).

In actual practice, the physical lottery is cumbersome. There may also be some question as to its randomness. Consider the 1970 draft lottery fiasco. There, the objective was to determine priorities of selection by date of birth. A physical method much like the raffle was used. The results appeared, however, to have a pattern; December birthdays had disproportionately low numbers, while January birthdays had predominantly high numbers. Investigation showed that the undesirable results were due to the fact that the capsules had not been mixed at all; since the December capsules were the last to be placed into the hopper, many December dates were drawn first. Because low dates implied almost certain draft eligibility, considerable controversy was created, and lawsuits were filed to invalidate the entire lottery.

Sample Selection Using Random Numbers

A more acceptable way to obtain a random sample is to use random numbers. *Random numbers* are digits generated by a process which allows for the equal probability that each possible number will be the next. For example, in a list of five-digit random numbers, each value between 00000 and 99999 has the same chance of appearing at each location. Appendix Table B is a list of five-digit random numbers. Notice that these numbers give the appearance of having been selected from a lottery. Indeed, in a special manner, they have been. They were generated by an electromechanical process at the Rand Corporation. These particular numbers have passed many tests and have been certified as pure by the scientific community. Had the Selective Service used such a table of random numbers to select its birthdays, it would have been guilt-free in the eyes of all save those receiving the low numbers.

There are no guarantees that a list of random numbers will not exhibit a pattern of some sort. For example, 8 may follow 7 more frequently than any other value. However, in a long list, a pattern of any sort is highly unlikely (8 should follow 7 about 10 percent of the time).

In order to use random numbers to select our ten conductors, first it will be necessary to assign a number to each member of the population. In this case, we can use two digits; the assigned numbers will therefore run from 00 to 99. It will make no difference how these numbers are assigned, but care must be taken to insure that the person making the assignments does not know which page of the random number table will be used. In Table 3-1, the 100 conductors have been arranged in alphabetical order and assigned the numbers 00 through 99 in sequence.

The ten conductors will be chosen by reading down the first column of the random numbers table until ten different values are obtained. Notice that we only require two-digit numbers, but the table has five-digit entries. This is no problem, for the extra digits may be ignored and only the first two digits of each entry may be used. If a value is obtained more than once before all ten conductors have been chosen, then that number will be skipped and the next one on the list used instead. The values obtained by starting at the top of the first column of Appendix Table B (page 552) and reading down the column are listed below,

TABLE 3-1 Symphony Conductors

00. Abbado	25. Golschmann	50. Mehta	75. Santini
01. André	26. Hannikainen	51. Mitropoulos	76. Sargent
02. Anosov	27. Hollingsworth	52. Monteux	77. Scherchen
03. Ansermet	28. Horenstein	53. Morel	78. Schippers
04. Argenta	29. Horvat	54. Mravinsky	**79.** Schmidt-Isserstedt
05. Barbirolli	30. Jacquillat	**55.** Newman	80. Sejna
06. Beecham	31. Jorda	56. Ormandy	**81.** Serafin
07. Bernstein	32. Karajan	57. Paray	**82.** Silvestri
08. Black	33. Kempe	58. Patanè	83. Skrowaczewski
09. Bloomfield	34. Kertesz	59. Pedrotti	84. Slatkin
10. Bonynge	35. Klemperer	60. Perlea	85. Smetáček
11. Boult	**36.** Kletzki	61. Prêtre	86. Solti
12. Cantelli	37. Klima	62. Previn	87. Stein
13. Cluytens	38. Kondrashin	63. Previtali	88. Steinberg
14. Dorati	39. Kostelanetz	64. Prohaska	89. Stokowski
15. Dragon	40. Koussevitzky	65. Rekai	**90.** Svetlanov
16. Erede	41. Krips	66. Reiner	91. Swarowsky
17. Ferencsik	42. Kubelik	67. Reinhardt	92. Szell
18. Fiedler	43. Lane	68. Rignold	93. Toscanini
19. Fistoulari	44. Leinsdorf	69. Ristenpart	94. Van Otterloo
20. Fricsay	45. Maag	**70.** Rodzinski	95. Van Remoortel
21. Frühbeck de Burgos	46. Maazel	71. Rosenthal	96. Vogel
22. Furtwängler	47. Mackerras	72. Rozhdestvensky	97. Von Matacic
23. Gamba	48. Markevitch	73. Rowicki	98. Walter
24. Giulini	49. Martin	**74.** Sanderling	99. Watanabee

along with the corresponding conductors, who now comprise our sample:

12651	Cantelli	**74**146	Sanderling
81769	Serafin	**90**759	Svetlanov
36737	Kletzki	**55**683	Newman
82861	Silvestri	**79**686	Schmidt-Isserstedt
21325	Frühbeck de Burgos	**70**333	Rodzinski

It does not matter how the random number table is read (from left to right, right to left, top to bottom, bottom to top, or diagonally), but it is important not to read the same number location more than once and not to look ahead before deciding in which direction to proceed. If the numbers are larger than required, the leading or trailing digits may be ignored. If a number on the list has no counterpart in the population (which may happen when the population is smaller than the possible number of random numbers), it may be skipped.

Unlike judgment and convenience samples, random samples are free from sampling bias. No particular elementary units are favored. This still does not guarantee that a random sample will be extremely representative of the population from which it came. The chance effect may cause it not to be. But it will be possible to assess the reliability of a sample in terms of probability. This can only be done with random samples. When sampling bias is present, an objective basis for measuring its effect does not exist. Therefore, *only the random sample has a theoretical basis for quantitative evaluation of its quality*. As we shall see in later chapters, probability is used for such evaluations.

Types of Random Samples

The preceding example is just one type of random sampling scheme. At times it may be desirable to modify the selection procedure without altering the essential features of the random sample. We will refer to the type of sample just discussed as a *simple random sample*, for which we provide the following

DEFINITION A ***simple random sample*** is obtained by choosing units in such a way that each unit in the population has an equal chance of being selected.

The Systematic Random Sample

A simple random sample is free of sampling bias. However, using a random number table to choose the elementary units can be a cumbersome procedure. If the sample is to be collected by a person untrained in statistics, then there is the danger that he might misinterpret instructions and make the selections improperly. To simplify the data collection, instead of using a list of random numbers, the elementary units may be chosen by selecting every 10th or 100th unit, for example, after the first unit has been chosen randomly. Such a procedure is called a *systematic random sample*, for which we have the following

DEFINITION A ***systematic random sample*** is obtained by selecting one unit on a random basis and then choosing additional units at even-spaced intervals until the desired number of units have been obtained.

Example 3-4 A telephone company is conducting a billing study in order to justify a new rate structure to the state's public utility commission. Data are to be collected by taking a random sample of individual telephone bills from cities and metropolitan areas classified according to population. Billing data are not available at a single location, but are spread throughout the state in several regional revenue accounting offices, where they are stored by telephone number. Time and cost constraints require that the sample information be obtained by the field offices and then forwarded to company headquarters for processing. A company statistical analyst has designed a procedure for each revenue accounting office to employ in collecting the samples.

One city has 10,000 telephones sharing the 825 prefix. The telephone numbers run from 825-0000 through 825-9999. A sample size of 200 is required. Dividing the number of telephones by 200, the interval width of 50 is obtained.* Thus the

* Allowances may have to be made for numbers not currently in service. Here we may assume they are all active.

bill for every 50th telephone number will be incorporated into the sample. The analyst chooses a number lying between 0 and 49 at random. Suppose he chooses 37. His instructions to the revenue accounting office specify that the billing figures for every 50th telephone number, starting with 825-0037, be forwarded to his office. Thus, he will receive the figures for 825-0037, 825-0087, 825-0137, 825-0187, and so on. It will then be a simple manual task for a clerk to pull every 50th telephone bill from the files.

A systematic random sample may be somewhat prone to sampling bias: some extraordinary property may be attributable to the elmentary units assigned particular numbers, and the initial random number chosen may tend to favor this property. For instance, it is well known that businesses prefer, for mnemonic ease, telephone numbers that have an easily recalled pattern (for example, 825-4455). Had the analyst in Example 3-4 chosen the random number 05, then half of the sample would have consisted of telephone numbers ending with 55, and it is likely that a disproportionate number of businesses would have been represented. If businesses tend to have unusually large telephone bills, then such a listing would provide an obvious source of bias. This bias cannot be remedied by choosing an innocuous starting value such as 09, because, for the same reason, the sample may include too few businesses. The sampling bias can be very subtle in systematic random sampling, but this is the price that must be paid for simplifying the data collection procedures.

The Stratified Sample

In certain instances it is desirable to insure that various subgroups of a population be represented in the sample in accordance with their respective prevalence in the population. In statistical parlance, these subgroups are referred to as *strata*. Ideally, the population itself should be split into separate entities when it is significantly heterogeneous. In practice, however, this may complicate the statistical analysis. Next best is to force the sample to reflect the population's lack of homogeneity as closely as possible, by guaranteeing that the sample have the same proportions of each subgroup that exist in the population itself. This is accomplished by taking a stratified sample, for which we provide the

DEFINITION A ***stratified sample*** is obtained by independently selecting a separate simple random sample from each population stratum.

Whether or not stratification of the population is desirable will depend upon the purpose of the study, but it is usually carried out because certain nonhomogeneities are present in the population. There are also other reasons why a stratified sample may be desirable. It may be more convenient, for example, to collect samples separately from regional centers, as in the preceding example. Or there may be some prevailing reason to divide the population—for instance, to compare strata.

Example 3-5 Voter preference polls have been taken prior to each presidential election in order to infer who would win the election if it were to be held at the time of the poll. Because of the expense involved, the large pollsters, working under contract to news agencies, draw a random sample of voters from the entire United States. Sampling is not conducted independently within each state.

The choice of who shall be President is determined by the electoral college. The membership of the college has historically voted as a block by state, with all votes going to the candidate receiving the plurality of that state's votes. American history shows that sometimes the man who receives the most votes from the public can still lose the election in the electoral college. The reason for this disparity is that the winner need only achieve small margins in several large states, even though he loses by large amounts in the rest. Thus he wins a majority of electoral votes while he loses the national plurality. This has happened twice since the Civil War: in 1876, Rutherford B. Hayes won over Samuel J. Tilden, who received the most popular votes, and in 1888, Benjamin Harrison beat the largest popular vote getter, Grover Cleveland. The situation is further distorted by the fact that electoral votes are rationed to each state in accordance to its representation in *both* houses of Congress. Thus Alaska, with about one-hundredth the voting population of California, has about one-fourteenth as many votes, making each Alaskan voter about seven times as powerful as a Californian.

A sampling scheme that attempts to predict the outcome of an election must treat each state as a separate stratum to adequately cope with the distortion of the electoral college. The sample results must be separately compiled for each state, and the winner of each determined. The results must then be combined in the same proportion as the electoral votes in order to present a distortion-free prediction.

The Cluster Sample

Yet another form of random sampling is used prevalently in consumer surveying, where it is more economical to interview several persons in the same neighborhood. For example, contrast the bother and expense of collecting a thousand responses in a city of 100,000 by simple random sampling with the relative ease of completely canvassing a few selected neighborhoods. In the former case, the chosen elementary units are scattered throughout the city; in the latter, the elementary units live next door to each other. Random sampling would therefore require considerable travel between interviews, while the door-to-door neighborhood canvass would minimize travel time. Because the interviewer will ordinarily be paid by the hour, the simple random sample would be far more expensive.

Neighborhood groupings of people are referred to as *clusters*. Populations may be divided into clusters according to criteria, such as geographical proximity, yielding groups that are easy to observe in their entirety. Most often, the clusters that are to be observed are chosen randomly, and hence we make the

DEFINITION A ***cluster sample*** is obtained by selecting clusters from the population on the basis of simple random sampling. The sample comprises a census of each random cluster selected.

The only justification for cluster sampling is economic. It is fraught with many dangers of sampling bias. For instance, in neighborhood surveys, similar responses may be obtained from the entire cluster, as people of similar age, family size, income, and ethnic and educational backgrounds tend to live in the same vicinity. There is thus a high risk that persons with certain backgrounds or predilections will not be represented in the sample because the clusters in which they predominate were not chosen; conversely, other groups may be unduly represented.

At this point, it might be helpful to emphasize the similarities and differences between cluster and stratified samples. Both separate the population into groups, but the basis for distinction is usually the heterogeneity of strata versus the accessibility of clusters. In stratified samples, all groups are represented, whereas the cluster sample comprises only a fraction of the groups. In a stratified procedure, a sample is taken from within each stratum, while in cluster sampling, a census is conducted within each group. The goal of stratified sampling is to eliminate certain forms of bias; the goal of cluster sampling is to do the job cheaply, which enhances the effects of bias.

Systematic, stratified, and cluster sampling are approximations to simple random sampling. To the extent that the statistical theory applicable to simple random sampling is applied to samples obtained via these other procedures, there will be erroneous conclusions. Throughout much of the remainder of this book, techniques will be developed primarily for analyzing simple random samples. The more complex sampling schemes require more complicated methodology.

EXERCISES

3-10 What types of samples do the following represent?

(a) Telephone callers on a radio "talk" show.

(b) The 30 stocks constitutiong the Dow-Jones Industrial Average.

(c) The record of a gambler's individual wins and losses from a day of placing one dollar bets on black in roulette.

(d) The oranges purchased by a housewife from the produce section of a market.

3-11 In the following situations, would you recommend the use of judgment, convenience, or random sampling, or some combination of these? Explain the reasons for your answer.

(a) A salesman for a large paper manufacturer is preparing a kit of product samples to carry on his next road trip to printing concerns.

(b) A teacher asks his students for suggestions as to how to improve curriculum. He plans to use their suggestions as a basis for a questionnaire concerning curriculum preferences.

(c) A city government wants to compare the wages of its clerical workers to persons holding comparable jobs in private industry.

(d) A utility company plans to purchase a fleet of large-sized cars from one of four different manufacturers. The criterion for selection will be economy of operation.

(e) A newspaper editor selects letters from the day's mail to print on the editorial page. (Answer from the editor's point of view.)

3-12 An oil company wishes to obtain a random sample of its customers in order to estimate the relative proportion of purchases made from other oil companies. An accountant on the operations staff has recommended that a random sample of credit card customers be selected. An independent survey firm would be retained to obtain detailed records of the following month's purchases, by brand and quantity, for each sample unit. Evaluate this procedure.

3-13 A medical research foundation wishes to obtain a random sample of expectant mothers in order to investigate postnatal developments in their newborn babies. Various metropolitan areas are to be represented. Suggest a procedure that might be used to select the sample in a particular city.

3-14 A certain U.S. Senator is reputed to follow the majority opinion of his constituents on important pending legislation. Over the years, he has accumulated a panel of 20 persons whom his secretary contacts before a major Senate vote. Panel members are replaced only when they die or change their state of residence. Usually the Senator's votes coincide with the majority of this panel. Comment on this policy.

3-15 A random sample is desired for each of the following situations. Indicate whether a cluster, stratified, or a combined sampling procedure should be used. Explain the reasons for your choice in each case.

(a) A large retail food chain wishes to determine its customers' attitudes toward trading stamps. Because there is a limited budget for taking the survey, the sample replies are to be obtained by interviewing customers on the store premises.

(b) An agronomist wants to take a sample to compare the effectiveness of two fertilizers. He wishes to test them on a variety of crops and under a range of climate and soil conditions.

(c) An airline is taking a random sample of its first-class and tourist-class passengers in order to obtain a quality assessment of its in-flight meals. The ratings must be taken just after the meals.

3-5 DECIDING STATISTICAL PROCEDURES

A complete discussion of statistical procedures will take up several of the later chapters in this book. All of these procedures make generalizations regarding a population when only sample data are available. We have called this process inductive or inferential statistics. The type of inference to be made and the procedures to be used depend upon the goals of the study and should be determined during the planning stage. There are many ways to do this.

The simplest procedure estimates a population's parameter, such as its mean, standard deviation, or proportion. The most complicated procedures test some hypothesis regarding the population's characteristics. Ordinarily, several testing schemes may be employed. In all such tests, the proper decision to be made is indicated directly from the sample results. Many tests actually compare two or more populations and require a separate sample from each.

In the fields of education, psychology, and medicine, the two-sample testing procedure is a popular way to decide which of two approaches works best. Often the status quo must be compared to a proposed improvement. For example, a new teaching method for reading might be evaluated in terms of comprehension test scores. The population of scores achieved by students under the current plan would be compared to the similar test results of another group of readers constituting the elementary units in the population of test scores by children exposed to the new program. Because the new program may or may not be adopted, the associated population does not yet exist. Nevertheless, a sample of children using the new program represents this potentially large group, which makes it a target population. The actual comparison would be made between separate samples from both populations.

The readers in the sample from the current program are referred to as the *control group*, because they are not subjected to anything new. Those children in the second sample constitute the *experimental group*, since these students are using a new procedure on a trial basis primarily to evaluate its merit. Control and experimental groups are encountered in similar testing situations, such as in evaluating a new drug or medical procedure, in selecting a more effective fertilizer, or in checking new safety devices.

The simplest procedure for conducting a two-sample test is to select the two groups *independently*, so that the readers in our control and experimental groups would be chosen using separate lists of random numbers. (But both samples would ordinarily be derived from a common frame of applicable students, because the experimental group represents only a target population.) One drawback to independent sampling is that any differences found in the groups' reading comprehension scores could be attributed to causes other than the particular reading programs, such as previous level of reading readiness or home background. The danger that some outside influence may camouflage the actual differences between the two reading programs could be minimized through a stratified sampling scheme, where two separate groups are selected for various categories of student background, or by using large enough groups so that each would provide a representative cross section.

A common method for choosing the control and experimental groups is to pair each elementary unit with one that closely matches it in terms of potential extraneous influences. One member of each pair would be assigned to the experimental group and the other member to the control group. This is called *matched-pairs sampling*. The procedure is epitomized by studies of identical twins, who are genetically the same and who normally share nearly the same environment as well. In our survey, one twin would be assigned to a class having the existing reading program and the other would participate in the new program. Any reading-score differences between sample groups could thereby be confidently isolated to differences in the reading programs and not to other causes. But in practice, twins are rare and are not usually used in such tests. Yet unrelated individuals could still be paired. For example, a new drug for heart patients could be tested with pairs matched in terms of age, sex, weight, marital status, smoking habits, medical histories, and so on, thus producing "near twins." Identical twins who were reared apart have been extensively studied in an attempt to isolate the effects of heredity and environment upon intelligence.*

Example 3-6 The experiences of one state during the ecology renaissance of the early 1970s illustrates the pitfalls of inadequate planning when conducting a statistical study. A new law was about to go on the books requiring that all soft drink and beer containers be returnable. This, it was hoped, would eliminate the unsightly refuse problem caused by pop-top cans. The old-fashioned bottle was to be used exclusively. To guarantee that the empty bottles were returned, a deposit was required by the seller, just as it had been throughout the United States 20 years before.

To assess the effectiveness of the beverage container law, a study was made to determine how it affected accumulations of roadside litter. A random sample of 30 one-mile stretches of highway was chosen, cleared of all beer and soda cans, and the left to the vagaries of normal public neglect and mistreatment for three months before the new law became effective. The 30 miles of highway served as the control group, representing state residents' littering habits when beverage containers were not dictated by law. At the end of three months, the litter was cleared and the cans were counted. For the initial three months under the new law, the same 30 miles—now the experimental group—were again left to accumulate beverage containers (which would only be bottles this time). The accumulations of the two groups were then compared.

This study shows how the same elementary units can sometimes serve in both the control and the experimental groups. In this case, each one-mile segment after the law became effective was matched with itself in a preceding time period. This way traffic conditions and local idiosyncrasies were alike and could not be partly responsible for any differences found between the two groups. However, using the same units in both groups is only valid if no bias is induced upon the second set of observations by the first set. Here is where this study fell apart. When the beverage litter was measured the second time around, precious little of it was found. But all other kinds of litter were practically absent, too. Furthermore, adjacent stretches of highway were abnormally clean. At first blush, environmental officials were ecstatic at the apparent widespread impact of their new law. But they became embarrassed when it was determined that the same pattern prevailed only near the test tracts and not throughout the state.

Who absconded with the litter and why? Perhaps environmentalists—well aware that the new law would be temporary unless it proved effective—got wind of the study and tried to help it along by cleaning up the test stretches. Maybe when the state had cleaned the one-mile parcels, local residents were shamed into picking up some of their own mess.

Upon the advice of a statistician, officials discarded the second set of results and matched the original 30, one-mile stretches of highway with 30 similar ones untainted by having been cleaned. Just one pass was made through the new areas, and only returnable bottles were counted (the old cans did not have to be).

The later chapters of this book will be concerned largely with the variety of statistical procedures available for conducting a sampling study: Chapter 7 is concerned with making estimates; Chapters 8–14 cover a variety of testing procedures, with the emphasis in Chapter 9 being placed on associative inferences.

* One well-known report on identical twins may be found in H. H. Newman, F. N. Freeman, and K. H. Holzinger, *Twins: A Study of Heredity and Environment* (Chicago: University of Chicago Press, 1937).

REVIEW EXERCISES

3-16 Starting in the fifth column of the random number table on page 552, select a simple random sample of ten symphony conductors from those listed in Table 3-1 (page 83). Use only the first two digits of each random number from the table. List the names selected.

3-17 List the names for a systematic random sample of ten symphony conductors. Choose your conductors from those in Table 3-1, assuming that the starting random number provides Sir John Barbirolli for the first name.

3-18 Suppose that each successive ten names in Table 3-1 provides a stratum. Select a stratified random sample containing one conductor from each group. Assume that the following random numbers apply in the respective strata:

6, 5, 5, 2, 1, 9, 8, 7, 7, 0

(Redesignate the names in each stratum, starting with 0 and ending with 9; maintain the original alphabetical sequence.) List the names selected.

3-19 Treat each successive five names in Table 3-1 as a cluster. A random cluster sample of two clusters from the 20 is to be selected. Use the random numbers 18 and 4 to find the sample. (Designate the successive clusters as 0, 1, 2, and so on.) List the names obtained.

4 Probability

Let us imagine . . . a person just brought forth into this world. . . . The Sun would, probably, be the first object that would engage his attention; but after losing it the first night, he would be entirely ignorant whether he should ever see it again. . . . But let him see a second appearance or one return *of the Sun, and an expectation would be raised in him of a second return, and he might know that there was an odds of 3 to 1 for* some *probability of this. This odds would increase, as before represented, with the number of returns. . . . But no finite number of returns would be sufficient to produce absolute or physical certainty.*

The Reverend Thomas Bayes (1763)

Probability is especially important in statistics because of the many principles and procedures based upon this concept. Indeed, probability plays a special role in all our lives, because we use it to measure uncertainty. We are continually faced with decisions leading to uncertain outcomes, and we rely on probability to help us make our choice. Think of the planned outdoor activities, such as picnics or boating, you canceled because the chance of bad weather seemed too likely. Remember those nights before examinations when you decided not to study some topics, because they probably would not be covered on the test?

A *probability* is a numerical value that measures the uncertainty that a particular event will occur. The probability of an event ordinarily represents the *proportion of times under identical circumstances that the outcome can be expected to occur*. We refer to this value as the event's *long-run frequency of occurrence*. The probability that the head side will show when a fair coin is tossed is 1/2. This can be verified experimentally by tossing a coin several times and observing that "heads" occur about one-half of those times.

Probability is a necessary part of statistical inference, because it measures the uncertainties involved in making generalizations from a sample. Suppose that 10,000 men are selected at random from all adult American males. If we know exactly how many seven-footers there are in the population, we may calculate the probability that our sample will contain any number of men taller than seven feet.

But records of the heights of all American men are not available, so the actual number of seven-footers remains unknown. If, however, the number of seven-footers in the sample is counted, then the number in the population may be estimated from the sample. But the accuracy of this estimate will remain uncertain. Common sense tells us that it would be unlikely for the sample to differ substantially from the population, so that the proportion of very tall men may be expected to be roughly the same. Probability is required to indicate just how likely the sample is to yield estimates that are in error by various amounts.

Probability was initially studied scientifically by several famous mathematicians more than 300 years ago in connection with gambling problems. The theory of probability has since evolved into one of the most elegant and useful branches of mathematics. Today devices ordinarily associated with gambling, such as dice and playing cards, are still useful in illustrating how to find probabilities.

4-1 FUNDAMENTAL CONCEPTS

The Random Experiment

Probabilities are obtained for the outcomes of situations which we conveniently call *random experiments.* We use the word "experiment," because the outcome is yet to be determined, whereas the adjective "random" signifies that any particular outcome is uncertain. In statistics, the most important random experiment is selecting the sample, whose elementary units are ordinarily chosen randomly from the population.

To illustrate the procedures for determining probabilities, we will begin by investigating other "experiments," notably gambling situations such as the flip of a coin or the toss of a die. Such situations not only vividly represent many sampling concepts, but also serve to explain how probability can relate to many "experiments" encountered by individuals and organizations. A person continually faces random experimental outcomes, such as the weather, the density of commuter traffic, or the quantity of mail received. Random experiments affect governments, in some measure determining future economic conditions, geopolitical attitudes, and the effectiveness of domestic programs. They also prevail in education and the social sciences, where the outcomes of innovative techniques and programs are subject to uncertainty, and in the biological, physical, and medical sciences, where the results of investigations are usually random experiments. Another area rich in uncertainty is the business world, where random experiments in some measure determine future sales levels, union wage settlements, and labor productivity.

The Event

The outcomes of a random experiment are called *events*. In the case of tomorrow's weather, the occurrence of precipitation would be one possible event. Even more specific events may be considered; for example, precipitation could be in the form of rain, hail, or snow. Or events may be quite detailed: 1.2 inches of rain may fall.

A preliminary step in finding an event's probability is to identify all of the possible outcomes of the random experiment with which the event is associated.

Sample Space

In cataloging a random experiment's possible events, it is convenient to discuss them in groupings or *sets*. A list of all possible outcomes or a complete enumeration of all possible events that may result from a random experiment is given the

DEFINITION The ***sample space*** for a random experiment is the set of all possible events that may occur. Exactly one and only one of the events will occur.

The sample space for tossing a coin contains just two events, head and tail, which may be conveniently expressed in set notation as:

$$\text{Sample space} = \{\text{head, tail}\}$$

Head and tail are the *elements* of the random experiment's sample space. One and only one of these events must occur. We will rule out such possibilities as the coin landing on its edge or being lost (perhaps falling down a gutter drain) by specifying that these are not legitimate outcomes of the random experiment; that is, in these instances, the toss is incomplete and must be repeated until either a head or a tail occurs.

Consider another random experiment—drawing a card from a fully shuffled deck of 52 ordinary playing cards. Here we have

$$\begin{aligned}&\text{Sample space}\\&\quad = \{\text{ace of spades, deuce of spades, \ldots, queen of diamonds, king of diamonds}\}\end{aligned}$$

It is sometimes convenient to show the sample space pictorially in the manner of Figure 4-1. Here, each point represents an element of the sample space, in this case the drawing of a particular card.

It would be possible to construct many different sample spaces for the same random experiment. For instance, if we were interested only in the suit of the

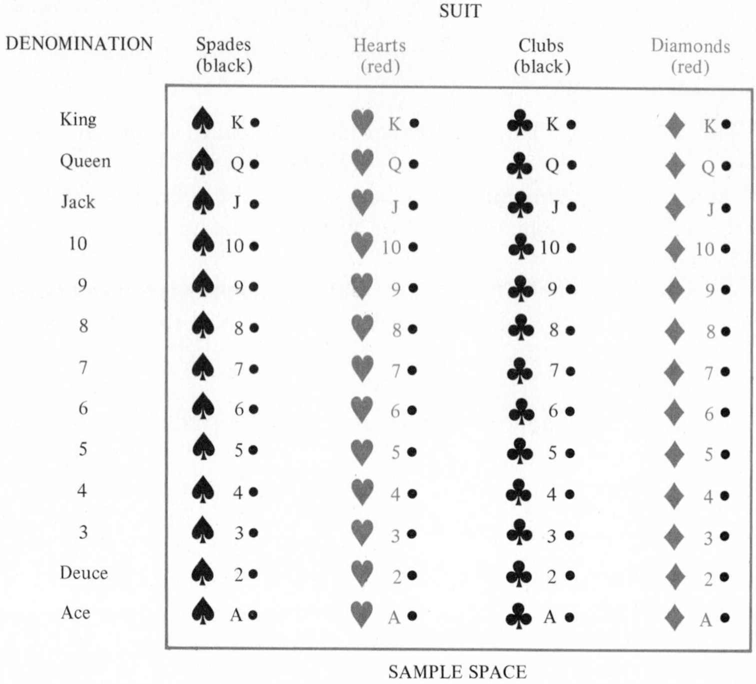

DENOMINATION	SUIT Spades (black)	 Hearts (red)	 Clubs (black)	 Diamonds (red)
King	♠ K •	♥ K •	♣ K •	♦ K •
Queen	♠ Q •	♥ Q •	♣ Q •	♦ Q •
Jack	♠ J •	♥ J •	♣ J •	♦ J •
10	♠ 10 •	♥ 10 •	♣ 10 •	♦ 10 •
9	♠ 9 •	♥ 9 •	♣ 9 •	♦ 9 •
8	♠ 8 •	♥ 8 •	♣ 8 •	♦ 8 •
7	♠ 7 •	♥ 7 •	♣ 7 •	♦ 7 •
6	♠ 6 •	♥ 6 •	♣ 6 •	♦ 6 •
5	♠ 5 •	♥ 5 •	♣ 5 •	♦ 5 •
4	♠ 4 •	♥ 4 •	♣ 4 •	♦ 4 •
3	♠ 3 •	♥ 3 •	♣ 3 •	♦ 3 •
Deuce	♠ 2 •	♥ 2 •	♣ 2 •	♦ 2 •
Ace	♠ A •	♥ A •	♣ A •	♦ A •

SAMPLE SPACE

FIGURE 4-1 Sample space describing the card selected randomly from a fully shuffled deck of 52 ordinary playing cards.

selected card, we might have used the set:

$$\{\text{spade, heart, club, diamond}\}$$

For our example, it will be convenient to keep the sample space for drawing a card in the more detailed form, making it possible to describe any type of outcome in terms of the more detailed elements—the individual cards. For this reason, we refer to the elements of the sample space shown in Figure 4-1 as *elementary events*.

Many random experiments are more elaborate. Consider the outcomes when three different coins—a penny, a nickel, and a dime—are tossed so that each will show either a head or a tail. The sample space is shown in Figure 4-2. Notice that the events are ordered triples such as (H, T, H), which represents the outcome "head for the penny, tail for the nickel, head for the dime." Even though three coins are tossed, this may be viewed as a single random experiment and the outcome (H, T, H) is a single elementary event. The events (H, H, T) and (T, H, H) are both different from (H, T, H), for in each case a different coin has the tail side showing. If we were describing the sample space for three tosses of the *same* coin, the elementary events would be represented in precisely the same manner—as eight ordered triples. In that case, the random experiment would

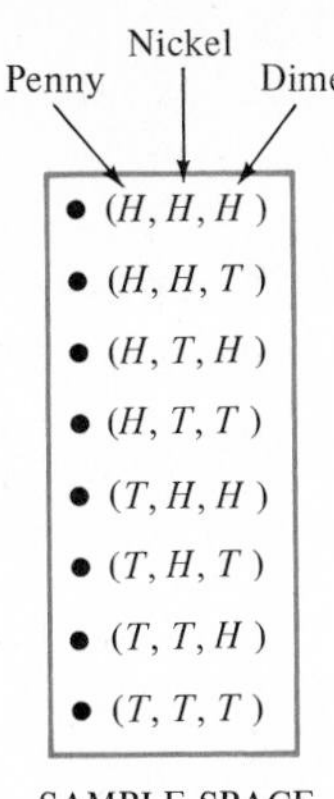

FIGURE 4-2 Sample space for the tossing of three coins.

have three *stages*, so that (H, T, H) would represent the event "head for the first toss, tail for the second toss, head for the third toss." (H, T, H) would be different from the event (T, H, H), because in each the tail appears on a different toss.

Event Sets

The elements of a random experiment's sample space are the simplest outcomes or elementary events. We may, however, be interested in more complex outcomes. For example, consider the sample space for the outcome of the toss of a six-sided symmetrical die:

$$\text{Sample space} = \{1, 2, 3, 4, 5, 6\}$$

where the elements represent the number of dots on the showing face. The outcome "an even-valued face" is a *composite event*, occurring whenever any one of the elementary events 2, 4, or 6 results. In keeping with our set representation, we denote the possible ways this event may occur by the following set:

$$\text{Even-valued face} = \{2, 4, 6\}$$

We will refer to such a set as an *event set*. Notice that it is a *subset* of the sample space: all of its elements also belong to the sample space. An event set for an elementary event, such as the king of spades, will contain a single element:

$$\text{King of spades} = \{\text{king of spades}\}$$

For most random experiments a great number of composite events are possible. Figure 4-3 shows a few that can be associated with drawing one card from a deck of 52 ordinary playing cards. Here, the respective event sets are pictured as groupings of those dots corresponding to the elementary events.

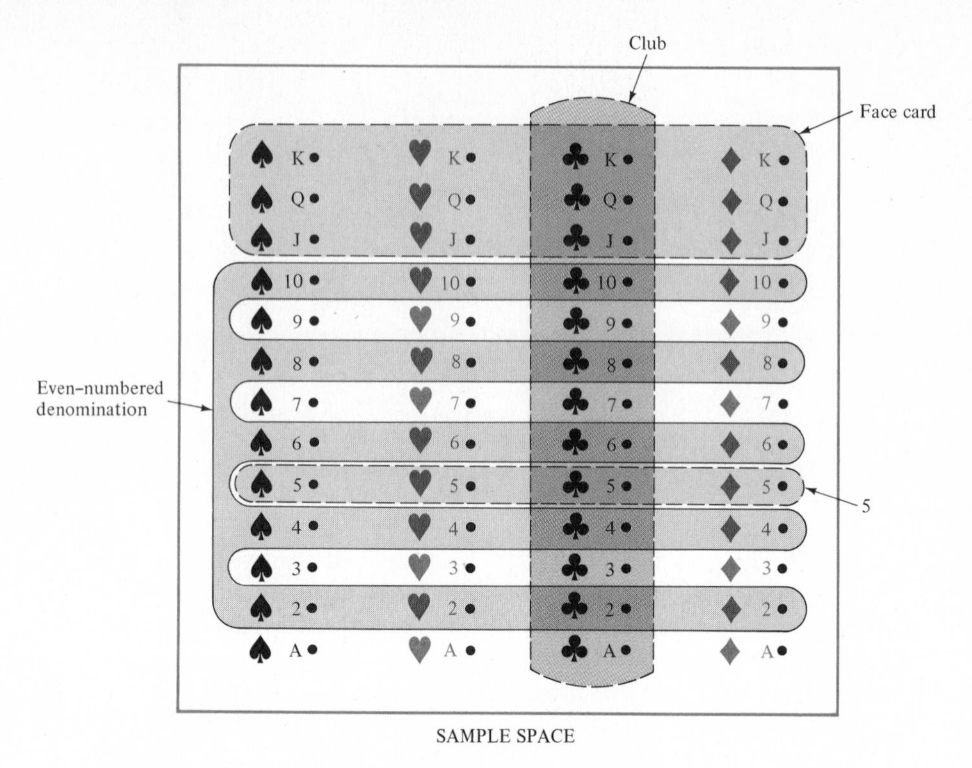

FIGURE 4-3 Composite events for drawing a card from a fully shuffled deck of 52 ordinary playing cards.

Basic Definitions of Probability

Classically, the probability of an event is defined as the relative frequency at which it occurs when the random experiment is repeated a large number of times. It is thus equal to the ratio of the number of times that the event occurs to the number of times that the random experiment is performed. The notation used to express the probability of an event is the bracketed form shown below:

BASIC DEFINITION

$$P[\text{event}] = \frac{\text{Number of times the event occurs}}{\text{Number of times the experiment is repeated}} \tag{4-1}$$

Expression (4-1) indicates that a probability is an empirically derived value, obtainable only after repeated experimentation. In practice, the actual performance of the experiments is unnecessary for a great many types of events. The probability of an event such as a head from the toss of a coin may be approximated by plausible reasoning with regard to the sort of outcomes that may be expected. Knowing that a coin has two sides, assuming that it is evenly balanced, and presuming that it is tossed with no bias toward one particular face, then it is reasonable to expect that a head should occur in about half of the tosses, and thus that:

$$P[\text{head}] = 1/2$$

In a similar manner, we may reason that the probability of drawing the ace of spades from a shuffled deck of 52 ordinary playing cards is 1/52, as there is no reason to suspect that certain of the 52 cards would be favored. Each card should appear with the same frequency in repeated trials. Each possible card is equally likely to turn up.

When all elementary events are equally likely, we may determine the probability of an event in the following manner:

PROBABILITY WHEN ELEMENTARY EVENTS ARE EQUALLY LIKELY When the sample space consists of elementary events reasoned to be equally likely, then

$$P[\text{event}] = \frac{\text{Number of elementary events in its event set}}{\text{Number of equally likely elementary events}} \tag{4-2}$$

Thus, since there is a single way to achieve a head (that is, since its event set is of size = 1, because head is an elementary event), expression (4-2) tells us that $P[\text{head}] = 1/2$, where the 2 represents the size of the sample space for one coin toss. The same procedure applies for composite events. For example, since the event "even-valued face" resulting from a die toss occurs whenever any one of the 3 elementary events 2, 4, or 6 is the outcome, and since there are exactly 6 possible elementary events, or showing faces:

$$P[\text{even-valued face}] = \frac{3}{6} = \frac{1}{2}$$

We may use this definition to find the probabilities of the composite events shown in Figure 4-3 for drawing a card from a fully shuffled deck of 52 playing cards:

$$P[\text{club}] = \frac{13}{52} = \frac{1}{4}$$

$$P[\text{face card}] = \frac{12}{52} = \frac{3}{13}$$

$$P[5] = \frac{4}{52} = \frac{1}{13}$$

$$P\left[\begin{array}{r}\text{even-numbered}\\ \text{denomination}\end{array}\right] = \frac{20}{52} = \frac{5}{13}$$

In each case, all cards (elementary events) are equally likely, and since there are 52 of them, the denominator in each case is 52 (the size of the sample space). The numerators are the sizes of the respective event sets (found by counting the number of elementary events—possible cards—in each set). The probabilities are therefore readily obtained by counting the number of elements in the respective sets and computing the ratios.

If the elementary events are not all equally likely, then the basic definition of probability must be used: the probabilities here must be estimated by repeating the random experiment many times. For example, if a die is shaved until it is asymmetrical, it becomes more likely to roll some sides than others. Logical reasoning cannot tell us what the probabilities ought to be for the faces of a shaved die. Only after many tosses can they be estimated from the results actually obtained. Perhaps a physicist could reason what the probabilities ought to be, but even he would be forced to rely upon empirically derived constants, such as the force of gravity.

From our basic definition of probability, it may be observed that a probability will always be a number between 0 and 1, inclusively. This is because the numerator in the probability fraction can never be negative nor can it be larger than the denominator. Two important observations follow from the definition. First, an event that is certain to occur will have the same value in both the numerator and the denominator, for the same events will result from all experiments (that is, with frequency = 1). Thus:

$$P[\text{certain event}] = 1$$

The event "the next President of the United States will be at least 35 years old" is a certain event and has a probability of 1, because the Constitution specifies a minimum age of 35. The event "food prices will rise, fall, or remain unchanged" likewise is certain and has a probability of 1.

At the other extreme, an impossible event's frequency ratio will always have a 0 in the numerator, for such an event will occur in none of the experiments. Thus:

$$P[\text{impossible event}] = 0$$

Those events that are not possible results of a random experiment are impossible events. For example, suppose that a pinochle card deck is used for our random experiment of drawing a card from a shuffled deck. A pinochle deck contains only 48 cards, all above 8 in denomination. The event space is therefore not the same as it is for a standard deck of 52 cards, there being no 8s, 7s, and so on. Thus, the event set of the composite event "drawing an 8 from a pinochle deck" has *no* elementary events; it is *empty*. We state this as:

$$\text{Drawing an 8 from a pinochle deck} = \{\ \ \}$$

The empty braces indicate that the event set is empty.

Since an 8 is impossible:

$$P[\text{drawing an 8 from a pinochle deck}] = 0$$

Alternative Expressions of Probability

Probabilities are not always expressed as ratios. They may be expressed as percentages, as odds, or as chances. These other forms are not inconsistent with our definition, because they may all be translated into the basic fraction or ratio form. For example, the probability of a head may be expressed in the following forms:

50 percent probability of a head. (Divide the percent by 100 to get the fraction 1/2.)

50-50 chance of a head. (This means an even chance, an *average* of 50 heads and 50 tails out of every 100 tosses.)

1 to 1 odds that a head will occur. (To express 1 to 1 as a fraction, add the two numbers $1+1=2$, and place the result in the denominator; then place the first number 1 in the numerator. The result is 1/2.)

EXERCISES

4-1 A smoker has eight pipes, two of which are meerschaums. One of his meerschaums has a curved stem. He has a total of four curved-stem pipes. He asks his son to bring him the curved-stem meerschaum. The boy, not knowing rose briar from ivory nor calabash from hookah, selects a curved-stem pipe at random. What is the probability that the father will get the pipe he wants?

4-2 A coin is tossed exactly three times in succession. Here, the sample space has the same form as Figure 4-2. List the elements of the event set for each of the following events and then determine each event's probability:

(a) Exactly two heads appear in the three tosses.
(b) The same side does not appear twice in succession.
(c) The toss sequence ends with a head.
(d) An odd number of tails is obtained.

4-3 A pair of six-sided dice, one red and one green, are tossed. The sides of each die cube exhibit the values 1 through 6:

(a) List the elementary events in the sample space.
(b) List the possible sums of the two showing face values.
(c) For each of your answers to (b), show which elementary events correspond to each possible sum value. Then determine the probability for each sum.

4-4 From your answers to Exercise 4-3, list the elementary events and then determine the probabilities for each of the following composite events:

(a) The sum value lies between 5 and 7, inclusively.
(b) The sum value lies between 8 and 12, inclusively.
(c) The sum value lies outside the range 4 through 9.
(d) The red and green dice have the same value.

(e) The red die has a value greater than the green die.
(f) The red die has a value greater than the green die, and the sum value is greater than or equal to 9.

4-5 Determine the probabilities of the following events:
(a) That a man chosen randomly from a group of ten men is a doctor, if the group contains two doctors.
(b) Winning a raffle with a single ticket out of 10,000.
(c) Getting both heads on one toss each of a dime and a penny. (First determine the sample space.)
(d) Getting a number greater than 2 for the value of the showing face from the toss of a six-sided die.

4-2 EVENTS AND THEIR RELATIONSHIPS

Events may be quite complex, for they can be expressed in terms of several other events. In order to develop tools for finding the probabilities of the more complex events, it is first necessary to scrutinize the relationships between events.

Events Having Components

To facilitate our discussion and to make precise the relationships between events, we will begin by considering an outcome expressed in terms of other events.

Union of Event Sets

A great many outcomes may be described in terms of the occurrence of any one of several events. For example, drawing a face card from an ordinary deck of playing cards occurs regardless of whether a king, queen, or jack is actually obtained. Thus, we may state the equivalence:*

Face card = king *or* queen *or* jack

Each of the events listed on the right-hand side of this equation are *component events*. "Face card" therefore comprises all the elementary events of each of its components, so that its event set contains the elements of all three component event sets, as shown in Figure 4-4. We may generalize the type of relation between these events by the following

DEFINITION The *union* of two sets A, B, denoted by

$$A \text{ or } B$$

is a set that contains all the elements of both sets.

* For simplicity, the Boolean symbols $\cap$ and $\cup$ are not used in this text. In their place, we use the italicized *and* and *or*.

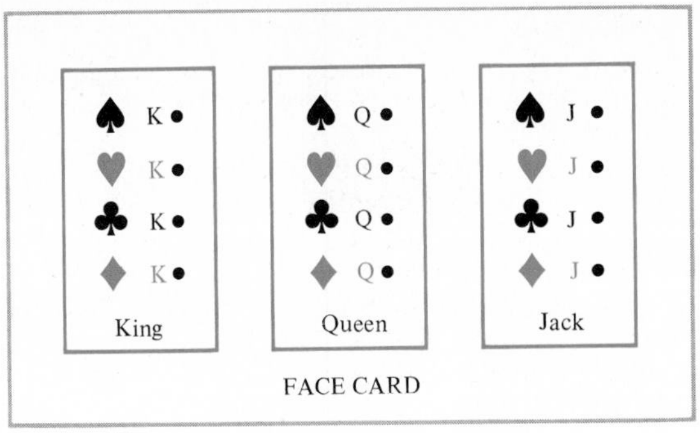

Face card = king *or* queen *or* jack

FIGURE 4-4 Portrayal of an event set comprising the union of three component events.

An event that occurs whenever any one or any combination of its component events occurs is the union *of those events.*

Intersection of Event Sets

Another type of complex outcome from a random experiment occurs only when *all* of several component events occur. For example, with the random experiment involving the tossing of a penny, a dime, and a nickel, the event "all heads" occurs only when both the events "dime is a head" and "all coins show same side" occur. Thus, we have

All heads = dime is a head *and* all coins show same side

"All heads" occurs when *both* of the component events occur. This can only happen if the elementary events common to both component events occur, so that:

$$\begin{aligned}\text{All heads} &= \{(H,H,H),(H,T,H),(T,H,H),(T,T,H)\}\ \textit{and}\ \{(H,H,H),(T,T,T)\}\\ &= \{(H,H,H)\}\end{aligned}$$

This is shown in Figure 4-5, where (H, H, H) is seen to be the element where the two event sets *intersect*. The relation between the foregoing events may be generalized by the following

DEFINITION The ***intersection*** of two sets A, B, denoted by

$$A \text{ and } B$$

is a set containing only those elements common to both.

Thus, the event set of an outcome that occurs only when all of its component events occur is the intersection *of those components. Such events are sometimes referred to as* joint events.

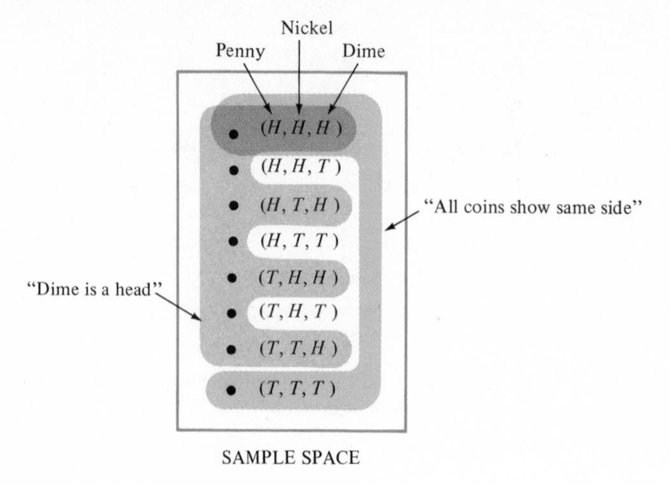

FIGURE 4-5 Portrayal of the intersection of two events.

Relationships Between Events

We will now consider two special relationships between events that are important to probability.

Mutually Exclusive Events

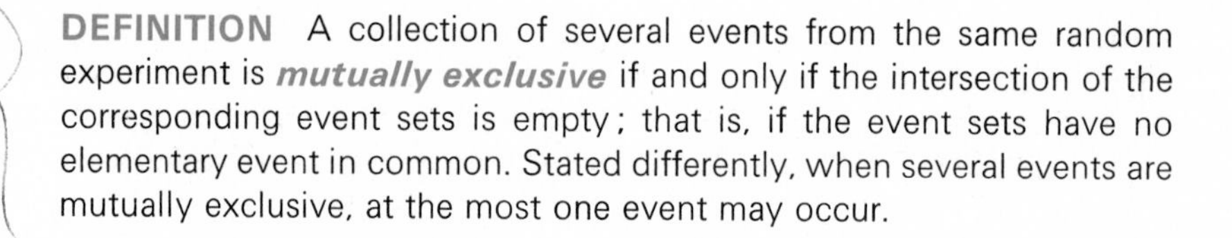

DEFINITION A collection of several events from the same random experiment is ***mutually exclusive*** if and only if the intersection of the corresponding event sets is empty; that is, if the event sets have no elementary event in common. Stated differently, when several events are mutually exclusive, at the most one event may occur.

It may be useful to consider just the two events A and B diagramed in Figure 4-6. Such representations are called Venn diagrams, after their originator John Venn.

A very simple example of a collection of mutually exclusive events is given by the coin toss. There are two possible events, a head or a tail. Since both events cannot occur on the same toss, they are mutually exclusive: *the occurrence of one event precludes the occurrence of the other.*

Example 4-1 Next year's annual rainfall for San Francisco is an uncertain quantity that may range from as low as 10 inches or less to a high of 40 inches or more. Any level of precipitation between these extremes is a possible event. However, since there will be just one rainfall level, all the possibilities are mutually exclusive events.

(a) Here, *A and B* = { }, so *A* and *B* are mutually exclusive.

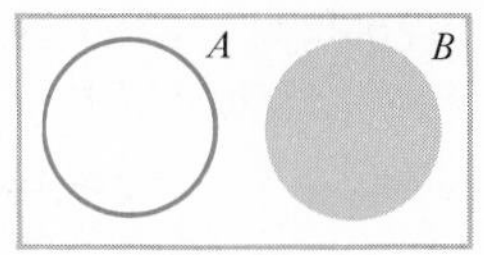

(b) Here, *A and B* = *B*, because *B* is a subset of *A*. *A* and *B* are not mutually exclusive.

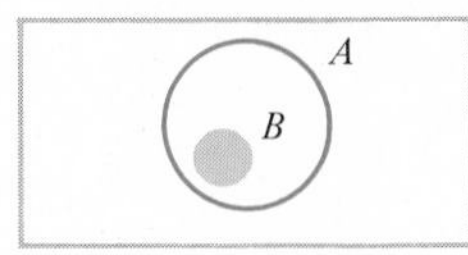

(c) Here, *A and B* is not empty, so *A* and *B* are not mutually exclusive.

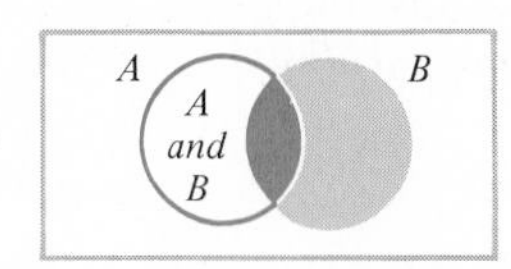

FIGURE 4-6 Venn diagrams showing the cases for the intersection of two events.

Example 4-2 A psychologist is testing a client's personality traits using the Minnesota Multiphasic Personality Inventory (MMPI). The subject will be rated on 13 different scales. The events—being "above normal in hypochondriasis" and being "below normal in schizophrenia"—are not mutually exclusive, since the client may simultaneously be overly anxious regarding his health and yet have a superior grasp of reality in his thinking.

Collectively Exhaustive Events

DEFINITION A collection of events resulting from the same random experiment is ***collectively exhaustive*** if the union of respective event sets comprises the entire sample space, or, in other words, if at least one of the events is certain to occur.

Again consider the example of the coin toss. The events head and tail are collectively exhaustive, since one of them must occur. The events in the preceding rainfall example are also collectively exhaustive. The two outcomes mentioned in the personality test illustration are not collectively exhaustive, as it is possible to score at other levels on either the hypochondriasis or the schizophrenia scales.

The notions of collective exhaustiveness and mutual exclusiveness are logically independent. That is, the presence or absence of one relationship in a collection of events does not imply the presence or absence of the other. The following example illustrates the point.

Example 4-3 A card is selected at random from a fully shuffled deck of 52 ordinary playing cards. Some events describing the characteristics of the card are categorized in the groupings below:

(1) *Mutually exclusive but not collectively exhaustive:*

odd-numbered denomination
face card

(because there are even-numbered denominations)

(2) *Collectively exhaustive but not mutually exclusive:*

red suit
black suit
spade

(because the spades are black cards)

(3) *Neither collectively exhaustive nor mutually exclusive:*

king
red suit

(since there are red kings and black cards that are not kings)

(4) *Mutually exclusive and collectively exhaustive:*

heart
diamond
club
spade

EXERCISES

4-6 The design engineer for a satellite power system has ordered a mock-up test on the ground. The system will run for six months, and will use a total of ten power cells. By the end of six months, anywhere between zero and ten cells will have expired, but the exact number is uncertain. List the elementary events in the sample space. Then find the respective event set for the following composite events for the number of dead cells:

(a) Not less than 3 nor more than 7.
(b) Greater than 7.
(c) At the most 4.
(d) Less than 5 or greater than 8.
(e) Less than or equal to 6.

4-7 A six-sided die is tossed. For the following events describing the values of the showing face, list the elementary events.

(a) at least 5; (b) at most 4; (c) 2 or greater; (d) 5 or less.

4-8 For the following random experiments, indicate whether or not the listed events are collectively exhaustive. If they are not, explain why.

(a) A box contains nine colored objects: three are red, three are blue, and three are white. Two objects are drawn from the box. Some possible resulting color combinations for the pair are: all red, all blue, red and white, all white, and blue and white.

(b) A new job applicant is: (1) either male (M) or female (F); (2) either a college graduate (G) or not (not G); and (3) either under 30 (U) or 30 or over (O). The following events describing the attributes of the applicant are of interest:

M, not G, *and* O | F, G, *and* O | F, G, *and* U
M, G, *and* U | F, not G, *and* U | M, G, *and* O
M, not G, *and* U

(c) A record disc will be inspected to determine its quality in terms of (1) high (H) or low (L) electric charge; (2) whether it is scratched (S) or not scratched (not S); and (3) whether it is warped (W) or flat (F). The joint events are

H, S, and W	*H,* not *S, and F*	*L,* not *S, and W*
H, S, and F	*L, S, and W*	*L,* not *S, and F*
H, not *S, and W*	*L, S, and F*	

4-9 For each of the following situations, indicate whether the events are mutually exclusive. If they are not, state why.

(a) Toss of a six-sided die: even-valued result, 1-face, 2-face, 5-face.

(b) Thermometers are inspected and rejected if any of the following is found: poor calibration, inability to withstand extreme temperatures without breaking, not within specified size tolerances.

(c) A manager will reject a job applicant for any of the following reasons: lack of relevant experience, slovenly appearance, too young, too old.

4-3 THE ADDITION LAW

There are several laws that can ease our task of determining the probabilities of a complex event. These follow the basic dichotomy of an event's composition: union versus intersection. They enable us to decompose a complicated probability problem into manageable pieces which may then be analyzed separately, using the basic definitions of probability.

Finding the probability of events that may be conveniently expressed as the union of component events can be simplified considerably by applying the *addition law*. This law is predicated upon the fact that, due to the logical nature of *or*, the number of ways in which the desired event may occur is the net sum of the ways in which the component events may occur. In its simplest form, we have the following expression for the

ADDITION LAW

$$P[A \text{ or } B] = P[A] + P[B] - P[A \text{ and } B] \qquad (4\text{-}3)$$

$P[A \text{ and } B]$ is subtracted to avoid double counting those situations in which the two events both occur.

Example 4-4 Consider the composite event "ace *or* heart," describing the properties of a card drawn from a completely shuffled deck of 52 ordinary playing cards. From the addition law:

$$\begin{aligned} P[\text{ace } or \text{ heart}] &= P[\text{ace}] + P[\text{heart}] - P[\text{ace } and \text{ heart (i.e., ace of hearts)}] \\ &= \frac{4}{52} + \frac{13}{52} - \frac{1}{52} \\ &= \frac{16}{52} \end{aligned}$$

Figure 4-7 shows that there are 16 elementary events in the sample space that are either ace or heart. Note that the event sets for ace and heart intersect at the ace of hearts. The sizes of the two sets are 4 and 13, respectively, and the ace of hearts is included in both. Therefore, the probability of ace of hearts must be subtracted from the sum of these probabilities in order to provide a result that is consistent with the basic probability definition. Since there are 16 cards in a deck of 52 that are either ace or heart, the correct probability must be 16/52.

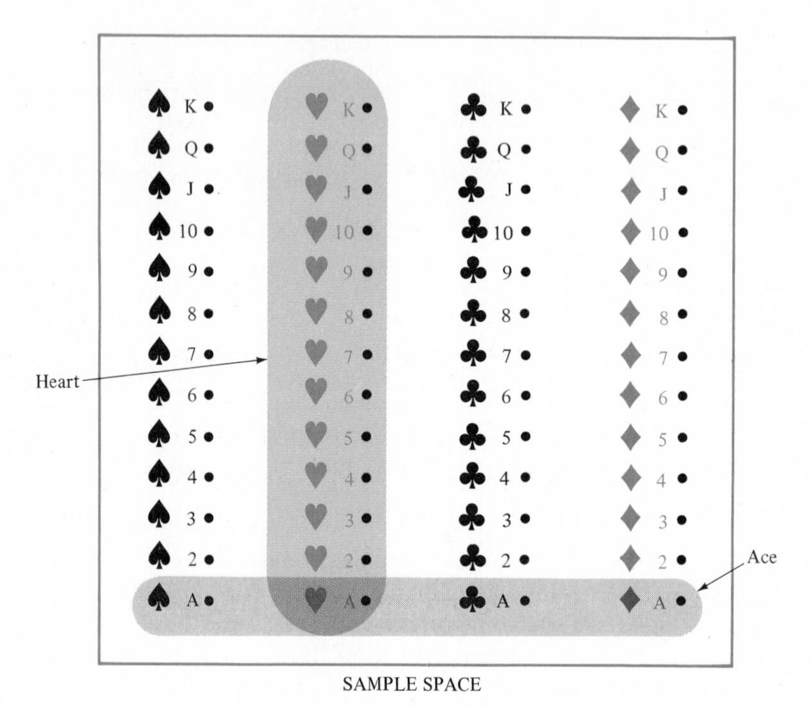

FIGURE 4-7 The union of events "heart" and "ace" is the entire colored and grey area.

When the events are *mutually exclusive*, a modified form of the addition law applies, since only one of the events can possibly occur. This makes the joint occurrence of events A and B impossible, so that $P[A \text{ and } B] = 0$. Thus, in this special instance, the addition law may be modified:

ADDITION LAW FOR MUTUALLY EXCLUSIVE EVENTS

$$P[A \text{ or } B] = P[A] + P[B] \tag{4-4}$$

The above law applies for the union of any number of events. As an illustration, we again consider the event "face card." We have established that

Face card = king *or* queen *or* jack

We notice that the events "king," "queen," and "jack" are mutually exclusive. Thus:

$$\begin{aligned} P[\text{face card}] &= P[\text{king } or \text{ queen } or \text{ jack}] \\ &= P[\text{king}] + P[\text{queen}] + P[\text{jack}] \\ &= \frac{4}{52} + \frac{4}{52} + \frac{4}{52} \\ &= \frac{12}{52} \end{aligned}$$

In this instance, the same probability result could have been found more simply by using the basic probability definition for equally likely events. Observing that 12 of the 52 cards in a deck are face cards, we can calculate the probability of drawing a face card to be 12/52. At this point, the reader may wonder why the addition law is ever needed if the probabilities can be found more easily another way. As we shall see, it is helpful when one encounters an event having a great many components or when elementary events are not equally likely (so that we cannot simply divide two numbers to obtain the answer).

When the component events are also *collectively exhaustive*, then, by definition, at least one is certain to occur. Hence, the composite event in this case is a certain event, and it therefore has a probability of 1. These facts may be used to make the addition law more specific:

ADDITION LAW FOR MUTUALLY EXCLUSIVE AND COLLECTIVELY EXHAUSTIVE EVENTS

$$P[A \text{ or } B] = P[A] + P[B] = 1 \qquad (4\text{-}5)$$

Again, this law applies to any number of events. For example, every card in an ordinary deck falls into one of the four suits: hearts, diamonds, clubs, or spades. Thus, the top card of a shuffled deck is certain to belong to one of these suits. It follows that:

$$P[\text{heart } or \text{ diamond } or \text{ club } or \text{ spade}] = 1$$

This may be verified from the addition law for mutually exclusive events, expression (4-5):

$$\begin{aligned} &P[\text{heart } or \text{ diamond } or \text{ club } or \text{ spade}] \\ &= P[\text{heart}] + P[\text{diamond}] + P[\text{club}] + P[\text{spade}] \\ &= \frac{13}{52} + \frac{13}{52} + \frac{13}{52} + \frac{13}{52} = \frac{52}{52} = 1 \end{aligned}$$

Application to Complementary Events

The addition law can be very useful in dealing with *complementary events* (opposites). An event and its complement are collectively exhaustive, because either one or the other must occur. For example, a person chosen at random must be either male or female (which may be expressed as not male). Since these events are obviously mutually exclusive, for any event A with a complement that is not A, the addition law provides:

$$P[A \text{ or not } A] = P[A] + P[\text{not } A] = 1$$

From this expression it follows that:

$$P[A] = 1 - P[\textit{not } A] \tag{4-6}$$

This principle can be useful when the event "not A" has a probability value that is easier to find than does A itself.

Suppose we wish to find the probability of getting at least one head in three tosses of a fair coin. "At least one" means "some," and the *opposite of some is none.* The complementary event is therefore "no heads." We have seen that only one of the eight equally likely outcomes (T, T, T) produces no heads. Thus:

$$\begin{aligned} P[\text{at least 1 head}] &= 1 - P[\text{no heads}] \\ &= 1 - P[(T, T, T)] \\ &= 1 - \frac{1}{8} = \frac{7}{8} \end{aligned}$$

This considerably shortens the work required to directly apply the addition law to "at least 1 head," which would have the component probabilities shown in Figure 4-8 and summarized below:

$$\begin{aligned} &P[\text{at least 1 head}] \\ &\quad = P[\text{exactly 1 head } \textit{or} \text{ exactly 2 heads } \textit{or} \text{ exactly 3 heads}] \\ &\quad = P[\text{exactly 1 head}] + P[\text{exactly 2 heads}] + P[\text{exactly 3 heads}] \end{aligned}$$

Substituting the probabilities from Figure 4-8, we have

$$P[\text{at least 1 head}] = \frac{3}{8} + \frac{3}{8} + \frac{1}{8} = \frac{7}{8}$$

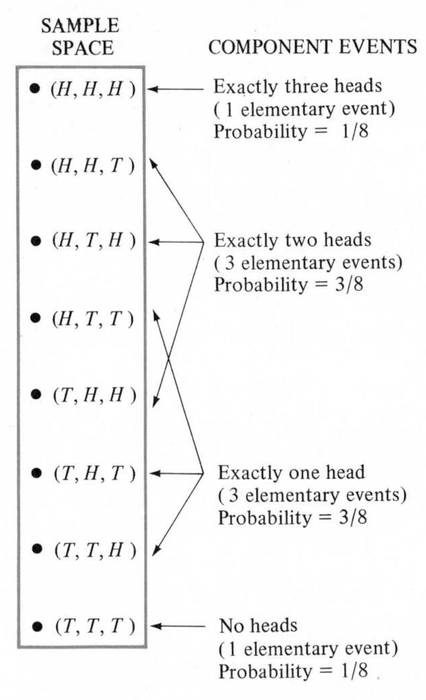

FIGURE 4-8 Component event probabilities for the outcome of three coin tosses.

EXERCISES

4-10 A researcher is studying albinism in guinea pigs. In a litter of four obtained from two colored parents, each having a recessive gene, the number of albino offspring may range from zero to four. The sample space for the number of albinos is therefore $\{0, 1, 2, 3, 4\}$.

(a) List the elements of the sets for the following events for the number of albinos: (1) An odd number; (2) At least 3; (3) More than 1 but less than 4; (4) Fewer than 4; (5) Exactly 3.

(b) List the elementary events in the event set for the complements of the outcomes in part (a).

4-11 A card is selected from a fully shuffled deck of 52 ordinary playing cards.

(a) Determine the following probabilities, treating the ace as the lowest denomination (one), the king and jack as odd denominations, and the queen as an even denomination:

(1) ace
(2) below 7
(3) odd
(4) even
(5) spade
(6) ace *and* spade
(7) odd *and* below
(8) even *and* spade
(9) even *and* below

(b) Using the values obtained above, apply the addition law to find the following probabilities:

(1) odd *or* below 7
(2) ace *or* spade
(3) even *or* below
(4) even *or* spade

4-12 An antique car parts supplier has determined the following probabilities for the number of annual orders for Locomobile fuel pumps:

Number of Orders	Probability
0	.3
1	.2
2	.1
3	.1
4	.1
5	.1
6	.1
7 or more	0
Total	1.0

Find the probability that there will be:
(a) Less than 4 orders.
(b) Between 2 and 6 orders.
(c) At least 1 order.
(d) Between 2 and 4 orders.
(e) At the most 2 orders.

4-13 Events A, B, C are mutually exclusive and collectively exhaustive, each having a probability of 1/3. Find:
(a) $P[A \text{ or } B]$
(b) $P[\text{not } C]$
(c) $P[\text{not } (A \text{ or } B)]$
(d) $P[A \text{ or } B \text{ or } C]$
(e) $P[\text{not } (A \text{ or } B \text{ or } C)]$

4-14 The antique car parts supplier in Exercise 4-12 has provided the following probabilities for the annual number of orders for Pierce-Arrow, Duesenberg, and Silver Ghost carburetors:

Number of Orders	Probabilities for Pierce-Arrow	Duesenberg	Silver Ghost
0	.5	.4	.3
1	.3	.2	.2
2	.2	.1	.2
3	.1	.1	.1
4	.1	.1	.1
5	.1	0	.1
6 or more	0	0	0

For each type of car, determine whether the probability values have been assigned properly in accordance with the addition law for mutually exclusive and collectively exhaustive events. If not, state why (verbally).

4-4 CONDITIONAL PROBABILITY AND THE JOINT PROBABILITY TABLE

Conditional Probability

Some random experiments can result in the joint occurrence of two or more events. This may be because several stages are involved, as in tossing a coin more than once, or it may be due to the simultaneous occurrence of several events, such as selecting a person at random who will be either male or female, married or single, and over or under 21 years old. In these cases, new questions arise regarding how the separate events of an outcome are related. In particular, do the conditions imposed by the occurrence of some events affect the probabilities of the other events? Thus, we may ask whether the probability of getting a head on the eleventh toss ought to be any different when the previous ten tosses had all been heads than if they had all been tails. Or we may question whether the probability that a female under 21 turns out to be married should be any different than the probability that an older woman turns out to be married. We refer to probability values obtained under the stipulation that some events have occurred or will occur as *conditional probabilities.*

Conditional probability may be illustrated for the outcome of drawing a card from a fully shuffled deck. Suppose another person draws a card from the deck without letting you see it, but in a brief glimpse you see enough to know it must be a face card. There are only 12 face cards in the deck. What is the probability that the card is a king? Although the deck contains 4 kings, our answer is not 4/52, since our surreptitiously gained information indicates that some of the 52 cards are impossible. The sample space has been restricted to the 12 face cards, so that the only remaining uncertainty is which of these 12 cards has been removed. In a sense, there is a new "sample space" having as elementary events the 12 face cards. Using basic concepts, we determine from expression (4-2) that the probability of a king is $4/12 = 1/3$. Thus, we may state that the conditional probability of a king *given* face card is

$$P[\text{king} \mid \text{face card}] = 4/12 = 1/3$$

where the vertical bar stands for "given."

Joint Probability Table and Marginal Probabilities

To further illustrate how probabilities may be obtained when there are several simultaneous events from a random experiment, let us consider the following situation.

The 100 psychology majors at a certain university have been classified in terms of sex and class level. The number of students in each category is provided in Table 4-1. Suppose that one student is chosen by lottery. The letters W, M, L, and U will be used to simplify the following discussion. From the basic definition for equally likely events, the following *joint probability* values are determined:

$$P[W \textit{ and } L] = \frac{10}{100} = .10 \qquad P[M \textit{ and } L] = \frac{40}{100} = .40$$

$$P[W \textit{ and } U] = \frac{20}{100} = .20 \qquad P[M \textit{ and } U] = \frac{30}{100} = .30$$

TABLE 4-1 Number of Psychology Majors, by Sex and Class Level

	Lower Division (*L*)	*Upper Division* (*U*)	Total
Woman (*W*)	10	20	30
Man (*M*)	40	30	70
Total	50	50	100

The marginal totals may be used to determine the probabilities that the student is a particular sex or is in a certain class level. For instance, the probability that a woman is chosen is

$$P[W] = \frac{30}{100} = .30$$

In a like manner, we can find $P[M]$, $P[L]$, and $P[U]$. Since only the numbers appearing in the margins of Table 4-1 are needed to compute these probabilities, they are sometimes called *marginal probabilities.*

Construction of the *joint probability table* shown in Table 4-2 may be helpful. The joint events represented by each cell in a joint probability table are mutually exclusive. Thus, the marginal probabilites may also be found by applying the addition law for mutually exclusive events [expression (4-4)]. For example, the event W has two mutually exclusive components: *W and L*, *W and U*. Hence, using the appropriate values from Table 4-2, we can determine that the probability of choosing a woman is

$$\begin{aligned} P[W] &= P[(W \text{ and } L) \text{ or } (W \text{ and } U)] \\ &= P[W \text{ and } L] + P[W \text{ and } U] \\ &= .10 + .20 = .30 \end{aligned}$$

TABLE 4-2 Joint Probability Table for Attributes of Randomly Selected Psychology Major

	Lower Division (*L*)	*Upper Division* (*U*)	*Marginal Probability*
Woman (*W*)	.10	.20	.30
Man (*M*)	.40	.30	.70
Marginal Probability	.50	.50	1.00

The probability values in the cells of the row (or column) corresponding to the event of the desired characteristic are added to obtain the respective marginal probabilities. Thus, we may calculate the other marginal probabilities using the addition law:

$$P[M] = P[M \text{ and } L] + P[M \text{ and } U] = .40 + .30 = .70$$

$$P[L] = P[W \text{ and } L] + P[M \text{ and } L] = .10 + .40 = .50$$

$$P[U] = P[W \text{ and } U] + P[M \text{ and } U] = .20 + .30 = .50$$

Computing Conditional Probability from Joint Probability

We are now ready to provide a formal definition of conditional probability when the joint probability of the two events is known.

DEFINITION Let A and B be two events. Then, the ***conditional probability*** of A given that B must occur, denoted by the expression $P[A|B]$, is

$$P[A|B] = \frac{P[A \text{ and } B]}{P[B]} \qquad (4\text{-}7)$$

Applying this definition to the random selection of a psychology major, the conditional probability that the student is a woman given that she is an upper-division student is

$$P[W|U] = \frac{P[W \text{ and } U]}{P[U]} = \frac{.20}{.50} = .40$$

Notice that the same result can be obtained from the data in Table 4-1 by finding the proportion of upper-division students who are also women:

$$\frac{20}{20+30} = \frac{20}{50} = .40$$

Thus, we see *the conditional probability of A given B is the proportion of times that A occurs out of all the times B occurs.* This explains why we divide the joint probability by the probability of the given event to obtain $P[A|B]$.

Figure 4-9 shows how the same result can also be obtained directly from a diagram of the sample space by noticing that the condition of being an upper-division student limits the outcomes to a new, smaller sample space of size 50. The probability of obtaining a woman is now found by observing that 20 events out of 50 provide this outcome.

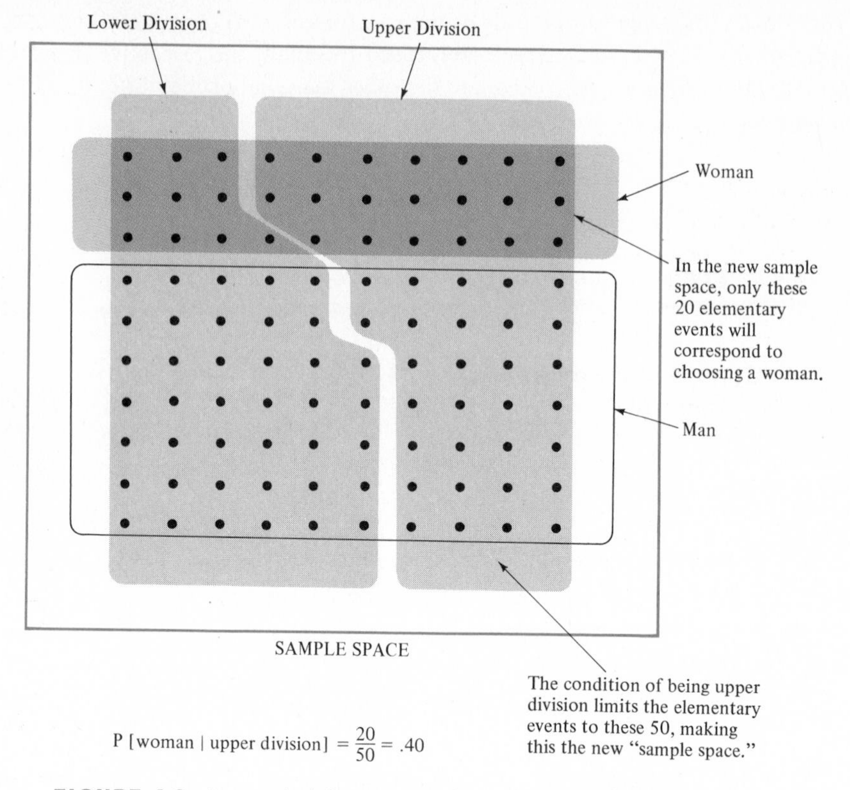

FIGURE 4-9 Portrayal of the concept of conditional probability.

We may use expression (4-7) to calculate the conditional probability found previously for a king, given face card:

$$P[\text{king}\,|\,\text{face}] = \frac{P[\text{king } \textit{and} \text{ face}]}{P[\text{face}]} = \frac{4/52}{12/52} = \frac{4}{12} = \frac{1}{3}$$

Notice that in both our illustrations the conditional probabilities are different from the corresponding probabilities when there are no stipulations, which we call the *unconditional probabilities.* We have

$$P[\text{king}\,|\,\text{face}] = \frac{1}{3} \neq \frac{1}{13} = P[\text{king}]$$

$$P[W\,|\,U] = .40 \neq .30 = P[W]$$

The unconditional and conditional probabilities are not always different, but the comparative values are used in establishing statistical independence, which will be discussed in Section 4-6.

Expression (4-7) is applicable to obtaining the conditional probability only when the joint probability of A and B is known. When $P[A \text{ and } B]$ is not known, then the conditional probability can be found only by applying the basic concepts of probability to those experiments which have resulted in B.

It may seem that the notion of conditional probability is an unnecessary embellishment. However, the principle underlying conditional probability is extensively used in our daily decision making; for example, when the sky is heavily overcast, enhancing the chance of rain, you carry an umbrella. Conditional probability is often used in decision making. Insurance companies, for example, use it to determine their rates.

Example 4-5 Life insurance companies charge a sizable premium for covering the lives of steeplejacks, miners, divers, and other members of occupational groups who are subject to greater hazards than most people. The mortality tables upon which insurance rates are based indicate that such persons have a shorter life expectancy than the population as a whole; in essence, their probability of dying in any year is higher. Additional information about an insurance applicant's occupation affects this probability; the likelihood of the event "untimely death" is affected by the occurrence of the event "applicant is a steeplejack."

EXERCISES

4-15 An employment agency specializing in clerical and secretarial help classifies candidates in terms of primary skills and years of experience. The skills are bookkeeping, switchboard, and stenography (we will assume that no candidate is proficient in more than one of these). Experience categories are one year or less, between one and three years, and three years or more. There are 100 girls currently on file, and their skills and experience are summarized in the following table.

	Skill			
Experience	*Bookkeeping*	*Switchboard*	*Stenography*	Total
Less than One Year	15		30	50
Between one and three Years	5	10		20
More than Three Years			10	
Total		30		100

One girl is chosen at random. Find the following:

(a) The missing numbers in the table.
(b) P [stenographer *or* bookkeeper]
(c) P [stenographer | more than three years experience]
(d) P [bookkeeper *or* less than one year experience]

4-16 You have drawn a card from a fully shuffled deck of 52 ordinary playing cards. Find:

(a) P [ace | red]
(b) P [ace of diamonds | red]
(c) P [diamond | red]
(d) P [face card | red]

4-17 For events X, Y, and Z, the following probability values hold:

$P[X \text{ and } Y] = 1/3$ $\quad$ $P[X \text{ and } Z] = 1/6$

$P[X] = 1/2$ $\quad$ $P[Z] = 4/9$

Answer the following:

(a) Find $P[Y|X]$.
(b) Find $P[Z|X]$.
(c) Find $P[Y]$, if Y and Z are mutually exclusive and collectively exhaustive.

4-18 A and B are mutually exclusive and collectively exhaustive. If $P[A] = 0.3$, find the following:

(a) $P[B]$ $\quad$ (b) $P[A \text{ and } B]$ $\quad$ (c) $P[A|B]$ $\quad$ (d) $P[A \text{ or } B]$

4-19 A new family with two children of different ages has moved into the neighborhood. Suppose that it is equally likely that either child will be a boy or girl. Hence, the following situations are equally likely:

Youngest	*Oldest*	
boy	boy	(B, B)
boy	girl	(B, G)
girl	boy	(G, B)
girl	girl	(G, G)

(a) Find P [at least one girl].
(b) If you spot one of the children and she is a girl, what is the conditional probability that the other child is a boy (that is, that the family has exactly one boy)?
(c) Given that one child is a girl, what is the conditional probability that the other child is a girl?

4-5 MULTIPLICATION LAW

We have seen how the addition law can be used to find the probability of the union of two events. We now introduce the *multiplication law*, which is used to find the probability of the joint occurrence or intersection of two or more events. The multiplication law is especially necessary when the component events have probability values that can be readily found but the joint probability cannot be directly obtained through the basic probability definitions. For instance, the necessary information may not be summarized conveniently in a joint probability table, as it was in the psychology major illustration in the previous section.

MULTIPLICATION LAW

$$P[A \text{ and } B] = P[A] \times P[B|A]$$

and $\quad$ (4-8)

$$P[A \text{ and } B] = P[B] \times P[A|B]$$

This law is an immediate consequence of the definition of conditional probability. (Each expression is obtained by multiplying both sides of the applicable conditional probability ratio by the given event's probability and canceling terms.)

We may continue with the psychology major illustration to show how the multiplication law can be applied. Recall that $P[W|U] = .40$ and also that $P[U] = .50$. Applying the multiplication law:

$$P[W \text{ and } U] = P[U] \times P[W|U] = .40(.50) = .20$$

This is the same joint probability value found earlier.

When the joint probabilities are already known, the multiplication law is not needed. But there are situations where only conditional and marginal probabilities are available, and the joint probabilities may be obtained only by using the multiplication law.

Example 4-6 A highway commissioner has found that half of all fatal automobile accidents in his state may be blamed on drunken drivers. Only 4 in 1,000 reported accidents have proved fatal, and 10 percent of all accidents in the state are attributable to drunken drivers. The commissioner wishes to summarize this information in a joint probability table relating to future accidents. Assuming that the present pattern prevails, the probability that a reported accident happens to be fatal is

$$P[F] = 4/1{,}000 = .004$$

while the probability that a drunken driver causes the accident (fatal or not) is

$$P[D] = .10$$

and the conditional probability that a drunken driver causes the accident given that it is fatal is

$$P[D|F] = .50$$

The joint probability table for cause and kind of accident events is provided below. Only the numbers shown in black are directly provided by the data given above. The colored probability values were obtained in the following manner.

The multiplication law provides the joint probability that a fatal accident is caused by a drunken driver:

$$P[D \text{ and } F] = P[F] \times P[D|F]$$
$$= .004(.50) = .002$$

TABLE 4-3 Joint Probability Table for Cause and Kind of Automobile Accident

	Fatal (F)	*Nonfatal (not F)*	*Marginal Probability*
Drunken Driver (D)	**.002**	**.098**	.100
Other Cause (O)	**.002**	**.898**	**.900**
Marginal Probability	.004	**.996**	1.000

The marginal probabilities for the causes and the types of accidents must sum, respectively, to 1. It therefore follows that:

$$P[O] = 1 - P[D] = 1 - .100 = .900$$

$$P[\text{not } F] = 1 - P[F] = 1 - .004 = .996$$

The remaining joint probabilities can be found by utilizing the fact that the joint probabilities in each row and column must sum to the respective marginal probability values. Thus, subtracting the known joint probability value from the marginal probability, we find the unknown joint probability values:

$$P[D \textit{ and } \text{not } F] = P[D] - P[D \textit{ and } F] = .100 - .002 = .098$$

$$P[O \textit{ and } F] = P[F] - P[D \textit{ and } F] = .004 - .002 = .002$$

$$P[O \textit{ and } \text{not } F] = P[O] - P[O \textit{ and } F] = .900 - .002 = .898$$

Joint Probabilities for More than Two Events

When we wish to find the joint probability of more than two events, the multiplication law may be extended:

$$P[A \textit{ and } B \textit{ and } C] = P[A] \times P[B|A] \times P[C|A \textit{ and } B] \qquad \text{(4-9)}$$

Suppose that we have a box containing five marbles, each of a different color: yellow, red, green, orange, and purple. Three marbles are selected from the box at random, one at a time, so that all marbles remaining in the box have an equal chance of being selected at each drawing. Once drawn, a marble is not replaced. The following events are designated:

A = first marble is yellow

B = second marble is green

C = third marble is orange

It is easy to establish that $P[A] = 1/5$. If a yellow marble is selected, then only four remain, and one of these is green. Thus, the conditional probability that the second marble will be green given that the first is yellow is $P[B|A] = 1/4$. Likewise, should the yellow and green marbles be selected, then the third marble is equally likely to be one of the three colors red, orange, or purple, so that $P[C|A \textit{ and } B] = 1/3$. The joint probability for A, B, and C may be determined by applying expression (4-9):

$$P[A \textit{ and } B \textit{ and } C] = \frac{1}{5} \times \frac{1}{4} \times \frac{1}{3} = \frac{1}{60}$$

For a large number of component events, the same approach applies.

Matching-Birthdays Problem

The following example nicely illustrates how the multiplication law may be applied to a large number of events.

Example 4-7 The problem is to determine the approximate probability that there is at least one matching birthday (day and month) among a group of persons. For simplicity, it will be assumed that each day in the year is equally likely to be a person's birth anniversary; February 29 will be combined with March 1. (This assumption is not strictly true; some months are more popular, so that our result will only be approximate.)

It will be simplest to find the probability of the complementary event—no matches—by using the multiplication law. Envision each member of a group of size n being asked in succession to state his birthday. We conveniently define our events as follows:

A_i = the ith person queried does not share a birthday with the previous $i-1$ persons.

Then $P[\text{no match}] = P[A_1 \text{ and } A_2 \text{ and} \ldots \text{and } A_{n-1} \text{ and } A_n]$, which is the probability of the event that no person shares a birthday with the preceding persons. Since there is no previous person for the first to match with, the event A_1 is certain, so that:

$$P[A_1] = 1 = \frac{365}{365}$$

Also, we obtain

$$P[A_2 | A_1] = \frac{364}{365}$$

because there are 364 days in which the second person may not match birthdays with the first. Continuing in this same manner, we finally obtain

$$P[A_n | A_1 \text{ and } A_2 \text{ and} \ldots \text{and } A_{n-1}] = \frac{365-n+1}{365}$$

since the nth person cannot have a birthday on the previously cited $n-1$ dates, leaving only $365-(n-1)$ days allowable for his birthday. Therefore, we may apply the multiplication law to obtain:

$$P[\text{no matches}] = \frac{365}{365} \times \frac{364}{365} \times \frac{363}{365} \times \cdots \times \frac{365-n+1}{365}$$

The probability of at least one match may be found from this:

$$P[\text{at least one match}] = 1 - P[\text{no matches}]$$

One interesting issue is finding the size of the group for which the probability exceeds 1/2 that there is at least one match; knowing this number, you can amaze your less knowledgeable friends and perhaps win a few bets. The "magical" group size turns out to be 23.

Why such a low group size? If it were physically possible to list all the ways (triples, sextuples, septuples, and so forth) in which 23 birthdays can match (there are several million, and for each of them a tremendous number of date possibilities), an intuitive appreciation as to why could be attained. Table 4-4 shows the matching probabilities for several group sizes. Note that for groups above 60 there is almost certain to be at least one match.

TABLE 4-4 Probabilities of at Least One Matching Birthday

Group Size	*P [no matches]*	*P [at least one match]*
3	.992	.008
7	.943	.057
10	.883	.117
15	.748	.252
23	.493	.507
40	.109	.891
50	.030	.970
60	.006	.994

EXERCISES

4-20 A perfectly balanced coin is tossed fairly twice in succession. Let H_1 and T_1 denote the possible outcomes of the first toss and H_2 and T_2 the possible results from the second toss.
(a) What is the value of $P[H_2 \mid T_1]$?
(b) What is the value of $P[H_2 \mid H_1]$?
(c) Use the multiplication law to find the probability of (1) H_1 *and* H_2, and (2) T_1 *and* T_2.

4-21 Given the following probability values:

$$P[A_2 \mid A_1] = .3; \quad P[A_1] = .6; \quad P[A_2] = .4;$$

$$P[A_3 \mid A_2] = .4; \quad P[A_3 \mid A_1 \text{ and } A_2] = .3;$$

use the multiplication law to find:

$$P[A_1 \text{ and } A_2]$$

$$P[A_2 \text{ and } A_3]$$

$$P[A_1 \text{ and } A_2 \text{ and } A_3]$$

4-22 A professor grading a statistics examination has promised to give an equal number of A's, B's, C's, and D's. If three students are chosen randomly:
(a) What is the probability that all three will receive different grades?
(b) What is the probability that at least two of the three students will receive the same grade?

4-23 A fruit inspector accepts or rejects shipments of bananas after performing tests on a few sample bunches. He rejects 15 percent of all shipments inspected. Thus far, he has rejected 95 percent of all bad shipments inspected, and 10 percent of all shipments have ultimately proved bad.
(a) Using the above experience as a basis, find the values for the probabilities regarding the outcome of any particular shipment handled by this particular inspector: P[reject], P[bad], P[reject | bad].

(b) Find P[reject *and* bad], using the multiplication law with the appropriate values found in part (a).

(c) Construct a joint probability table showing the joint probabilities and the marginal probabilities for the inspector's actions (accept or reject) and the quality (good or bad) of the banana shipment.

4-6 STATISTICAL INDEPENDENCE

In the psychology student lottery illustrated in Section 4-4, there is a difference between the prevalence of women at the lower- and upper-division levels. Referring to Table 4-1 on page 114, only 10 out of 50, or 20 percent, of the lower-division students are women, whereas 20 out of 50, or 40 percent, of the upper-division students are women. Thus, for the students in this example, being a woman is more common at the upper-division level than at the lower-division level. We may conclude that the events "woman" and "upper-division student" are *dependent*, and we can see that there is a dependency between sex and class level.

On the other hand, we ordinarily think of the outcomes from tossing two coins as *independent*. Unless the coins are unfairly tossed, getting a head on the first toss should not influence the outcome of the second toss. It is very important to determine whether or not events are independent, for in such cases procedures for finding probabilities may often be streamlined. We therefore make the following

DEFINITION Two events A and B are ***statistically independent*** if the chance of one is unaffected by the occurrence of the other; that is, if

$$P[A \mid B] = P[A]$$

Whenever the above equality holds, so must the following:*

$$P[B \mid A] = P[B]$$

With the psychology majors, the event "woman" (W) is not independent of (and therefore is dependent upon) the event "upper division" (U). This is indicated by the fact that

$$P[W \mid U] = .40$$

does not equal

$$P[W] = .30$$

so that the requirement of statistical independence is violated.

* Several events may be *collectively* independent. If they are, every possible conditional probability for all combinations of events must equal the corresponding unconditional probability.

As another example, a card is drawn from a fully shuffled deck of 52 ordinary playing cards. The events "jack" and "face card" are dependent, since

$$P[\text{jack} \mid \text{face}] = \frac{P[\text{jack } \textit{and} \text{ face}]}{P[\text{face}]}$$

$$= \frac{4/52}{12/52} = \frac{1}{3}$$

and this differs from

$$P[\text{jack}] = \frac{4}{52} = \frac{1}{13}$$

That "jack" and "face" are dependent is evident from the fact that knowing a card is a face card increases the probability that the card will be a jack. Knowing the occurrence of one event changes the probability that the other event will occur from what the probability would have been without the knowledge of the first event.

Simplification of the Multiplication Law

When events are independent, the joint probabilities may be determined from the multiplication law without using conditional probabilities. *The "unconditional" probabilities thus have the same values as the conditional probabilities.*

MULTIPLICATION LAW APPLIED TO INDEPENDENT EVENTS

$$P[A \textit{ and } B] = P[A] \times P[B] \tag{4-10}$$

Returning to the playing card illustration, *suit* (heart, club, spade, or diamond) and *denomination* (ace, deuce, three, and so on) are statistically independent. This is because the probability of obtaining an ace is the same (1/13), regardless of whether or not its suit is known—in all cases, exactly one-thirteenth of the cards can be an ace. Thus, the probability that a card is ace *and* a heart (the ace of hearts) is the product of the respective *unconditional* probabilities for the component events, or:

$$P[\text{ace } \textit{and} \text{ heart}] = P[\text{ace}] \times P[\text{heart}]$$

$$= \frac{1}{13} \times \frac{1}{4} = \frac{1}{52}$$

This fact may be extended to any number of independent events. For example, it is often presumed that a voter's sex, political affiliation, and marital status are statistically independent; that is, knowledge of any one or any combination of two of these attributes does not affect the chances that a particular voter will have the third attribute. Under these circumstances, the probability that the 812th name on the voter registration list will be a married man who is a Democrat could be found by multiplying together the unconditional probabilities for each attribute:

$$P[\text{man } \textit{and} \text{ married } \textit{and} \text{ Democrat}] = P[\text{man}] \times P[\text{married}] \times P[\text{Democrat}]$$

Independence and Random Sampling

The selection of a sample may be viewed as a random experiment having several stages, where some elementary units are removed from a population in such a way that with each successive selection all remaining units have an equal chance of being chosen. In other words, the selections are made *randomly*, and since the particular elementary units to be chosen are not known in advance, the sampling process may be considered a random experiment.

An illustration from manufacturing quality control will help us to set the stage. Suppose that 5 percent of all items made by a machine prove to be defective. Let's assume that the incidence of defectives is sufficiently erratic that the attributes of successive items are independent events. In deciding whether or not to readjust the machine, common practice would be to periodically take a sample of items to determine if the number of defectives happens to be excessive. Suppose that three successive items are chosen for this purpose. The multiplication law for independent events could be used to determine the probability that the sample contains all defectives, all good items, or both defective and good items. In order that all possible outcomes can be considered, it is sometimes helpful in such a situation to obtain the entire sample space and to find the probabilities for every elementary event.

Probability Tree Diagram

In describing a sample space, it sometimes is convenient to explain the sampling process in terms of a *probability tree diagram.* Consider the tree diagram provided in Figure 4-10 for the machine being sampled above. Here, each successive sample observation outcome is represented by a branch, and the probability of an outcome is indicated alongside its branch. Here we use .05 as the probability of a defective item and .95 for the probability of a good item. Since there are two complementary events possible for each item observed, there are two separate branches for each item—one for defective and one for good. There is a different branching point or *fork* for each observation. Two forks are thus required for the second item—a different one for each possible attribute of the first observation. Because four distinct outcomes are possible for

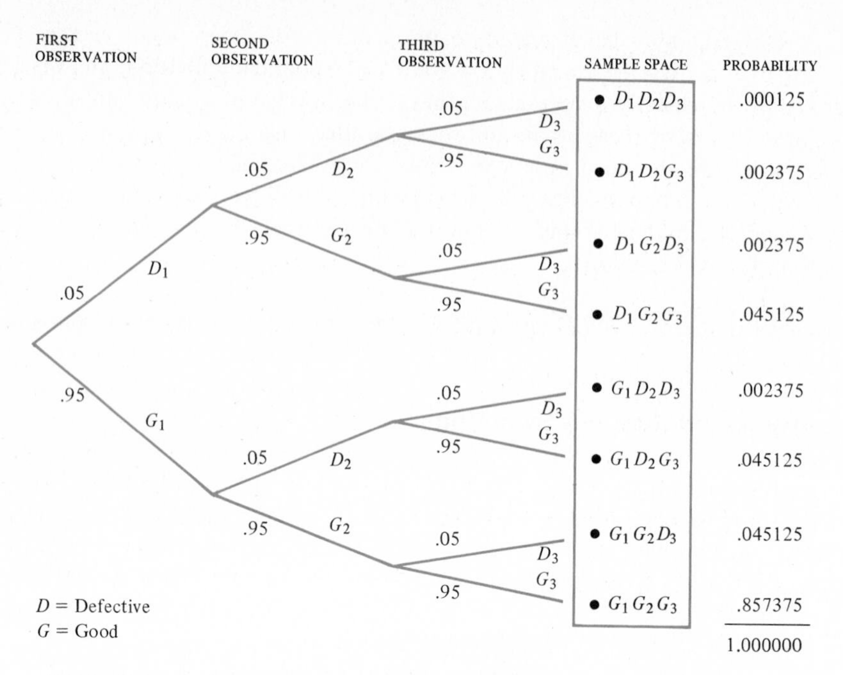

FIGURE 4-10 Probability tree diagram for selecting a quality control sample of three items from a production process.

the quality of the earlier items, four forks are needed for the third observation. To distinguish the outcomes for each item, we use the subscripts 1, 2, and 3. For instance, D_1 means that the first item will be defective; G_2 means that the second will be good. Altogether there are eight paths through the tree, each representing different sample outcomes. Each path leads to a different elementary event for this random experiment, so that the eight end positions provide the sample space for the final sample results.

The probabilities for each elementary event may be found in the manner used earlier, by multiplying together the probability values for the branches on the path leading to that particular event. For instance, the multiplication law may be applied to find the probability that all three items are defective:

$$\begin{aligned} P[D_1 \textit{ and } D_2 \textit{ and } D_3] &= P[D_1] \times P[D_2] \times P[D_3] \\ &= (.05)(.05)(.05) \\ &= .000125 \end{aligned}$$

Since the events D_1, D_2, and D_3 are independent, we use unconditional probabilities, and in all cases, the probability of obtaining a defective item happens to be the same for every observation to be made. The other outcome probabilities in Figure 4-10 were found in the same way.

Sampling With and Without Replacement

Let us slightly modify our example. Suppose that a population of 100 items is made on the same machine, of which 5 percent are defective. A three-item sample is *randomly* selected and set aside, and the quality of each item is determined.

The probability tree diagram of this experiment is provided in Figure 4-11. The results of each selection are *not independent*, because after a selection the composition of the remaining items changes. The proportion of defective items remaining increases or decreases, depending upon the quality of the prior selection. Thus, if the first item proves to be defective, so that D_1 occurs, then there are only four defectives left out of 99 remaining items, and the probability that the second item is defective D_2 is 4/99. But if G_1 is the first event, then D_2 has a probability of 5/99, because any one of five remaining defective items could be chosen. These values are *conditional probabilities*, because D_2 has a different chance of occurring in each case. Returning to our original multiplication law, we find the probability of getting all defectives in this revised sampling situation:

$$P[D_1 \text{ and } D_2 \text{ and } D_3] = P[D_1] \times P[D_2 | D_1] \times P[D_3 | D_1 \text{ and } D_2]$$

$$= \left(\frac{5}{100}\right)\left(\frac{4}{99}\right)\left(\frac{3}{98}\right)$$

$$= .0000618$$

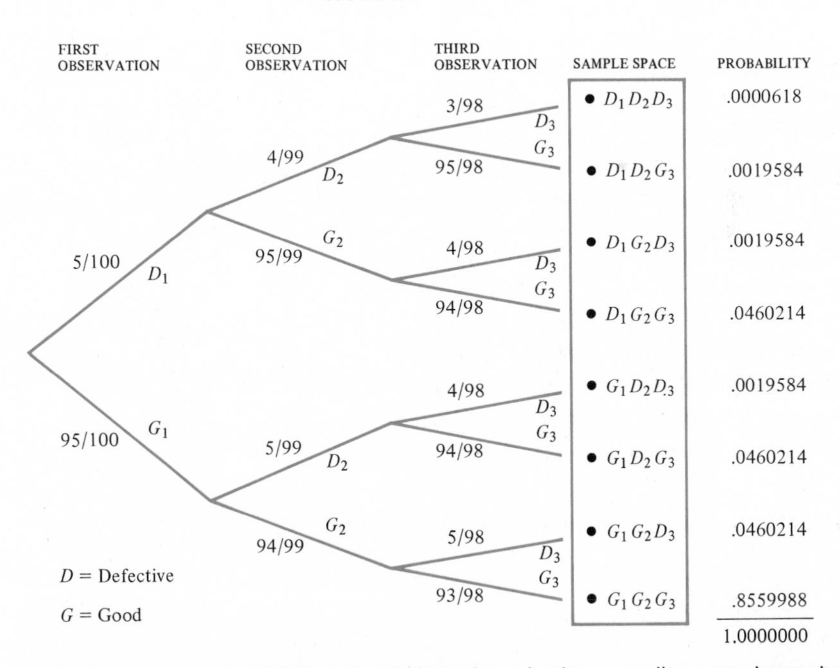

FIGURE 4-11 Probability tree diagram for selecting a quality control sample of three items without replacement from a population of size 100.

Notice that this result differs from the one obtained when successive observations are independent.

When the population units are set aside after each selection, the sampling is done *without replacement*. If, instead, each inspected item were replaced in the group and again allowed the same chance of being chosen as any of the other items, then the outcomes would have identical probabilities for each observation, as in Figure 4-10. Such a procedure is called *sampling with replacement*. Although intuitively wasteful, sampling with replacement makes for simpler probability calculations. More will be said about this in later chapters.

Independence and the Law of Large Numbers

Another example may prove useful to illustrate a very important property of random experiments with several stages, each having outcomes independent of the others.

Suppose a fair coin is tossed 20 times, and each time a head appears. The probability of this happening is less than one-millionth.* Does this mean that in a long series of future tosses 20 extra tail outcomes must occur in order to balance the long-run frequency of occurrence ratio and to coincide with "heads" having a probability of 1/2? Also, does this increase the probability of a tail next time?

The answer to both questions is "no." First, there are no guarantees that that any particular sequence of results will exhibit heads even close to one-half of the time. This presents no contradiction here, for there is a nonzero probability that any number of heads—say, 10,000—will occur in sequence. All that can be stated is that it is not very likely there will be many more (or fewer) heads than half the number of tosses. A law of probability, the *law of large numbers*, in essence states that the probability that the result will deviate significantly from that indicated by the theoretical long-run frequency of occurrence becomes small as the number of experiments (tosses) increases. Thus, if the coin tosses were continued several thousand more times, the effect of the 20 "extra" heads on the resulting frequency would not be very noticeable.

If the coin-tossing process is a fair one (that is, if there are no biases in the tossing process in favor of heads), then it may be inferred that the probability of a head should be 1/2. Also, if the process is fair, then a head would be no more likely to follow a previous head than would a tail; the events of succeeding and preceding tosses are independent. Hence:

$$P[\text{head} \mid \text{head previously}] = P[\text{head}] = 1/2$$

and

$$P[\text{tail}] = 1 - P[\text{head}] = 1/2$$

* This provides rather strong empirical evidence that the coin may indeed be unfair. But—although a rare outcome—this is a possible result even with a fair coin.

The probability of a tail should not increase or decrease if the fact is accepted that fairness is present. After obtaining 20 heads in a row, human nature may lead to the inference that the tossing mechanism is indeed unfair. The seemingly aberrant result does not, however, constitute proof of unfairness.

EXERCISES

4-24 A box contains 100 marbles, 60 red and 40 green. There are 30 striped; 70 are solid. There are 10 green-striped marbles. One marble is chosen at random from the box.
(a) Construct a joint probability table summarizing the events describing the properties (color, solid or striped) of the selected marble; include the marginal probabilities.
(b) Find P[marble is solid | marble is green].
(c) Is the event "marble is solid" independent of the event "marble is green"? Why?

4-25 The following probabilities have been obtained for the results of a random experiment: $P[R \text{ and } S] = .10$; $P[R \text{ and } T] = .3$; $P[S \text{ and } T] = .2$; $P[R] = .50$; $P[S] = .4$; and $P[T] = .50$. Specify which of the following pairs of events are statistically independent: R, S; R, T; S, T. State the reasons for your answers.

4-26 Ten percent of the ball bearings in a lot of 50 are known to be overweight. Three bearings are randomly selected, one at a time, and weighed. Each in turn is returned to the lot and allowed the same chance of being selected as the unweighed items. Construct a probability tree diagram for this situation. Find the following probabilities describing the final results of the sampling procedure: (a) No overweight items are selected; (b) All the items are found to be overweight; (c) Exactly one overweight ball bearing is found.

4-27 Repeat Exercise 4-26 if the successively weighed items are not replaced.

4-7 COMMON ERRORS IN APPLYING THE LAWS OF PROBABILITY

Some of the most prevalent errors committed in the determination of probability values are due to the improper use of the laws of probability. A few of the common mistakes are listed here:

(1) Using the addition law to find the probability of the union of several events when they are *not* mutually exclusive, without correcting for the double counting of possible occurrences.

Example 4-8 Suppose that casualty insurance underwriters have established that the probabilities of a city experiencing one of the following natural disasters in the next decade are

tornado	.5
flood	.3
earthquake	.4

We cannot say that the probability of suffering one of these acts is $.5 + .3 + .4 = 1.2$. Clearly, two or more of these disasters may occur over a ten-year period, and some may occur more than once.

(2) Using the addition law when the multiplication law should be used, and conversely. Remember that *or* signifies addition and that *and* signifies multiplication.

For example, the probability of drawing a red face card is the same as that for the event "red *and* face." Recall that:

$$P[\text{red}] = \frac{26}{52} \qquad \text{and} \qquad P[\text{face}] = \frac{12}{52}$$

If we add these values:

$$\frac{26}{52} + \frac{12}{52} = \frac{38}{52}$$

we obtain a meaningless result. Since red and face are independent events:

$$P[\text{red } \textit{and} \text{ face}] = \frac{26}{52} \times \frac{12}{52}$$
$$= \frac{6}{52}$$

(3) Using the multiplication law for independent component events when the events are dependent.

For example, a common commission of this error occurs when replacement is mistakenly assumed in calculating the probability of obtaining a particular sample result. As we have seen, removal of an item from a group changes the group's composition and hence the probabilities that future selections will be of a certain type.

(4) Improperly identifying the complement of an event. For example, *the complement of none is some* (which may be expressed as "one or more," or as "at least one").

The following example, which actually happened, dramatically illustrates how ludicrous results may be obtained through the incorrect application of probability laws.*

Example 4-9 An elderly woman was mugged in the suburb of a large city. A couple was convicted of the crime; the evidence upon which the prosecution rested its case was largely circumstantial. Probability was used to demonstrate that an extremely low probability of one twelve-millionth existed that any specific couple could have committed the crime. The probability value was determined by

* For a detailed discussion, see "Trial by Mathematics," *Time* (April 26, 1968), 41.

using the multiplication law for independent events. The events—the characteristics ascribed by witnesses to the couple who actually did the deed—are listed below, along with the assumed probabilities:

Characteristic Event	*Assumed Probability*
Drives yellow car	1/10
Interracial couple	1/1000
Blonde girl	1/4
Girl wears hair in ponytail	1/10
Man bearded	1/10
Man black	1/3

Multiplying the above values, the probability was obtained that any specific couple, chosen at random from the city's population, had all six characteristics:

$$\frac{1}{10} \times \frac{1}{1000} \times \frac{1}{4} \times \frac{1}{10} \times \frac{1}{10} \times \frac{1}{3} = \frac{1}{12{,}000{,}000}$$

Since the defendants had all six characteristics and the jury was mystified by the overwhelming strength of the probability argument, they were convicted.

The Supreme Court of the state heard the appeal of one of the defendants. The defense attorneys, after obtaining some good advice on probability theory, attacked the prosecution's analysis on two points: (1) the rather dubiously assumed event probability values, and (2) the invalid assumption of independence implicit in using the multiplication law in the above manner (for example, the proportion of black men having beards may have been greater than the proportion of the population as a whole having beards; also, "interracial couple" and "black man" are definitely not independent events). The judge accepted the arguments of the defense and noted that the trial evidence, even allowing for its incorrect assumptions, was misleading on another score—namely, a high probability that the defendants were the only such couple should really have been determined in order to demonstrate a strong case. Using the prosecution's original figures and its assumptions of independence, it can be demonstrated that the probability is large that at least one other couple in the area had the same characteristics.

4-8 COUNTING TECHNIQUES

There are random experiments whose sample spaces are so large that we cannot list all of their elementary events. One can envision events for which the space contained in all the world's books would not be enough for a complete listing of the possibilities. For example, there are about 400 trillion trillion ways to sequence the 26 letters of the alphabet. Assuming 500,000 entries per book, 100 books per person, 6 billion persons, we have space for only 300,000 trillion entries. Fortunately, probabilities for such events may still be calculated by the use of the techniques developed in this section. These enable us to count possibilities by large multiples and to develop rules for employing logical principles to determine what a count should be by taking shortcuts and thereby giving us the capability to deal with astronomical quantities.

Underlying Principle of Multiplication

The most difficult sample spaces to physically enumerate arise from random experiments with multifaceted outcomes. Each elementary event can be categorized in a number of different ways. As we have seen, this kind of an experiment may have several stages, such as successive coin tosses. Or it may have only one stage but several dimensions, like selecting an object which may belong to several categories. For example, we may consider a particular order received by an automobile assembly plant as a single random experiment with a complex outcome involving the options requested; the number of possible combinations is so staggering that it is an extreme rarity for two independently ordered cars coming from the same plant to be just alike. In either case, it will be convenient to treat the experiment as one having several stages. Thus, we may envision the car order as one involving first a specification of color, then size of engine, followed by body style (sedan, convertible, or whatever), and so on.

When the number of possibilities in a particular stage will be the same no matter what the outcomes of the preceding stages are, our problem is quite simple. We count only the number of possibilities in each stage and then multiply these together.

Example 4-10 Consider the number of possible ways a man could choose to dress himself when selecting items from his wardrobe. Suppose that he has 5 suits, 8 shirts, 3 belts, 15 pairs of socks, 6 pairs of shoes, and 20 ties. Figure 4-12 shows how we may break this situation into stages. The total number of possible

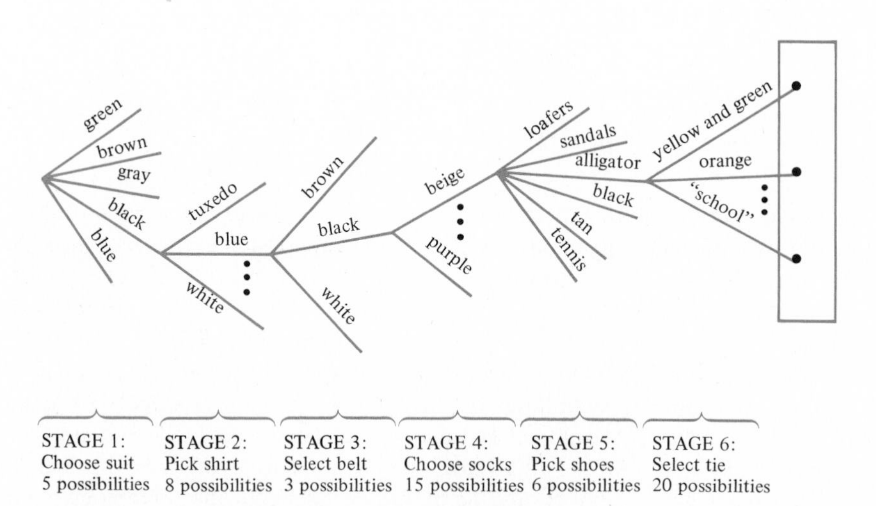

FIGURE 4-12 Diagram of possible wardrobe selections.

choices would then be

$$5 \times 8 \times 3 \times 15 \times 6 \times 20 = 216{,}000$$

Notice that it is impracticable to list all of the possibilities. This would be unnecessary here, anyway. However, if our man is particular about the potential for clashing combinations (if he never wears his white belt with his black suit or his socks with his sandals) then we would have no choice but to laboriously list all of the possibilities, taking shortcuts whenever possible.

Example 4-11 A deck of 52 ordinary playing cards is thoroughly shuffled and the five top cards are removed. The probability that these cards may be placed in a denominational sequence to form a straight is to be determined. It does not matter, in what sequence the cards appear when drawn.

Here, our sample space comprises all possible five-card results. As these are equally likely, we may find the probability of a straight by dividing the number of results involving straights by the size of the sample space. To find the number of five-card straights,* one may begin by listing the possibilities:

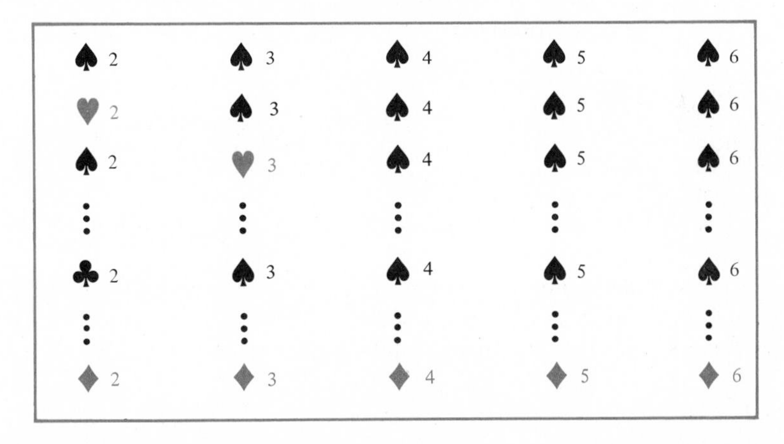

This list, as will be shown, has 1,024 entries. Lists of equal length would have to be constructed for straights beginning with 3, with 4, and so on, and with 10 (ace being a high card only). Nine such lists would be required. The total number of five-card straights possible is therefore 9,216. More than 200 pages of this book would be required to show a complete listing.

How did we obtain 9,216? First, we observe that there are nine denominations for the lowest card. All of these denominational groupings are of materially the same form, each containing the same number of possible five-card arrangements. Therefore, if straights beginning with the deuce are investigated first, then the results obtained will be applicable to straights of any denomination. What may be observed from the above list? Note that each possible outcome differs from the others only in the choice of suits for the cards of the five denominations. For the deuce, any one of four suits (hearts, diamonds, clubs, or spades) would be acceptable, and likewise for the 3, 4, 5, and 6. The following list summarizes the possible suit arrangements:

* Poker players will recognize that this list would normally include straight flushes (wherein all cards belong to the same suit); these may be eliminated here, thereby reducing the number of possibilities.

$$\underbrace{\text{deuce of}}_{\substack{\text{hearts}\\ \text{diamonds}\\ \text{clubs}\\ \text{or spades}}}, \quad \underbrace{\text{3 of}}_{\substack{\text{hearts}\\ \text{diamonds}\\ \text{clubs}\\ \text{or spades}}}, \quad \underbrace{\text{4 of}}_{\substack{\text{hearts}\\ \text{diamonds}\\ \text{clubs}\\ \text{or spades}}}, \quad \underbrace{\text{5 of}}_{\substack{\text{hearts}\\ \text{diamonds}\\ \text{clubs}\\ \text{or spades}}}, \quad \underbrace{\text{6 of}}_{\substack{\text{hearts}\\ \text{diamonds}\\ \text{clubs}\\ \text{or spades}}},$$

The total number of suit assignments is found by the following multiplication:

$$\begin{bmatrix}\text{number of}\\ \text{deuce suits}\end{bmatrix} \times \begin{bmatrix}\text{number of}\\ \text{3 suits}\end{bmatrix} \times \begin{bmatrix}\text{number of}\\ \text{4 suits}\end{bmatrix} \times \begin{bmatrix}\text{number of}\\ \text{5 suits}\end{bmatrix} \times \begin{bmatrix}\text{number of}\\ \text{6 suits}\end{bmatrix}$$

which equals

$$4 \times 4 \times 4 \times 4 \times 4 = 4^5 \text{ or } 1{,}024$$

Duplicating the above argument for the eight other denominational categories of straights, we arrive at:

$$\underbrace{9}_{\substack{\text{total number}\\ \text{of possible}\\ \text{denominational}\\ \text{categories}}} \times \underbrace{4^5}_{\substack{\text{total number}\\ \text{of suit}\\ \text{arrangements}\\ \text{for each}}} = \underbrace{9{,}216}_{\substack{\text{total number}\\ \text{of five-card}\\ \text{straights}}}$$

The number of equally likely situations in the denominator must now be determined before the five-card straight probability can be calculated; this will be done later in this chapter.

The principle used in the foregoing examples is a general one. It may also be applied to certain types of situations of a different nature.

The multiplication principle is applicable even when the same physical object is used over and over again. For example, it tells us that the number of possible outcomes from a sequence of five tosses of the same coin is

$$2 \times 2 \times 2 \times 2 \times 2 = 2^5 = 32$$

Observe that an identical number of outcomes are possible when five different coins are each tossed once.

Number of Ways of Sequencing Objects: The Factorial and Permutations

A large class of situations involves the enumeration of sequences in which objects may occur. The following example illustrates how this may be accomplished.

Example 4-12 The editor of *Psychomation Magazine* wants to determine the sequence in which to place five articles (A, B, C, D, and E) in next month's issue. How many different sequence versions are possible?

The question may be answered by considering each possible position in succession. Position one may be any one of the five articles; thus, there are five choices. Choosing one will then leave four articles available for position two—four choices; again, one of these is to be selected. For the first two positions, there are already 20 possible choices. These are partially listed below (verify this to your satisfaction by completing the list).

Possible Positions of A B C D *and* E

Position One	*Position Two*	*Choices for Position Three*
A	B	C, D, E
A	C	B, D, E
A	D	B, C, E
A	E	B, C, D
B	A	C, D, E
⋮	⋮	⋮

Suppose that article A is chosen for position one and article B is chosen for position two. This leaves articles C, D, and E to be assigned to positions three, four, and five. The following six positions are possible:

Position Three	*Position Four*	*Position Five*
C	D	E
C	E	D
D	C	E
D	E	C
E	C	D
E	D	C

A similar list of equal length may be prepared for any of the other 19 assignments to the first two positions; all such lists would contain six entries (verify this to your own satisfaction). For the magazine, 20×6 or 120 different article-to-position sequences are therefore possible. This figure could also have been obtained from the following multiplication:

$$\underbrace{5}_{\substack{\text{number of}\\ \text{choices for}\\ \text{position}\\ \text{one}}} \times \underbrace{4}_{\substack{\text{number of}\\ \text{choices for}\\ \text{position}\\ \text{two}}} \times \underbrace{3}_{\substack{\text{number of}\\ \text{choices for}\\ \text{position}\\ \text{three}}} \times \underbrace{2}_{\substack{\text{number of}\\ \text{choices for}\\ \text{position}\\ \text{four}}} \times \underbrace{1}_{\substack{\text{number of}\\ \text{choices for}\\ \text{position}\\ \text{five}}}$$

A multiplication of the type used in the foregoing example is called a factorial product, or simply a *factorial*, and is denoted by placing an exclamation point after the highest number. Thus, a 5 factorial is

$$5! = 5 \times 4 \times 3 \times 2 \times 1 = 120$$

In general, when n is any positive whole number, n factorial ($n!$) may be determined by the following multiplication:

$$n! = n \times (n-1) \times (n-2) \times \cdots \times 2 \times 1 \tag{4-11}$$

We define

$$1! = 1$$

$$0! = 1$$

The fact that 0! is 1 is usually a perplexing notion. If you think of 0! as representing the number of different sequences of assigning no (0) articles to positions in a magazine with no (0) positions, you may observe that there is only one way of doing this; that is, there is only one way to sequence nothing.

Factorial values literally become astronomically large for modest values of n. For instance:

$$10! = 3{,}628{,}800$$

$$20! = \text{approximately 2.4 billion billion}$$

$$100! = \text{approximately 9 followed by 157 zeros}$$

It is sometimes useful to determine the number of ways in which a number of objects can be selected from a given group by considering the order of selection. Suppose we have n objects from which r objects are to be removed. The number of ways in which it is possible to select r objects may be determined from the following product:

$$\underbrace{n}_{\substack{\text{number of}\\ \text{choices for}\\ \text{first object}}} \times \underbrace{(n-1)}_{\substack{\text{number of}\\ \text{choices for}\\ \text{second object}}} \times \underbrace{(n-2)}_{\substack{\text{number of}\\ \text{choices for}\\ \text{third object}}} \times \cdots \times \underbrace{(n-r+1)}_{\substack{\text{number of}\\ \text{choices for}\\ r\text{th object}}}$$

Note that when $r-1$ objects have been removed, $n-r+1$ [that is, $n-(r-1)$] objects remain, leaving this many choices for selection of the rth object. In Example 4-12, $r = 5$ articles were selected, one at a time, from $n = 5$ for inclusion in the magazine. There, each possibility differed only in *sequence*, so that $n = r$. But when r is smaller than n (if, for example, we took four of the articles for the magazine), each possible group of selections would also differ with respect to *which particular articles were chosen*. Each such possibility is called a *permutation*. A permutation can involve a particular arrangement of fewer objects than the total or a particular sequence of the entire collection. If we multiply the product displayed earlier by $(n-r)!/(n-r)!$, we see that it can be expressed more simply:

$$\frac{n(n-1)\cdots(n-r+1)(n-r)!}{(n-r)!} = \frac{n!}{(n-r)!}$$

This enables us to make the following statement regarding the

NUMBER OF PERMUTATIONS The number of possible permutations of r objects from a collection of size n denoted by P_r^n, is

$$P_r^n = \frac{n!}{(n-r)!} \tag{4-12}$$

For the number of permutations of $r = 3$ articles out of $n = 5$ in our magazine example:

$$P_3^5 = \frac{5!}{(5-3)!} = \frac{5!}{2!} = 5 \times 4 \times 3 = 60$$

When all of the objects in the collection are taken:

$$P_n^n = \frac{n!}{(n-n)!} = \frac{n!}{0!} = n!$$

so that, in this case, the number of permutations is the same as the number of ways to sequence n objects.

Number of Combinations of Objects

The number of permutations P_r^n considers not only which r objects are selected but also their sequence or arrangement. Sometimes we do not care how the r objects are arranged; we merely wish to determine the number of possible *combinations* of objects.

Example 4-13 Suppose that, due to size limitations, *Psychomation Magazine* must restrict the number of articles in next month's issue to three; thus, only three of the original five may appear. How many combinations of 3 objects out of 5 are there? Each of the ten possible combinations are listed here:

ABC	ABE	ACD	BCD	BDE
ABD	ACE	ADE	BCE	CDE

Again, there is an alternative approach to determining the answer. Imagine choosing the articles one position at a time. We have 5 choices for the first position. This leaves 4 choices for the second, and only 3 articles for our final choice. Then the following multiplication describes the number of choices:

$$5 \times 4 \times 3$$

But the result of the above product is 60, not 10. The same situations have been accounted for several times each for instance, the combination ABC has been accounted for in all of the following sequences:

ABC BAC CAB ACB BCA CBA

But these sequences differ only in the order in which the items appear; the same three articles are involved in all cases. Since the order in which an article is chosen is not now of interest (we are only interested in whether it is selected), the redundant accounting must be corrected. This can be accomplished by dividing the earlier product by the number of sequences in which the three articles ultimately chosen could have been selected; that is, by 3!, or $3 \times 2 \times 1 = 6$:

$$\frac{5 \times 4 \times 3}{3!} = 10$$

The fraction in the above example may be transformed into a version which will later prove useful in analyzing other situations. Multiplying the numerator and denominator by 2! (thereby leaving the ratio's value unchanged) yields:

$$\frac{(5\times4\times3)\times2!}{3!\times2!},$$

which may be rewritten

$$\frac{5\times4\times3\times2\times1}{3!\times2!}=\frac{5!}{3!\times2!}$$

Another representation is

$$\frac{5!}{3!\times(5-3)!}$$

From this final representation, an easily generalized form for the number of combinations may be inferred:

> **NUMBER OF COMBINATIONS** The number of combinations of r objects taken from n objects, denoted by C_r^n, may be determined from:
>
> $$C_r^n=\frac{n!}{r!(n-r)!} \tag{4-13}$$

Notice that this expression is similar to expression (4-12) for the number of permutations. However, permutations are distinguished by order, while combinations are not. Since there are $r!$ ways to order (or sequence) r objects, the number of combinations is the number of permutations divided by $r!$, so that:

$$C_r^n=\frac{P_r^n}{r!} \tag{4-14}$$

The number of combinations of size $n-r$ may also be determined from expression (4-13) by the following line of reasoning. Whenever there are $n-r$ selected objects, there are r unselected objects; hence, the number of combinations of unselected objects must be the same as the number of combinations of selected objects. In the magazine problem in Example 4-13, there must be ten combinations of 2 (or $5-3$), because for each combination of size 3 listed, there is one combination of size 2 not on the list.

As the expression for the number of combinations is composed of factorials, the numbers may be quite large. Table 4-5 provides some examples of combination sizes.

All of the tools necessary to finish answering the problem in Example 4-11 have now been developed. The probability of a five-card straight can be found by dividing the number of ways that a five-card straight can occur (which we found to be 9×4^5) by the number of equally likely events. Any combination of five cards is equally likely (since each of the 52 cards in the deck has an equal

TABLE 4-5 Number of Combinations of r Objects Taken From n

n	r	$C_r^n = \frac{n!}{r!(n-r)!}$
6	3	20
8	5	56
12	7	792
20	5	15,504
100	50	≈ 100,000 trillion trillion

chance of being on the top of a completely shuffled deck). Therefore, the number of equally likely events is the same as the number of combinations of $r = 5$ cards taken from $n = 52$ cards:

$$C_5^{52} = \frac{52!}{5!(52-5)!}$$

In evaluating this type of fraction, do not try to calculate the denominator and numerator separately, because the numbers become far too large to handle easily and the amount of work required is unwarranted. Instead, follow this procedure.

First, express the fraction in the following manner:

$$\frac{52!}{5!47!}$$

Now, 52! may be factored as:

$$52! = 52 \times 51 \times 50 \times 49 \times 48 \times 47!$$

so that

$$\frac{52!}{5!47!} = \frac{52 \times 51 \times 50 \times 49 \times 48 \times 47!}{5!47!}$$

The 47! terms may then be canceled, which gives us:

$$\frac{52 \times 51 \times 50 \times 49 \times 48}{5!}$$

Factoring 5!, we obtain:

$$\frac{52 \times 51 \times 50 \times 49 \times 48}{5 \times 4 \times 3 \times 2 \times 1}$$

Further cancellation yields:

$$\frac{52 \times 51 \times \overset{\overset{5}{\cancel{10}}}{\cancel{50}} \times 49 \times \overset{\overset{4}{\cancel{12}}}{\cancel{48}}}{\cancel{5} \times \cancel{4} \times \cancel{3} \times \cancel{2} \times 1} = 52 \times 51 \times 5 \times 49 \times 4 = 2{,}598{,}960$$

The resulting figure is the number of possible five-card combinations. Dividing our earlier result for the number of straight combinations by this number, we obtain:

$$P[\text{five-card straight}] = \frac{9 \times 4^5}{\dfrac{52!}{5!(52-5)!}}$$

$$= \frac{9{,}216}{2{,}598{,}960}$$

$$= .00355$$

$$\approx \frac{1}{282}$$

After a large number of attempts at drawing five cards from a deck, you would, on the average, expect to encounter a straight only once in 282 tries; that is, the odds against drawing a five-card straight are about 281 to 1.

EXERCISES

4-28 Calculate the following:

(a) $6!$ (b) $\frac{14!}{3!11!}$ (c) $8!$ (d) $\frac{7!}{4!3!} \times \frac{11!}{7!2!}$

(e) $\frac{37!}{4!(37-4)!}$

(*Hint* for (e): Express the numerator as a product of numbers and 33 factorial; then cancel.)

4-29 The following famous nursery rhyme contains a counting problem.

As I was going to St. Ives,
I met a man with seven wives.
Every wife had seven sacks,
Every sack had seven cats,
Every cat had seven kits,
Kits, cats, sacks, and wives,
How many were going to St. Ives?

Find (1) the number of cats and (2) the number of kits encountered on the journey.

4-30 A cafeteria has 3 types of salad (carrot, tossed, and bean); 4 entrees (ham, chop suey, meat loaf, and tacos); 3 vegetables (corn, peas, and stewed tomatoes); 6 drinks (cola, iced tea, hot tea, coffee, lemonade, and orange juice); and 4 desserts (ice cream, apple cobbler, apple sauce, and yogurt). For two dollars, you may have 1 salad, 2 different entrees, 2 different vegetables, 1 drink, and 2 different desserts.

(a) How many ways are there to obtain 2 entrees? (*Hint:* Ignore order of selection.)

(b) How many ways are there to obtain 2 vegetables?
(c) How many ways are there to obtain 2 desserts?
(d) How many combinations of two-dollar dinners are allowable when the full quota of items is selected and when there are no duplications? (*Hint:* Do not attempt to diagram the outcomes.)

4-31 A traveling salesman must visit ten cities in one trip. In how many sequences may he make his stops?

4-32 If the top 13 cards are drawn from a deck of 52 ordinary playing cards which has been fully shuffled, they are the equivalent of a bridge hand. How many possible hands are there when the order in which they are obtained does not matter? How many hands are possible when the order does matter? (Answers in terms of factorials are sufficient.)

4-33 Determine the number of possible occurrences for each of the following situations:
(a) A coin is tossed ten times.
(b) Six dice are tossed simultaneously.
(c) One die is tossed six times.
(d) Six objects are selected from ten distinguishable items (order of selection is not considered).

4-34 Seating assignments are being made for the guests at an awards banquet. There are 20 guests (ten couples) to be seated along a table with ten chairs on each side. A name card is to appear at every place setting.
(a) In how many different ways is it possible to place the name cards?
(b) If no man is to sit aside or across from another man, how many arrangements are possible?
(c) If each couple is to sit side by side, how many arrangements are possible when (1) the sexes are alternated, and (2) members of the same sex may sit beside each other?

REVIEW EXERCISES

4-35 There are ten applicants for the position of recreational director in a city. Some characteristics of the candidates are provided in the table below.

Name	*Age*	*College Graduate*	*Marital Status*	*Previous Experience*
Mr. Braun	28	No	Single	Yes
Mrs. Charles	37	No	Married	Yes
Mr. Feeley	42	No	Single	No
Mr. Gordon	53	Yes	Single	Yes
Ms. Kish	28	Yes	Married	No
Mr. Lambert	35	No	Married	Yes
Mr. Minsky	45	Yes	Married	No
Miss Olivera	33	Yes	Single	Yes
Mr. Snyder	39	No	Single	Yes
Mr. Wasserman	35	No	Married	Yes

The file of one applicant is chosen at random. For each of the following events, indicate the elementary events of its event set and its probability.
(a) College graduate.
(b) Older than 35.
(c) Same age as another applicant.
(d) No previous experience.
(e) Married.

4-36 Suppose that a card is selected from a fully shuffled deck of 52 ordinary playing cards. (For the sample space, refer to Figure 4-1 on page 96.)

(a) List the cards (K-H, 7-D, and so on) that are the elements in the following event sets: (1) Heart; (2) ace; (3) below 4; (4) 8 through 10.

(b) Using your answers to part (a), list the elements in the following event sets: (1) Heart *and* ace; (2) 8 through 10 *or* ace; (3) 8 through 10 *and* below 4; (4) below 4 *and* ace; (5) below 4 *or* ace.

4-37 Pinochle is a card game played with a special deck consisting of 48 cards. All of the cards are above 8 in denomination, and the deck includes aces. A pinochle deck may be compiled from two regular decks by setting aside the denominations 2 through 8. Suppose that this is done with one regular "Bee" deck and one standard "Bicycle" deck.

(a) Draw a sketch of the sample space for one card taken from the shuffled deck. Use a dot to represent each card in the pinochle deck, and make your sketch similar to the one on page 96. (*Hint:* Use two columns for each suit.)

(b) On your sketch, encircle the group of elementary events in the following event sets and write each set's identity beside the corresponding area: king, club, face card, nine, and so on. Then determine the probabilities of these events.

4-38 A General Motors plant manager once remarked, "I've never seen two cars exactly alike come off our assembly line." This is because the number of possible options is so large and the likelihood that two persons will order exactly the same car is so very small.

As an example, consider the abbreviated illustration below, where various options for a hypothetical model car are considered.

Option	*Number of Possibilities*
Body type (four-door, convertible)	4
Exterior color	8
Interior color	6
Engine size	4
Transmission	3
Suspension	2
Power accessory combinations	4
Sound system combinations	10
Air conditioning	2
Tire type/size	10
Cosmetic combinations	20
Window types	3
Automatic auxilary features	6

How many possible different cars might be made?

4-39 A box contains four marbles, each of a different color: (*R*)ed, (*Y*)ellow, (*W*)hite, and (*B*)lue. Three marbles are selected randomly from the box, one at a time, and set aside.

(a) Construct a three-stage probability tree diagram for this situation. Each fork must contain a branch for every color possible. (Remember, there is one less marble possible at each successive stage.) Label each branch with the corresponding color letter (subscripted with a 1, 2, or 3, depending upon whether it is the first, second, or third marble drawn) and the probability value for that branch.

(b) Determine the joint probability corresponding to each end position.

(c) Applying the addition law to the values obtained in (b), find the probability for each of the following composite events: (1) the red marble is selected before the yellow marble; (2) the blue marble is not selected;

(3) both the white and the yellow marbles are chosen; (4) the red marble is selected on the first draw and the white marble on the last draw.

4-40 The events listed below pertain to a card selected from a fully shuffled deck of 52 ordinary playing cards. Determine in each case the appropriate relationship (mutually exclusive, collectively exhaustive, both, neither).
(a) Heart, diamond, 10.
(b) 10, queen, ace.
(c) Face, nonface.
(d) Face, red, club, spade.

4-41 The following joint probability table represents the particular characteristics of a randomly selected retired military person.

	College Education		
Rank	*Yes*	*No*	*Marginal Probability*
Officer	.21	.13	____
Enlisted	.06	.60	____
Marginal Probability	____	____	1.00

(a) Find the missing marginal probability values.
(b) Find the joint probability that the selected retiree was both an officer and college educated.
(c) Find the conditional probability that the selected person was an officer, given that he graduated from college.
(d) Determine the percentage of college-educated persons who were officers.

4-42 An experiment is conducted using three boxes, each containing a mixture of ten (R)ed and (W)hite marbles. The three boxes have the following compositions:

Box A	Box B	Box C
6 R 4 W	4 R 6 W	7 R 3 W

Two marbles are selected randomly. The first is selected from Box A. If it is red, the second marble is to be picked from Box B, but if the first marble is white, the second marble will be taken from Box C. Using R_1 and W_1 to represent the color of the first marble and R_2 and W_2 to represent the color of the second:
(a) Find the following probabilities:

$$P[R_1] \qquad P[W_1]$$
$$P[R_2 | R_1] \qquad P[R_2 | W_1]$$
$$P[W_2 | R_1] \qquad P[W_2 | W_1]$$

(b) From your answers to part (a), use the multiplication law to determine the following joint probabilities:

$$P[R_1 \text{ and } R_2]$$
$$P[R_1 \text{ and } W_2]$$
$$P[W_1 \text{ and } R_2]$$
$$P[W_1 \text{ and } W_2]$$

(c) Determine the probabilities that:
(1) One red and one white marble will be chosen.
(2) Either two red or two white marbles will be chosen.

4-43 Have you ever wondered why a royal flush beats four of a kind in straight, five-card poker? Recall that a royal flush is an A, K, Q, J, ten straight of cards in the same suit, while four of a kind consists of four cards of the same denomination plus any fifth card.

Determine the number of five-card hands possible (ignoring the order in which a hand is filled) for obtaining a royal flush and for obtaining four of a kind. Since the rarer hand wins, which one is the best?

4-44 A train detection device for an automated rail network is 99.9 percent reliable; that is, 99.9 percent of the time when there is a stalled train between stations the device detects it, and it indicates that no train is present 99.9 percent of the time when there is not one. The probability that any particular departing train will stall before reaching the next station is .005.

(a) Find the probability that a hazardous situation will arise and a stalled train will go undetected.

(b) How many such situations does the law of large numbers imply that we can expect in one million station departures?

4-45 A horse race is to be run. There are nine qualified entries. Does this mean that after buying a "win" ticket on a randomly chosen horse in the race that you then have a probability of 1/9 of winning? Briefly explain your answer.

4-46 Given the probability data below:

$P[A] = 1/3$ $\quad P[B] = 1/2$ $\quad P[C] = 1/4$

$P[A \text{ and } B] = 1/8$ $\quad P[B|C] = 1/3$ $\quad P[C|A] = 1/5$

$P[A|B \text{ and } C] = 1/2$ $\quad P[A \text{ or } B \text{ or } C] = 17/20$

(a) Determine the following probabilities:

(1) $P[A \text{ and } C]$

(2) $P[B \text{ and } C]$

(3) $P[A \text{ and } B \text{ and } C]$

(4) $P[A|B]$

(5) $P[A|C]$

(6) $P[C|B]$

(7) $P[B \text{ and } C|A]$

(8) $P[A \text{ or } B]$

(9) $P[A \text{ or } C]$

(10) $P[B \text{ or } C]$

(b) Answer the following:

(1) Are the events in any letter pair mutually exclusive?

(2) Do A, B, and C form a collectively exhaustive collection of events?

(3) Which event pairs are *not* statistically independent?

4-47 A quality control inspector accepts only 5 percent of all bad items and rejects only 1 percent of all good items. Overall production quality of items to later be inspected is such that only 90 percent are good.

(a) Using the above percentages as probabilities for the next item inspected, find:

$P[\text{accept}|\text{bad}]$
$P[\text{reject}|\text{good}]$
$P[\text{good}]$

(b) Determine the values missing in the following joint probability table:

	Inspector Action		
Quality	*Accept*	*Reject*	*Marginal Probability*
Good			
Bad			
Marginal Probability			

(c) What is the probability that the inspector will accept or reject the next item incorrectly?

4-48 Four strangers wish to compare birth *months*. Assume that all 12 months are equally likely to contain each person's birthday. Then find the probability that there is at least one matching birth month.

4-9 OPTIONAL TOPIC: REVISING PROBABILITIES USING BAYES' THEOREM

In this section, we will introduce a procedure whereby probabilities can be revised when new information pertaining to a random experiment is obtained. The notion of revising probabilities is a familiar one, for all of us—even those with no previous experience in calculating probabilities—have lived in an environment ruled by the whims of chance and have made informal probability judgments. We have also intuitively revised these probabilities upon observing certain facts, and have changed our actions accordingly. For example, think how many times you have left home in the morning with no raincoat, only to look up and notice a menacing cloud cover which sends you back for some protection in case it rains. On first charging outdoors, you behaved in a manner consistent with your judgment that the probability of rain was small. But the presence of clouds caused you to revise this probability significantly upward.

Our concern for revising probabilities arises from a need to make better use of experimental information. We begin here by establishing a fundamental principle that follows immediately from the laws of probability developed earlier in this chapter. This is referred to as *Bayes' Theorem*, after the Reverend Thomas Bayes, who proposed in the eighteenth century that probabilities be revised in accordance with empirical findings.

Such empirical findings may result from very elaborate experiments; in science, for example, many man-years of effort and much expensive equipment are devoted to such activities as studying physical laws by observing solar eclipses. Empirical findings may also arise from a very minor effort, such as asking a person questions in order to become better acquainted. No matter what their scope, they have one feature in common: *experimental results provide information*. This information may serve to realign uncertainty. Data obtained by observing a solar eclipse can lend support to hypotheses regarding the effect of the sun's gravity on stellar light rays. A person's responses to your questions may help you decide whether or not it is worthwhile to become friends. Scientists can become more certain that the sun's gravity has an effect on light rays, and their observations can lead to finer choices for further experiments. The answers to your questions can tell you that an ugly person is really beautiful or that a physically attractive person is vain and selfish.

Most information is not conclusive. Any empirical test can camouflage the truth. For instance, some potentially good students will do badly on college entrance examinations and some poor students will score highly. A good example of how such information may be unreliable is illustrated by the oil wildcatter's seismic test. Not completely reliable, a seismic survey can deny the presence of oil in a field already producing it and can confirm the presence of oil under a site already proved dry. Still, such imperfect findings can be valuable. An unfavorable test result can increase the chance of rejecting a poor prospect—college applicant or drilling lease—or a favorable result can enhance the likelihood of selecting a good one.

The information obtained will affect the probabilities of those events which determine the consequences of each act. We can revise the probabilities of these events upward or downward, depending upon the evidence obtained. Thus, the geologist increases the probability that he will find oil if he obtains a favorable seismic survey analysis, and he decreases the probability if he obtains an unfavorable survey.

Bayes' Theorem

Consider a random experiment having several possible mutually exclusive events, $E_1, E_2, \ldots, E_n$. Suppose that the probabilities of each event, $P[E_1], P[E_2], \ldots, P[E_n]$ have been obtained. These probabilities are referred to as *prior probabilities*, because they represent the chances of the events *before* the results from the empirical investigation are obtained. The investigation itself may have several possible outcomes, each statistically dependent upon the Es. For any particular result, which we may designate by the letter R, the conditional probabilities $P[R \mid E_1], P[R \mid E_2], \ldots, P[R \mid E_n]$ are often available. The result itself serves to revise the event probabilities upward or downward. The resulting values are called *posterior probabilities*, since they apply *after* the informational result has been learned.

The posterior probability values are actually conditional probabilities of the form $P[E_1 \mid R], P[E_2 \mid R], \ldots, P[E_n \mid R]$ that may be found according to

BAYES' THEOREM The posterior probability of event E_i for a particular result R of an empirical investigation may be found from:

$$P[E_i \mid R] = \frac{P[E_i]P[R \mid E_i]}{P[E_1]P[R \mid E_1] + P[E_2]P[R \mid E_2] + \cdots + P[E_n]P[R \mid E_n]} \tag{4-15}$$

The principle underlying Bayes' Theorem may best be explained in terms of the following example.

Example 4-14 A box contains four fair dice and one crooked die with a leaded weight which makes the six-face appear on two-thirds of all tosses. You are asked to select one die at random and toss it. If the crooked die is indistinguishable from the fair dice and the result of your toss is a six-face, what is the probability that you tossed the crooked die?

The events in question are

$$E_1 = \text{fair die}$$

$$E_2 = \text{crooked die}$$

The empirical investigation here is the toss itself, so we use:

$$R = \text{six-face}$$

Since 1 out of 5 dice is crooked, the prior probabilities for the type of die tossed are

$$P[E_1] = 4/5$$

$$P[E_2] = 1/5$$

Because all sides of a six-sided fair die are equally likely, the probability of obtaining a six-face when tossing a fair die is

$$P[R \mid E_1] = 1/6$$

If the crooked die is tossed, we are told that

$$P[R \mid E_2] = 2/3$$

Applying expression (4-15), the posterior probability that the die you tossed is crooked is

$$\begin{aligned} P[E_2 \mid R] &= \frac{P[E_2]P[R \mid E_2]}{P[E_1]P[R \mid E_1]+P[E_2]P[R \mid E_2]} \\ &= \frac{(1/5)(2/3)}{(4/5)(1/6)+(1/5)(2/3)} \\ &= \frac{2/15}{4/15} = \frac{1}{2} \end{aligned}$$

Here, we see that the probability of tossing a crooked die must be revised upward from the prior value of 1/5 that applies when there is no information to 1/2 after we know that the toss resulted in a six-face.

Posterior Probability as a Conditional Probability

Although it has a special interpretation, a posterior probability is merely a conditional probability when some relevant result is given, and it can be found in the same manner:

$$\begin{matrix}\text{Posterior} \\ \text{probability} \\ \text{of event}\end{matrix} = P[\text{event} \mid \text{result}] = \frac{P[\text{event } \textit{and} \text{ result}]}{P[\text{result}]}$$

A straightforward procedure for calculating an event's posterior probability is first to find the joint probability that the event will occur with the given result and then divide by the probability of that result. This is exactly what expression (4-15) accomplishes. The numerator is the joint probability found by using the multiplication law. The denominator is the probability of obtaining the particular empirical result and is basically the sum of all those joint probabilities for potential outcomes where that result might occur. In practice, expression (14-15) can be cumbersome to use when the data are given in such a way that the posterior probabilities may be found more directly.

For example, if a statistics class contains just as many men as women, then the prior probability that an examination paper chosen at random will belong to a man (M) would be 1/2. Now suppose that the exams have been graded and 20 percent of the papers received a mark of "C or better" (C) *and* were written by men; a total of 60 percent of the exams were scored "C or better." If a randomly selected test sheet was graded "C," we have sufficient information to calculate the posterior probability that it was written by a man:

$$P[M \mid C] = \frac{P[M \text{ and } C]}{P[C]} = \frac{.20}{.60} = \frac{1}{3}$$

Here, the information regarding the paper's grade causes us to revise downward the probability that it belongs to a man.

Typically, the probability values needed to make a simple calculation such as this are not immediately available. When we use evidence or empirical results to revise probabilities, our knowledge of the various events involved is usually structured so that some preliminary work is necessary to obtain the needed probability values. To accomplish this, it may help to construct a joint probability table first.

Obtaining Posterior Probabilities from the Joint Probability Table

Example 4-15 A noted lawyer specializing in defending his corporate clients in personal injury suits thinks that getting a sympathetic jury is half the battle. In large measure, he can winnow a jury panel down to a largely sympathetic group by preemptory challenges. Since a potential juror's leaning in a particular case remains largely concealed during the selection interview, superficial characteristics must be relied upon in accepting or rejecting jury candidates. The lawyer has found that mature and stable persons, those who "made it on their own," are the most likely to be sympathetic to the defendant, while young people tend to have a "social worker attitude" which makes them likely to favor the plaintiff. From post-trial talks with jurors over the years, the lawyer has managed to identify their attitudes toward his clients. He has found that 65 percent of the sympathetic jurors have been older persons. A special bar study in his county has shown that only 30 percent of jury panel members are sympathetic to the defendant in a personal injury suit.

In a negligence suit resulting from an elevator accident, all but one juror has been chosen and both lawyers have no more challenges left. Of the available jurors, six are younger and four are older. Depending upon alphabetical sequence, one of these ten will be the last juror.

Table 4-6 is a joint probability table constructed regarding this last juror. The probability values obtained directly from the data provided above are shown in black. The colored numbers were obtained from these data by first noting that:

$$P[U] = 1 - P[S] = .70$$

The fact that we have been told the following conditional probability:

$$P[O \mid S] = .65$$

enables us to find the joint probability of obtaining an older, more sympathetic juror:

$$P[O \text{ and } S] = P[O \mid S]P[S] = .65(.30) = .195$$

The remaining values follow from this one and from the fact that the joint probabilities must sum to the respective marginal values.

The prior probability that a sympathetic juror will be obtained is only .30 (which is also the marginal probability for this event). Suppose that an older juror is chosen. The posterior probability that he will also be sympathetic is

$$P[S \mid O] = \frac{P[O \text{ and } S]}{P[O]} = \frac{.195}{.40} = .4875$$

The above example shows how we can calculate posterior probabilities by relying upon basic concepts, instead of using the complicated expression of Bayes' theorem. Nevertheless, as Table 4-6 shows, essentially the same steps are required in either case.

TABLE 4-6 Joint Probability Table used to Illustrate Posterior Probability Calculation

	Older (*O*)	*Younger* (*Y*)	
Sympathetic (*S*)	.195	**.105**	.300
Unsympathetic (*U*)	**.205**	**.495**	**.700**
	.400	**.600**	1.000

OPTIONAL EXERCISES

4-49 A local television weatherman makes a daily forecast indicating the probability that it will rain tomorrow. On one particular evening, he announces an 80 percent chance of rain (R) the next day. The manager of the city golf courses has established a policy that he will only water the greens if the probability of rain is less than 90 percent. Using the local TV forecast as his prior probability, the manager also relies upon his father-in-law's rheumatism: historically, he gets a "rain pain" (P) on 90 percent of all days followed by rain, but he also gets a pain on 20 percent of the days not followed by rain. The following probabilities therefore apply:

$$P[R] = .80 \qquad P[P \mid R] = .90 \qquad P[P \mid \text{not } R] = .20$$

(a) Assuming that the golf course manager's father-in-law is currently receiving pain signals, find the posterior probability that it will rain tomorrow. Should the manager water the greens?

(b) If the manager's father-in-law feels just fine, what is the posterior probability of rain tomorrow?

4-50 A marketing researcher wishes to determine from a given response to a question, whether a randomly chosen person will choose BriDent when he next purchases toothpaste. The question will reveal whether the selected person recalls the name BriDent, an event we will designate R. From previous testing, it has been established that 99 percent of those persons buying BriDent had previously recalled the name. It has also been found that only 10 percent of the nonbuyers of BriDent recalled this particular brand name. Since BriDent now has 30 percent of the toothpaste market, the researcher chooses .30 as the prior probability that the person will buy BriDent. Denoting this event by B, we have the following probabilities:

$$P[B] = .30 \qquad P[R|B] = .99 \qquad P[R|\text{not } B] = .10$$

(a) Suppose that the chosen person replies that he remembers BriDent. What is the posterior probability that he will make BriDent his next purchase?

(b) Suppose that he does not remember BriDent. Find the posterior probability that he will pick BriDent when next buying toothpaste.

4-51 An artist has finished a portrait. She believes that the client will have either a warm (W) or cold (C) response, with prior probabilities $P[W] = .90$ and $P[C] = .10$. In the past, from trying her ideas out on her husband, who either likes (L) or dislikes (D) a piece of work, she has found $P[L|W] = .80$ and $P[D|C] = .90$.

(a) Calculate the missing values in the following table:

Prior Probabilities	*Conditional Probabilities*	*Joint Probabilities*	*Posterior Probabilities*
$P[W] = .90$	$P[L\|W] = .80$	$P[L \text{ and } W] = .72$	$P[W\|L] = ?$
$P[C] = .10$	$P[L\|C] = ?$	$P[L \text{ and } C] = ?$	$P[C\|L] = ?$
		$P[L] = ?$	
$P[W] = .90$	$P[D\|W] = ?$	$P[W \text{ and } D] = ?$	$P[W\|D] = ?$
$P[C] = .10$	$P[D\|C] = .90$	$P[C \text{ and } D] = .09$	$P[C\|D] = ?$
		$P[D] = ?$	

(b) What is the posterior probability that the client will respond warmly if the artist's husband likes her portrait?

4-52 An oil wildcatter has assigned a .50 probability to striking oil on his property. He orders a seismic survey that has proved only 80 percent reliable in the past: given oil it predicts favorably 80 percent of the time; given no oil, it augurs unfavorable with a frequency of .8.

(a) Given a favorable seismic result, what is the probability of oil?

(b) Given an unfavorable seismic result, what is the probability of oil?

4-53 Your friend places two coins, identical in all respects except that one is two-headed, into a box. Without looking, you select one coin from the box and lay it on the table. A head side shows.

(a) What is the prior probability that the coin has two heads?

(b) What is the posterior probability that the coin has two heads?

4-54 A life insurance company gives an aptitude test to new salesmen in order to predict their success at selling. Like most aptitude tests, this one is imperfect. Some high scorers will be poor salesmen, lacking the necessary motivation. A few low scorers will turn out to be excellent salesmen. Sacrificing justice for efficiency, the company has established a score of 80 as the minimum for hiring. This score was determined after a year of testing new hires, who were divided into two categories—satisfactory and unsatisfactory. The results of the tests are summarized below:

	Proportion of Men Who		
Performance	*Scored Below 80*	*Scored 80 or Above*	*Total*
Satisfactory	.10	.40	.50
Unsatisfactory	.30	.20	.50
Total	.40	.60	1.00

Since the men tested represent a typical cross section of applicants, the above figures have been judged acceptable as probabilities for future candidates.

(a) What is the prior probability that a candidate will be satisfactory?

(b) What is the posterior probability that a candidate will perform unsatisfactorily if he is hired after scoring 80 or above?

(c) What is the posterior probability that a candidate *would have been* satisfactory (were he hired) after scoring below 80?

5 Probability Distributions and Expected Value

The Curve described by a simple molecule of air or vapor is regulated in a manner just as certain as the planetary orbits; the only difference between them is that which comes from our ignorance.

Marquis de Laplace (1820)

The preceding chapter provided the fundamental concepts of probability. We are now ready to see how these concepts may be applied to two broad categories of problems. The traditional application encountered in statistical investigations uses probability to qualify generalizations made from samples. In recent years, probability has been incorporated into new tools that are used to analyze decision making under uncertainty.

The outcomes from a sampling study will usually be treated as uncertain events. We will make use of the population frequency distribution to find the probability that a particular sample result will be obtained. The sample outcome will be treated as a variable, which, because it is determined by chance, is referred to as a *random variable*. The values of a random variable will be treated as uncertain events occurring with a set of probabilities that we call a probability distribution.

It is convenient to summarize the information in a probability distribution in terms of one or two key values. Like populations, probability distributions exhibit measures of central tendency and variance. Calculated in a similar manner, an "average" figure for a random variable may be obtained. Such a number is referred to as an *expected value*. The expected value concept may be used as a decision-making tool for comparing probability distributions. Its major role in sampling is that it enables us to analyze the kinds of results that might be obtained from a sampling study. For example, the height of a man randomly chosen from a population having a mean height of 5′10″ would be expected to be 5′10″. As we will see, the expected value principle can be beneficial in trying to estimate an unknown population mean solely from sample results.

5-1 DISCRETE PROBABILITY DISTRIBUTIONS

The Random Variable Concept

Let's consider the decision problem of a school superintendent who must select materials to include in the reading curriculum. One choice may be oriented to computerized instruction. Various measures may be used to compare the attractiveness of this approach with competing methods. One common gauge is the ranking of aggregate reading comprehension scores achieved by the district's students on a statewide test, relative to all other school districts.

The actual ranking achieved lies in the future and is therefore subject to uncertainty. Since the reading rank may assume any one of a number of possible values, we may view it as a *variable*.

We refer to a variable whose value is determined by chance as a *random variable*. Thus, the reading rank to be achieved from computerized instruction is a random variable. In general, we can make the following

DEFINITION A ***random variable*** is a numerical quantity whose value is determined by the outcome of a random experiment.

Before the random experiment is conducted, we must treat the outcome as a variable. The random variable assumes an actual numerical value only after the experiment's outcome is known.

A single random experiment may have many different kinds of random variables. Instead of reading comprehension rank in the foregoing situation, vocabulary levels or reading speed might have been the random variable. Each of these values would be obtained in a different manner, but from the same data. The elementary events of the sample space would remain unchanged.

Example 5-1 Roulette, the internationally popular casino game, vividly illustrates the relationship between elementary events and random variables. The procedure is for the player to commit himself to a particular bet. A wheel having 38 slots is then spun and a ball is set in motion. The outcome is determined by the slot into which the ball drops. The sample space is therefore one of the 38 slots; 36 of these are numbered from 1 to 36 (half of these are red, the others black) and two green slots are numbered 0 and 00.

There are several ways of placing a roulette bet. For instance, a player may place a dollar on a particular number (say, 7). If the ball drops into the 7-slot, the player receives a 35-dollar payoff;* otherwise he forfeits his bet. Another way to

* Since there are 38 equally like slots, the probability of getting 7 is 1/38. To be a fair gamble, the payoff should be 37 to 1, but the house does not pay on 0 or 00, giving it an edge and a built-in source of long-run profits.

play is to bet a dollar on the red or the black field (say, red). If the ball drops into a red slot, the player receives a dollar; the bet is lost if a black or a green number appears. In either type of gamble, the sample space for the outcome of the spin of a roulette wheel comprises 38 equally likely elementary events.

For each player, the random variable of prime interest is his winnings. The random variable of the player betting on the 7 has two values: +\$35 (if 7 comes up), and −\$1 (if the ball drops into any of the other slots). The less venturesome red-player's random variable also has two values: +\$1 (if the ball lands in a red slot), and −\$1 (if the ball drops into a black or a green slot). For the random experiment, we have two distinctly different random variables, *depending upon the type of gamble made.*

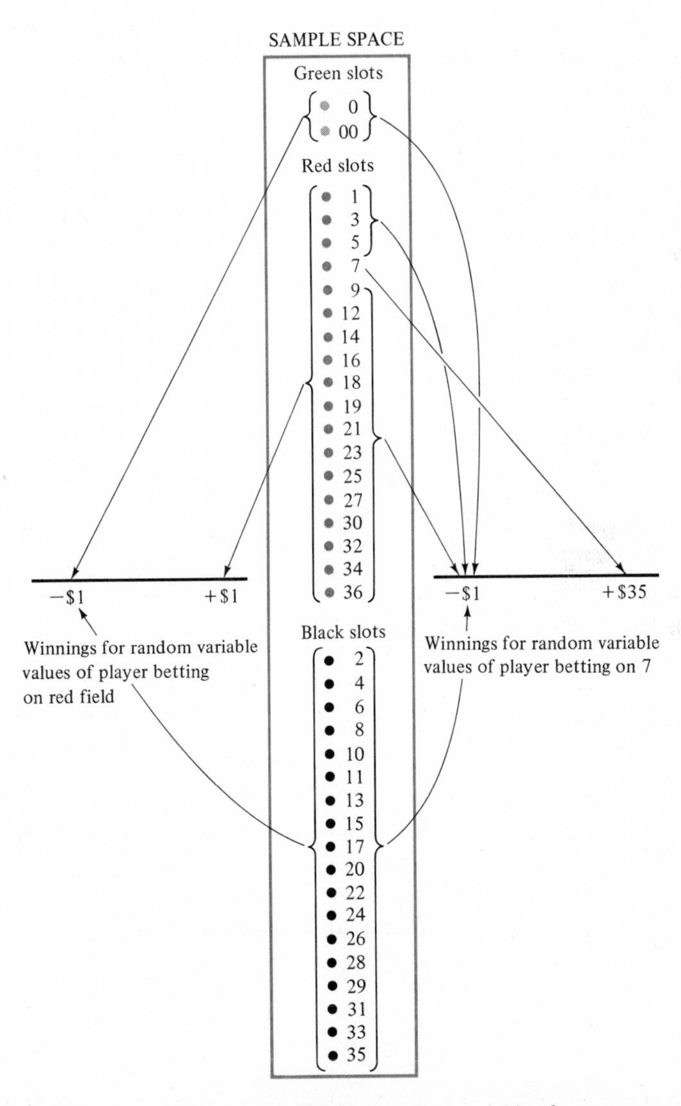

FIGURE 5-1 Portrayal of two different random variables for two types of roulette gambles.

Figure 5-1 shows how each of the random variables in Example 5-1 is defined in terms of the same sample space. The values of the respective random variables are determined on each play by which elementary event occurs. The arrows show how each elementary event is matched to a particular point on a numerical scale, depending on the choice of random variable. A mathematical interpretation of a random variable therefore is that it is a function that matches or maps the elementary events onto their corresponding points on a numerical scale.

A random variable must assume numerical values. For example, the outcomes of a coin toss—head and tail—are non-numerical. We can associate a random variable with coin tossing only by assigning numbers to these outcomes, as would be the case in a wager where a head results in winnings of $1 and tail in a loss of $1. In this case, the winnings rather than the side showing would be the random variable, with possible values of $1 and −$1. As we will see in Section 5-2, an expected value for the random variable can be obtained by an averaging procedure. Thus, we can determine an average level of winnings from the coin toss. But there is no way to average the head and the tail, which are non-numerical.

The Probability Distribution of a Random Variable

The notion of a random variable is fundamental to the application of probability concepts to decision making. It is also necessary to establish a body of statistical theory that can help us interpret sample evidence. In both cases, we need techniques for finding the probabilites for possible random variable values. The relation between a random variable's values and their probabilities is summarized by the *probability distribution.*

Each possible random variable value corresponds to a particular composite event in the underlying sample space. Thus, we may use the procedures introduced in Chapter 4 to obtain the probabilities for each possible value. For example, again consider the random variable of Example 5-1: winnings from a gamble at the roulette table. For the single-number bettor, winnings equal to $35 occur whenever the event "the ball drops into the slot of the number played" does. Likewise, winnings equal to −$1 (that is, losing $1) occur when the complementary event "some other slot receives the ball" occurs. There are 37 out of 38 equally likely possibilities that the second event will occur. Thus, the probabilities of our random variable values are equal to the probabilities of the corresponding events. In this instance, denoting the single-number player's winnings by W:

$$P[W = \$35] = 1/38$$

$$P[W = -\$1] = 37/38$$

The probability values assigned to each value correspond to events that are *mutually exclusive*. This is because we specify that each outcome corresponds to exactly one random variable value. In the case of roulette, each slot of the wheel corresponds to a win or to a loss, but not to both. The associated events are also *collectively exhaustive*, meaning that some value for the random variable will always be achieved. In roulette, this corresponds to the fact that the wager must be either lost or won.

Expressing the Probability Distribution

We are now ready to define the probability distribution.

DEFINITION The ***probability distribution*** of a random variable X provides a probability for each possible value. These probabilities must sum to 1, and they are denoted by

$$P[X = x]$$

where x represents any one of the possible values that the random variable may assume.

It is advantageous to use a letter as a symbol to denote the random variable. (This is strictly a convenience, as it is in algebra, to keep expressions uncluttered.) Symbols will make it easier for us to deal with random variables whose probability values can be determined by mathematical formulas.

Many probability distributions can be expressed in terms of a table like Table 5-1 for the reading rank to be achieved by the school district children when computerized reading instruction is provided. Here, we represent this rank by the letter X (but any letter would suffice). Notice that the probabilities

TABLE 5-1 Probability Distribution of Reading Comprehension Ranking of School District

Possible Rank x		$P[X=x]$
1		.05
2		.10
3		.15
4		.17
5		.12
6		.08
7		.09
8		.06
9		.05
10		.05
11		.04
12		.04
	Total	1.00

sum to 1, because the events $X = 1, X = 2, \ldots$ are mutually exclusive and collectively exhaustive.

Not all probability distributions will be expressed by a table. Sometimes they will be described by an algebraic formula. This is one reason why we use the lower case x to represent the possible values. As the following example demonstrates, there may simply be too many possibilities to conveniently list them all in a table.

Example 5-2 A raffle is to be conducted with 10,000 tickets, numbered consecutively from 1 to 10,000. One ticket is to be selected from a barrel after a thorough mixing. As each ticket is equally likely, the probability that a particular one will be selected is 1/10,000. We define a random variable for this random experiment: the number of the selected ticket, to be denoted by Y. This enables us to shorten the statement "The number of the selected ticket will be 7,777" to read:

$$Y = 7{,}777$$

Thus, we can express the probability of this event by the following:

$$P[Y = 7{,}777] = \frac{1}{10{,}000}$$

As there are 10,000 outcomes, all having the same probability, we seek a better way to express the probability distribution than by means of a table. We may use a statement of the form

$$P[Y = y] = \frac{1}{10{,}000} \quad \text{for values of } y = 1, 2, \ldots, 10{,}000$$

which is equivalent to duplicating the previous probability expression 10,000 times, once for each ticket number. The lower case y is a "dummy" variable, representing one particular ticket number.

It is important to keep in mind that the capital letter Y represents the number of the ticket that *will be* selected and so has no determined value until the random experiment is completed. The lower case letter y is a surrogate or stand-in for each ticket number and is a device to avoid 10,000 repetitious statements. We will find that this shorthand representation can be of great value.

Discrete and Continuous Random Variables

Special difficulties arise when random variables do not assume a discrete number of possible values.

Example 5-3 For use in large motors precision one-inch-diameter ball bearings must be machined to within a tolerance of .01 of an inch. Each ball bearing is assumed to be so nearly perfectly round that no measuring device commercially available is capable of perceiving otherwise. One million of these ball bearings are used annually.

Each ball bearing sold has been inspected at uniform temperature by measurement, using a pair of "go–no go" gauges (stands with holes through which a bearing being tested is dropped). The diameter of one hole is set for 1.01 inches; the other for .99 inch. Each ball bearing must be small enough to go through the hole in the large gauge, but not so small that it falls through the hole in the small gauge. All oversized and undersized bearings are set aside for salvage.

How many of the million bearings produced annually are precisely one inch in diameter? To determine this, a more accurate measuring system must be established. Suppose that a pair of gauges could determine if bearings were within a hundred-thousandth (.00001) of an inch of being exactly one inch in diameter. Only about 1,000 ball bearings would pass this test (assuming nearly equal frequency for all values between .99 and 1.01 inches). Those bearings passing this second stage of testing could be measured again, this time with special gauge blocks accurate to within a millionth of an inch. About 100 of these would pass the test. Additional tests using optical equipment could filter out the majority of these bearings, leaving a handful to be measured even more accurately in a fourth stage. But how many would pass the fourth test? Would greater precision of measurement eliminate all of the ball bearings?

We cannot be certain whether one or more of the ball bearings would be precisely one inch in diameter, as the standard inch is presently defined and within our capability to measure it. But we can conclude that such ball bearings would be extremely rare.

Consider a random experiment involving the random selection of one of these ball bearings. As our random variable, we take the diameter of the ball bearing. We can conclude that the probability that this will be exactly one inch is so tiny as to be zero for all practical purposes.

The random variable illustrated in the foregoing example can assume any value on a continuous scale. Such a random variable is therefore given the

DEFINITION A ***continuous random variable*** is one that may assume any numerical value on a continuous scale.

Since there are an infinite number of numerical values possible for a continuous random variable, the probability that a particular single value will be attained is zero.

In contrast to the continuous random variable, another class of random variables, called *discrete random variables*, may assume a finite or countable number of numerical values. Examples 5-1 and 5-2 involved discrete random variables.

The dichotomy of discrete versus continuous random variables is important in probability theory because of the type of mathmatics required to describe the probability distributions of each. Much of the mathematics needed to handle continuous random variables requires calculus, so that the extensive development of many properties of these variables is beyond the scope of this book. However, continuous random variables share many properties with the discrete class, so that the more significant results attributable to both may be explored using discrete examples.

EXERCISES

5-1 Describe a circumstance which would create a random variable from each of the following random experiments:

(a) Tomorrow's weather.
(b) The sex of an unborn child.
(c) The winner of the World Series.

5-2 A gambling game involves a coin toss. The *price* to play is $1 (this is not returned). If the coin lands head-side up, the bettor receives a payment of $2. If a tail occurs, there is no payoff. Letting W represent the net winnings (payoff – cost) from one play of the gamble, determine the probability distribution of W.

5-3 From the following probability distributions for the receipts and expenses of a charity carnival, determine the probability distribution for the net proceeds (Receipts – Expenses).

Receipts	*Probability*		*Expenses*	*Probability*
$30,000	1/3		$30,000	1/3
40,000	1/3		40,000	1/3
50,000	1/3		50,000	1/3
	1			1

5-4 "Craps" is a favorite gambling game in which a pair of six-sided dice is tossed. Like roulette, there are many different ways to place a bet. One of these involves "playing the field," where the bettor indicates that he wishes to make a bet with a complicated payoff, depending on which faces of the dice show. If a "field" number, defined by a sum value of 2 through 4 or 9 through 12, occurs, the player wins. Should the roll of dice yield any other total, he loses. A field gamble is further complicated by varying payoffs: 1 to 1 for all field numbers but 2 or 12; 2 to 1 on a 2; and 3 to 1 on a 12. If a bettor wins, he keeps his original bet and is also paid his winnings. If he loses, he forfeits the amount of his wager.
(a) For a bet of $1, find the probability distribution of W, the gambler's net winnings from one field bet in craps. (You may wish to refer to your solution to Exercise 4-3 on page 101 to obtain the probabilities for the possible dice sums.)
(b) Solve (a) for a bet of $2.

5-2 EXPECTED VALUE AND VARIANCE

Expected Value of a Random Variable

The probability distribution of a random variable is similar to the frequency distribution of a quantitative population because it tells us what the long-run frequency of occurrence will be for each possible outcome when the random experiment is repeated. In Chapter 2 we saw that the population mean is a desirable summary, either for comparing populations or for making decisions regarding the population. Similarly, it is useful to find the average of the random variable values to be achieved from repeated random experiments. Because the outcomes are in the future, the average result is called an *expected value*.

An average, as we saw in Chapter 2, may be calculated in a number of different ways. The computation that we will use here for the expected value is analogous to the computation for finding the mean of grouped data.

DEFINITION The ***expected value*** of a discrete random variable X, denoted by $E(X)$, is the weighted average of that variable's possible values, where the respective probabilities are used as weights. The expected value may be determined from

$$E(X) = \sum xP[X = x] \tag{5-1}$$

To illustrate how we may calculate the expected value for a random variable, let X represent the number of dots on the face showing after a six-sided die is tossed. The probability distribution of X is provided in the first two columns of Table 5-2. Multiplying each possible result by its corresponding probability

TABLE 5-2 Expected Value Calculation for Die Toss Outcome

(1) *Possible Value* x	(2) $P[X=x]$	(3) *Weighted Value* (1) × (2) = $xP[X=x]$
1	1/6	1/6
2	1/6	2/6
3	1/6	3/6
4	1/6	4/6
5	1/6	5/6
6	1/6	6/6
	6/6 = 1	$E(X)$ = 21/6 = 3.5

produces a weighted value. Summing these results gives the expected value $E(X) = 3.5$. On the average, the number of dots obtained for the showing faces in a large number of die tosses will be 3.5. Because it is the average value achieved by the random variable X, we sometimes refer to $E(X)$ as the *mean of X*.

Meaning of Expected Value

The expected value has many uses. In a gambling game, it tells us what our long-run average losses per play will be. Sophisticated gamblers know that slot machines pay poorly in relation to the actual odds and that the average loss per play will be less in roulette or dice games. A mathematician, Edward Thorp, caused quite a stir in the early 1960s when he demonstrated that various betting strategies in playing the card game of blackjack will result in positive expected winnings.*

* *See* Edward Thorp, *Beat the Dealer*, Second Edition (New York: Random House, 1966). Unlike other gambling games, blackjack allows bets to be placed when the odds are in a player's favor. This is because the card deck may not be reshuffled after each stage of play. By significantly raising his bets at these times, a player will make a profit on the average.

The meaning of the expected value may be explicitly illustrated in terms of long-run frequency. Once again, consider the roulette gambler betting $1 on number 7. We may calculate his expected winnings $E(W)$ as follows:

Possible Winnings w	$P[W=w]$	$wP[W=w]$
+$35	1/38	+$35/38
−$1	37/38	−$37/38
		$E(W)$ = −$2/38 = −$.053

The expected winnings from a single gamble is therefore a loss of 5.3 cents. This means that if the gambler keeps making $1 bets indefinitely, he will lose an average of 5.3 cents on each. Of course, on an individual gamble, he will either win $35 or lose $1. But—again, on the average—he will win $35 in only 1 out of 38 gambles, while he will lose $1 in 37 out of 38.

Notice that in both of the preceding illustrations the expected values of the random variables are not in themselves outcomes of the random experiments. Getting 3.5 dots in a single die toss is impossible, and the roulette player will always be dealing in whole dollars. In either case, the expected value is an average result only. Although it may happen in some random experiments, there is no reason why the expected outcome should turn out to be a possible outcome from a single experiment.

Some Properties of Expected Value

One variable may sometimes be converted into another by means of a mathematical expression, such as $d = \frac{1}{2}gt^2$ for the distance d an object falls by t seconds after its release, or $a = \pi r^2$ for the area of a circle of radius r. In these cases, we say that d is a *function* of t and a is a *function* of r, or, in symbols, $d = f(t)$ and $a = f(r)$. The function concept is a useful one, because, in working with random variables, it is sometimes necessary to compute the expected value of a function of the original variable.

Several important properties of expected values permit us to take computational shortcuts in evaluating functions of random variables.

1. The expected value of a constant c is equal to the constant:

$$E(c) = c$$

This can be justified immediately by a short example. Suppose a random variable has a single value c, as would be the case if c represents the age (to the nearest day) of a person randomly chosen from a group of people all born on the same date. No matter who is selected, c will be the age obtained. Hence, c is also the expected age.

2. ***The expected value of the product of a constant c and a random variable X is equal to the constant times the expected value of the random variable:***

$$E(cX) = cE(X)$$

To see why this is so, consider a random variable X whose possible values are expressed in feet. Clearly, if the expected value in inches were desired instead, it could be obtained by multiplying the expected value in feet by 12 [that is, $c \times E(X)$, with $c = 12$], since one foot is equal to 12 inches. The result obtained would be the same as if the original measurements had been converted to inches and used to find the expected value. [This would be denoted by $E(cX)$, again with $c = 12$, for each value of X would have to be multiplied by 12 before averaging to obtain the expected value.]

3. ***The expected value of the sum of a constant c and a random variable X is the sum of the constant and the expected value of the random variable:***

$$E(c+X) = c + E(X)$$

This property is also easily justified by a simple example. Suppose that bushels of fruit are weighed, one at a time, using the same basket. Then the average weight of bushel baskets of fruit will be the same as the average weight of the bushels without the basket plus the weight of the basket.

These three properties may be combined and summarized as

PROPERTY 5-1 Let a and b be constants that remain unchanged for all possible values of X. Then:

$$E(a+bX) = a + bE(X) \tag{5-2}$$

The Variance of a Random Variable

Just as the expected value of a random variable is analogous to the weighted mean, the variability of random variables may be measured in much the same way as the variability in a population or a sample. The measure we will consider is the variance. Like the mean, the variance represents the same thing that it does for the population: the average of the squared deviations from the mean or expected value. The variance is itself an expected value of a function—the squared deviation—so that we make the following

DEFINITION The ***variance*** of a random variable X, denoted by $\sigma^2(X)$, is the expected value of the squared deviations from the expected value of the variable, and may be determined from the following expression:

$$\sigma^2(X) = E([X-E(X)]^2) = \sum[x-E(X)]^2 P[X=x] \tag{5-3}$$

For computational ease, the following eqiuvalent expression may be used to calculate the ***variance of a random variable***

$$\sigma^2(X) = \sum x^2 P[X = x] - E(X)^2 \qquad (5\text{-}4)$$

Example 5-4 A six-sided symmetrical die is rolled. We denote the number of dots on the showing face by the random variable X. The probability distribution of X is provided by the numbers in columns (1) and (2) of Table 5-3. Column (3) lists the values of x^2. Column (4) contains the weighted products of columns (2) and (3), their sum being

$$\sum x^2 P[X = x] = 15.167$$

The expected number of dots showing was previously found to be $E(X) = 3.5$. Thus, from expression (5–4):

$$\sigma^2(X) = 15.167 - (3.5)^2 = 2.917$$

Two useful properties of the variance are important.

1. ***The variance of the product of a constant and a random variable is equal to the constant squared times the variance of the random variable:***

$$\sigma^2(cX) = c^2\sigma^2(X)$$

To prove that this is true, consider X in units of feet. To express the random variable in inches, we multiply X by 12, to obtain the new random variable $12X$ (here, $c = 12$). As the units in which variance is expressed are the square of those units expressing the random variable [for example, if X is in feet, then $\sigma^2(X)$ is in square feet], then the units expressing $\sigma^2(cX)$ must be c^2 times those for $\sigma^2(X)$ itself; that is, as there are 12^2 square inches in a square foot, then $\sigma^2(12X) = 12^2\sigma^2(X)$.

2. ***The variance of the sum of a random variable and a constant equals the variance of the random variable since the added constant has no effect on the variance.***

TABLE 5-3 Calculation of Expected Value for Square of Dots on Die

(1) Possible Values of X x	(2) $P[X=x]$	(3) x^2	(4) $x^2P[X=x]$
1	1/6	1	1/6
2	1/6	4	4/6
3	1/6	9	9/6
4	1/6	16	16/6
5	1/6	25	25/6
6	1/6	36	36/6
	6/6 = 1		91/6 = 15.167

This may be justified by considering a random variable representing the measured heights of people. Suppose that these people are measured while standing on a block one-inch thick. Since the variance expresses dispersion in the heights, the thickness of the block should not matter.

These two properties together establish:

PROPERTY 5-2 Let a and b be two constants. Then:

$$\sigma^2(a+bX) = b^2\sigma^2(X) \tag{5-5}$$

The *standard deviation* of a random variable X, which we will abbreviate here as $\sigma(X)$, will have the same meaning as that of the population. By taking the square root of the variance, we calculate the ***standard deviation of a random variable***

$$\sigma(X) = \sqrt{\sigma^2(X)} \tag{5-6}$$

Expected Value and Sampling

The sampling experiment of randomly selecting an elementary unit from a population is of special interest. Here, the random variable would be the observed value of the chosen unit. Because the long-run average result of repeated sampling experiments should be about the same as the mean of the population, the expected value of the sample observation should equal the mean of the population. A parallel result should hold for the variance, so that we can have the following

RELATION The value of an elementary unit chosen randomly from a quantitative population is a random variable. Its expected value equals the population mean and its variance equals the population variance.

Example 5-5 The number of children in the 1,000 families residing in a particular town has the frequency distribution presented in Table 5-4. Using the grouped approximation procedures discussed in Chapter 2, the mean number of children μ and the variance σ^2 may be calculated:

$$\mu = \frac{\sum fX}{N} = \frac{2{,}833}{1{,}000} = 2.833$$

$$\sigma^2 = \frac{\sum fX^2 - N\mu^2}{N} = \frac{11{,}101 - 1{,}000(2.833)^2}{1{,}000} = 3.075$$

(We use N rather than $n-1$ as the divisor because the data pertain to the entire *population.*)

One family is to be chosen at random. The probability distribution for the selected family's number of children may be obtained from the population frequency distribution in Table 5-4. For example, to find the probability that $X=2$, we divide the number of two-child families by 1,000: $P[X=2] = 223/1{,}000 = .223$. The probability distribution is provided in Table 5-5, where $E(X)$ and $\sigma^2(X)$ are calculated. Notice how similar these calculations are to those used earlier to find the population parameters. We use the probabilities in the same manner as class frequencies, and they have corresponding values. We find that $E(X) = 2.833$ and $\sigma^2(X) = 3.075$; these are the same values as the corresponding population parameters, so that $E(X) = \mu$ and $\sigma^2(X) = \sigma^2$.

TABLE 5-4 Frequency Distribution of Number of Children in 1,000 Families in a Town

(1) Number of Children X	(2) Number of Families f	(3) fX	(4) X^2	(5) fX^2
0	63	0	0	0
1	185	185	1	185
2	223	446	4	892
3	207	621	9	1,863
4	153	612	16	2,448
5	87	435	25	2,175
6	53	318	36	1,908
7	21	147	49	1,029
8	5	40	64	320
9	1	9	81	81
10	2	20	100	200
Totals	1,000	2,833		11,101

TABLE 5-5 Probability Distribution for Number of Children in Randomly Selected Family

(1) Possible Number of Children x	(2) $P[X=x]$	(3) $xP[X=x]$	(4) x^2	(5) $x^2P[X=x]$
0	.063	.000	0	.000
1	.185	.185	1	.185
2	.223	.446	4	.892
3	.207	.621	9	1.863
4	.153	.612	16	2.448
5	.087	.435	25	2.175
6	.053	.318	36	1.908
7	.021	.147	49	1.029
8	.005	.040	64	.320
9	.001	.009	81	.081
10	.002	.020	100	.200
Totals	1.000	2.833		11.101

$E(X) = \sum xP[X=x] = 2.833$

$\sigma^2(X) = \sum x^2P[X=x] - E(X)^2 = 11.101 - (2.833)^2 = 3.075$

EXERCISES

5-5 A coin is tossed three times. Let X represent the number of heads obtained. Then determine the probability distribution for X and calculate $E(X)$.

5-6 Determine the probability distribution of the winnings W of the roulette bettor in Example 5-1 when he places a $1 bet on the red field. Calculate $E(W)$. Describe the meaning of your answer verbally.

5-7 The following is the probability distribution for the number of daily requests a medical laboratory recieves for blood-typing tests. Find the expected daily number of blood-typing requests.

Number of Requests	*Probability*
0	.1
1	.2
2	.3
3	.2
4	.1
5	.1

5-8 A scientist studying sleep behavior is interested in the number of rapid-eye-movement (rem) phases achieved by his subjects during eight hours of sleep. The following probability distribution applies for any person not yet studied:

Number of rem Phases	*Probability*
0	.1
1	.2
2	.3
3	.3
4	.1

Find the expected value and variance for the number of rem phases.

5-9 A laboratory uses a particular breed of rabbit for experiments. In investigating the impact of diet upon litter size, the following probability distributions apply:

Standard Diet		*Special Diet*	
Litter Size	*Probability*	*Litter Size*	*Probability*
5	.1	5	0.0
6	.2	6	.1
7	.2	7	.2
8	.3	8	.3
9	.1	9	.3
10	.1	10	.1

(a) Find the expected litter sizes using each diet.

(b) Assuming the diet will be chosen which is expected to produce the greatest number of offspring per year, which diet will be used if the expected number of litters per year is 4 with the standard diet and 3.5 with the special diet? (*Hint:* Assume that the number of litters per year is unaffected by the sizes of the individual litters.)

5-10 An investor wishes to buy a stock to be held for one year in anticipation of capital gain. He has narrowed his choice down to High-Volatility Engineering and Stability Power. Both stocks currently sell for $100 per share and yield $5 dividends. The probability distributions for next year's price has been judgmentally assessed for each stock. These are given below, where

P_1 = price of High-Volatility Engineering and P_2 = price of Stability Power

High-Volatility Engineering s	$P[P_1 = s]$	Stability Power s	$P[P_2 = s]$
$ 25	.05	$ 95	.10
50	.07	100	.25
75	.10	105	.50
100	.05	110	.15
125	.10		1.00
150	.15		
175	.12		
200	.10		
225	.12		
250	.14		
	1.00		

(a) Determine the expected prices for a share of each stock.
(b) Should the investor select the stock with the highest expected value? Discuss.

5-3 THE BINOMIAL DISTRIBUTION

In Chapter 2, we saw that populations could be classified as members of basic family groupings, according to the nature of the frequency distribution. The same thing has been done with many random variables encountered in statistics. As the first case, we describe a widely used probability distribution that treats sample outcomes from qualitative populations. In Chapter 6, the normal distribution will be described. In subsequent chapters, additional probability distributions will be introduced in conjunction with various statistical applications.

Many statistical applications involve situations which have only complementary non-numerical outcomes. Consider these three examples: (1) When deciding to place a magazine advertisement, an advertiser must ultimately consider whether or not the reader will be influenced enough to buy his product. (2) A politician retains a polling agency to find out whether or not he is preferred by a majority of the voters. (3) A medical researcher wishes to determine if his new drug does or does not provide patients with some degree of relief. In each of these cases, the relevant answers to be sought are, respectively, the number of persons who will buy, the number of voters favoring the candidate, and how many patients will respond favorably to treatment. Because a great many persons may be involved in all of the foregoing situations, any evidence used to find the answers to questions such as these is usually provided by a sample.

A person selected for the sample will provide one of two opposite responses: purchase the product or not, prefer the candidate or not, or respond to treatment or not. In sampling from such qualitative populations, the only numerical result ordinarily of interest is how many times or in what proportion a particular attribute occurs.

The *number of* persons providing the desired response (such as the number of voters who prefer a candidate) is the key issue in each of our illustrations. Knowing the probability distribution of this random variable should facilitate deciding what course of action to take. For example, the medical researcher would be less prone to promote his drug if only a small number of the sample patients responded favorably to treatment, for the probability of such an outcome would be small if the drug were truly effective.

We may expect each situation to have a different probability distribution.

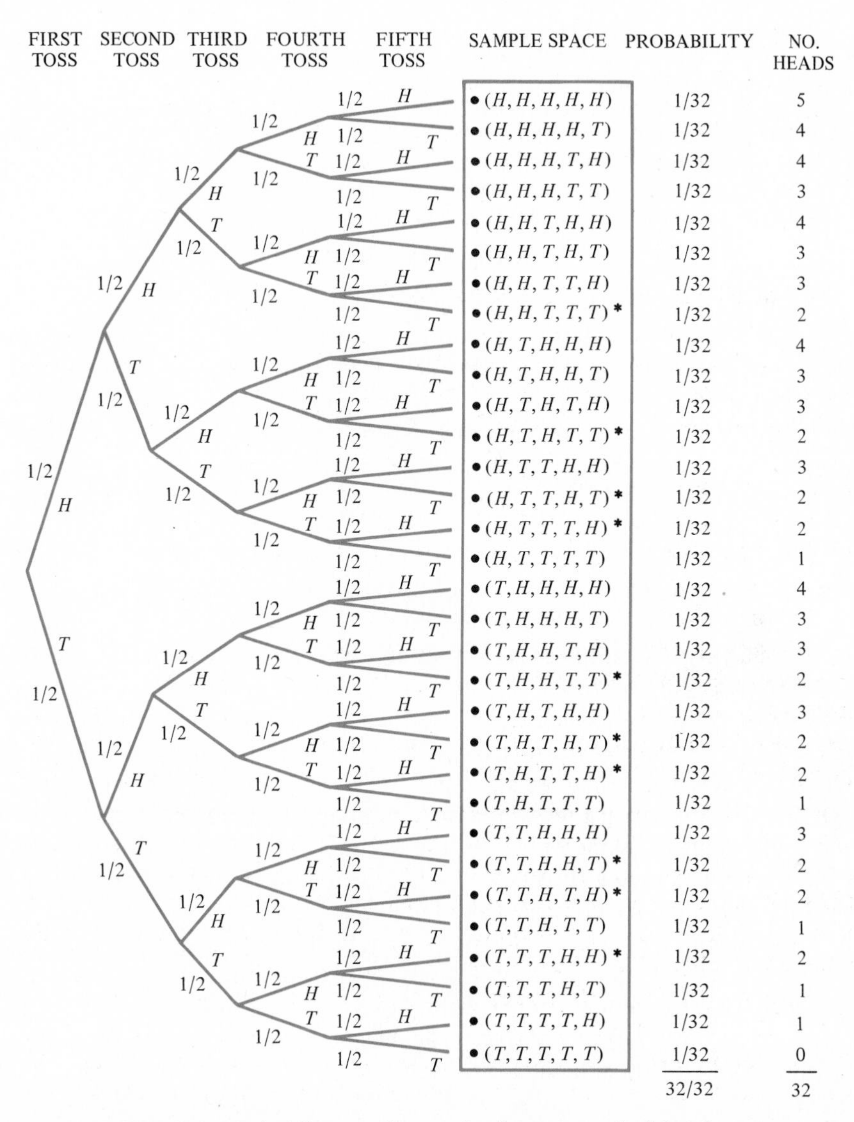

FIGURE 5-2 Probability tree diagram for five tosses of a fair coin.

There is no reason (except coincidence) why the probability that 50 voters will prefer the candidate should be the same as the probability that 50 patients will respond favorably to the drug. Yet, there are similarities between these two cases that prove advantageous in finding their respective probability distributions: they have characteristics in common that place them in the same family. Members of this family have a probability distribution referred to as a *binomial distribution*. The binomial distribution is epitomized by coin tossing and can be explained in terms of the following example.

Example 5-6 An evenly balanced coin is fairly tossed five times. This may be viewed as a five-stage experiment having the probability tree diagram shown in Figure 5-2, where the sample space is also listed. Our initial problem is to find the probability of obtaining exactly two heads. As each of the 32 outcomes is equally likely, the basic definition of probability allows us to find the answer in the following manner. Count the number of elementary events involving two heads and then divide this result by the total number of equally likely elementary events. The sample space contains 32 elementary events, 10 of which are two-head outcomes (see asterisks in Figure 5-2). Thus, we can determine that:

$$P[\text{exactly two heads}] = 10/32$$

As we saw in Chapter 4, it is impracticable to list all possible outcomes unless there are only a few of them. For instance, if 10 tosses were to be considered, then the analogous list would contain 1,024 (2^{10}) entries. For even longer toss sequences, it would be necessary to use the short cuts and probability laws already developed to ease computation. Before discussing a procedure to simplify finding such probabilities, it will be helpful to relate coin tossing to a similar class of situations.

The Bernoulli Process

A sequence of coin tosses is one example of a *Bernoulli process*.* A great many situations fall into the same category. All involve a series of experiments (such as tosses of a coin), which are referred to as *trials*. With each trial, *there are only two possible complementary outcomes*, like head or tail. Usually one outcome is referred to as a *success* and the other as a *failure*. Success is a broad term which covers the kind of outcome for which we wish to find the probability that a particular number of the same outcomes will occur. In the preceding illustration, getting a head would be a success. (This designation is arbitrary; in some other context, a head might be construed as a failure.)

Other examples include: the single childbirths in a maternity hospital, where each birth is a trial resulting in a boy or girl; the canning of a vegetable, where each trial is a full can that is slightly overweight or underweight (cans of precisely the correct weight are so improbable, we may ignore them); and the keypunching of numerical data, where each card completed is a trial which will

* After Jacob Bernoulli (eighteenth century), who proved a version of the law of large numbers.

either contain errors or be correct. In all cases, only two opposite trial outcomes are considered.

What further distinguishes these situations as Bernoulli processes is that *the success probability remains constant* from trial to trial. Thus, the probability of getting a head is the same, regardless of which toss is considered—and this must be the case for delivering a boy for any successive birth in the maternity hospital, picking up an overweight can of vegetables, and getting a correctly punched card each time. (The last condition would not hold if a keypuncher tires over time, so that the probability could be larger that an earlier card would be correct than that a later card would be.)

A final characteristic of a Bernoulli process is that *successive trial outcomes must be independent events.* Like a fairly tossed coin, the probability of getting a success (head) must be independent of what occurred in previous trials (tosses). The births in a *single family* might violate this requirement if the parents use certain recent techniques to obtain a second child whose sex is the opposite of their first child. Or a keypuncher's errors may occur in batches, due to fatigue, so that once an error is made it is more likely to be followed by another.

The results of sampling to determine response to advertising, voter preference, or favorable response to drug treatment may all be classified as Bernoulli processes. In order not to violate the requirements of independence and constant probability of success, we must sample with replacement (see Chapter 4, p. 127). This type of sampling allows each person the same chance of being selected each time, and perhaps of being chosen more than once.* The probability of a trial success would in each case be the proportion of persons in the respective population who would provide the desired response.

The Binomial Formula

When the trial outcomes are the results of a Bernoulli process, the number of successes is a random variable having a binomial distribution. The following expression may then be used to find the probability values. It is referred to as the ***binomial formula***

$$P[R=r] = \frac{n!}{r!(n-r)!}\pi^r(1-\pi)^{n-r} \tag{5-7}$$

where R = number of successes achieved
n = number of trials
π = trial success probability
$r = 0, 1, \ldots, n$

* When sampling without replacement, a different probability distribution, the *hypergeometric*, should be used. In practice, when the population is large, the conditions of a Bernoulli process are very nearly met and the binomial distribution is acceptable.

Expression (5-7) can be used to determine the probability found earlier for getting $R = 2$ heads in $n = 5$ tosses of a fair coin. In this case, $\pi = P[H] = \frac{1}{2}$ and $1 - \pi = P[T] = \frac{1}{2}$, so that

$$P[R = 2] = \frac{5!}{2!(5-2)!}\left(\frac{1}{2}\right)^2\left(1-\frac{1}{2}\right)^{5-2} = \frac{5!}{2!3!}\left(\frac{1}{2}\right)^2\left(\frac{1}{2}\right)^3$$

$$= 10\left(\frac{1}{2}\right)^5 = \frac{10}{32}$$

The factorial terms in the above product provide the number of outcomes involving exactly 2 heads, which is equal to 10 and represents the number of combinations of 2 particular tosses that may result in heads out of a total of 5 tosses made. The product involving $\frac{1}{2}$ represents the probability of getting any one of the 10 two-head sequences that are represented by the end positions of the probability tree diagram in Figure 5-2. Each of these positions is reached by traversing a particular path of 2 head and $5 - 2 = 3$ tail branches. The probability of doing this may be obtained by applying the multiplication law. Since a two-head result can happen in any one of 10 equally likely ways, the addition law of probability indicates that we add 10 of the identical product terms together, or, more simply, multiply by 10.

The entire binomial distribution that corresponds to the number of heads R resulting from $n = 5$ fair tosses of an evenly balanced coin is given in Table 5-6, where the probability values are found by applying the binomial formula for all possible r values.

TABLE 5-6 Binomial Distribution for the Number of Heads Obtained in Five Coin Tosses

Possible Number of Heads r	$P[R=r]$
0	$\frac{5!}{0!5!}\left(\frac{1}{2}\right)^0\left(\frac{1}{2}\right)^5 = \frac{1}{32} = .03125$
1	$\frac{5!}{1!4!}\left(\frac{1}{2}\right)^1\left(\frac{1}{2}\right)^4 = \frac{5}{32} = .15625$
2	$\frac{5!}{2!3!}\left(\frac{1}{2}\right)^2\left(\frac{1}{2}\right)^3 = \frac{10}{32} = .31250$
3	$\frac{5!}{3!2!}\left(\frac{1}{2}\right)^3\left(\frac{1}{2}\right)^2 = \frac{10}{32} = .31250$
4	$\frac{5!}{4!1!}\left(\frac{1}{2}\right)^4\left(\frac{1}{2}\right)^1 = \frac{5}{32} = .15625$
5	$\frac{5!}{5!0!}\left(\frac{1}{2}\right)^5\left(\frac{1}{2}\right)^0 = \frac{1}{32} = .03125$
	Total 1.00000

Because the probabilities obtained depend upon the value of r used, the right-hand side of expression (5-7) is a mathematical function in r. Because it provides nonzero probabilities only at specific points between 0 and n, to use a physical analogy, it concentrates its "mass" on these. (Referring to Table 5-6, no "mass" is concentrated on any point between 1 and 2, for example, as the probability is zero for any such impossible value of r.) Thus, a function such as expression (5-7) is occasionally referred to as a *probability mass function.*

The designation of which attribute is the "success" is completely arbitrary, but care must be taken to insure that the appropriate value of π is used. An interesting feature of the binomial formula is that it may also be used to obtain the probability that some number of failures will occur. For instance, the probability of obtaining exactly three tails in five tosses of the coin has the same value as the probability of obtaining two heads, because whenever there are two heads there must be three tails. In general, *when there are r successes, there must be $n-r$ failures.*

As we have already noted, different Bernoulli processes will have different probability values. But the number of successes from each process are random variables belonging to the binomial distribution family. Note that the probabilities for all possible values of R depend upon the value of π. Different sizes for n will result in a larger or smaller number of possible values for R and will also affect each probability value. *For purposes of calculating probabilities, one Bernoulli process differs from another only by the values of π and the sizes of n.*

Example 5-7 The administrator of a large airport is interested in the number of aircraft departure delays that are attributable to inadequate control facilities. A random sample of 10 aircraft takeoffs is to be thoroughly investigated. (In Chapter 7, we will discuss the criterion for deciding how large a sample ought to be taken.) If the true proportion of such delays in all departures is .40, what is the probability that 4 of the sample departures are delayed because of control inadequacies?

The sample investigations may be considered as trials in a Bernoulli process. Letting a control-caused delay be a success, the trial success probability is equal to the proportion of such outcomes, so that $\pi = .40$. To obtain the probability of $R = 4$ control-caused delays (successes) using the binomial formula [expression (5-7)] with $\pi = .40$, $n = 10$, and $r = 4$, we have:

$$\begin{aligned} P[R=4] &= \frac{10!}{4!(10-4)!}(.4)^4(1-.4)^{10-4} \\ &= 210(.4)^4(.6)^6 \\ &= 210(.0256)(.046656) \\ &= .2508 \end{aligned}$$

The Proportion of Successes

Ordinarily, the number of successes is a less useful random variable than the *proportion of successes*. The proportion of control-caused aircraft delays would be more meaningful to the administrator in Example 5-7 than the number of delays. The administrator could use this proportion to determine the total number of delays that would be encountered for all prognosticated levels of traffic.

The ratio of the number of successes to the number of trials, denoted by P, is the ***proportion of successes***

$$P = \frac{R}{n}$$

The probabilities of various possible values for P may also be calculated from the binomial formula. In so doing, we notice that $R = nP$. Therefore, when $R = r$, $nP = r$ or $P = r/n$. Therefore, we have:

$$P\left[P = \frac{r}{n}\right] = \frac{n!}{r!(n-r)!}\pi^r(1-\pi)^{n-r} \tag{5-8}$$

The probability distribution of the proportion of control-caused aircraft delays is shown in Table 5-7. Each probability value is calculated from expression (5-8), using $n = 10$ and $\pi = .4$.

TABLE 5-7 Binomial Distribution for the Proportion of Aircraft Delays

r	$\frac{r}{n} = \frac{r}{10}$	$P\left[P = \frac{r}{n}\right]$
0	0.0	.0060
1	.1	.0404
2	.2	.1209
3	.3	.2150
4	.4	.2508
5	.5	.2007
6	.6	.1114
7	.7	.0425
8	.8	.0106
9	.9	.0016
10	1.0	.0001
	Total	1.0000

The Binomial Distribution Family

To illustrate the concept of a family of binomial distributions, we construct Table 5-8, showing the binomial probability distributions for different trial success probabilities π. Each entry is obtained from expression (5-8).

In each case, a sequence of $n = 5$ trials applies. The same possible outcomes exist but, as shown in Table 5-8, the probabilities for the values of P differ due to the different values of π. When $\pi = .9$, a higher proportion of successes is

TABLE 5-8 Binomial Probability Distributions When $n = 5$

$\frac{r}{5}$	$P\left[P = \frac{r}{5}\right]$				
	$\pi = .1$	$\pi = .3$	$\pi = .5$	$\pi = .7$	$\pi = .9$
.0	.59049	.16807	.03125	.00243	.00001
.2	.32805	.36015	.15625	.02835	.00045
.4	.07290	.30870	.31250	.13230	.00810
.6	.00810	.13230	.31250	.30870	.07290
.8	.00045	.02835	.15625	.36015	.32805
1.0	.00001	.00243	.03125	.16807	.59049

more likely than when $\pi = .1$. Figure 5-3 shows graphs of these binomial probability distributions, constructed for the values in Table 5-8.

When $\pi = .1$, the probability distribution is quite positively skewed with the probabilities of small values of P being greatest. As π becomes larger, the

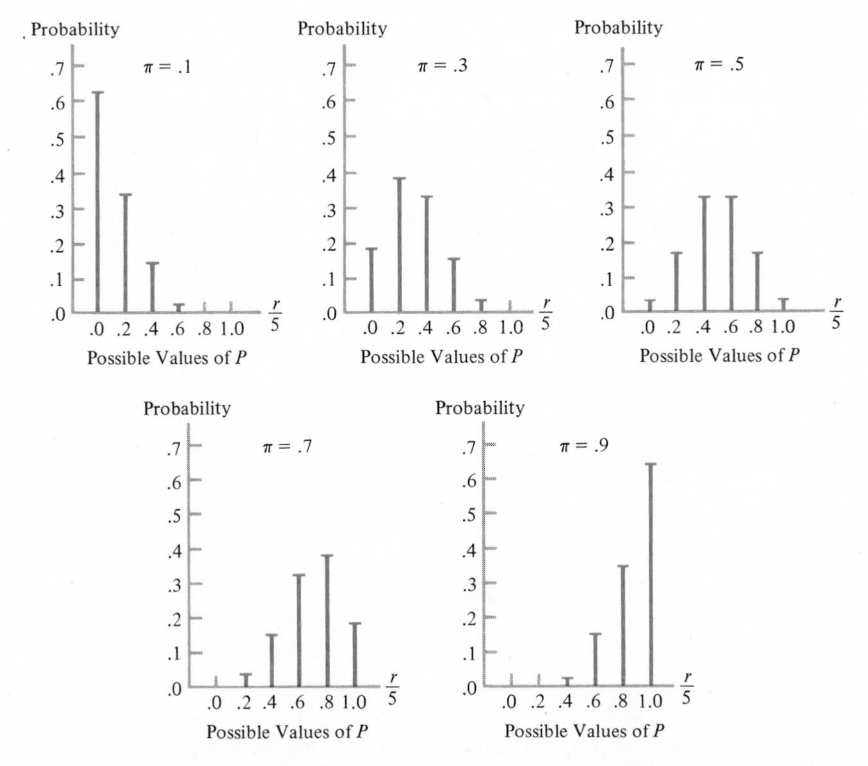

FIGURE 5-3 Binomial probability distributions for several values of π, with $n = 5$.

skew becomes less pronounced. When $\pi = .5$, the distribution is symmetrical, so that the probability of a particular P value result is equal to the probability of $1 - P$. When π is larger than .5, the distributions become negatively skewed: low values of P have the smaller probabilities and the intensity of the skew

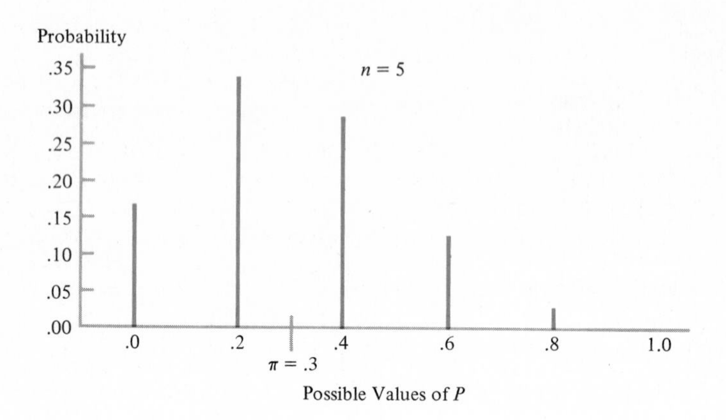

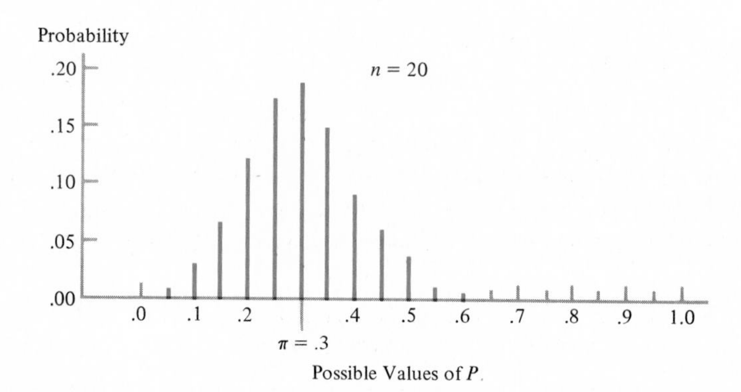

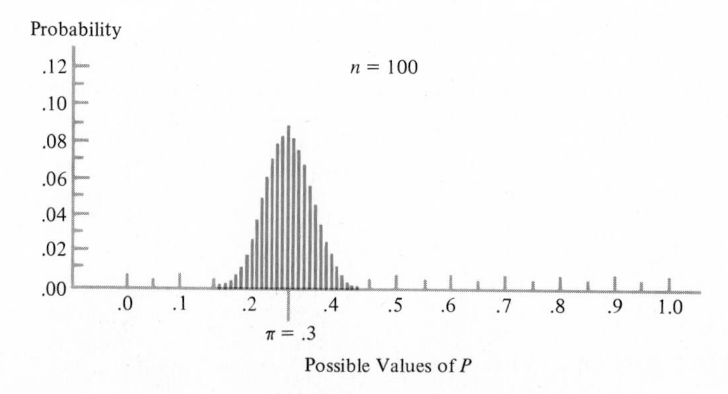

FIGURE 5-4 Binomial distributions for different levels of n, with $\pi = .3$.

increases with π. Notice that the probabilities for $\pi = .1$ are identical, but in *reverse* sequence, to those for $\pi = .9$. This will hold for every complementary pair of π values, such as .3 and .7 or .01 and .99.

The graphs in Figure 5-4 illustrate another interesting feature of the binomial distribution. Here, we fix π at .3 and vary n. Starting with $n = 5$, the number of trials is increased first to 20 and then to 100. With n trials, there are $n+1$ possible values of P (corresponding to $0, 1, 2, \ldots,$ or n successes). For larger n values, we cannot show all of the possibilities as spikes on the graphs, for some of the probabilities are extremely tiny. (For example, when $\pi = .3$ and $n = 20$, the probability of all successes, or $P[P=1]$, is less than 40 trillionths.)

Note that the number of spikes increases with n and that the spikes become more closely bunched as n increases. A very significant feature is the tendency for the spikes to assume the "bell shape." In Chapter 6, we will make use of this property to approximate the binomial distribution by the normal distribution. This fundamental property holds for any value of π as n becomes large. But it does not hold if $\pi = 0$ or $\pi = 1$; in either case, there is only one possible value for P. When $\pi = 0$, no trial can be a success, so P must always equal zero regardless of the size of n. Likewise, when $\pi = 1$, all trials must be successes, so $P = 1$ always.

The Cumulative Probability Distribution

Just as it is sometimes convenient to deal with cumulative frequencies, which are readily determined from the population frequency distribution, we may use *cumulative probabilities* for random variables. These are very simply obtained from the probability distribution in Table 5-7 by creating an additional column (4) for $P[P \leqslant r/n]$, as shown in Table 5-9. The values in this column

TABLE 5-9 Cumulative Probability Distribution for the Proportion of Aircraft Delays

(1) r	(2) $\frac{r}{n} = \frac{r}{10}$	(3) $P\left[P = \frac{r}{n}\right]$	(4) $P\left[P \leqslant \frac{r}{n}\right]$
0	0.0	.0060	.0060
1	.1	.0404	.0464
2	.2	.1209	.1673
3	.3	.2150	.3823
4	.4	.2508	.6331
5	.5	.2007	.8338
6	.6	.1114	.9452
7	.7	.0425	.9877
8	.8	.0106	.9983
9	.9	.0016	.9999
10	1.0	.0001	1.0000
	Total	1.0000	

are obtained by adding all the preceding entries in the values of $P[P = r/n]$. Thus:

$$P[P \leqslant 0] = P[P = 0] = .0060$$

and

$$\begin{aligned} P[P \leqslant .1] &= P[P = 0] + P[P = .1] \\ &= .0060 + .0404 \\ &= .0464 \end{aligned}$$

and

$$\begin{aligned} P[P \leqslant .2] &= P[P = 0] + P[P = .1] + P[P = .2] \\ &= .0060 + .0404 + .1209 \\ &= .1673 \end{aligned}$$

Table 5-9 is thus constructed cumulatively from the individual probability values, so that the values obtained constitute the *cumulative probability distribution* of the random variable P.

The probability mass function for P is graphed in Figure 5-5. The cumulative probability distribution function for P is also shown. The cumulative probability value corresponding to any particular proportion value is obtained from the *highest point* on the "stairway" directly above. For instance, the cumulative probability for values .5 or less is .8338, and not .6331, which belongs to the lower "step." Notice that the size of each step is the same as the height of the respective spike of the mass function. Thus, the underlying probability distribution may be obtained from the cumulative probability distribution by finding these step sizes. For example, to find the probability that $P = .5$ we find the difference

$$\begin{aligned} P[P = .5] &= P[P \leqslant .5] - P[P \leqslant .4] \\ &= .8338 - .6331 = .2007 \end{aligned}$$

Using Binomial Probability Tables

Because calculating binomial probabilities involves working with large factorial values and with small numbers raised to large powers, it is convenient to have them computed once and for all. Appendix Table C provides cumulative binomial probability values computed for various sizes of n, with a separate tabulation for each of several π's. Although binomial probability tables may be obtained for individual terms as well, cumulative values are easier to apply in most cases.

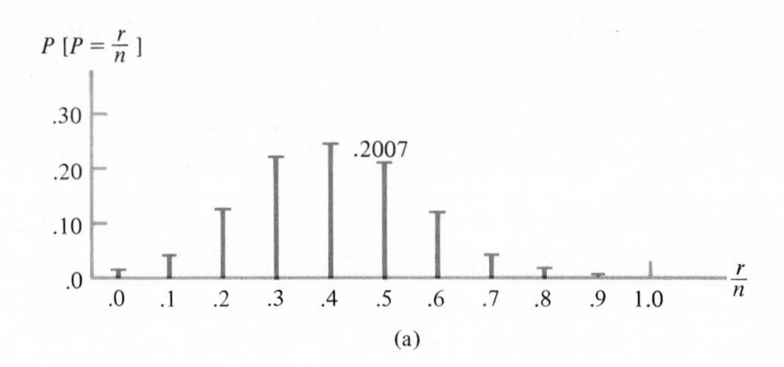

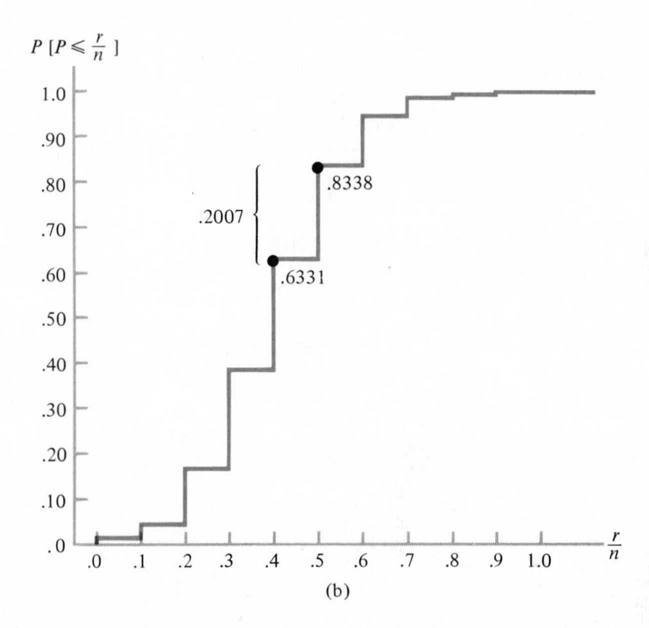

FIGURE 5-5 (a) Binomial probability mass function and (b) cumulative probability distribution function for the proportion of successes from a Bernoulli process with $n = 10$ and $\pi = .4$.

It is possible to use this table to compute probabilities for the number or proportion of successes in the variety of situations described below. For purposes of illustration, assume that we are interested in finding probabilities regarding the proportion of $n = 100$ patients who will respond favorably (success) to treatment with a drug that has proved successful 30 percent of the time, so that $\pi = .30$. We will use the portion of Appendix Table C that begins on page 562.

1. Obtaining a result less than or equal to a particular value. The probability that 40 percent or fewer patients will respond favorably is a cumulative probability value that may be read directly from the table when $r = 40$ successes:

$$P[P \leqslant .40] = P[P \leqslant 40/100] = .9875$$

2. Obtaining a result exactly equal to a single value. Recall that cumulative probabilities represent the sum of individual probability values and are portrayed graphically as a stairway (see Figure 5-5). A single value probability may be obtained by determining the size of the step between two neighboring cumulative probabilities by finding the difference:

$$P\left[P = \frac{r}{n}\right] = P\left[P \leqslant \frac{r}{n}\right] - P\left[P \leqslant \frac{r-1}{n}\right] \tag{5-9}$$

For example, the probability that exactly $r = 32$ percent of the patients will respond favorably would be

$$P\left[P = \frac{32}{100}\right] = P\left[P \leqslant \frac{32}{100}\right] - P\left[P \leqslant \frac{31}{100}\right]$$
$$= .7107 - .6331$$
$$= .0776$$

3. Obtaining a result strictly less than some value. In this case, we need only observe that

$$P\left[P < \frac{r}{n}\right] = P\left[P \leqslant \frac{r-1}{n}\right] \tag{5-10}$$

For example, the probability that fewer than $r = 30$ percent successes are achieved is the same as the probability that exactly $r-1 = 29$ percent or less successes are obtained, or

$$P\left[P < \frac{30}{100}\right] = P\left[P \leqslant \frac{29}{100}\right] = .4623$$

4. Obtaining a result greater than or equal to some value. Here, we use the fact that

$$P\left[P \geqslant \frac{r}{n}\right] = 1 - P\left[P < \frac{r}{n}\right] = 1 - P\left[P \leqslant \frac{r-1}{n}\right] \tag{5-11}$$

In finding the probability that at least $r = 20$ percent of the patients will respond favorably, we look up the cumulative probability that $r-1 = 19$ or less will respond and subtract this value from 1, or

$$P\left[P > \frac{20}{100}\right] = 1 - P\left[P \leqslant \frac{19}{100}\right]$$
$$= 1 - .0089 = .9911$$

Note that a related problem involves a situation where the result must be *strictly greater than* some value. In such a case, we find the cumulative probability for r itself and subtract this from 1. For example, the probability that more than $r = 20$ patients will respond to treatment is 1 minus its complementary probability that 20 or fewer patients respond, or

$$P\left[P > \frac{20}{100}\right] = 1 - P\left[P \leqslant \frac{20}{100}\right]$$
$$= 1 - .0165 = .9835$$

5. Obtaining a result that lies between two values. Suppose we want to find the probability that the proportion of successes will lie somewhere between .25 and .35, inclusively; that is, we want to determine that

$$P\left[\frac{25}{100} \leqslant P \leqslant \frac{35}{100}\right]$$

In this case, we obtain the difference between two cumulative probabilities:

$$P\left[P \leqslant \frac{35}{100}\right] - P\left[P \leqslant \frac{24}{100}\right] = .8839 - .1136 = .7703$$

The underlying rationale is that the first term above represents all outcomes with 35 or fewer successes, but we do not want to include those outcomes with successes numbering 24 and fewer. By subtracting the second cumulative probability from the first, we account only for those outcomes that are between 25 and 35 percent successful.

6. Finding probabilities when the trial success probability exceeds .5. For brevity, our binomial table stops at $\pi = .5$. Suppose we wish to find probabilities for the number or proportion of successes when the trial success probability is larger (say, .7). To do this, we can still use Appendix Table C by letting R represent the number of failures and π the trial failure probability.

Consider a keypuncher who correctly punches cards 99 percent of the time. Suppose we want to find the probability that at least 95 percent of $n = 100$ cards have been punched correctly. Here, a success represents a correct card and $\pi = .99$. We observe that more than 95 percent correct is the same as

5 percent or less incorrect (failures). For the moment, we use $\pi = 1 - .99 = .01$ as the trial failure probability, and using the portion of Table C on page 562, we find

$$P\left[P \leqslant \frac{5}{100}\right] = .9995$$

This illustrates that for every success event there is a corresponding failure event with the same probability.

Mean and Variance of Binomial Distribution

The expected value of P, the proportion of successes from a Bernouill process, may be determined from probabilities obtained from the binomial formula for a particular n and π. Intuitively, however, we would expect the proportion of successes, on the average, to be the same as the trial success probability. For instance, the proportion of heads resulting from a sequence of coin tosses should be 1/2, on the average, the same as the probability of obtaining a head from a single toss. We therefore have the ***expected value of P***

$$E(P) = \pi \tag{5-12}$$

The variance of P may be caluclated from the probability distribution in the manner of expression (5-4), but such calculations can be quite tedious. The following expression, applicable to the binomial distribution, provides the same answer more directly for the ***variance of P***

$$\sigma^2(P) = \frac{\pi(1-\pi)}{n} \tag{5-13}$$

The expressions just given for $E(P)$ and $\sigma^2(P)$ follow mathematically from the properties of the binomial distribution. Aside from their computational simplicity, an added advantage is that there is no need to compute the probabilities for the possible values of P first. Thus, knowing π and n allows us to quickly find $E(P)$ and $\sigma^2(P)$, making out later task of approximating the binomial distribution a simple one.

Note that since $\sigma^2(P)$ depends upon both the value of π and the number of trials n, if π is held fixed, $\sigma^2(P)$ will decrease as the number of trials increases.

Example 5-8 An inspector removes cans of soup from production at randomly chosen times. The proportion of all cans overweight is .05. Thus, the probability that any particular can is overweight will be $\pi = .05$. Letting P represent the proportion of cans the inspector will find overweight, or $E(P) = \pi = .05$, and n represent the number of cans he selects, Table 5-10 shows the values obtained for

$\sigma^2(P)$. Since $\sigma^2(P)$ measures dispersion in P, the greater the number of cans selected n, the closer, on the average, the proportion of overweight cans will be to the expected value $E(P)$.

TABLE 5-10 Variance of Proportion of Overweight Cans Selected When $\pi = .05$

Number of Cans n	$\sigma^2(P) = \frac{.05\,(.95)}{n}$
2	.02375
5	.0095
10	.00475
100	.000475

Suppose that $n = 10$ cans are selected from production runs of different canned food products and that each run has different proportions π of overweight cans. $\sigma^2(P)$ will differ with the type of food, as shown in Table 5-11. Notice that $\sigma^2(P)$ is largest when $\pi = .5$. This will always be true, regardless of the size of n.

TABLE 5-11 Variance of P When $n = 10$

Proportion of Production Overweight π	$\sigma^2(P) = \frac{\pi(1-\pi)}{10}$
.05	.00475
.25	.01875
.50	.025
.75	.01875
.95	.00475

By taking the square root of the variance, we may use the following to calculate the ***standard deviation of P***

$$\sigma(P) = \sqrt{\frac{\pi(1-\pi)}{n}} \tag{5-14}$$

As we will see in Chapter 6, the standard deviation of P is used in many statistical applications.

Now recall that the binomial distribution can be applied to either the proportion of successes P or to the number of successes R. The expected value and variance of R will not be the same as they are for P, but since $R = nP$, the properties of expected value and variance may be used to find $E(R)$ and $\sigma^2(R)$ directly from the shortcut expressions (5-12) and (5-13).

The expected value of R is determined by applying Property 5-1 (page 163), with $a = 0$ and $b = n$, so that $E(nP) = nE(P)$. Thus:

$$E(R) = E(nP) = n\pi \tag{5-15}$$

The variance of R is similarly obtained from Property 5-2 (page 165),

$$\sigma^2(R) = \sigma^2(nP) = n^2\sigma^2(P) = n^2\left(\frac{\pi(1-\pi)}{n}\right)$$

Canceling n terms simplifies the above to:

$$\sigma^2(R) = n\pi(1-\pi) \tag{5-16}$$

The standard deviation of R follows directly:

$$\sigma(R) = \sqrt{n\pi(1-\pi)} \tag{5-17}$$

EXERCISES

5-11 Can each of the following situations be classified as a Bernoulli process? If not, state why.

(a) Childbirths in a hospital, the relevant events being the sex of each newborn child.

(b) The outcomes of successive rolls of a die, considering only the events odd or even.

(c) A crooked gambler has rigged his roulette wheel so that whenever the player loses, a mechanism is released which gives him better odds; likewise, when a player wins, his chance of winning on the next spin is somewhat smaller than before. Consider the outcomes of successive spins.

(d) The measuring mechanism which determines how much dye to squirt into paint being mixed to customer order occasionally violates the required tolerances. The mechanism is very reliable when it is new, but with use it continually wears, becoming less accurate with time. Consider the outcomes (within or not within tolerance) for successive mixings.

(e) A machine produces items which are sometimes too heavy or too wide to be used. The events of interest express the quality of each successive item in terms of both weight and width.

5-12 An evenly balanced coin is fairly tossed seven times.

(a) Determine the probabilities of obtaining: (1) exactly two heads; (2) exactly four heads; (3) no tails; (4) exactly three tails.

(b) What do you notice about your answers to (2) and (4)? Why is this so?

5-13 n parts are randomly chosen from a production process which yields 5 percent defectives.

(a) What is the expected proportion of defectives?

(b) How many defectives are expected when $n = 5$? When $n = 10$? When $n = 100$?

5-14 A political polling agency has contacted a random sample of $n = 50$ persons. The actual proportion of all persons favoring a new law is $\pi = .30$. Assuming that the binomial distribution applies, use Appendix Table C to find the probabilities for the following outcomes relating to the number of favorable responses R: (a) $R \leqslant 11$; (b) $R = 15$; (c) $R < 18$; (d) $R \geqslant 20$; (e) $R > 25$; (f) $17 \leqslant R \leqslant 23$.

5-15 A fair coin is tossed 20 times in succession. Using Appendix Table C, determine the probability that the proportion of heads obtained is:

(a) Less than or equal to .4.

(b) Equal to .5.

(c) Less than .75.

(d) Greater than or equal to .6.

(e) Greater than .65.

(f) Between .4 and .7, inclusively.

5-16 Using the probability distribution for P provided in Table 5-8 with $\pi = .1$, calculate $E(P)$. Use the definition of expected value expression (5-1), with r/n in place of x and P in place of X. Your answer should equal π, which here is .1.

5-17 From the probability values in Table 5-8, construct the cumulative probability distribution of P when $n = 5$ and $\pi = .7$.

5-18 A form of malnutrition occurs in 10 percent of all persons. Determine the probabilities of the following malnutrition outcomes for five randomly chosen persons.

(a) All have it.

(b) None have it.

(c) At least one has it.

(d) At least 40 percent have it.

(e) Between 40 and 80 percent, inclusively, have it.

5-19 The chief engineer in a chemical plant has established a policy that five sample vials be drawn from the final stage of a chemical process at random times over a four-hour period. If one or more vials (20 percent or more) have impurities, all the settling tanks are cleaned. Find the probability that the tanks must be cleaned when:

(a) The process is so clean that the probability of a dirty vial is $\pi = .01$.

(b) Same as (a), but with $\pi = .05$.

(c) Same as (a), but with $\pi = .20$.

(d) Same as (a), but with $\pi = .50$.

(*Hint:* Use the fact that $P[\text{at least one dirty vial}] = 1 - P[\text{no dirty vials}]$.)

5-20 A production process produces defective parts at rate of .05. If a random sample of five items is chosen, what is the probability that at least 80 percent of the sample will be defective?

5-21 A lot of 100 items in which 5 percent are defective is sampled *without replacement*. Five items are chosen.

(a) Does the binomial distribution apply here? Explain your answer.

(b) What is the probability that not one of the five sample items is defective?

5-22 A lopsided coin provides a 60 percent chance of a head on each toss. If the coin is tossed 18 times, find the probability that the number of heads obtained is:

(a) Less than or equal to 8.

(b) Equal to 9.

(c) Less than 15.

(d) Greater than or equal to 12.

(e) Greater than 13.

(f) Between 8 and 14, inclusively.

5-4 CONTINUOUS PROBABILITY DISTRIBUTIONS

To begin our discussion of continuous probability distributions, let us consider a random experiment involving the selection of a sample from a population.

The frequency distribution obtained from grouping the years of company service of each of the 22,000 employees of the Wheeling Wire Works is given in Figure 5-6. The height of each bar of this histogram represents the relative frequency for durations of employment in the interval covered by the bar. The random variable of interest, which we will denote by the letter X, is the length of service of one person chosen randomly from the population. The time an employee has spent working for the company may be measured on a continuous scale, so that this time may be expressed to any desired fraction of the year. This makes X in our example a *continuous random variable.*

The probability that X lies within a particular class interval will be the same as the relative frequency of employees having salaries in that range.

Recall that the bars of the histogram for a relative frequency distribution have areas proportional to the relative frequency of the variate values in the corresponding interval. Since the relative frequencies must sum to 1, we may

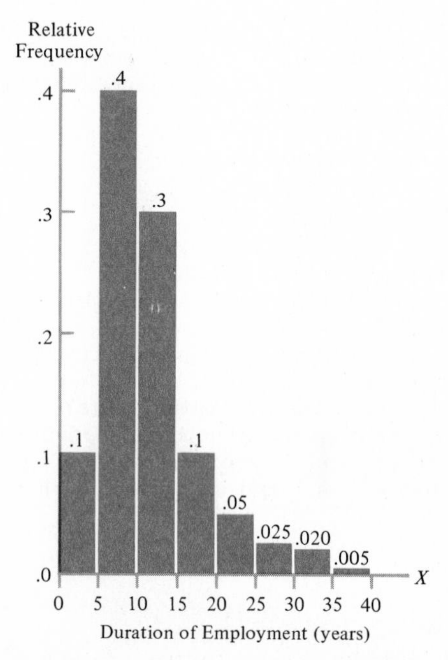

FIGURE 5-6 Frequency distribution of years of employment.

consider the total area under the histogram as being equal to 1. There, each class interval is one 5-year unit wide. Thus, the probability that X lies inside a particular class interval equals the area of the bar covering that interval. Therefore, the probability that our employee served between 10 and 15 years must be .3—the area under the bar or rectangle of the histogram that covers these values, or width × height = $1 \times .3 = .3$.

But what if we want to consider an event that is more specific? For example, we may wish to find the probability that our selected person was employed between 10.95 and 11.05 years—a range of values lying wholly within a single class interval. In this case, the frequency distribution does not provide sufficiently detailed information for us to find the probability directly.

Smoothed Curve Approximation

Our problems arise from the fact that the histogram artificially forces us to treat population variates in discrete lumps. Each value is arbitrarily placed

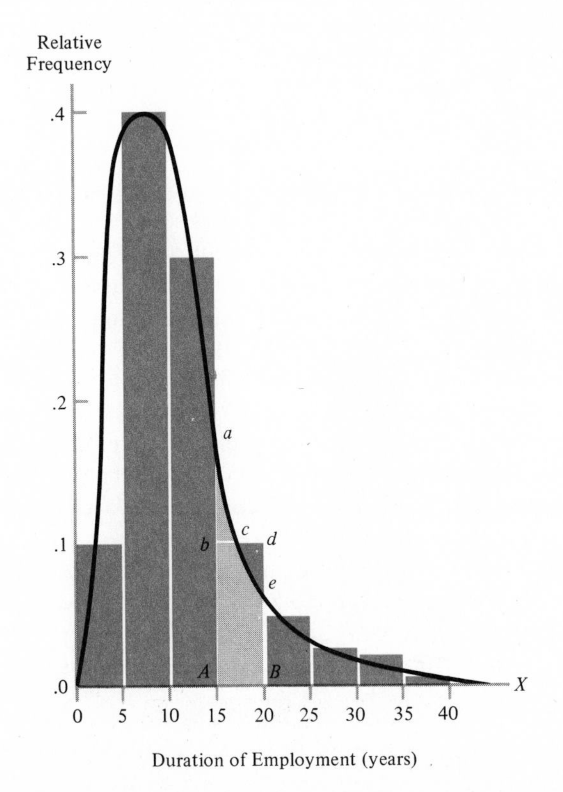

FIGURE 5-7 Frequency distribution from Figure 5-6 approximated by a continuous curve.

into one of a very few class intervals. To accurately consider intervals of any width, we may construct a smoothed, continuous approximation to the histogram like the one in Figure 5-7, where the curve is superimposed on the histogram from Figure 5-6. This may then be used to generate any probability values desired.

If the smoothed curve is drawn properly, the area under portions of the curve will correspond closely to the area of the histogram bars for that part of the horizontal axis, and the total area under the curve can also be taken as unity. Thus, in Figure 5-7, the shaded area under the portion of the curve covering the points between A and B is nearly the same as the corresponding area under the bar covering the values from A to B. Here, the curve has been drawn so that the wedge-shaped area described by the points a, b, c is nearly the same as the area described by c, d, e. The sum of the areas for portions that are cut off the histogram by the curve should equal the sum of those areas the curve adds by smoothing the corners.

The relative frequency of the population values lying between A and B in Figure 5-7 may be approximated by the area under the portion of the curve covering these values. We may therefore conclude that ***the probability that a continuous random variable assumes some value in an interval is represented by the area under the portion of the continuous curve covering the interval.***

Probability Density Function

As we saw in Chapter 2, population frequency distribution graphs can be categorized according to their general shapes. We can construct smoothed curve approximations for any of these distributions. The process of selecting the curve is in essence one of choosing the appropriate shape. Fortunately, a great deal is known about curves having shapes that fit the more common types of frequency distributions.

Curves of a particular shape can be defined by an equation or function. Thus, we can place population frequency distributions into various groups, according to which mathematical function yields a curve of the shape that best matches that of the population histogram. If we know the function defining the curve, then it is possible to mathematically compute the area over an interval to obtain the required probability.

A unique set of probabilities is obtained from a particular curve. Consider, for example, the two curves in Figure 5-8 for two random variables, X and Y, whose values range over the same scale. The areas covering the values between points A and B are different for the two curves because the curve shapes are different. Thus:

$$P[A \leqslant X \leqslant B] \neq P[A \leqslant Y \leqslant B]$$

so that X and Y have different probability distributions. Note that the thickness or density of the two curves is distributed differently. For this reason, the mathematical expression describing such a curve is sometimes called a *probability density function*. In general, we denote this function by $f(x)$, where the lower case x represents the possible values of the random variable X. (Since density is measured on the same scale, we will continue to use "relative frequency" to designate the vertical axis.) The probability density function therefore describes the

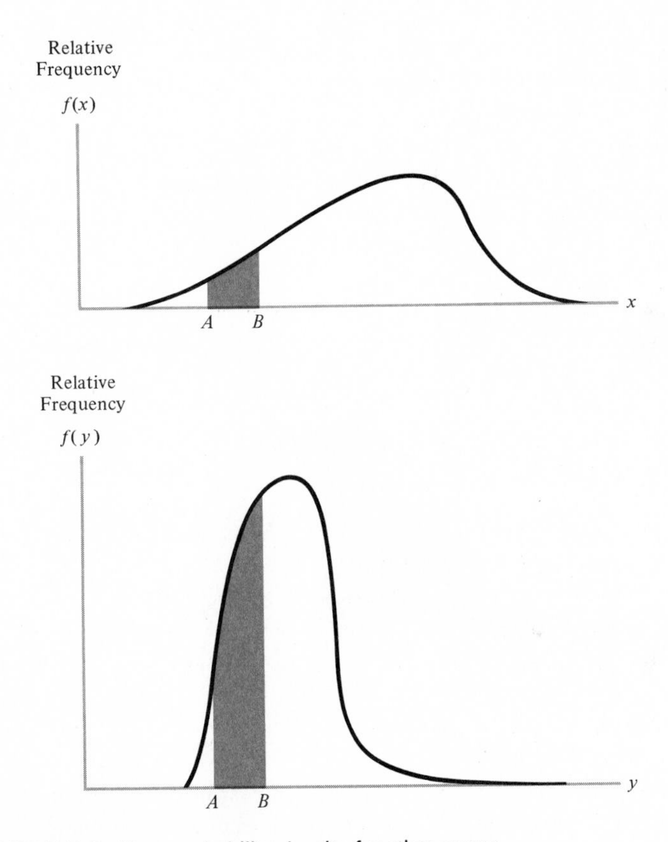

FIGURE 5-8 Two probability density function curves.

probability distribution of a continuous random variable, so that a random variable may be categorized by the form of its function. The probability density function is the continuous counterpart to the discrete random variable's probability mass function. But the latter expresses a probability directly, while *it is the area under the density function that yields probability values.*

There is zero area under the portion of the curve that covers a single point, since there is no width. This reflects the fact that for a continuous random variable X, $P[X = x] = 0$, which was established in Example 5-3 (page 158).

Cumulative Probability Distribution

In Section 5-3, we defined the cumulative distribution for the discrete random variable. Analogously, we can make the following

DEFINITION The ***cumulative probability distribution*** of a continuous random variable X is the set of all values expressing $P[X \leqslant x]$, where x is a particular number. For a given x, the cumulative probability is the total area under the density function covering points to the left of x.

The cumulative probability distribution means precisely the same thing for discrete and continuous random variables. In the former, the cumulative prob-

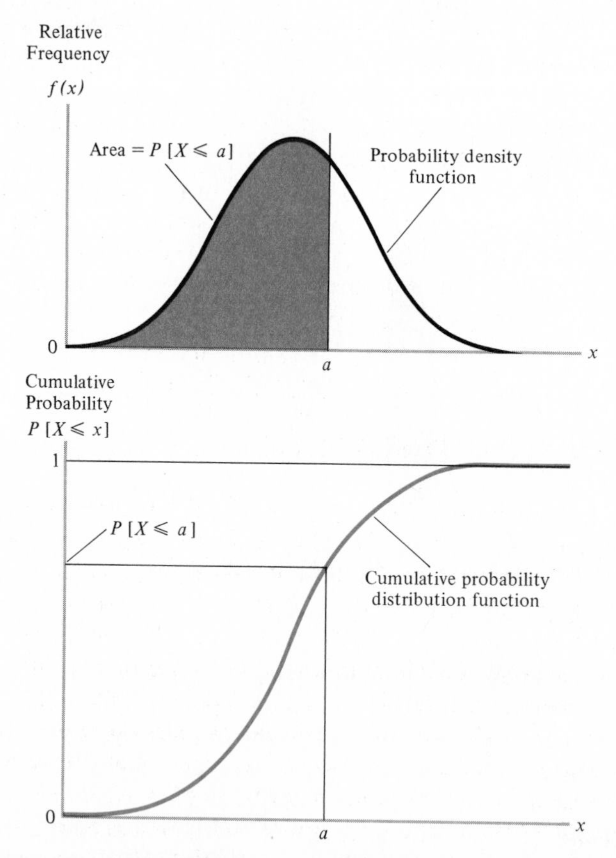

FIGURE 5-9 Graphical expressions of the probability distribution of a continuous random variable.

ability graph resembles a stairway (see Figure 5-5) instead of a smoothed curve. In either case, it may be expressed algebraically, as a graph, or as a table of values. The graphical expressions of the probability distribution of a continuous random variable are presented in Figure 5-9.

A property of the cumulative probability distribution of a continuous random variable is that

$$P[X \leqslant x] = P[X < x]$$

because $P[X=x] = 0$, as we have already seen. (This is also true in the case of the discrete random variable, except when x lies directly beneath a stepping point on the cumulative probability stairway.)

The Expected Value and Variance

As with discrete random variables, we may view the expected value of a continuous random variable as the long-run average result from many repeated random experiments. The variance is also analogous. Both may be calculated from the probability density function, but to do so requires mathematics beyond the scope of this book. For the common distributions we will encounter, the values have already been determined.

Many random variables may have the same basic density function and differ only by the values of the parameters specifying the particular function. They are then said to belong to the same *distribution family*. In Chapter 6, we will discuss one of these families—the normal distribution. Other continuous probability distributions will be introduced in later chapters.

5-5 THE SAMPLING DISTRIBUTION

We are now ready to set the stage for a smooth transition from probability theory to statistical planning and analysis. To do this, we will concern ourselves largely with deductive statistics. Thus, for the time being, it will be assumed that many details regarding the population under investigation are known, and we will study the kinds of *possible* results that may be obtained for samples taken from these populations. It is important to emphasize that the deductive viewpoint is not usually encountered in a real sampling situation. In actual applications, the sample results are known and the population details are unknown. After all, the reason we take a sample is to draw conclusions about a population. The purpose served by reversing our vantage point is that, knowing how sample results are generated for known pouplations and knowing what values are likely to occur, we are better able to answer the more difficult inductive questions involved in generalizing about an unknown population from a known sample.

One important type of inference to be made from a sample is the estimation of a population parameter such as the mean μ or proportion π. Usually, we would employ the sample statistic counterparts ($\bar{X}$ for μ and P for π) in making the estimate.

In the planning stage of a sampling study, before the data are collected, we can speak of $\bar{X}$ and P only in terms of probability. Their values are yet to be determined and will depend upon which particular elementary units happen to be randomly selected from the population. Thus, before the sample results are obtained, $\bar{X}$ and P are random variables. To determine the amount of chance sampling error and to qualify the accuracy of the resulting estimate, we must determine the probability distribution of the statistic used.

We emphasize that $\bar{X}$ and P are viewed as random variables at one stage and, at a later phase, as statistics whose value can be calculated from observed data. The probability distributions for $\bar{X}$ and P are of key importance. They are called *sampling distributions*, a special term distinguishing probabilities regarding $\bar{X}$ and P from those for other random variables not connected with sampling situations. The sampling distribution of $\bar{X}$ or P is determined in the same way as the probability distribution for any other random variable.

Sampling Distribution of Sample Mean and Proportion

To illustrate how we might find sampling distributions for $\bar{X}$ and P, consider the population of Law School Aptitude Test (LSAT) scores achieved by four hypothetical Slippery Rock seniors who have applied to Harvard that is

TABLE 5-12 Hypothetical Population of LSAT Scores

Name	*LSAT Score*
Bates	600
Jones	500
O'Hara	700
Smith	600

$$\mu = \frac{600+500+700+600}{4} = 600$$

$$\sigma = \sqrt{\frac{(600-600)^2+(500-600)^2+(700-600)^2+(600-600)^2}{4}}$$

$$= \sqrt{5{,}000} = 70.71$$

given in Table 5-12. To stretch our point a bit, suppose that the Harvard Law School will admit exactly two applicants from Slippery Rock and that on the basis of total credentials all four of these candidates are viewed equally well. The admissions committee will therefore draw straws, so that the scores of the lucky two will be a random sample from the population of four.

Table 5-13 shows the six sample results that are possible and the corresponding values of the sample mean and the proportion of LSAT scores greater than 600. From this information, we can determine the probabilities that the chosen sample will yield possible values for each statistic by finding the number of ways that each value can possibly occur. For example, there are two potential

TABLE 5-13 Possible Sample Results for Randomly Selecting Two LSAT Scores Without Replacement

Applicants Selected	*LSAT Scores*	*Sample Mean* ($\bar{x}$)	*Sample Proportion* (p)
Bates, Jones	600, 500	550	0
Bates, O'Hara	600, 700	650	1/2
Bates, Smith	600, 600	600	0
Jones, O'Hara	500, 700	600	1/2
Jones, Smith	500, 600	550	0
O'Hara, Smith	700, 600	650	1/2

samples where the mean LSAT score is 600, so the probability of this result is $2/6 = 1/3$. The sampling distributions for $\bar{X}$ and P are shown in Table 5-14.

Using the procedures outlined earlier in this chapter, we can calculate the expected value, variance, and standard deviation of $\bar{X}$ and P. As the remainder of this chapter will show, these values can be found directly from the population parameters μ, σ, and π, when these parameters are known.

TABLE 5-14 Sampling Distribution of $\bar{X}$ and P for LSAT Scores Selected Without Replacement

$\bar{x}$	$P[\bar{X}=\bar{x}]$	$\bar{x}P[\bar{X}=\bar{x}]$	p	$P[P=p]$	$pP[P=p]$
550	1/3	550/3	0	1/2	0
600	1/3	600/3	1/2	1/2	1/4
650	1/3	650/3		1	$E(P) = 1/4$
	1	$E(\bar{X}) = 600$			

The basic concepts underlying the sampling distribution of $\bar{X}$ are shown in Figure 5-10. This distribution may be viewed as roughly analogous to the results that would be obtained if many different samples were taken and the sample mean $\bar{X}$ were calculated each time.

IMAGINARY EXPERIMENT

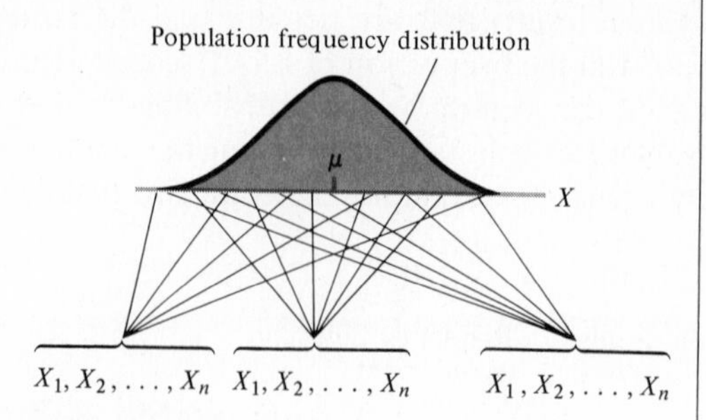

Many samples of size n are taken from the same population. Means are calculated from each sample and grouped into a frequency distribution.

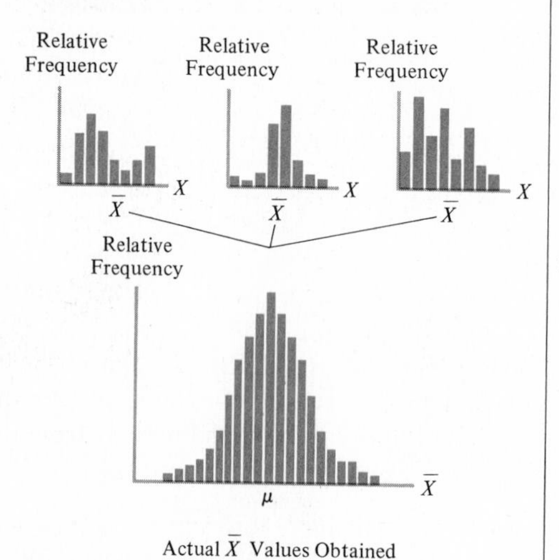

Actual $\bar{X}$ Values Obtained

THEORETICAL COUNTERPART

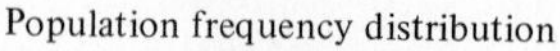

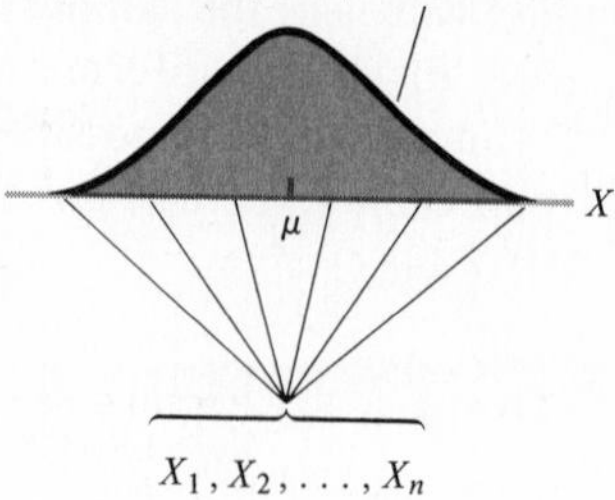

A single sample of size n is *to be* taken from the population, and then $\bar{X}$ is to be calculated.

The sampling distribution of possible $\bar{X}$ values is determined, using only the principles of probability theory.

Relative Frequency

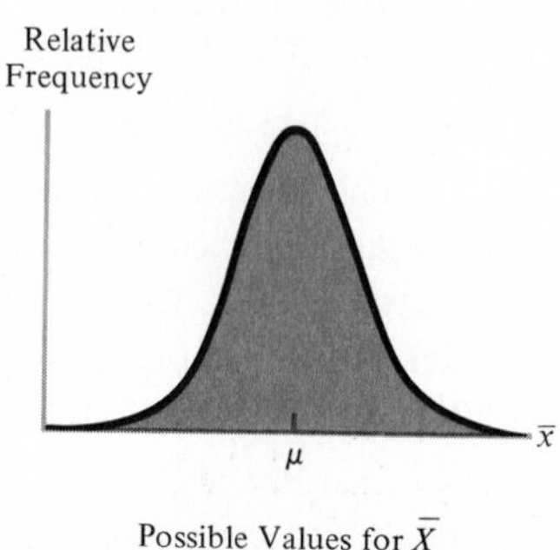

Possible Values for $\bar{X}$

FIGURE 5-10 Interpretation of a theoretically derived sampling distribution. The procedure outlined at the left is never necessary but shows the concepts underlying the theory. The histogram at bottom left is an experimental representation of the sampling distribution at bottom right.

Expected Value of the Sample Mean

The mean of the population of the four LSAT scores in our present illustration is $\mu = 600$, while the population proportion of scores greater than 600 is $\pi = .25$. Unlike this example, the full set of data needed to calculate population parameters is ordinarily unavailable, so the values of μ and π are normally estimated from the partial sample results instead. Usually the sample mean $\bar{X}$

and the sample proportion P are calculated and used for this purpose. We will consider how well these estimates perform in Chapter 7. In laying the necessary groundwork, we will now examine two further properties of the sampling distribution: its mean, or expected value, and its standard deviation.

The expected value of $\bar{X}$ is calculated in Table 5-14. Notice that $E(\bar{X}) = 600$, which is equal to the population mean μ. In general, this will be true for any random sampling situation, so we may conclude the following relation for the ***expected value of*** $\bar{X}$

$$E(\bar{X}) = \mu \tag{5-18}$$

This fact is quite plausible. In effect, it says that the long-run average value of sample means is the same as the mean of the population from which the sample observations are taken.

The rationale for this is simple. If $\bar{X}$ is calculated over and over again from different samples taken from the same population, then the successive sample means will tend to cluster about the mean of the population. If a sample of 100 men is taken from a population where the mean height is $\mu = 5'10''$, then the sample mean $\bar{X}$ is expected to be $5'10''$ as well. This does not imply that $\bar{X}$ cannot turn out to be $5'9\frac{1}{2}''$ or $5'10\frac{1}{2}''$, but *on the average* $\bar{X}$ will equal $5'10''$.

The above relationship can be formally justified by looking at the individual sample observations themselves as random variables, each with its own probability distribution. Every successive observation may be represented symbolically by X_1 for the first, X_2 for the second, and so forth. The sample mean is the sum of these observations divided by the sample size n. Multiplying the sample mean by the sample size, we obtain:

$$n\bar{X} = X_1 + X_2 + \cdots + X_n \tag{5-19}$$

The expected value of $n\bar{X}$ is therefore the same as the expected value of the sum of the individual random variables. It can be shown that the expected value of the sum will always be equal to the sum of the individual expected values:

$$E(n\bar{X}) = E(X_1 + X_2 + \cdots + X_n) = E(X_1) + E(X_2) + \cdots + E(X_n) \tag{5-20}$$

Since any single sample observation has an expected value equal to the population mean, the right-hand side of expression (5-20) is equal to $n\mu$. Therefore $E(\bar{X}) = \mu$, since $E(n\bar{X}) = nE(\bar{X})$ and the n's cancel.

Standard Deviation of Sample Mean

The standard deviation of the sample mean may be found in a similar manner. In determining it, however, we again encounter the distinction mentioned in Chapter 4 (page 127) regarding sampling with or without replacement:

sample outcomes are statistically independent when sampling with replacement and statistically dependent when sampling without replacement. As we will see, whether or not sample outcomes are independent will affect the value of the standard deviation for $\bar{X}$.

TABLE 5-15 Possible Sample Results for Randomly Selecting Two LSAT Scores with Replacement

Applicants Selected	*LSAT Scores*	*Sample Mean* ($\bar{x}$)	*Sample Proportion* (p)
Bates, Bates	600, 600	600	0
Bates, Jones	600, 500	550	0
Bates, O'Hara	600, 700	650	1/2
Bates, Smith	600, 600	600	0
Jones, Bates	500, 600	550	0
Jones, Jones	500, 500	500	0
Jones, O'Hara	500, 700	600	1/2
Jones, Smith	500, 600	550	0
O'Hara, Bates	700, 600	650	1/2
O'Hara, Jones	700, 500	600	1/2
O'Hara, O'Hara	700, 700	700	1
O'Hara, Smith	700, 600	650	1/2
Smith, Bates	600, 600	600	0
Smith, Jones	600, 500	550	0
Smith, O'Hara	600, 700	650	1/2
Smith, Smith	600, 600	600	0

Table 5-15 shows the possible sample results for randomly selecting two of the 4 Slippery Rock LSAT scores when sampling with replacement. Notice that there are $4 \times 4 = 16$ equally likely outcomes (because different orders of selection are now counted) and that the same name can be selected both times. Although this may seem peculiar, we should keep in mind that sampling is usually not done with replacement.

The sampling distributions of P and $\bar{X}$ are summarized in Table 5-16.

TABLE 5-16 Sampling Distributions of $\bar{X}$ and P for LSAT Scores Selected with Replacement

$\bar{x}$	$P[\bar{X}=\bar{x}]$	p	$P[P=p]$
500	1/16	0	9/16
550	4/16	1/2	6/16
600	6/16	1	1/16
650	4/16		1
700	1/16		
	1		

The expected value and the standard deviation for $\bar{X}$ are calculated in Table 5-17. Notice that, as earlier, $E(\bar{X}) = 600$. The standard deviation of $\bar{X}$ is $\sigma(\bar{X}) = 50$. In many statistical applications, the standard deviation of a statistic is referred to as its *standard error*. Because we use it so often, the standard error of $\bar{X}$ is represented by the special symbol $\sigma_{\bar{X}}$. Here, $\sigma_{\bar{X}} = 50$.

TABLE 5-17 Calculation of Mean and Standard Deviation for $\bar{X}$ for LSAT Scores Selected with Replacement

$\bar{x}$	$P[\bar{X}=\bar{x}]$	$\bar{x}P[\bar{X}=\bar{x}]$	$\bar{x}^2$	$\bar{x}^2P[\bar{X}=\bar{x}]$
500	1/16	500/16	250,000	250,000/16
550	4/16	2,200/16	302,500	1,210,000/16
600	6/16	3,600/16	360,000	2,160,000/16
650	4/16	2,600/16	422,500	1,690,000/16
700	1/16	700/16	490,000	490,000/16
		$E(\bar{X}) = 600$		362,500

$$\sigma^2(\bar{X}) = 362{,}500 - (600)^2 = 2{,}500$$

$$\sigma_{\bar{X}} = \sigma(\bar{X}) = \sqrt{2{,}500} = 50$$

We have found that $E(\bar{X}) = \mu$. It is interesting to see if $\sigma_{\bar{X}}$ similarly relates to the population standard deviation. We may use the following expression to calculate the ***standard error of $\bar{X}$ when sampling with replacement***

$$\sigma_{\bar{X}} = \frac{\sigma}{\sqrt{n}} \tag{5-21}$$

This says that the standard error of $\bar{X}$ relates directly to the population standard deviation σ and is inversely proportional to the square root of the sample size. Thus, both population variability and sample size influence the variability in possible values of $\bar{X}$. In Table 5-12, we found that the population of LSAT scores had a standard deviation of $\sigma = 70.71$. In this illustration, $n = 2$, so that

$$\sigma_{\bar{X}} = \frac{70.71}{\sqrt{2}} = \frac{70.71}{1.414} = 50$$

which is the same value for $\sigma_{\bar{X}}$ found in Table 5-17.

To explain why this relation holds, we first note that variability is additive, so that *in sampling with replacement*

$$\sigma^2(n\bar{X}) = \sigma^2(X_1+X_2+\cdots+X_n) = \sigma^2(X_1) + \sigma^2(X_2) + \cdots + \sigma^2(X_n) \tag{5-22}$$

This says that the variance of the two selected test scores combined is equal to the sum of the individual score variances, or twice the variance of a single score. There is no canceling effect. Looked at another way, the range also measures variability. Each of the selected LSAT scores may range from 500 (Jones) to 700 (O'Hara), and the sum of any two scores can range from 1,000 (Jones, Jones) to 1,400 (O'Hara, O'Hara)—an interval twice as wide. Here we see that the variability of the sum of *two* variables is *twice* as large as the variability of one variable alone.

Expression 5-22 applies for random variables where the outcomes for any one variable have probabilities that are not influenced by the particular results for the other variables. For example, in the popular casino game of "craps," two dice are tossed and the sum of both dice determines the gambler's reward. The variance of the sum of the showing faces is the sum of the individual variances (found on page 164 to be 2.917), or 2(2.917) = 5.834. In the parlor game Yahtzee, five dice are tossed, so the variance of their sum would be 5 times 2.917. In both cases, this is because the individual die outcomes are independent events. The respective numerical results are therefore referred to as *independent random variables.* Likewise, whenever random sampling is done with replacement, the results are referred to as an *independent random sample.*

Since $\sigma^2(n\bar{X}) = n^2\sigma^2(\bar{X})$ by Property 5-2 (page 165), it follows that the variance of $\bar{X}$ is the sum of the individual observation variances divided by n^2. Because each of the n observations has the same variance σ^2 as the population, the following is true:

$$\sigma^2(\bar{X}) = \frac{n\sigma^2}{n^2} \tag{5-23}$$

Canceling the n's and noting that $\sigma^2_{\bar{X}} = \sigma^2(\bar{X})$, we obtain:

$$\sigma^2_{\bar{X}} = \frac{\sigma^2}{n} \tag{5-24}$$

Taking the square roots of expression (5-24) gives us expression (5-21).

Influence of Sampling Without Replacement

When sampling without replacement, the sample outcomes are not independent. Independence between observations means the same thing as independence between events. For example, suppose that the heights of all adult American males comprise the population of interest. Then, if the first man randomly selected to be in the sample happens to be over seven feet tall, independence requires that this fact will not influence the probability distribution for the second man's height, and so forth for all men selected. However, if the selection is made without replacement, choosing a seven-footer first will have an influence on the second man's height probability. In such cases, the determination of $\sigma_{\bar{X}}$ may become more involved than we have indicated here.

The difficulty arises because, without replacement, variability will no longer be additive and the variance of the sum will be smaller than $n^2\sigma^2$, so that $\sigma_{\bar{X}}$ will be less than $\sigma/\sqrt{n}$ due to a canceling effect. This is because the early selection of an extremely small or large value from a population will reduce the variability of the remaining selections, while the early choice of a middling value will make later observations more varied than they normally would be.*

No practical difficulty exists unless the population is small. For instance, if our population contains 1,000 men, exactly one of whom is over seven feet tall, then independence is violated when sampling without replacement. For a very large population, the degree of dependence will be so slight that it can be ignored. For the United States as a whole, the removal of one seven-footer from the sample would not appreciably change the remaining proportion of persons who are this tall. But in a small population, this could seriously affect the sampling outcome.

In Chapter 6, we will investigate the practical significance of this fact upon sampling without replacement. Independence or dependence between random variables is a very important area of statistics; it will be investigated in later chapters.

Mean and Standard Error of *P*

For simple random samples, it is true that the expected value of the sample proportion is the same as the population proportion, or $E(P) = \pi$, which says that the long-run average value of P is equal to π. For the LSAT scores in our illustration, the proportion greater than 600 is $\pi = 1/4$, which is equal to $E(P)$ calculated in Table 5-14.

When sampling with replacement, the sampling distribution of P is the binomial for the population proportion π (which also represents the trial success probability) and the sample size n which represents the number of trials. As we saw earlier, the standard deviation for P may be calculated from π and n alone. Again, we will sometimes refer to this as the standard error of P, which we denote by σ_P. We may use the following expression to calculate the ***standard error of P when sampling with replacement***

$$\sigma_P = \sqrt{\frac{\pi(1-\pi)}{n}} \tag{5-25}$$

When the random sample is a dependent one due to sampling without replacement, expression (5-25) overstates the value for σ_P. As with the standard error for $\bar{X}$, we will further investigate this in Chapter 6.†

* These same principles also apply between sample observations representing two distinct populations and between dependent random variables. The difference between the variance of the sum and the sum of the variances may be summarized in terms of the *covariance*, which is half this difference and which may be positive or negative.

† The sampling distribution of P then becomes the hypergeometric distribution, which is similar to and is often approximated by the binomial.

EXERCISES

5-23 A sample of size 2 is to be randomly selected *without replacement* from five persons having the following IQ's:

Identity	*IQ*
Mr. A	100
Mrs. D	120
Mr. J	120
Ms. P	100
Miss R	90

(a) Find the sampling distribution of the sample mean.
(b) Find the sampling distribution of the sample proportion of females.

5-24 Suppose instead that the sample in Exercise 5-23 was chosen *with replacement.*
(a) Find the sampling distribution of the sample mean.
(b) Calculate the expected value, variance, and standard deviation of the sample mean.

REVIEW EXERCISES

5-25 You are offered the following proposition. A coin is tossed twice. For each tail you must forfeit one dollar, and for each head you will receive two dollars. Determine the probability distribution for your net winnings W. (*Hint:* It will help if you determine which outcomes [for example, (H, T)] apply to each possible amount.)

5-26 During the early 1950s, the card game of canasta was popular throughout the world. In canasta, points are assigned to cards in the following manner: red 3 = 100 points; joker = 50 points; ace or deuce = 20 points; 8 through king = 10 points; black 3s and 4 through 7 = 5 points. A canasta deck is comprised of two ordinary decks of playing cards, each containing 52 cards and 2 jokers. For the first card dealt from a shuffled canasta deck, determine the probability distribution for its corresponding points.

5-27 Calculate the mean, variance, and standard deviation for your net winnings W in Exercise 5-25, using your probability distribution from that exercise.

5-28 Calculate the mean, variance, and standard deviation for the points of the first card dealt from the canasta deck in Exercise 5-26, using your probability distribution from that exercise.

5-29 Suppose that the proportion of adult U.S. citizens who approve of the way the President does his job is $\pi = .50$. A random sample of $n = 5$ persons is chosen. Using the binomial formula, determine the probability that: (a) exactly 5 approve; (b) none approve; (c) exactly 3 approve.

5-30 Find the standard deviation for the proportion of successes P found in n Bernoulli trials for the following situations:
(a) $n = 100, \pi = .8$ (b) $n = 25, \pi = .5$ (c) $n = 100, \pi = .10$.

5-31 Forty percent of the voters in a city are black. A jury of 12 persons has been impaneled for a trial. Assuming that every registered voter has an equal chance of being chosen, use Appendix Table C to find the following probabilities for the representation of blacks on the jury:

(a) At least 5 black jurors.
(b) No black jurors.
(c) All black jurors.
(d) At most 10 black jurors.

5-32 The ages of the straight-A students in a certain statistics class are:

Mr. C	24
Miss H	19
Mr. J	21
Mr. M	20
Mrs. T	23

(a) A random sample of two persons is taken *without replacement*. Determine the sampling distribution of the mean age $\bar{X}$, and then determine the value of the standard error of $\bar{X}$.
(b) Calculate the standard deviation of the *population* of straight-A students. Using this value, determine the standard error of $\bar{X}$ if the random sample of two persons in (a) had been taken *with replacement*.

6 The Normal Distribution

I know of scarcely anything so apt to impress the imagination as the wonderful form of cosmic order expressed by the "Law of Frequency of Error." The Law would have been personified by the Greeks and deified, if they had known of it. It reigns with serenity and in complete self-effacement amidst the wildest confusion. The hugher the mob and the greater the apparent anarchy, the more perfect is its sway. It is the supreme law of Unreason. Whenever a large sample of chaotic elements are taken in hand and marshaled in the order of their magnitude, an unsuspected and most beautiful form of regularity proves to have been latent all along. The tops of the marshaled row form a flowing curve of invariable proportions; and each element, as it is sorted into place, finds, as it were, a preordained niche accurately adapted to fit it. If the measurement at any two specified Grades in the row are known, those that will be found at every other Grade, except towards the extreme ends, can be predicted in the way already explained and with much precision.

Sir Francis Galton

The normal distribution is perhaps the most important distribution encountered in statistical applications. One very good reason for this is that so many physical measurements and natural phenomena have actual observed frequency distributions that closely resemble the normal distribution. These include not only distributions for the physical measurements of height and weight of both persons and things, but also other human characteristics, such as IQ. Relative frequencies of these and many more populations closely resemble the normal curve in Figure 6-1 when their histograms are represented graphically

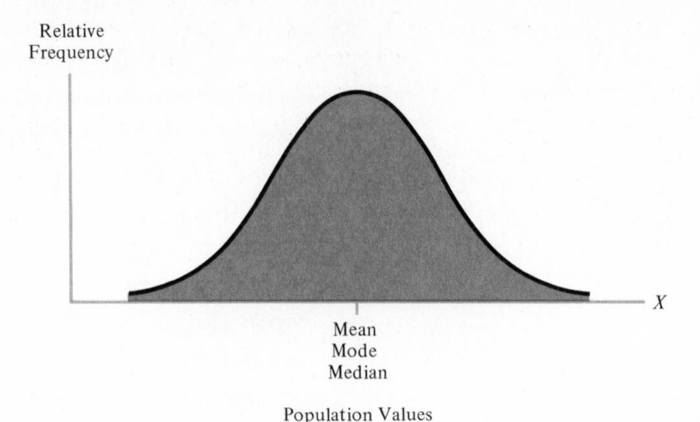

FIGURE 6-1 Frequency curve for the normal distribution.

by a smoothed curve. But there is a more fundamental reason why the normal distribution is so important to statistics. A theoretical property of the sample mean allows us to use it to find probabilities for various sample results. Thus, the normal curve has a basic role to play in situations where inferences are made regarding the value of the population mean when only the sample mean can be calculated directly.

6-1 CHARACTERISTICS OF THE NORMAL DISTRIBUTION

Several interesting features are apparent from Figure 6-1. Notice that the curve is shaped very much like a bell with a single peak, making it *unimodal*. It is *symmetrical* about its center. A normally distributed population's mean lies at the center of its normal curve. Because of symmetry, the *median* and the *mode* of the distribution also occur at the curve's center, so that the mean, median, and mode all have the same value. Although it is impracticable to show this on the graph, the tails extend indefinitely in both directions, never quite touching the horizontal axis.

We say that a population having a frequency distribution approximated by the shape of the normal curve is *normally distributed*. If the random variable X is the value of an elementary unit chosen randomly from a normally distributed population, then we also say that X is normally distributed. Furthermore, the same curve serves as a representation of the probability density function for X.*

* The function denoting the height of the curve is

$$f(x) = \frac{1}{\sqrt{2\pi\sigma^2}} e^{-[(x-\mu)^2/2\sigma^2]}$$

where π is the ratio of the circumference to the diameter of a circle (3.1416), and e is the base of natural logarithms (2.7183).

The normal curve depends upon only two parameters: the population mean μ and the standard deviation σ. No matter what the values of μ and σ, the total area under the normal curve is always 1.

The mean, as we have seen, is a measure of central tendency or location. When normal curves for populations having different means are graphed on the same axes, they will therefore be located at different positions along the horizontal axis. Figure 6-2 illustrates this for three different populations with means of 30, 80, and 120.

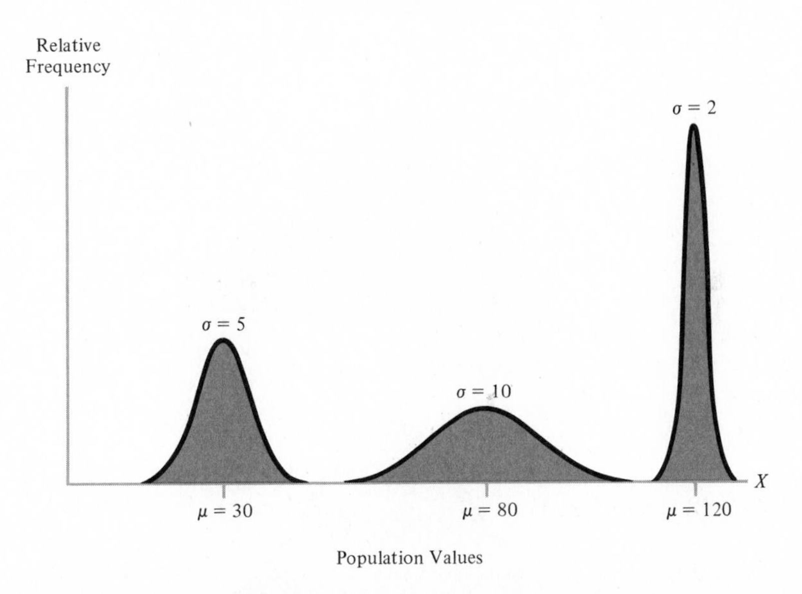

FIGURE 6-2 Three different normal distributions graphed on a common axis.

Figure 6-2 also illustrates that the shape of a normal curve is determined by the population's standard deviation. Distributions with small standard deviations have narrow, peaked "bells," and those with large σ's have flatter curves with less pronounced peaks. The three populations shown in Figure 6-2 have standard deviations of 5, 10, and 2, respectively. A very large class of populations belong to the normal family, and each member differs only by its mean and its standard deviation.

The probability that a normally distributed random variable assumes values within a particular interval equals the area of the portion of the curve covering that interval. The areas can be found by referring to a table, but before we describe how to use this table, note the very useful property of any normal distribution that is illustrated in Figure 6-3. *The area under the normal curve covering an interval symmetrical about the mean depends only upon the distance, measured in standard deviations, that separates the end points from the mean.* For instance, about 68 percent of the population has values lying within one standard

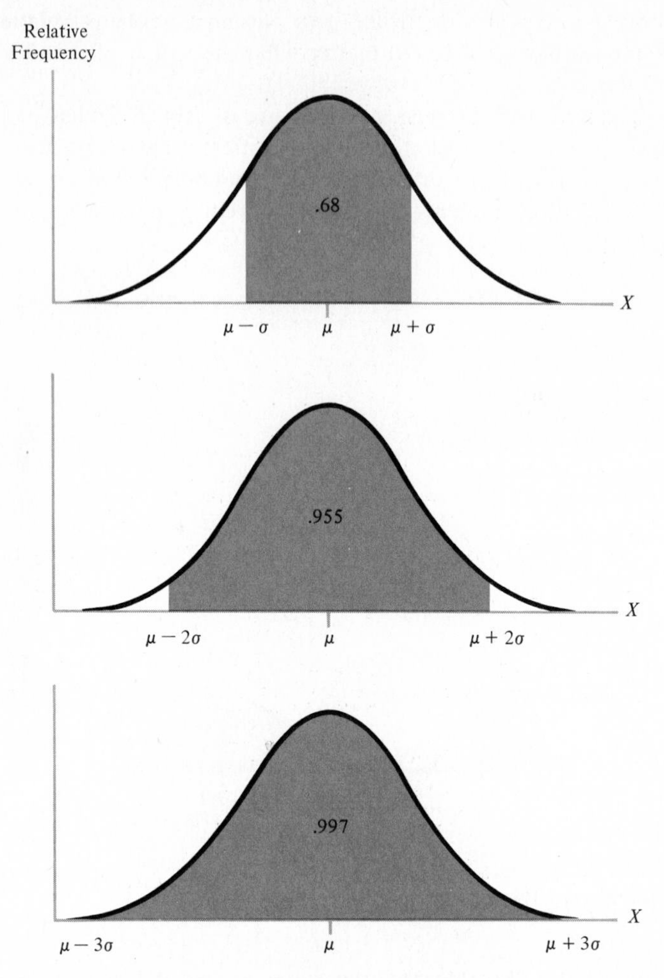

FIGURE 6-3 Relationship between the area under the normal curve and the distance from the mean, expressed in standard deviation units.

deviation in either direction of the mean; that is, the area under the curve over the interval $\mu - \sigma$, $\mu + \sigma$ is .68. This is true no matter what the values of μ and σ happen to be. It is also true that about 95.5 percent of the population has values lying within two standard deviations of the mean, and approximately 99.7 percent fall within three standard deviations.

Example 6-1 The commander of an Army division has wagered with his brother-in-law, who commands a Marine division, that his Army troops are taller. To verify this, the Army commander orders his aide to compile the heights of his soldiers from their medical records. The aide calculates the mean height to be 70 inches and finds that the standard deviation is 2 inches. He also constructs a histogram, which exhibits an almost perfect bell shape. Thus, the aide believes that the heights of the Army soldiers may be described in terms of the normal distribution.

The Marine commander arrives at similar conclusions. The mean height of his men is 69.5 inches, with a standard deviation of 2.5 inches. The two populations are compared in Table 6-1. Figure 6-4 shows the frequency curves for these two populations.

TABLE 6-1 A Summary of Normally Distributed Heights from Two Populations

	Height (inches)	
	Army	*Marine*
Mean (μ)	70	69.5
Standard deviation (σ)	2	2.5
68 percent lie between $\mu - \sigma$, $\mu + \sigma$	68, 72	67, 72
95.5 percent lie between $\mu - 2\sigma$, $\mu + 2\sigma$	66, 74	64.5, 74.5
99.7 percent lie between $\mu - 3\sigma$, $\mu + 3\sigma$	64, 76	62.5, 77

The Army commander claims that he wins the bet, as the mean height of his men is .5 inches greater than the mean Marine height. The Marine commander objects, noting that his division contains a greater percentage of very tall men. (Notice that the upper tail of the Marines' frequency curve lies above the Army division's curve.) The Army commander agrees, but retorts that there is also a higher percentage of shorter Marines. (Observe the lower tails of the two frequency curves.)

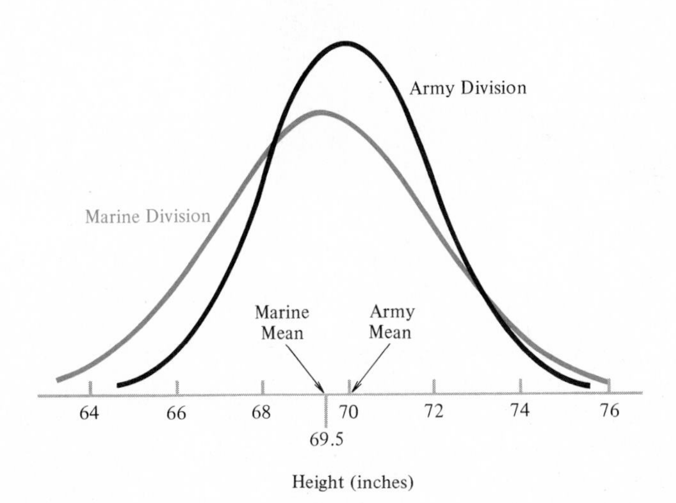

FIGURE 6-4 Normal curves for the heights of the Marine and Army divisions.

This example illustrates a difficulty in comparing populations by using a measure of location, such as the mean. The fact that some Marines are taller than the tallest soldiers is due to the difference in population variabilities. The

Marine division is a more disperse group with a larger standard deviation (2.5 inches compared to 2.0 inches for the Army division).

6-2 FINDING AREAS UNDER THE NORMAL CURVE

In order to obtain probability values for normally distributed random variables, we must first find the appropriate area lying under the normal curve. This can be accomplished by using Appendix Table D (page 564).

We will illustrate how to find the desired areas for the time taken by a particular typesetter to compose one page of standard type. We will assume that the population of times is normally distributed, with a mean of $\mu = 150$ minutes and a standard deviation of $\sigma = 30$ minutes. The time for any given page, such as the next one to be composed, represents a randomly chosen time from this population.

The probability that between 150 and 175 minutes will be required is represented by the colored area under the normal curve in Figure 6-5. As we

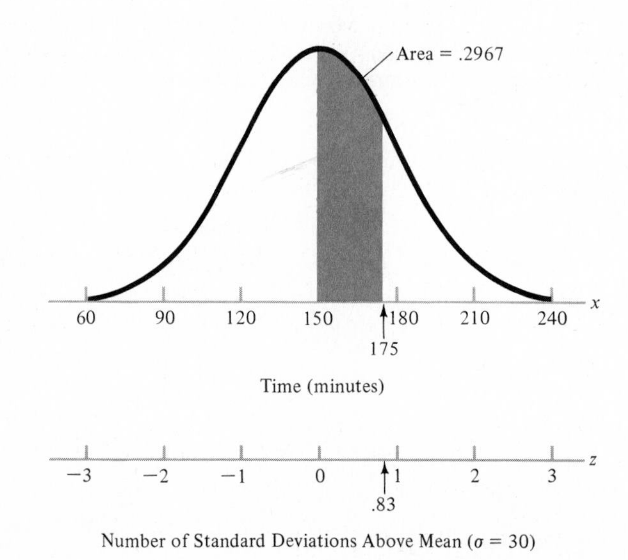

FIGURE 6-5 Determining the area under a normal curve.

have seen, the area beneath the normal curve between the mean and a certain point depends only upon the number of standard deviations separating the two points. We see that 175 minutes is equivalent to a distance above the mean of .83 standard deviation. This figure is determined by observing that 175 minutes minus the mean of 150 minutes is equal to 25 minutes. Since the standard deviation is 30 minutes, 25 minutes is only a fraction, 25/30 = .83, of the standard deviation.

Appendix Table D has been constructed for a special curve, called the *standard normal curve*, which provides the area between the mean and a point above the mean some specified distance measured in standard deviations. Because the distance will vary with the situation, it is treated as a variable and denoted by the letter z. Sometimes the value of z is referred to as a *normal deviate*. The distance z that separates a possible normal random variable value x from its mean may be determined from the following expression for the ***normal deviate***

$$z = \frac{x - \mu}{\sigma} \tag{6-1}$$

A negative value will be obtained for z when x is smaller than μ.

The first column of Table D lists values of z to a single decimal place. The second decimal place value is located at the head of one of the remaining 10 columns. The area under the curve between the mean and z standard deviations is found at the intersection of the correct row and column. For example, when $z = .83$, we find the area of .2967 by reading the entry in the .8 row and the .03 column. The area under the normal curve for a completion time between 150 and 175 minutes is thus .2967, which represents the probability that the next 500 lines of print will take this long to set.

Using Normal Curve Table

Appendix Table D only provides areas between the mean and some point above it. But this table may be employed to find areas encountered in other common probability situations, such as those in Figure 6-6. Each of these is described below.

(a) ***Area between mean and some point lying below it.*** To find the probability that the completion time lies between 125 and 150 minutes, first we calculate the normal deviate from expression (6-1):

$$z = \frac{125 - 150}{30} = -.83$$

Here, z is negative because 125 is a point lying below the mean. Since the normal curve is symmetrical about the mean, this area must be the same as it would be for a positive value of z of the same magnitude (in this case .2967, as before). It is therefore unnecessary to tabulate areas for negative values of z. The area between the mean and a point lying below it will equal the area between the mean and a point equally distant above it.

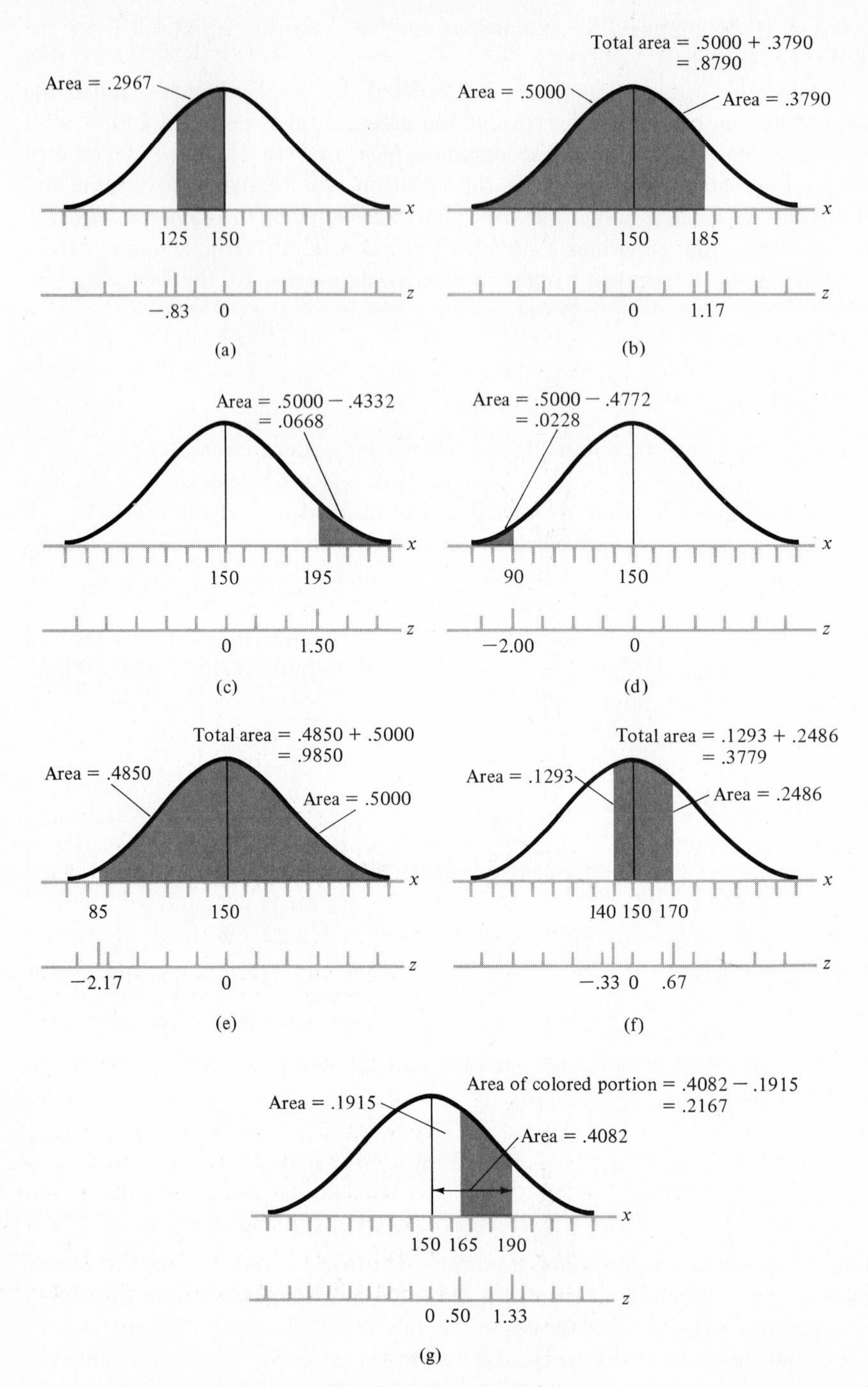

FIGURE 6-6 Various areas under the normal curve. x = completion time in minutes; z = standard deviations.

(b) *Area to the left of a value above the mean.* To find the probability that the type can be set in 185 minutes or less, we have to find the entire colored area below 185. Here, we must consider the lower half of the normal curve separately. Since the entire normal curve area is 1, the area under the half to the left of 150 must be .5. The area between 150 and 185 is found from Table D, with $z = (185-150)/30 = 1.17$, to be .3790. The entire colored portion is the sum of the two areas, or $.5000+.3790 = .8790$.

(c) *Area in upper tail.* We find the probability that the number of minutes required exceeds 195 by first finding the area between the mean and 195. The normal deviate is $z = (195-150)/30 = 1.50$, for which Table D yields the area .4332. Since the area in the upper half of the normal curve is .5, our desired area is found by subtracting the unwanted portion, or $.5000-.4332 = .0668$.

Because the total area under the normal curve is 1, we may use the value of the upper tail area to calculate the area to the left of 195. This may be found by subtracting the upper tail area of .0668 from 1, giving $1.0000-.0668 = .9332$ as the area to the left of 195 (or the probability that the time will be less than 195 minutes). Similarly, when the area to the left of a point is known, the area to its right can be found by subtracting this value from 1. For example, in (b) above, we found the area to the left of 185 to be .8790. Thus, the area in the upper tail to the right of 185 must be $1-.8790 = .1210$.

(d) *Area in lower tail.* We find the probability that 90 minutes or less will be taken to set the type by two steps similar to those in (c) above. First the area between 90 and 150 is found. Using $z = (90-150)/30 = -2.00$, we obtain .4772 from Table D. This value is subtracted from .5, yielding $.5000-.4772 = .0228$.

(e) *Area to the right of a value below the mean.* To find the probability that the completion time will be equal to or greater than 85 minutes, the area between 85 and the mean is added to the area to the right of the mean, which is .5. Here, we calculate $z = (85-150)/30 = -2.17$, so that, from Table D, the area is .4850. Adding this to .5, the combined area is $.5000+.4850 = .9850$.

From this value, the area in the lower tail below 85 may be found by subtracting .9850 from 1: $1-.9850 = .0150$. Likewise, we may find the area to the right of 90 by subtracting the lower-tail area found in (d) above from 1, or: $1-.0228 = .9772$.

(f) *Area under portion overlapping the mean.* In order to find the probability that the time taken will be between 140 and 170 minutes, we need only add the portion of the colored area that lies below the mean to the one above it. The respective normal deviate values are calculated to be $z = (140-150)/30 = -.33$ and $z = (170-150)/30 = .67$. From Table D, the lower area is .1293 and the upper area is .2486, so that the combined area is $.1293+.2486 = .3779$.

Suppose that we wish to find the probability that between 120 and 180 minutes will be taken. The normal deviates are $z = (120-150)/30 = -1$ and $z = (180-150)/30 = 1$. From Table D, .3413 is the same area for both sides, so that the combined area is $.3413+.3413 = .6826$. Because z expresses the number of standard deviation units from the mean, we see that .6826 is a more precise value for the area between $\mu \pm \sigma$ than the value used in Figure 6-3.

(g) *Area between two values lying above or below the mean.* To find the probability that the composition time lies between 165 and 190 minutes, first we have to determine the areas between the mean and each of these two values. The respective normal deviates are therefore $z = (165-150)/30 = .50$ and $z = (190-150)/30 = 1.33$. From Table D, we see that the area between the mean and 190 is .4082, while the area between the mean and 165 is only .1915. Thus, the colored area is found by subtracting the smaller area from the larger one, or: $.4082-.1915 = .2167$.

A similar procedure could be applied for an area lying below the mean. It is also possible to find the value for a complementary situation—that the time will be either below 165 minutes or greater than 190 minutes—by subtracting the colored area from 1, or : $1-.2167 = .7833$.

Inequalities and Normal Curve Areas

The normal curve represents values lying on a continuous scale, such as height, weight, and time. There is zero probability that a specific value, such as 129.40 minutes, will occur (there is zero area under the normal curve covering a single point). Thus, in finding probabilities, it does not matter whether we use "strict" inequalities, such as the composition time is "less than" ($<$) 129.40 minutes, or an ordinary inequality, such as the time is "less than or equal to" ($\leqslant$) 129.40 minutes. Using $z = (129.40-150)/30 = -.69$, the area is the same in either case: $.5000-.2549 = .2451$.

Cumulative Probabilities and Percentiles

Cases (b) and (d) in our discussion of normal curve areas show how all the values for a normal cumulative probability distribution may be determined. We have seen that the probability of a time of 185 minutes or less is .8790, while the probability of 90 minutes or less is .0228. For values above the mean, when z is positive, Table D directly provides the cumulative probabilities by simply adding .5 to the tabulated areas. For values below the mean, when z is negative, the area given by the table is subtracted from .5 to obtain the cumulative probability.

It is frequently necessary to find a normally distributed population's percentile values. Recall from Chapter 3 that a percentile is the population value below which a certain percentage of the population lies. There, we saw that a percentile or fractile value could be obtained directly from the cumulative relative frequency distribution.

To find a population percentile, we must read Table D in *reverse*, since the specified percentage represents an area under the normal curve. For example, the 90th percentile for the number of minutes taken is a particular number of minutes. The area under the normal curve to the left of this value will be .90. Thus, the area between the mean and the number of minutes to be found will be $.90-.50 = .40$.

Searching through the body of the table, we select the area that lies closest to this figure. The closest value is .3997. Since .3997 is in the $z = 1.2$ row and the

.08 column, the corresponding normal deviate is $z = 1.28$. This means that our desired time is 1.28 standard deviations or $30 \times 1.28 = 38.4$ minutes above the mean. Adding this to the mean, the 90th percentile is $150 + 38.4 = 188.4$ minutes.

In general, the procedure for finding a percentile is begun by reading Table D in reverse to find z. The corresponding population value x may then be calculated from the following expression, which provides the ***percentile for a normal population***

$$x = \mu + z\sigma \tag{6-2}$$

Note that below the 50th percentile the lower-tail area is used to find z, which in these cases will be negative.

Example 6-2 Human engineers (whose specialty is the problem of integrating human factors into the design of man–machine systems) are designing the cockpit of a new jet aircraft and want to arrange the positions of certain controls so that 95 percent of all pilots can reach them while seated. They want to arrange a mockup so that the reach of a small man can be used to compare design alternatives. In order to select the man, the required reach radius must be determined. This will involve finding the maximum reach radius exceeded by 95 percent but not by 5 percent of all pilots. Thus, the engineers must find the reach that corresponds to the fifth percentile.

The maximum reach radii of airline pilots are assumed to be approximately normal, with a mean of $\mu = 48$ inches and a standard deviation of $\sigma = 2$ inches. They seek the point below which the area under the normal curve is equal to .05. This means that Table D, must be searched to find the area closest to $.50 - .05 = .45$. It just so happens that two areas are equally close, .4495 and .4505, corresponding to normal deviates of 1.64 and 1.65. The desired figure actually lies somewhere in between 1.64 and 1.65. For simplicity, the engineers choose 1.64, as they ordinarily wish to err on the side of a larger rather than a smaller tail area.* Since the fifth percentile is below the 50th, they are dealing with a lower-tail area and their normal deviate is negative, so that $z = -1.64$. (The fifth percentile lies below the mean and is smaller, so that a distance of 1.64 standard deviations must be subtracted from the mean.) From expression (6-2), the fifth percentile is therefore:

$$x = 48 - 1.64(2) = 44.72 \text{ inches}$$

Standard Normal Random Variable

The area under any normal curve may be found by using the standard normal curve. This curve provides the probability distribution for the *standard normal random variable* Z. Here, our notation of using capital letters for random variables means that the lower case z represents a point on the scale of possible values for Z. Because z expresses distance in standard deviations above or below

* An interpolation procedure would be more accurate, but to keep our discussions less complicated, throughout this text we will always choose the nearest table value and break ties arbitrarily.

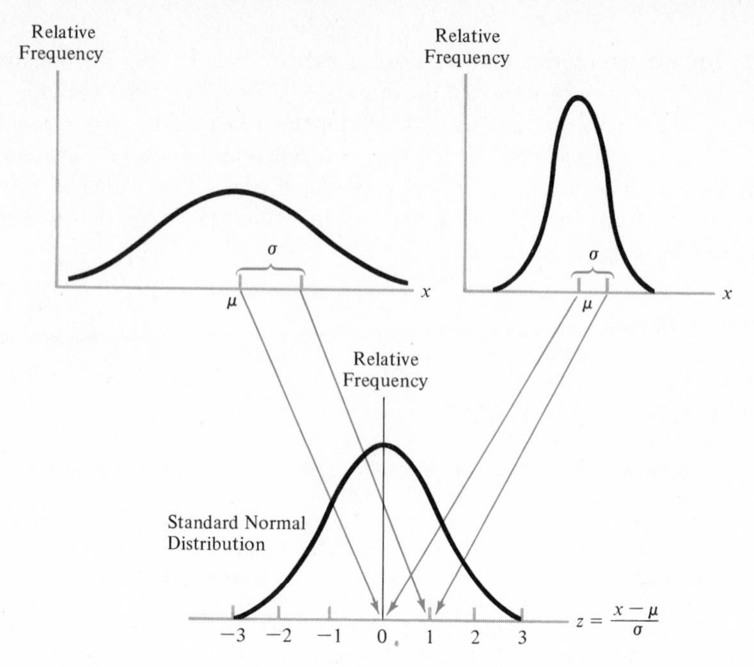

FIGURE 6-7 Illustration of the linear transformation of normal random variables into the standard normal distribution.

the mean, the center of the standard normal curve lies at zero, so that the expected value of Z is zero. Likewise, the standard deviation of Z is equal to 1.

In the type-composition time illustration, we in essence transformed the original random variable X, the time to complete 500 lines, into the standard normal random variable Z whenever we used Table D to find areas. Such a transformation may be accomplished physically by a procedure consisting of two parts: shifting the center of the curve and then stretching or contracting it. To shift the original curve so that its center lies above the point $x = 0$, subtract μ from each point on the x axis. Then the repositioned curve may be stretched or squeezed so that the scale on the horizontal axis matches the scale for the standard normal distribution. This may be accomplished by dividing all values of the random variable by its standard deviation. The resultant transformed curve will have the same shape as the standard normal curve illustrated in Figure 6-7. The net effect will always be the same, no matter what the values of μ and σ are. The horizontal scale may be either expanded or contracted.

Fortunately, a physical transformation of the original random variable X into the standard normal random variable is unnecessary. Instead, we may algebraically manipulate X itself, according to the following expression to make the ***transformation of X into the standard normal variable***

$$Z = \frac{X - \mu}{\sigma} \qquad (6\text{-}3)$$

Example 6-3 IQ tests are designed so that the scores achieved by a cross section of persons is normally distributed for all practical purposes. One of the most popular is the Stanford-Binet test, which has a mean of $\mu = 100$ points and a standard deviation of $\sigma = 16$. Letting X represent the IQ score achieved by a randomly chosen person, it is possible to determine various probabilities for the value achieved.

For example, we may find the probability that this person's IQ is at most 140:

$$P[X \leqslant 140] = P\left[\frac{X-\mu}{\sigma} \leqslant \frac{140-100}{16}\right]$$

Here we subtract $\mu = 100$ from both sides of the original inequality and then divide by $\sigma = 16$. This change transforms X into the standard normal random variable Z and 140 into the normal deviate value 2.50, so that our original problem is equivalent to evaluating $P[Z \leqslant 2.50]$.

From Appendix Table D, we know that the area between zero and $z = 2.50$ is .4938. Thus

$$P[X \leqslant 140] = P[Z \leqslant 2.50] = .5000 + .4938 = .9938$$

We see that there is a better than 99 percent chance that this person's IQ falls below 140 on the Stanford-Binet scale.

Different IQ tests have standard deviations other than the Stanford-Binet, although the means are generally the same: 100 points. Now suppose we ask the same question for another test with a standard deviation of $\sigma = 10$ points. In this case

$$P[X \leqslant 140] = P\left[\frac{X-\mu}{\sigma} \leqslant \frac{140-100}{10}\right] = P[Z \leqslant 4.0]$$

From Table D, the area between zero and $z = 4.0$ is .49997. Thus

$$P[X \leqslant 140] = .50000 + .49997 = .99997$$

Notice that IQ's above 140 are considerably rarer with this test than with the Stanford-Binet.

The fact that different types of IQ tests are scaled differently makes it hard to compare IQ scores. For this reason, various intelligence-test results are often transformed into *standard scores*—normal deviate values which can be calculated from expression (6-1). In this example, 140 points on the Stanford-Binet provides a standard score of 2.50, while 140 points on the other test corresponds to a standard score of 4.0, indicating that a person scoring 140 points on the second test would exhibit a much higher level of intelligence. (To obtain a standard score equal to 4.0 on the Stanford-Binet test, a person would have to be 4 standard deviations above the mean, so that his IQ would be $100+4(16) = 164$ on that scale.)

Areas for Large z

Suppose that σ was even smaller than the 10 used in the above example. z would then be so large in absolute value that Table D would not provide the area. Our table has to stop somewhere, for the tails of the normal curve extend indefinitely. In general, whenever z exceeds 3.09, the area between the mean and z is so very close to .5 that for practical purposes .5 would be used in such cases. The upper- and lower-tail areas would be so close to zero that they would be *negligible*. For example, the area below $z = 5$ would be approximately 1, as would the area above $z = -5$, while the areas above $z = 5$ and below $z = -5$ would each be approximately zero.

Concluding Remarks

Recall that the tails of the normal curve never touch the horizontal axis. This implies that there is a probability—although perhaps an infinitesimal one—that the random variable can exceed any value. Consider the distribution of heights of men, for instance. The normal curve quite literally assigns a probability to the event that a man will be born who will grow to be more than one mile tall. Most people would accept that this is impossible, but according to the normal curve the probability of this event is not zero—although it is quite remote. Worse yet, the normal curve says that there are probabilities of men being negatively tall. Absurd as these implications seem, they are merely by-products of employing a convenient mathematical expression to describe a particular curve which happens to fit some empirical frequency distributions very nicely, and they should not detract from the utility of the normal distribution. From the table of areas under the normal curve, only about .13 percent of the area lies beyond a distance of 3σ above the mean. Areas in the tails of the theoretical normal distribution are quite tiny for greater distances. (At 4σ, the tail area is only .003 percent.) Any discrepancy between reality and the theoretical normal curve occurring beyond four standard deviations can thus be safely ignored. (A mile-tall man would be more than 40,000 standard deviations from the mean.)

EXERCISES

6-1 The measurement errors for the height of a weather satellite above a ground station are normally distributed, with a mean of zero and a standard deviation of one mile. These errors will be negative if the measured altitude is too low and positive if the altitude is too high. Find the probability that for the next orbit the error will be

(a) Between zero and +1.55 miles.
(b) Between −2.45 and zero miles.
(c) +.75 miles or less.
(d) Greater than +.75 miles.
(e) −1.25 miles or less.
(f) Greater than −1.25 miles.
(g) Between +.10 and +.60 miles.
(h) Between +1 and +2 miles.

6-2 The lifetime of a particular model of a stereo cartridge is normally distributed, with a mean of $\mu = 1{,}000$ hours and a standard deviation of $\sigma = 100$ hours. Find the probability that one of these cartridges will last:

(a) Between 1,000 and 1,150 hours.
(b) Between 950 and 1,000 hours.
(c) 930 hours or less.
(d) More than 1,250 hours.
(e) 870 hours or less.
(f) Longer than 780 hours.
(g) Between 700 and 1,200 hours.
(h) Between 750 and 850 hours.

6-3 The time required by a nurse to inject a shot of penicillin has been observed to be normally distributed, with a mean of $\mu = 30$ seconds and a standard deviation of $\sigma = 10$ seconds. Find the following percentiles: (a) 10th; (b) 25th; (c) 50th; (d) 75th; (e) 90th.

6-4 The heights of the male students at a particular university were found to be normally distributed, with a mean of $\mu = 5'10''$ and a standard deviation of $\sigma = 2.5''$. Find the height below which the following percentages of men lie: (a) 1; (b) 5; (c) 30; (d) 60; (e) 95; (f) 99.

6-5 An architect is designing the interior doors in a men's gymnasium. He wants to make them high enough so that 95 percent of the men using the doors will have at least a one-foot clearance. Assuming that the heights will be normally distributed, with a mean of 70 inches and a standard deviation of 3 inches, how high must the architect make the doors?

6-6 The quality control manager shuts down an automatic lathe for corrective maintenance whenever a sample of parts it produces has an average diameter either greater than 2.01 inches or smaller than 1.99 inches. The lathe is designed to produce parts with a mean diameter of 2.00 inches, and the sample averages have a standard deviation of .005 inches. Using the normal distribution:

(a) What is the probability that the quality control manager will stop the process when the lathe is operating as designed, with $\mu = 2.00$ inches?

(b) Suppose a part within the lathe wears out and it begins to produce parts that on the average are too wide, with $\mu = 2.02$ inches. What is the probability that the lathe will continue to operate?

(c) Suppose that an adjustment error causes the lathe to produce parts that on the average are too narrow, so that $\mu = 1.99$ inches. What is the probability that the lathe will be stopped?

6-3 SAMPLING DISTRIBUTION OF THE SAMPLE MEAN FOR A NORMAL POPULATION

The normal curve describes the frequency distribution of a great many populations. Notable among these are measurements of persons and things. Some important questions can be answered by applying the normal curve. For instance, a new testing procedure may be analyzed by a counseling center seeking a replacement for their present test. A sample of persons can be given the new test in order to estimate the mean score to be achieved by all those who may ultimately take the test. Because scores obtained with the current test closely fit the normal distribution, there is reason to believe that the normal curve will apply to the new one as well.

We will show how the probability distribution for the sample mean can be determined. It has been established mathematically that under certain circumstances, when the population frequencies are described by the normal curve, *the sampling distribution of $\bar{X}$ itself is also normally distributed*.

Mean and Standard Deviation of $\bar{X}$

In order to find probabilities for the value of $\bar{X}$, we must determine which particular normal curve applies. Recall that the shape of the applicable normal curve is determined by the mean (expected value) and standard deviation.

Suppose that a random sample of size n is selected from a population with a mean of μ and a standard deviation of σ. In Chapter 5, we showed that the following applies for the ***expected value of the sample mean***

$$E(\bar{X}) = \mu \tag{6-4}$$

This fact is true regardless of the shape of the underlying population distribution.

The standard deviation of $\bar{X}$ depends partly upon the variability of the population, but it also depends partly upon the sample size n. A large sample should be more accurate than a small one; thus, when n is large, $\bar{X}$ values will tend to cluster more tightly about μ than they would if n were small. Both the standard deviation of the population and the size of the sample therefore affect the standard deviation of the sample mean, with large values of n making the sample mean smaller.

We have referred to the standard deviation of $\bar{X}$ as the *standard error of* $\bar{X}$. One reason for this is that the variability in possible sample mean values provides an indication of the accuracy of estimates to be made. It allows us to quantify chance-caused sampling errors. We may use the following expression to calculate the ***standard error (deviation) of*** $\bar{X}$

$$\sigma_{\bar{X}} = \frac{\sigma}{\sqrt{n}} \tag{6-5}$$

This result is true for *any* population whenever the sample observations are independent.

Knowing the values of μ and $\sigma_{\bar{X}}$, we can determine the appropriate normal curve to apply in finding the probability values for $\bar{X}$. The shape of this normal curve is determined by its standard deviation $\sigma_{\bar{X}}$, and its center lies at μ.

Finding Probabilities for $\bar{X}$

We may summarize the foregoing discussions by stating the following

PROPERTY If $\bar{X}$ is the mean of a random sample taken from a normally distributed population having a mean of μ and a standard deviation of σ, then its sampling distribution is also normal, its mean is also μ, and its standard deviation is $\sigma_{\bar{X}} = \sigma/\sqrt{n}$. This is true no matter what the size of the sample happens to be.*

As an illustration, suppose that $n = 100$ people are chosen as a random sample from a normally distributed population of IQ test scores having an unknown mean of μ and a standard deviation of $\sigma = 10$ points. The sample

* As we will see in Chapter 7, when σ is *also* of unknown value and n is small, it is necessary to use the Student t statistic instead of $\bar{X}$. This statistic has a different sampling distribution.

mean $\bar{X}$ is normally distributed with the same mean and has the standard deviation:

$$\sigma_{\bar{X}} = \frac{10}{\sqrt{100}} = \frac{10}{10} = 1$$

The probability may be found that the sample mean differs from the unknown population mean by a level of precision of no more than one point in either direction. We find that:

$$P[-1 \leqslant \bar{X} - \mu \leqslant +1] = P\left[\frac{-1}{1} \leqslant \frac{\bar{X}-\mu}{\sigma_{\bar{X}}} \leqslant \frac{+1}{1}\right]$$
$$= P[-1 \leqslant Z \leqslant +1] = .6826$$

Here, $\bar{X}$ was transformed in the usual manner into the ***standard normal random variable***

$$Z = \frac{\bar{X}-\mu}{\sigma_{\bar{X}}} \tag{6-6}$$

For areas under the normal curve, we find from Table D that the area between the mean and $z = 1$ is .3413. This gave us $2(.3413) = .6826$ for the above probability.

Role Played by the Standard Error of $\bar{X}$

$\sigma_{\bar{X}}$, the standard error of $\bar{X}$, equals the population standard deviation σ divided by the square root of the sample size n. Thus, $\sigma_{\bar{X}}$ will be smaller than σ whenever n exceeds 1. This simply reflects the fact that values for $\bar{X}$ are more alike than individual population values are. The normal curve for the population will therefore be flatter than the one we use to calculate probabilities for $\bar{X}$.

Since the standard error of the sample mean is inversely proportional to the square root of the sample size, the larger the sample is, the smaller $\sigma_{\bar{X}}$ will be. As $\sigma_{\bar{X}}$ becomes smaller, the possible values of $\bar{X}$ cluster more tightly about μ. This is reflected in the shape of the normal curve, which is more peaked when $\sigma_{\bar{X}}$ is small.

As an illustration, the normal curves for the possible values for the sample mean to be calculated from the IQ tests are shown in Figure 6-8 for two different sample sizes: $n = 100$ and $n = 400$. Although only one value of n will actually be used, we can compare the normal curves in order to see which would be more accurate. The curve for the larger sample size is much denser near μ. This means that when $n = 400$, it is more likely that $\bar{X}$ will turn out to be close to μ. The shaded portions represent the probability that $\bar{X}$ will lie within one point of μ.

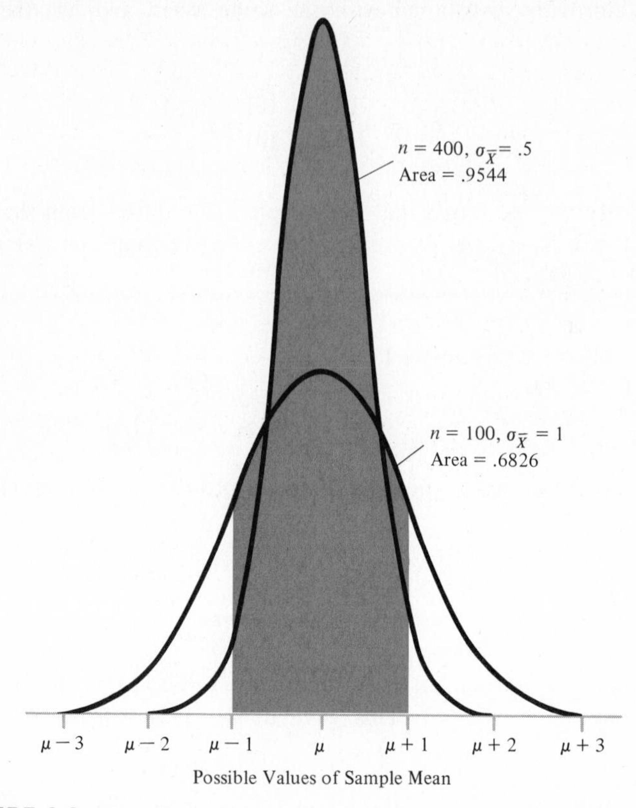

FIGURE 6-8 The effect of increased sample size upon the sampling distribution of $\overline{X}$.

For $n = 100$, we calculated this probability to be .6826. When $n = 400$, the standard error of $\overline{X}$ is

$$\sigma_{\overline{X}} = \frac{10}{\sqrt{400}} = .5$$

The analogous probability is then:

$$\begin{aligned} P[-1 \leqslant \overline{X} - \mu \leqslant +1] &= P\left[\frac{-1}{.5} \leqslant \frac{\overline{X} - \mu}{\sigma_{\overline{X}}} \leqslant \frac{+1}{.5}\right] \\ &= P[-2 \leqslant Z \leqslant +2] \\ &= 2(.4772) = .9544 \end{aligned}$$

We see that the area covering plus or minus one point from μ is larger for $n = 400$. This fact indicates that a larger sample size provides a more reliable estimate of μ for the same level of precision—plus or minus one point. As the larger sample size is reflected by a smaller value of $\sigma_{\bar{X}}$, we can conclude that the smaller the value of $\sigma_{\bar{X}}$, the more accurate the sample result will be.

The standard error of $\bar{X}$ is also affected by the population standard deviation σ. Thus, the size of $\sigma_{\bar{X}}$ depends upon how highly disperse the population values are. This means that the standard error of $\bar{X}$ is a single gauge of sample accuracy that incorporates the effects of both the size of the sample and the variability of the population itself.

We illustrate the effect of population dispersion on sample accuracy by comparing two samples, taken from different populations.

Example 6-4 The mean gasoline mileages of compact and intermediate cars of the same manufacturer are to be estimated by a consumer testing service. It will be assumed that the two car sizes represent different normally distributed populations, with standard deviations of 2 miles per gallon (mpg) for the compacts and 4 mpg for the intermediates. The mileages of intermediate cars are thus a more disperse population.

The standard errors of the sample mean for each type of car are calculated below, assuming that a separate random sample of size $n = 100$ is taken from each population.

$$\text{For compacts:} \qquad \sigma_{\bar{X}} = \frac{2}{\sqrt{100}} = .2$$

$$\text{For intermediates:} \quad \sigma_{\bar{X}} = \frac{4}{\sqrt{100}} = .4$$

Figure 6-9 illustrates the normal curves obtained for the respective sampling distributions of $\bar{X}$. The two curves are drawn on separate graphs to emphasize the fact that different populations are involved, with the compacts having the smaller mean. The colored areas provide the probabilities that each sample mean

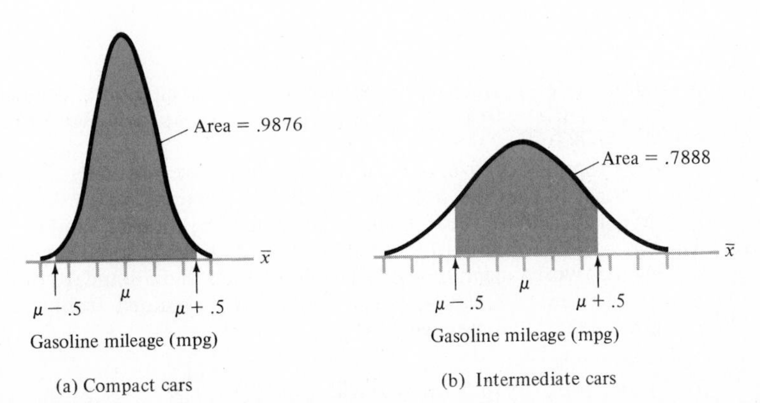

FIGURE 6-9 Sampling distributions for sample means taken from populations having different standard deviations.

will lie within $\pm .5$ mpg of the respective unknown population means. These are

$$P[-.5 \leqslant \bar{X}-\mu \leqslant +.5] = P\left[\frac{-.5}{.2} \leqslant \frac{\bar{X}-\mu}{\sigma_{\bar{X}}} \leqslant \frac{+.5}{.2}\right]$$
$$= P[-2.5 \leqslant Z \leqslant +2.5]$$
$$= 2(.4938) = .9876 \quad \text{for compacts}$$

and

$$P[-.5 \leqslant \bar{X}-\mu \leqslant +.5] = P\left[\frac{-.5}{.4} \leqslant \frac{\bar{X}-\mu}{\sigma_{\bar{X}}} \leqslant \frac{+.5}{.4}\right]$$
$$= P[-1.25 \leqslant Z+1.25]$$
$$= 2(.3944) = .7888 \quad \text{for intermediates}$$

A more reliable result is obtained for the compacts, whose normal curve for possible sample means provides a tighter clustering of $\bar{X}$ about μ. This is because the standard error of the sample mean is smaller for the compact cars.

Distinguishing s and $\sigma_{\bar{X}}$

The reader is cautioned against confusing $\sigma_{\bar{X}}$ with s. Both are standard deviations, but they represent entirely different phenomena—another reason why we usually call $\sigma_{\bar{X}}$ the standard error of $\bar{X}$. Recall that s is the standard deviation of a particular sample result and is actually calculated from the observed sample values at the final stage of sampling; s measures the variability between the observations actually made.

In contrast, $\sigma_{\bar{X}}$ measures the variability of the *possible* $\bar{X}$ values that *might be* obtained and is mainly used in the planning stage. Even though only one numerical value will be achieved for $\bar{X}$, depending upon the sample results, we have demonstrated the need to know the pattern of variation among the possible values that $\bar{X}$ might assume. It is this variation that is summarized by the standard error of the sample mean.

EXERCISES

6-7 A random sample of size n is selected from a normally distributed population. Find $\sigma_{\bar{X}}$ when $\sigma = 10$ for the following values of n: (a) 4; (b) 9; (c) 25; (d) 100; (e) 400; (f) 2,500; (g) 10,000.

6-8 Random samples of size 100 are selected from five normally distributed populations, having the following standard deviations σ: (a) .1; (b) 1.0; (c) 5.0; (d) 20; and (e) 100. Calculate the value of $\sigma_{\bar{X}}$ for each sample.

6-9 The times taken to administer blood-pressure tests at a hospital are normally distributed, with a mean time of $\mu = 1.50$ minutes and a standard deviation of $\sigma = .25$ minutes. A sample of $n = 25$ patients is obtained, from which the mean test time $\bar{X}$ will be calculated.

(a) Find the value of $\sigma_{\bar{X}}$.

Determine the probability that $\bar{X}$ will

(b) Lie between 1.40 and 1.50 minutes.

(c) Exceed 1.35 minutes.

(d) Fall at or below 1.45 minutes.
(e) Lie between 1.52 and 1.58 minutes.
(f) Fall above 1.65 minutes.

6-10 The city public health department closes down a certain beach when the concentration of *E. coli* bacteria, present in raw sewage, becomes too high. Assume that on a particular day the contamination-level index is normally distributed, with a mean of $\mu = 160$ and a standard deviation of $\sigma = 20$ for each liter of water. A sample of $n = 25$ liters is taken, and the mean index value $\bar{X}$ is found.
(a) Calculate $\sigma_{\bar{X}}$.

Determine the probability that $\bar{X}$ will
(b) Lie between 150 and 160.
(c) Exceed 148.
(d) Fall at or below 153.
(e) Lie between 165 and 170.
(f) Fall above 162.

6-11 The population of lifetimes of guinea pigs injected from birth with a carcinogenic substance is normally distributed, with a mean of $\mu = 800$ days and a standard deviation of $\sigma = 100$ days. A random sample of $n = 25$ animals is chosen.
(a) What is the probability that the sample mean will differ from the population mean by more than 20 days in either direction?
(b) What is the probability that $\bar{X}$ will fall below 740 days?
(c) Suppose that the dosage is increased and the population mean lifetime is undetermined. Assuming that the population standard deviation remains at $\sigma = 100$ days and that $\bar{X} = 700$, what is the probability that this value or less could have been obtained if (1) $\mu = 750$; (2) $\mu = 700$; (3) $\mu = 650$; or (4) $\mu = 600$?

6-4 SAMPLING DISTRIBUTION OF $\bar{X}$ WHEN POPULATION IS NOT NORMAL

Even when a sample is selected from a population whose frequency distribution is not normal, for large sample sizes the sampling distribution of the mean will still be approximately normal. Before discussing this result, called the *central limit theorem*, we will consider how the sample size affects the sampling distribution of $\bar{X}$. In doing this, we will use the two facts established in Section 6-3, which hold regardless of the population characteristics: (1) the expected value of $\bar{X}$ is always equal to the population mean μ; and (2) for independent observations, the standard error of $\bar{X}$ is equal to the population standard deviation divided by the square root of the sample size, or $\sigma_{\bar{X}} = \sigma/\sqrt{n}$.

Effect of the Sample Size n upon the Sampling Distribution of $\bar{X}$

Figure 6-10 illustrates the sampling distributions of $\bar{X}$ for samples of various sizes. Here, $\bar{X}$ represents the mean number of bedrooms in a sample of randomly selected homes in a certain city. The same frequency distribution in Table 6-2 was used to generate all the cases shown. Note that as the sample size increases, the number of possible values of $\bar{X}$ grows and the sampling distributions become more symmetrical about μ. Although the number of possible $\bar{X}$ values increases with n, the total variability declines, as evidenced by the tendency of the values farther from μ to decrease in probability while the number of spikes of significant height near μ grows. This tendency is reflected in the standard error of $\bar{X}$, which becomes smaller as n becomes larger. Even more

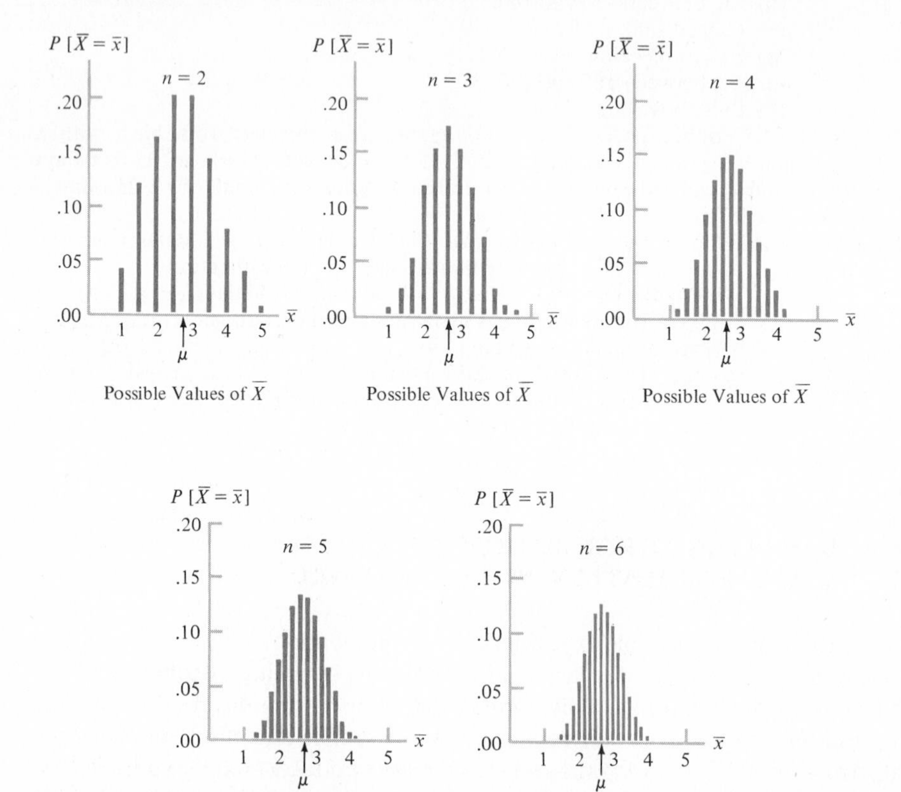

FIGURE 6-10 Comparative sampling distributions of $\bar{X}$ for different sample sizes.

interesting is the pattern presented by the spikes themselves, resembling the bell shape that we ordinarily associate with the normal distribution. This tendency is more pronounced with the larger n values.

This illustrates a crucial point of statistical theory, which is substantiated by the central limit theorem to be discussed next. The tendency of the sampling distribution of $\bar{X}$ to assume the bell shape indicates how we can obtain the sampling distribution of $\overline{X}$ even when the population is not normally distributed.

TABLE 6-2 Relative Frequency Distribution for the Number of Bedrooms in Residential Units

Number of Bedrooms X		Relative Frequency
1		.2
2		.3
3		.2
4		.2
5		.1
	Total	1.0

The Central Limit Theorem

The foregoing illustration suggests that the normal distribution may be used as a basis for approximating the sampling distribution of $\overline{X}$. Fortunately, it has been established mathematically that this is indeed the case. It is this fact that has made the normal distribution of such fundamental importance in statistics.

We now state the

CENTRAL LIMIT THEOREM As the sample size n becomes large, when each of the observations is independently selected from a population having a mean of μ and a standard deviation of σ, the sampling distribution of $\overline{X}$ tends toward a normal distribution with a mean of μ and a standard deviation of $\sigma_{\overline{X}}$.

Note that the central limit theorem is applicable regardless of the shape of the population frequency distribution. It is valid for populations having the skewed, bimodal, uniform, or exponential frequency distributions discussed in Chapter 2. It may be used whether the observation random variable is discrete or continuous. Figure 6-11 shows the sampling distributions of $\overline{X}$ obtained mathematically for samples taken from populations having frequency distributions of various shapes. Note that in every case, as the sample size increases, the sampling distribution of $\overline{X}$ becomes more bell shaped. This occurs more quickly for unimodal, symmetrical populations, which yield bell shapes for small samples; for larger sample sizes, all have shapes of the same form.

The utility of the central limit theorem lies in the fact that it enables us to make an inference about a population without knowing anything more about its frequency distribution than we can glean from a sample, although we must still presume a specific value for σ.

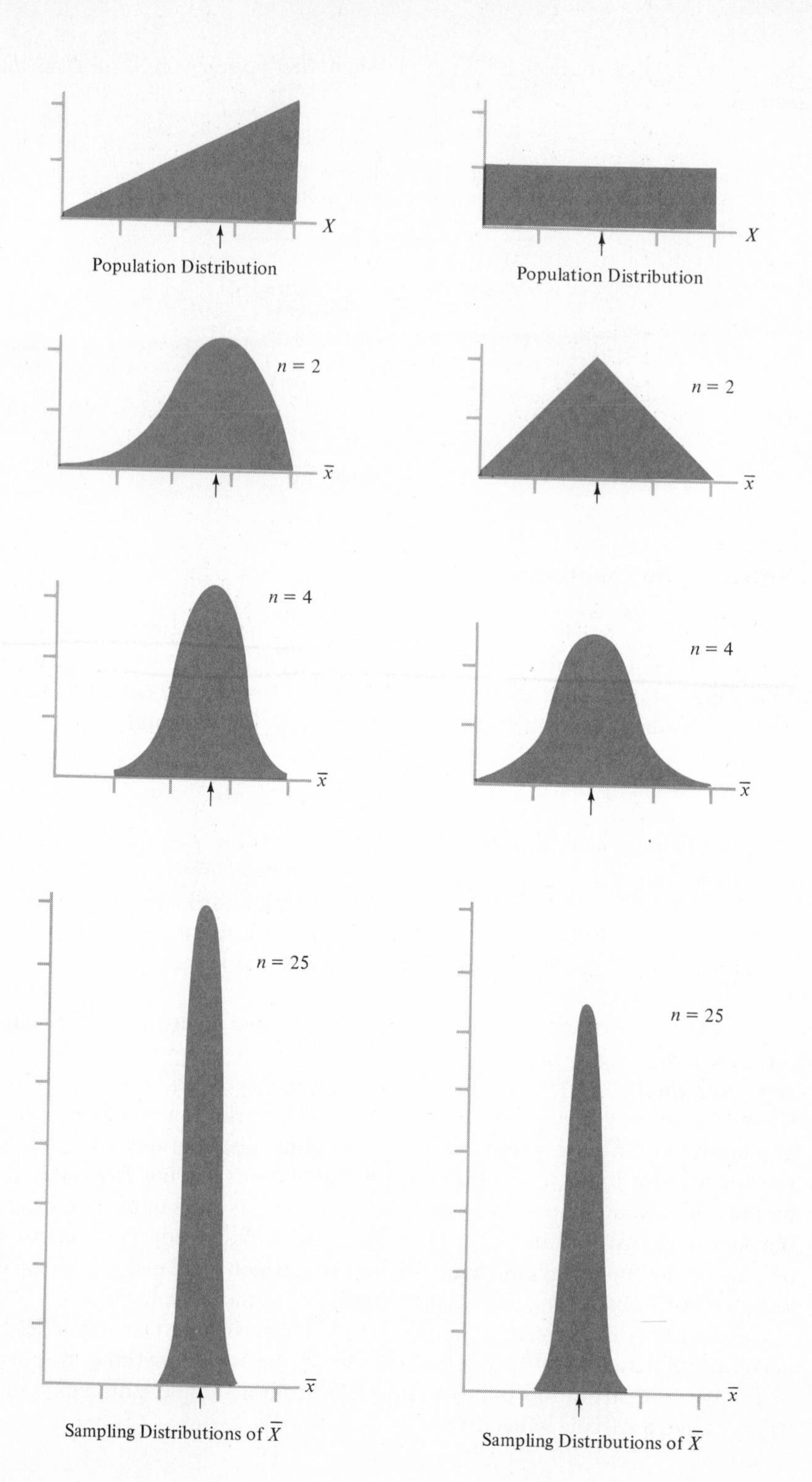
X
Population Distribution
X
Population Distribution
n = 2
$\bar{x}$
n = 2
$\bar{x}$
n = 4
$\bar{x}$
n = 4
$\bar{x}$
n = 25
$\bar{x}$
Sampling Distributions of $\bar{X}$
n = 25
$\bar{x}$
Sampling Distributions of $\bar{X}$

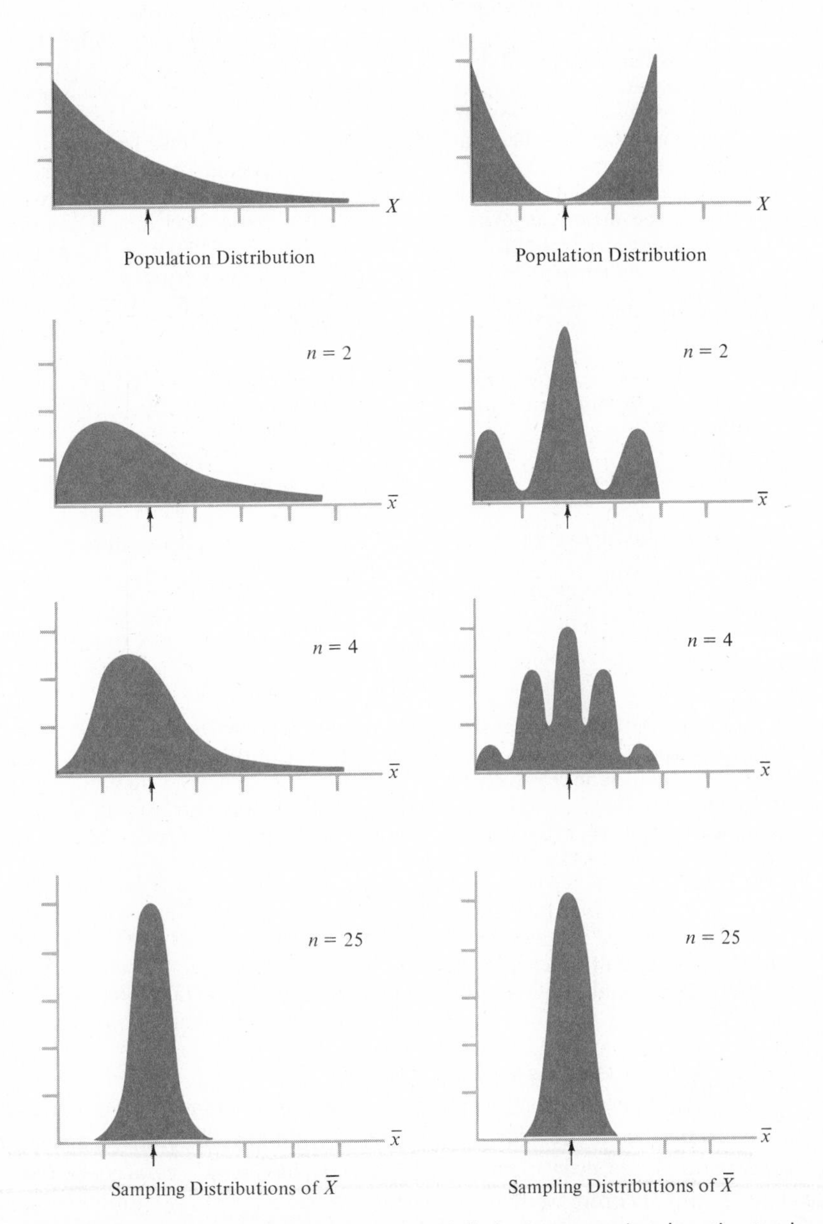

FIGURE 6-11 Illustration of the central limit theorem, showing the tendency toward normality in the sampling distribution of $\overline{X}$ as n increases, for various populations.

Example 6-5 Much scientific research regarding human health and psychology is rooted in initial studies of rat physiology and behavior. Over the years, stable populations of these rodents have been established in laboratories around the world. Much criticism has been made of using such animals in any kind of testing, as various laboratories have their own genetic strains which have been cultured for hundreds of generations in sterile, isolated environments. Thus, two laboratories may obtain entirely different results by performing an identical experiment on their respective races of rats, while a third kind of reaction might be expected when "ill-behaved" wild rats are used.

One researcher has been attempting to break the prevailing rat-research syndrome by duplicating certain medical experiments using only first- and second-generation wild rats captured in the cities. One of his studies involved estimating the mean longevity of rats fed a synthetic food diet. The original laboratory experiment indicated that the population of lifetimes was positively skewed, with a mean of $\mu = 24$ months and a standard deviation of $\sigma = 6$ months. Assuming that the lifetimes of these wild rats have the same distribution, what is the probability that a sample of size $n = 100$ will have a mean lifetime of 25 months or more?

Even though the population is skewed, the central limit theorem indicates that the sampling distribution is normal, with a mean μ of 24 months and a standard deviation of

$$\sigma_{\bar{X}} = \frac{\sigma}{\sqrt{n}} = \frac{6}{\sqrt{100}} = .6$$

The probability that $\bar{X}$ will be at least 25 months may be calculated by:

$$P[25 \leqslant \bar{X}] = P\left[\frac{25-24}{.6} \leqslant \frac{\bar{X}-\mu}{\sigma_{\bar{X}}}\right]$$

$$= P[1.67 \leqslant Z]$$

$$= .5000 - .4525 = .0475$$

Thus, if the same mean and standard deviation apply for the wild rats, then $\bar{X}$ would exceed 25 months in only about 5 percent of these experiments. Such an outcome ought to be relatively rare.

Suppose that $\sigma = 6$ months is indeed the case for wild rats as well. Also, assume that the experimenter actually obtains a mean longevity of $\bar{X} = 30.53$ months with his hardy wild animals! This would make the actual sample mean for wild rats $30.53 - 24 = 6.53$ months greater than the mean for laboratory rats, or more than 10 standard deviations ($10\sigma_{\bar{X}}$) beyond $\mu = 24$. In other words, this outcome is so rare that the upper-tail area under the normal curve will be quite tiny—far too small to even be reflected in our table of areas. The obvious conclusion is that wild rats live substantially longer than laboratory strains fed the same diet, and the actual value of μ is considerably larger than 24 months.

Applicability of the Central Limit Theorem

The central limit theorem is perhaps the most important result of statistical theory. It allows us to develop procedures for making inferences about the means of populations *even when little more than the sample results are known.* The central limit theorem applies universally, whether the population is discrete or continuous.* Thus, the normal curve plays a fundamental role in a great

* One restriction is that the population must have a finite variance. This is a theoretical limitation of no practical significance to populations ordinarily encountered.

portion of statistical theory. In Chapter 7, we will make use of this result in estimating population means. In Chapter 8, the theorem will be applied in describing procedures for making decisions based upon the sample mean, which may tend to confirm or deny an assumption regarding the true value of the population mean. In the next section of this chapter, we will see that the central limit theorem applies to samples taken from small as well as from large populations. Section 6-6 will then show that the same result allows us to use the normal curve to approximate the binomial distribution.

EXERCISES

6-12 A sample of size $n = 49$ is taken from a population having a mean of $\mu = 100$ and a standard deviation of $\sigma = 14$. Find the probability that $\bar{X}$ will lie between 96 and 104.

6-13 An economist serving as a consultant to a large teamsters' local wishes to obtain an estimate of the mean annual earnings μ of the membership. He will use the mean $\bar{X}$ of a sample of $n = 100$ drivers as an estimate. Assuming that the standard deviation of the membership's annual earnings is $\sigma = \$1{,}500$, answer the following:

(a) Find the probability that the estimate will lie within $200 of the actual population mean.

(b) Let n be increased to 400. Find the probability that the estimate will lie within $200 of the actual population mean. What is the percentage increase in probability over the probability found in (a)?

(c) Determine the probability that the estimate lies within $100 of the true population mean when $n = 100$. What is the percentage reduction in this result over the probability found in (a)?

6-14 A public transportation agency wishes to estimate the mean mileage that can be obtained on a new type of radial tire. Due to operating methods, a tire may be used on several buses during its useful life. A separate mileage log must therefore be kept on each tire in the sample. Since this procedure is costly, only $n = 100$ tires are to be used. In a previous study on another tire type, a standard deviation of $\sigma = 2{,}000$ miles was determined. It is assumed that the same figure will apply with the new tires.

(a) What is the probability that the sample mean will be within 500 miles of the population mean?

(b) The maintenance superintendent has stated that the new tires may have greater mileage variability than the tires now used. Assuming that the standard deviation is changed to $\sigma = 4{,}000$ miles, find the probability that the sample mean differs from the population mean by no more than 500 miles.

(c) Comparing your answers to parts (a) and (b), what can you conclude regarding the effect of increased population variability upon the reliability of the sample mean as an estimator?

(d) Suppose that a sample of size $n = 200$ tires is used instead. If a standard deviation of $\sigma = 2{,}000$ miles is determined, find the probability that the sample mean differs from the population mean by no more than 500 miles.

(e) Comparing your answers to parts (a) and (d), what may you conclude about the effect of increased sample size upon the reliability of the sample mean as an estimator?

6-15 In order to make a decision regarding the optimal number of toll booths to open during various times of the week, the operations manager of a port authority has ordered an extensive study. One unanticipated finding is that the mean time to collect a toll decreases as the traffic becomes heavier. For example, on late Friday afternoon, the collection times were found to have a mean of $\mu = 10$ seconds, with a standard deviation of $\sigma = 2$ seconds. On slower Wednesday mornings, the mean was $\mu = 12$ seconds and the standard deviation was $\sigma = 3$ seconds. A consistency check is now to be made to determine whether the season of the year affects efficiency. Random samples of $n = 25$ cars are taken on Wednesday mornings and Friday afternoons. Assuming that the above results are true population parameters, answer the following:

(a) What is the probability that the Wednesday sample mean will differ by more than 1 second from the assumed mean?

(b) What is the same probability for Friday?

(c) Why do the probabilities found in (a) and (b) differ?

6-5 SAMPLING DISTRIBUTION OF $\bar{X}$ WHEN POPULATION IS SMALL

We have established that for independent sample observations, the standard error for $\bar{X}$ may be obtained from the expression, $\sigma_{\bar{X}} = \sigma/\sqrt{n}$. Observations will always be independent when sampling is done with replacement. But, ordinarily, sampling is done *without replacement*. When populations are large in comparison to the sample size, the probability distributions for successive observations are changed imperceptibly by the removal of earlier items. In these cases, we can generally draw the same conclusions that we would if the population were infinite, and thus we can assume independence between observations.

But when the population is small in comparison to the sample size, we must reflect this fact in computing $\sigma_{\bar{X}}$. ***The standard error of $\bar{X}$ for small populations is***

$$\sigma_{\bar{X}} = \frac{\sigma}{\sqrt{n}}\sqrt{\frac{N-n}{N-1}} \tag{6-7}$$

where σ = population standard deviation, N = population size, and n = sample size.

Finite Population Correction Factor

The term $\sqrt{(N-n)/(N-1)}$ is referred to as the *finite population correction factor*. When n is small in relation to N, then the factor is very near to 1, so that the standard error obtained from a sample without replacement is close in value to one obtained with replacement. Note that the numerator in the expression will never be greater than the denominator, so that $\sigma_{\bar{X}}$ will be smaller when a sample is taken without replacement. *In practice, the finite population correction is usually ignored whenever n is less than 10 percent of N.*

The normal distribution approximately describes the sampling distribution of $\bar{X}$, even though the observations are not independent with samples from small populations.

Example 6-6 Suppose that a survey is to be taken of incomes in two cities: city A has a population of $N = 10{,}000$; city B has a population of $N = 2{,}500$. It will be assumed that the same standard deviation of $\sigma = \$1{,}500$ is common to both populations. A sample of size $n = 1{,}000$ is taken from each.

The following calculations, using expression (6-7), provide the standard error of $\bar{X}$ for city A:

$$\sigma_{\bar{X}} = \frac{\$1{,}500}{\sqrt{1{,}000}}\sqrt{\frac{10{,}000 - 1{,}000}{10{,}000 - 1}} = \$45.00$$

and, analogously, for city B:

$$\sigma_{\bar{X}} = \frac{\$1{,}500}{\sqrt{1{,}000}}\sqrt{\frac{2{,}500 - 1{,}000}{2{,}500 - 1}} = \$36.75$$

Notice that the standard error of $\bar{X}$ for city B is smaller than that for city A. This is to be expected, since B is a smaller city. But notice that the standard error for city A is only about 23 percent greater, although its population is four times as large.

The probability that $\bar{X}$ *exceeds* the mean by 100 or more may be calculated for each city. For A:

$$\begin{aligned} P[100 \leqslant \bar{X} - \mu] &= P\left[\frac{100}{45.00} \leqslant \frac{\bar{X} - \mu}{\sigma_{\bar{X}}}\right] \\ &= P[2.22 \leqslant Z] \\ &= .5000 - .4868 = .0132 \end{aligned}$$

and for B,

$$\begin{aligned} P[100 \leqslant \bar{X} - \mu] &= P\left[\frac{100}{36.75} \leqslant \frac{\bar{X} - \mu}{\sigma_{\bar{X}}}\right] \\ &= P[2.72 \leqslant Z] \\ &= .5000 - .4967 = .0033 \end{aligned}$$

In either case, the probability of obtaining a sample mean that exceeds the respective mean by \$100 or more is relatively small. Notice, though, that the same sample size provides about the same protection against obtaining extremely large sample values in *both* cities, regardless of the size of either.

In general, we can make the following

CONCLUSION The probability that the sample mean is unusually small or large is about the same for both small and large populations having the same standard deviation, as long as the sample size is relatively small in relation to the population size.

When n is small in relation to N, the finite population correction factor gets close to 1. Thus, when a sample of $n = 1{,}000$ is used, it would yield almost the same finite population factor for a population of $N = 100{,}000$ as it would for one of $N = 1{,}000{,}000$ (.9949 vs .9995). Thus, if the standard deviations of each are identical, then there would be an imperceptible difference in the respective values of $\sigma_{\bar{X}}$ and the probabilities of possible sample results would be nearly identical. In this instance, a sample of size $n = 1{,}000$ will be almost as reliable in a large population as in one that is one-tenth its size. This is why we concluded in Example 3-5 (page 86) that a sample of Alaskan voters in a presidential election poll would have to be about the same size a sample of Californian voters for the sample results to be equally reliable.

EXERCISES

6-16 Suppose that a finite population has a standard deviation of $\sigma = 10$. Calculate the standard error of $\bar{X}$, $\sigma_{\bar{X}}$, for each of the following value pairs for the population and sample size:
(a) $N = 1{,}000$; $n = 500$
(b) $N = 1{,}000$; $n = 50$
(c) $N = 10{,}000$; $n = 5{,}000$
(d) $N = 10{,}000$; $n = 1{,}000$
(e) $N = 10{,}000$; $n = 500$

6-17 A customs official wishes to estimate the mean weight for each of three batches of copper ingots obtained from different smelters. $N = 500$ ingots have arrived from Chile, $N = 1{,}000$ from Bolivia, and $N = 700$ from Arizona. Samples of $n = 100$ ingots from each batch are weighed. It will be assumed that each smelter produces batches of ingots having a mean weight of $\mu = 100$ pounds, with a standard deviation of $\sigma = 5$ pounds.
(a) Find the probability that each respective batch yields a sample whose mean lies between 99 and 101 pounds.
(b) Find the analogous probability values when the sample means are obtained from batches 10 times as large (using the same sample size). Do your answers differ significantly between smelters? Explain.

6-18 Suppose that the customs official in the preceding exercise establishes a rule to enable him to decide when an entire batch of ingots should be individually weighed. Since weighing is a time-consuming procedure, he only wants to do this when there is strong sample evidence that the ingots are underweight.

Assume that an acceptable batch of $N = 1{,}000$ ingots has a mean weight of $\mu = 100$ pounds, but that the official does not know this value. He takes a sample of $n = 200$ and calculates $\bar{X}$. Suppose that the batch standard deviation is $\sigma = 5$ pounds. Three decision rules are to be considered:

1. Accept batch (do not weigh all 1,000 ingots) if $\bar{X} \geq 100$ pounds.
2. Accept batch if $\bar{X} \geq 99.5$ pounds.
3. Accept batch if $\bar{X} \geq 99.0$ pounds.

Find the probability under each of the above decision rules that the customs official will accept the batch without weighing all the ingots. If he prefers the rule that maximizes the probability of not weighing all 1,000 ingots, which rule should he choose?

6-6 SAMPLING DISTRIBUTION OF P AND THE NORMAL APPROXIMATION

In Chapter 5, we established that the proportion of successes obtained from a Bernoulli process is a random variable having the binomial distribution. A random sample selected from a population of two complementary attributes may be construed to be a Bernoulli process whenever the population is infinite or whenever the sampling is done with replacement. In other words, the binomial distribution is applicable whenever the sample observations are taken independently.

Advantages of Approximating the Binomial Distribution

The calculation of probabilities from the binomial formula can be quite a chore. In many cases, the task may seem insurmountable. Consider, for example, calculating the probability that $R = 324$ convictions are obtained in the next $n = 2{,}032$ criminal cases brought to trial in a state, when the average rate of convictions is $\pi = .537$. This would involve evaluating

$$\frac{2{,}032!}{324!(2{,}032-324)!}(.537)^{324}(1-.537)^{2{,}032-324}$$

The computations would require working with both extremely large and minuscule numbers, making the use of logarithms or other approximations almost mandatory.

Of course, with a high-speed digital computer, the task might be manageable. Binomial probabilities can be calculated and tabulated. But a new problem then arises: for what values of π and n should tables be constructed? Obviously, not all π values can be accommodated, for they are infinite in number (all possible values between 0 and 1). Also, the probability table for a particular π would be quite long for even a moderately large n. Since no table can be constructed that will be satisfactory for all envisionable contingencies, and because one applicable to a moderate number of situations would require a great deal of space, tables are no panacea.*

* Of course, if a high-speed digital computer could be queried whenever the need arose, there would be no need for tables. With the growing accessibility to time-sharing terminals that connect the user directly to a large computer, it may eventually be most convenient to use such a procedure and thus avoid the approximations described in this section.

Normal Distribution as an Approximation

If a satisfactory approximation to the binomial distribution can be used, then the difficulties just discussed can be avoided. Recall that a graph of the binomial distribution tends to be bell shaped as n becomes large (see Figure 5-4, page 176). This suggests that for large sample sizes, the binomial approaches the shape of the normal curve. Indeed, it may be established that the central limit theorem also applies to the proportion of successes P.* This fact proves quite advantageous, because normal curve areas are conveniently tabulated.

In Chapter 5, we established expressions for the mean and the standard deviation of P. The mean of P is

$$E(P) = \pi \tag{6-8}$$

Retaining the terminology in the preceding section, the standard deviation of the possible values for the statistic P, calculated from the following expression, is the ***standard error of P***

$$\sigma_P = \sqrt{\frac{\pi(1-\pi)}{n}} \tag{6-9}$$

Assuming that P has a sampling distribution shaped like a normal curve, P may be transformed into the standard normal random variable by the following:

$$Z = \frac{P-\pi}{\sigma_P} \tag{6-10}$$

The suitability of the approximation may be illustrated by considering the two distributions when $n = 10$ and $\pi = .5$. The actual sampling distribution is plotted in Figure 6-12. The probability values were obtained from Appendix Table C (page 555). In order to compare this discrete distribution to the continuous normal distribution, the probabilities are presented as bars rather than spikes. The height of each bar represents the probability that P assumes the value at the midpoint of the bar's base. Since the base of each bar may be considered to be of unit width (letting each unit represent a .10 increment on the scale of p), the area of a bar also represents the probability of the P value at the midpoint. The area .3770 of all the bars above the point $P = .6$ therefore represents the probability that the value of P obtained will be greater than or equal to .6. This is represented by the gray area in Figure 6-12.

From the above relation, we can obtain the normal approximation to the

* Letting each sample success represent a "score" of $X = 1$ and each failure a "score" of $X = 0$, the sample mean "score" $\bar{X}$ is the number of successes divided by n, or $\bar{X} = R/n$. This equals P, the sample proportion of successes.

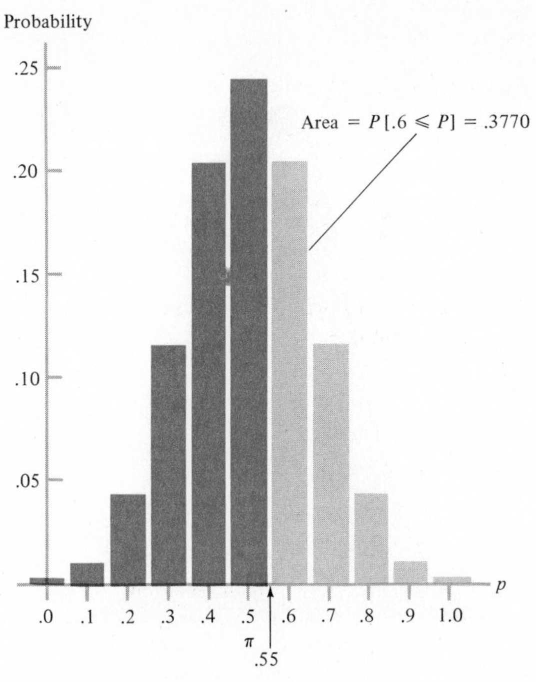

FIGURE 6-12 Binomial sampling distribution of *P*.

sampling distribution of P. Here, the mean and the standard error are

$$\pi = .5$$

$$\sigma_P = \sqrt{\frac{.5(1-.5)}{10}} = .158$$

The corresponding normal curve is graphed in Figure 6-13. Treating P as an approximately normal random variable, we may find the probability that P lies above .6 as follows:

$$P[.6 \leqslant P] = P\left[\frac{.6-.5}{.158} \leqslant \frac{P-\pi}{\sigma_P}\right]$$
$$= P[.63 \leqslant Z]$$
$$= .5000 - .2357 = .2643$$

This probability value is represented by the gray area under the approximating normal curve shown in Figure 6-13.

Note that the area under the normal curve above .6 only approximates the true probability indicated by the gray area under the bars in Figure 6-12. The true probability is .3770, while the normal curve provides a probability of .2643.

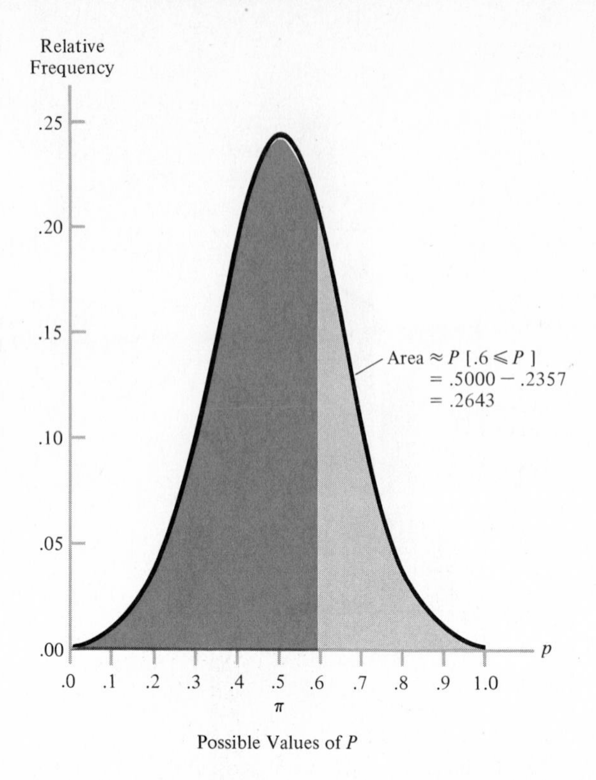

FIGURE 6-13 Normal approximation to the sampling distribution of P, when $n = 10$, $\pi = .5$.

The discrepancy between the binomial probabilities and those found with the normal curve will be negligible for larger sample sizes. We may better assess the nature of the approximation in Figure 6-14, where the graph for the binomial distribution for P is superimposed on the normal curve. Here, a continuity correction has been applied to obtain a better approximation.*

* From this graph, we can see that a value could have been obtained from the normal curve area above .55 rather than .60:

$$
\begin{aligned}
P[.55 \leq P] &= P\left[\frac{.55 - .50}{.158} \leq \frac{P - \pi}{\sigma_P}\right] \\
&= P[.32 \leq Z] \\
&= .5000 - .1255 = .3745
\end{aligned}
$$

The value .3745 is much closer to the true binomial probability of .3770. Using .55 instead of .60 applies a continuity correction of amount $.5/n = .5/10 = .05$. The need for a continuity correction arises from the fact that the *discrete* binomial distribution is being approximated by a *continuous* normal distribution. Thus, we may improve our normal curve approximation by subtracting $.5/n$ from the lower limit for P and adding $.5/n$ to the upper limit. Throughout the remainder of this book, the continuity correction will be ignored, making discussions conceptually simpler. For most sample sizes n where the normal approximation will be made, n is large enough so that $.5/n$ is relatively tiny and the continuity correction only slightly improves the approximation.

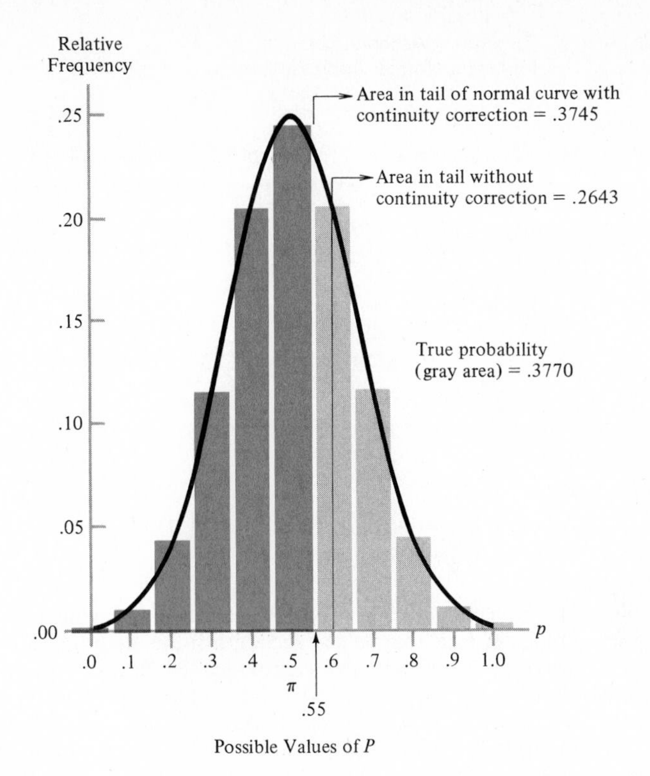

FIGURE 6-14 Comparison of binomial sampling distribution of P to its normal approximation.

Table 6-3 gives the commonly accepted guidelines for using the normal approximation. These guidelines have been constructed according to the popular rule that the normal approximation to the binomial is adequate whenever both of the following hold:

$$n\pi \geqslant 5 \tag{6-11}$$

$$n(1-\pi) \geqslant 5$$

Some statisticians insist upon even larger sample sizes than those listed in Table 6-3 before the approximation is acceptable. Note that, in some cases, a very large sample size ought to be used. This is because the skew of the binomial distribution is so pronounced for large or small π that the bell shape is assumed by the binomial distribution only for very large n.

TABLE 6-3 Commonly Accepted Guidelines for Using Normal Approximation to the Binomial

Whenever π Equals	*Use the Normal Approximation Only If n Is No Smaller Than*
.5	10
.40 or .60	13
.30 or .70	17
.20 or .80	25
.10 or .90	50
.05 or .95	100
.01 or .99	500
.005 or .995	1,000
.001 or .999	5,000

An Example

We may illustrate the usefulness of the normal approximation to the binomial by means of the following example.

Example 6-7 A candidate for mayor wants to entice voters to switch their support to him. He feels that he can accomplish this by promising more recreational services than the incumbent has provided. He therefore wants to gauge the potential attraction of more services by estimating the proportion of voters who now use city services. If 50 percent of the voters are *assumed* to require city recreational services, what is the probability that 40 percent or fewer voters in a sample of size $n = 100$ will use recreational services?

We assume that the sample proportion of users has a binomial sampling distribution (because the population is large). Here $\pi = .50$, and we wish to evaluate the probability that $P \leqslant .40$. Referring to Table 6-3, we see that $n = 100$ is sufficiently large for making the normal approximation. We therefore calculate:

$$\sigma_P = \sqrt{\frac{.5(1-.5)}{100}} = .05$$

Thus

$$P[P \leqslant .40] = P\left[\frac{P-\pi}{\sigma_P} \leqslant \frac{.40-.50}{.05}\right]$$

$$= P[Z \leqslant -2.00]$$

$$= .5000 - .4772 = .0228$$

So the probability is quite small that 40 percent or fewer of the sample will use city recreation services. If a sample result of $P = .40$ were actually obtained, this would be a rare outcome when the actual proportion of users is assumed to be $\pi = .50$. Such sample evidence would strongly indicate that the unknown value of π is smaller than .50.

For $n = 100$, we are fortunate to have the actual cumulative binomial probability values in Appendix Table C. For $\pi = .50$ and $n = 100$, we find that:

$$P[P \leqslant .40] = P\left[\frac{R}{n} \leqslant \frac{40}{100}\right] = .0284$$

In this case, the normal approximation provides a value quite close to the actual one.

If a sample of $n = 5$ were used, then the above probability calculation could not have been obtained using the normal approximation, because Table 6-3 indicates that n would be too small for $\pi = .50$. This desired probability would therefore have to be obtained directly from the binomial distribution.

Distribution of *P* for Samples Taken from a Small Population Without Replacement

As in the case of the sample mean, sampling without replacement requires a different sampling distribution for P. This distribution is called the *hypergeometric distribution*, and it has the same mean as the binomial, or:

$$E(P) = \pi$$

As with $\bar{X}$, the variability in values of P is smaller when there is no replacement. This is reflected in σ_P. *For small populations, the standard error of P is*

$$\sigma_P = \sqrt{\frac{\pi(1-\pi)}{n}}\sqrt{\frac{N-n}{N-1}} \qquad (6\text{-}12)$$

Expression (6-12) is the standard deviation of the binomial distribution multiplied by the same finite population correction factor that is used with $\bar{X}$.

The normal approximation may be used here in the same manner as it is in sampling with replacement, except that for sampling without replacement the standard error of P is obtained from expression (6-12). Again, we usually ignore the finite population correction when n is less than 10 percent of N.

Example 6-8 A small, professional society has $N = 5{,}000$ members. The president has mailed $n = 500$ questionnaires to a random sample of members asking whether they wish to affiliate with a larger group. Assuming that the proportion of the *entire* membership favoring consolidation is $\pi = .7$, find the probability that the sample proportion P differs from this by no more than .05.

Here we observe that $E(P) = \pi = .7$, and we calculate the standard error from expression (6-12):

$$\sigma_P = \sqrt{\frac{.7(1-.7)}{500}}\sqrt{\frac{5{,}000-500}{5{,}000-1}} = .0194$$

The desired probability is then found:

$$P[-.05 \leq P-\pi \leq .05] = P\left[\frac{-.05}{.0194} \leq \frac{P-\pi}{\sigma_P} \leq \frac{.05}{.0194}\right]$$

$$= P[-2.58 \leq Z \leq 2.58]$$

$$= 2(.4951) = .9902$$

An Application of the Normal Approximation to Acceptance Sampling

We may illustrate the usefulness of the normal approximation to the binomial by an example involving *acceptance sampling*. A manufacturer dependent upon an outside supplier for raw materials or components for his own production will want to accept only shipments of some set minimum quality. Because of the expense involved, the decision to accept or reject an incoming shipment is based on a sample of the items received. The procedure for accomplishing this is an acceptance sampling plan. This approach is common in quality control.

Consider a certain automobile assembly plant that obtains its headlamps from a supplier in batches of $N = 500$. The supplier has designed its production operation in such a way that, for various reasons, about 5 percent of its output is defective. The assembly plant quality control manager inspects a sample of size $n = 100$ (without replacement) from each shipment. If 7 or fewer (that is, 7 percent or fewer) items in the sample are found to be defective, he will accept the shipment; otherwise, the batch will be rejected and returned to the supplier.

The number 7 is referred to as the *acceptance number*. It forms the basis for a *decision rule*, which specifies the action to be taken regarding a particular batch. We may determine what the implications are to both the consumer and the producer by using such a rule to calculate various probabilities.

Suppose that a batch of headlamps is received that contains $\pi = .10$ defectives. If the quality control manager knew the π value, he would reject the shipment. But since only a sample will be taken, he may end up accepting this poor batch. What is the probability that the manager will accept the shipment?

The shipment will be accepted if the sample proportion defective P is $\leqslant .07$. Employing the normal approximation, we first find the standard error of P, using expression (6-12):

$$\sigma_P = \sqrt{\frac{.10(1-.10)}{100}}\sqrt{\frac{500-100}{500-1}} = .027$$

The probability of accepting the bad batch is therefore:

$$\begin{aligned} P[P \leqslant .07] &= P\left[\frac{P-\pi}{\sigma_P} \leqslant \frac{.07-.10}{.027}\right] \\ &= P[Z \leqslant -1.11] \\ &= .5000 - .3665 = .1335 \end{aligned}$$

Thus, there is a .1335 probability that a poor batch of headlamps will be accepted. Since accepting a bad lot is an incorrect or erroneous decision that will hurt some users of a product, such a probability value is sometimes referred to as the *consumer's risk*.

The supplier can also be hurt by another kind of erroneous decision. Suppose that an acceptable lot having $\pi = .05$ defectives is shipped. If, by chance, more than 7 defectives are selected, the good shipment will be rejected. To find the probability that this will occur, we again use the normal approximation. The standard error of P is

$$\sigma_P = \sqrt{\frac{.05(1-.05)}{100}}\sqrt{\frac{500-100}{500-1}} = .0195$$

The probability of rejecting the good batch is therefore:

$$\begin{aligned} P[.07 \leqslant P] &= P\left[\frac{.07-.05}{.0195} \leqslant \frac{P-\pi}{\sigma_P}\right] \\ &= P[1.03 \leqslant Z] \\ &= .5000 - .3485 = .1515 \end{aligned}$$

Such a probability that an acceptable shipment will be erroneously rejected is referred to as the *producer's risk*.

The producer's risk and consumer's risk are illustrations of the more general Type I and Type II errors of hypothesis testing to be discussed in Chapter 8. There, we will describe how a decision rule may be constructed to balance both risks and to keep the probabilities of erroneous decisions at desired levels.

EXERCISES

6-19 Find the probability that the proportion P of successes in $n = 100$ trials of a Bernoulli process with trial success probability $\pi = .5$ lies between each of the following pairs of values:
(a) .40, .60; (b) .35, .65; (c) .50, .65; (d) .50, .90.

6-20 Find the probabilities that the proportion P of successes obtained from a Bernoulli process with trial success probability $\pi = .5$ will lie between .4 and .6 when the number n of trials is 64, 100, 400. What relationship do you notice between the probability values obtained and the trial size n? How do you explain this relationship?

6-21 A cannery accepts a shipment of tomatoes whenever the proportion that is ripe P in a sample of $n = 100$ is $\geqslant .90$. Assuming that shipments are sufficiently large that the sampling process may be represented (with only minor error) as a Bernoulli process, answer the following:
(a) What is the probability of accepting a poor shipment in which the proportion of ripe tomatoes is $\pi = .8$?
(b) What is the probability of rejecting a good shipment in which the proportion of ripe tomatoes is $\pi = .95$?
(c) How do you account for the fact that these probabilities indicate that some shipments of poor quality will be accepted while some shipments of good quality will be rejected?

6-22 Random samples of voter preferences are obtained for the candidates to the U.S. Senate. One is taken from a precinct having $N = 500$ registered voters; the other, from a precinct having $N = 1{,}000$. In each case, the sample size is $n = 100$. There are just two candidates, a Republican and a Democrat, and, for (a) and (b) below, each voter in the sample favors one of the candidates.

(a) Assuming that the proportion of Republicans in both precincts is $\pi = .40$ and that the voters will adhere strictly to party lines, find the probability for each sample that the majority of the voters polled will favor the Republican candidate.

(b) Same as (a), but with samples of size $n = 49$.

(c) Suppose that some of the voters are "undecided" when polled. Can any sample which includes undecided's be truly reflective of the votes made by the precinct population on election day? Explain.

REVIEW EXERCISES

6-23 As closely as possible, find the following percentiles for a normal distribution of women's heights with a mean of $\mu = 66$ inches and a standard deviation of $\sigma = 1.5$ inches:

(a) 50th (b) 15th (c) 95th (d) 75th (e) 33rd

6-24 A large rancher has established that the gestation period of his cows may be represented by the normal distribution with a mean of $\mu = 281$ days and a standard deviation of $\sigma = 5$ days. Assuming that the length of pregnancies for the next 100 births is a random sample from this population, find the probabilities that the mean gestation period will

(a) Exceed 282 days.

(b) Lie between 280 and 282 days.

(c) Be less than 282.5 days.

(d) Be less than 280.5 days.

6-25 The weights of "one-ounce" gold ingots casted by a rare-metals refinery are normally distributed with a mean of $\mu = 28$ grams and a standard deviation of $\sigma = 1$ gram. A shipment of 25 ingots is to be made.

(a) What is the probability that the mean ingot weight will be less than 27.5 grams?

(b) Find the probability that the entire shipment will weigh more than 705 grams.

6-26 The annual incomes of surgeons constitute a highly positively skewed population. Nevertheless, it is still possible—according to the central limit theorem—to find the sampling distribution of the mean of a random sample of these incomes by using the normal distribution. Suppose that the population has an unknown mean of μ and a standard deviation of $\sigma = \$10{,}000$. An estimate of μ will be made using the value of the sample mean $\bar{X}$. It is desired that this estimate fall within $\pm \$1{,}000$ of the true mean.

(a) If $n = 100$ incomes, find the probability that the estimate meets the desired accuracy.

(b) If $n = 625$ incomes are used in the sample instead, find the probability that the estimate meets the desired accuracy.

6-27 The IQ scores achieved by the first-grade students in a certain state are normally distributed, with a mean of $\mu = 105$ and a standard deviation of $\sigma = 10$ points. A random sample of n scores is to be selected. Find the standard error of the sample mean and the probability that this mean IQ will exceed 106 points, when: (a) $n = 100$ (b) $n = 25$ (c) $n = 900$

6-28 A parapsychologist is testing the extrasensory perception of a purported clairvoyant. He has a deck of cards, half of which are red and half of which are black. He selects the top card and asks his subject (who is in another room) to identify its color. The procedure is repeated 100 times. Each successive card is replaced in the deck, and the deck is shuffled before the next card is drawn. A subject with no powers of ESP should give the correct color 50 percent of the time, so that $\pi = .5$. Under this assumption, what is the probability that a correct response will be obtained for 65 or more of the 100 cards? (Use the normal approximation.)

6-29 For each of the following sampling situations, indicate (by yes or no) whether the sampling distribution of the sample proportion may be approximated by the normal curve. In each case, assume that the population is quite large in relation to the sample size.

(a) $\pi = .50$ $n = 9$
(b) $\pi = .10$ $n = 36$
(c) $\pi = .83$ $n = 100$
(d) $\pi = .45$ $n = 20$
(e) $\pi = .01$ $n = 1{,}000$

6-30 A political-polling firm believes that the proportion of persons favoring a certain presidential policy is $\pi = .4$. Suppose that a sample of $n = 100$ persons is selected at random from the entire electorate. The proportion favoring the presidential policy may be approximated by the normal distribution.

(a) If the assumed parameter value holds, what is the probability that 50 percent or more of the persons queried will favor the policy?
(b) Suppose that the actual parameter value is only $\pi = .37$. What is the probability that 50 percent or more of the sample will favor the policy?

6-31 A student marks an examination consisting of 36 true-or-false questions by tossing a coin. For each question, he answers true for a head and false for a tail. Assuming that half of the correct answers should be marked true, find the probability that the student will pass the examination by marking at least 75 percent of the answers correctly. (Use the normal approximation.)

6-32 If the population of members of the U.S. Army has a mean of 70 inches and a standard deviation of 3 inches, determine the probabilities that the mean height for a random sample of 100 soldiers will be

(a) Between 70 and 70.5 inches.
(b) Less than 69.5 inches.
(c) Greater than 72 inches.
(d) Less than 68 inches.
(e) Between 69.4 and 70.8 inches.

6-33 A random sample of 100 soldiers is chosen without replacement from a regiment of 900 men. If the mean regiment height is 71 inches, with a standard deviation of 2.5 inches, determine the probabilities that the sample mean height will be

(a) Between 71 and 71.4 inches.
(b) Less than 70.5 inches.
(c) Greater than 71.6 inches.
(d) Between 70.7 and 71.5 inches.

6-34 A state government agency has a policy of measuring the quantities in all bulk products purchased. For instance, on the average, 1,000-foot rolls of transparent tape ought to be close to the specified length. In evaluating 10,000 roll shipments, the purchasing director has specified the following quality control policy. A random sample of 100 rolls of tape is to be selected, and the mean length determined. If this mean is greater than or equal to 990 feet, the entire shipment is to be accepted; otherwise, the shipment is to be rejected and returned to the manufacturer.

(a) Suppose that a shipment has a mean roll length of only 980 feet, with a standard deviation of 50 feet. Find the consumer's risk or probability of accepting these inferior-quality tapes.
(b) Suppose that a shipment has mean roll length of 1,005 feet, with a standard deviation of 60 feet. Find the producer's risk or probability of rejecting these good-quality tapes.

7 Statistical Estimation

The more numerous the number of observations and the less they vary among themselves, the more their results approach the truth.

Marquis de Laplace (1820)

This chapter is concerned with the kinds of estimates available from sample data and with the procedures for making these estimates. Problems of estimation are crucial in practically every statistical application. This is true in the sciences, where generally only sample results are available for establishing values for a wide variety of parameters, including those physical constants (like $g=32$ ft/sec/sec for the acceleration due to the earth's gravity) that are so fundamental to many theoretical models. In the biological and medical sciences, the effects of chemicals—whether for controlling pests or saving lives—must be estimated from sample data obtained from experiments. In the behavioral sciences and education, statistical estimates are crucial for evaluating programs and procedures. Administrators in both business and government, at all levels, require estimates that can only be made from samples. Estimation problems are perhaps most difficult in economic planning, where the impact of policies can affect the lives of literally every person in an entire nation.

7-1 ESTIMATORS AND ESTIMATES

Using samples to estimate the value of a population parameter is one of the more prevalent forms of statistical inference. A sample statistic used for this purpose is referred to as an *estimator*. An important segment of statistical theory is concerned with finding statistics that are appropriate to use as estimators. For instance, the sample mean is a particularly good estimator for the population mean. We shall see why this is so.

The Estimation Process

Parameter estimation requires a great deal of planning. The first step is to choose an estimator. Having done that, a major concern is to select appropriate procedures so that sampling error may be controlled. The choice of sample selection method—judgment or convenience sampling, or some version of random sampling—will affect this. As we have seen, only certain random samples are free from sampling bias, so our discussion will center on these. We have also seen that sampling error may be controlled by using a sufficiently large sample, as small samples tend to be less reliable. A practical consideration will be to balance the level of reliability that is influenced by sampling error against the costs of obtaining the sample.

Due to chance, the sample statistic actually obtained may not be very close in value to the population parameter, but the only way to be certain of the parameter's value would be to take a census of the population. Therefore, methodology must be developed that will enable the statistician to assess the reliability of his result. This is tricky business when there is only a theoretical basis for knowing what values are reasonable.

Estimates fall into two categories: point and interval estimates. Differing circumstances dictate that an estimate will be in one form or the other. For example, a commonly quoted figure for the mean height of adult American males is 5′9½″. Such a single value is called a *point estimate*. The major difficulty with a number like 5′9½″ is that it gives us the false impression that it is the true mean value, right down to the smallest fraction of an inch. This is not so. Rather, such a number is the mean of a relatively small sample of heights, and, as we have seen, one sample mean can vary considerably from another—even when both samples are selected from the same population. Reflecting this chance variation, the *interval estimate* specifies a range of values, such as 5′9″–5′10″, within which the population parameter is believed to lie. This type of estimate is advantageous, since it acknowledges the potential for sampling error by indicating that the parameter lies between two numbers.

Whether a point or an interval estimate is to be made depends upon the objective of the statistical study. Interval estimates of needed quantities would be unsuitable to a manufacturer as the basis for placing orders for raw materials, for the supplier cannot accept an order for "something between 2,000 and 3,000 units." Likewise, knowing only that the wattage of a stereo amplifier lies between 50 and 80 makes it difficult to compare that model to other brands. But the superiority of an interval estimate is evident in many kinds of planning or reporting. For example, knowing that the next year's enrollment might lie between 10,000 and 11,000 students, a university president is in a better position to make a budget when the estimate indicates the presence of uncertainty as to the actual enrollment; he can visualize the best and worst cases, and thereby obtain a better basis for evaluating the risks involved.

We are presently concerned only with estimates based upon samples taken from already existing populations or from identifiable target populations. This excludes an entire class of predictions that may be based upon other kinds of information. The most notable of these would be economic forecasts based upon time-series data. For example, every year the news media provide a plethora of forecasted levels of the gross national product. Although sampling may be involved in this type of estimate, the major role is played by other factors, such as judgment, current trends, contemplated changes in government policies, and the state of international affairs. Some useful procedures that can be helpful in making such predictions will be discussed in Chapter 9. Another category of excluded predictions is that regarding the parameter of some future population. For example, the median annual family income for the United States in 1984 cannot be estimated by means of the usual sample, because the elementary units cannot be presently defined and the variate values themselves lie in the future. Although the population may be imagined, it cannot be directly observed until 1984.

EXERCISES

7-1 For each of the following situations, indicate whether or not a sample may be selected from some population to make a current estimate. Explain your answers.

(a) An estimate is to be made of the mean annual lifetime earnings of recent college graduates.

(b) The taste preferences of the potential buyers of a new product that is not currently on the regular market are to be determined.

(c) The response of heart patients to a new drug, untried on humans, is to be measured.

(d) The Federal Power Commission wishes to estimate the kilowatt-hour usage by customers at the end of the coming decade.

7-2 For each of the following parameters, indicate whether you would use a point or an interval estimate and explain your choice.

(a) The mean age of product buyers, to be used in comparing alternative advertising plans.

(b) The proportion of an evening's television viewing population watching a particular program, to be presented to advertisers as evidence of the program's drawing power.

(c) The mean pH value (a measure of acidity) of intermediary ingredients in a chemical process, so that a formula may be developed indicating the amount of neutralizing agent to apply.

(d) The median income of doctors, to be published in a government report.

7-2 POINT ESTIMATES

Often several alternative statistics can be used as estimators. For instance, to estimate the population mean, we could use any one of three sample statistics: the mean, the median, or the mode. We can guess that the sample mean may be the most suitable estimator, but how can we substantiate our choice?

The most desirable feature of an estimator is that it has a value close to the unknown value of the population parameter. We will need to develop criteria for comparing alternatives in terms of their capability to match this parameter value. The criteria themselves must necessarily be of a theoretical nature, because ordinarily we will have no experimental basis for comparing alternative estimators. Not knowing the population mean, for example, it is not possible to choose between the sample mean or median by their values alone.

Under certain circumstances, the sample mean, median, and mode will each have a value lying close to the population mean. The basic questions are therefore: Which statistic will be the most reliable estimator? Which will require the least expenditure of resources in terms of sample size? An ancillary question, also quite important, is: Do we have an easily applied theoretical procedure for determining the probability that the value obtained for the estimator statistic will lie sufficiently close to the parameter?

Criteria for Statistics Used as Estimators

Three criteria have been developed that may be used to compare statistics in terms of their capability to estimate a parameter. One of these determines whether there is a tendency, on the average, for the statistic to assume values close to the parameter in question. Another considers the overall variablity in the sampling distribution of the statistic as summarized by the standard error, which, as we have seen, affects the reliability of a sample. A final consideration is how the reliability of the statistic as an estimator is affected by the size of the sample; it is desirable to have an estimator statistic that becomes more reliable as the sample size increases. Each of these criteria is defined and discussed below.

Unbiased Estimators

DEFINITION An ***unbiased estimator*** is a statistic that has an expected value equal to the population parameter being estimated.

The notion of bias, as used here, differs from that previously discussed in that it has nothing to do with the tendency to favor selection of certain population elementary units for the sample; an estimator can be biased even when sampling bias is absent. This is a theoretical type of bias—a statistic that does not, on the average, tend to yield values equal to the parameter is said to be biased. The sample mean $\bar{X}$ is an unbiased estimator of the population mean μ, since we have established that

$$E(\bar{X}) = \mu$$

We may now indicate why we chose to define the sample variance s^2 in the way we did in Chapter 2 (page 54):

$$s^2 = \frac{\Sigma(X-\bar{X})^2}{n-1} \tag{7-1}$$

The $n-1$ divisor is used instead of n. However, it may be proven that s^2 (as defined with the $n-1$ divisor) is an unbiased estimator of σ^2, since

$$E(s^2) = \sigma^2$$

An intuitive reason why $n-1$ is used as the divisor is that we obtain a somewhat larger value for the sample variance than the one we obtain by dividing by n. The need for the larger estimator provided by s^2 reflects the fact that ordinarily a sample has less diversity than its population, for the part is rarely more disperse than the whole. (s^2 may turn out to be larger than σ^2, if a disproportionately large number of extreme values are selected for the sample; however, on the average, s^2 values will tend to equal σ^2.)

Efficient Estimators

DEFINITION One statistic is a more ***efficient estimator*** than another if its standard error is smaller for the same sample size.

To demonstrate this relationship, we compare the mean and the median of a sample taken from a large, symmetrical population. It may be shown that the sample median is an unbiased estimator of μ when the population is normally distributed. For a large sample size the sample median m has a standard error of

$$\sigma_m = 1.2533\frac{\sigma}{\sqrt{n}}$$

and we have shown that the standard error of $\bar{X}$ is

$$\sigma_{\bar{X}} = \frac{\sigma}{\sqrt{n}}$$

Thus, it follows that the standard error of the sample median is 1.2533 times as great as that of the sample mean. This implies that the sample mean is more efficient than the sample median as an estimator of μ. Medians of samples taken from normally distributed populations tend to be more unlike each other than means of identical samples. From this fact, it follows that it is more likely that an estimate close in value to the population mean will be obtained when the sample mean is used.

Consistent Estimators

DEFINITION A statistic is a ***consistent estimator*** of a parameter if the probability that the value of the statistic is very near that of the parameter becomes closer and closer to unity with increasing sample size.

The sample mean is a consistent estimator of the population mean. This is evident because $\sigma_{\bar{X}}$, which expresses the variability in the possible values of $\bar{X}$, becomes smaller as $\sqrt{n}$ grows larger. We also know that the potential values of $\bar{X}$ cluster around μ, so that less variability in these values implies a greater probability that the value actually obtained will be close to μ. As we have seen, $\sigma_{\bar{X}}$ is inversely proportional to $\sqrt{n}$, so that when n becomes quite large, $\sigma_{\bar{X}}$ gets very small; in the limiting case, n is so large ($n = N$) that $\sigma_{\bar{X}}$ becomes zero. This implies that $\bar{X}$ can then assume only one possible value, μ, with a probability of 1. In general, *a statistic whose standard error becomes smaller as n gets larger will be consistent.*

When an estimator is consistent, it then becomes more reliable with larger samples. Consistency alone does not guarantee reliable sample results; this is achieved only by increasing the sample size. But for a larger sample to be more reliable, consistency is a necessary condition. The net effect is that the use of such an estimator allows the statistician to buy more reliability by paying the price to obtain a larger sample. This is not to say that the benefit of increased reliability is worth the price. As with any resource utilization, there is a point of diminishing returns, to borrow from the language of economics, when an expenditure level is reached at which the benefit of greater reliability no longer has a greater value than the money saved by limiting the size of the sample.

Choosing the Estimator

We may apply the criteria for evaluating the characteristics of the many types of statistics to the comparison of estimators. If we wish to estimate μ, we will achieve better results by using $\bar{X}$ as our estimator than we will by using m. (We would never consider using m to estimate μ, except when the population is believed to be symmetrical.) The reader may verify that both statistics are consistent and unbiased. However, since $\bar{X}$ is more efficient than m, it will be a more reliable estimator of μ for a given expenditure of resources on sampling.

The criteria do not always provide a clear-cut determination of which statistic is the best one to use. Each candidate estimator must be analyzed as a separate case, in light of the criteria, since there is no general rule to follow when the best choice is not immediately apparent. Usually, we can say that the better of two estimators that are both unbiased and consistent is the one that is more efficient. But if the better statistic has a sampling distribution that is cumbersome or theoretically difficult to handle, then the second choice will generally be used. (The sample mode has a distribution that can be so categorized; thus, we will have little to say about the mode as an estimator.)

Commonly Used Estimators

For the Population Mean

As we have seen, $\bar{X}$ is the most desirable estimator of μ. It is unbiased, consistent, and more efficient than other candidate estimators such as the sample median. It also has a readily obtainable *normal* sampling distribution when the sample is sufficiently large. Knowing the sampling distribution, as we will see in the remainder of this chapter, enables us to choose a sample large enough to achieve a desired level of reliability in our estimate. It also allows us to qualify the estimate actually obtained when the sample results have been collected and tabulated.

TABLE 7-1 Sample Results of Words Per Page

Number of Words X	X^2		*Number of Words* X	X^2
383	146,689		175	30,625
325	105,625		278	77,284
411	168,921		351	123,201
416	173,056		423	178,929
395	156,025		327	106,929
372	138,384		381	145,161
293	85,849		317	100,489
361	130,321		362	131,044
216	46,656		338	114,244
431	185,761		411	168,921
406	164,836		371	137,641
394	155,236		393	154,449
402	161,604		388	150,544
376	141,376		295	87,025
268	71,824		421	177,241
		Totals	10,680	3,915,890

$$\bar{X} = \sum X/n = 10{,}680/30 = 356 \text{ words}$$

$$s = \sqrt{(\sum X^2 - n\bar{X}^2)/(n-1)} = \sqrt{[3{,}915{,}890 - 30(356)^2]/29} = \sqrt{3{,}924.48} = 62.65 \text{ words}$$

Example 7-1 A publisher needs to determine the number of words in one of his books. Rather than having someone count all the words, which would take several hours of painfully boring work, an estimate of the mean number of words per page μ is obtained by counting the number of words on each of 30 randomly chosen pages. The sample mean number of words per page $\bar{X}$ is then calculated. Since there are 600 pages in the book, the estimated number of words is obtained by multiplying $\bar{X}$ by 600. The sample results are shown in Table 7-1. The sample mean obtained is

$$\bar{X} = 356 \text{ words}$$

Thus, the point estimate of μ is 356 words. This results in an estimate of

$$356 \times 600 = 213{,}600$$

words in the entire book.

For the Population Standard Deviation

The sample variance s^2, as we have shown, is an unbiased and consistent estimator of the population variance σ^2. We may take the square root of s^2 to estimate the standard deviation, since σ is the square root of σ^2.

For the previous example, the standard deviation of the number of words per page was calculated in Table 7-1 to be $s = 62.65$ words. This may be used as an estimate of σ, the standard deviation of words per page for the entire book.

For the Population Proportion

The proportion π of elementary units in the population having a particular attribute may be estimated by the corresponding sample proportion P. As we have seen, $E(P) = \pi$, so that P is an unbiased estimator of π. It is also consistent, since the standard error of P

$$\sigma_P = \sqrt{\frac{\pi(1-\pi)}{n}}$$

decreases with increasing n. Thus, the variability in possible P values decreases, so that the probability that the value obtained for P will be very close to π increases as the sample size increases. In the limiting case, when n is the same size as the population, $P = \pi$ with certainty.

Example 7-2 A consumer information service has been asked to measure the tendency of the average housewife to purchase whatever she picks up, whether it is needed or not. A random sample of shoppers is selected for an experiment. Each is asked to prepare a detailed shopping list before embarking on the week's food forage. During her visit to the store, each sample shopper is "tailed" and a record is kept of each item she touches (except for packaged meats and produce). At the end of the shopping, the items on her original list are compared to those she bought. The number of items bought but not on the the list is added to the number of items touched but not bought. The total obtained represents the number of items touched or bought on impulse (that is, not included in the original shopping plans).

The procedure is repeated for several weeks, and separate records are kept on each shopper. The proportion P of impulse items bought is then estimated for every shopper in the sample experiment. One woman has accrued a total of 132 impulse items; of these, she purchased 48. Her value of P is therefore determined to be

$$P = \frac{48}{132} = .364$$

which may be taken as a point estimate of the actual proportion of impulse items she would buy under similar circumstances.

The P's for all the shoppers were determined in this manner. Their mean was then used to estimate the proportion of impulse purchases for the shopping population as a whole. The average proportion of such purchases was estimated to be .55. The consumer service therefore advised its clients to keep their hands on their carts and rely on their initial grocery-buying decisions.

EXERCISES

7-3 A survey is taken of medium-sized city libraries to determine annual book expenditures. A random sample of $n = 10$ libraries has been taken, resulting in the following figures (in thousands of dollars, rounded): 15, 5, 2, 7, 25, 19, 11, 9, 13, 42.

(a) Find a point estimate of the mean expenditures for all libraries as a group.

(b) Find a point estimate of the standard deviation of expenditures for the population.

(c) Estimate the proportion of expenditures more than $10,000.

(d) Estimate the proportion of expenditures less than $5,000.

7-4 The random experiment of tossing a die twice in succession may be viewed as analogous to taking a random sample of size $n = 2$ with replacement from a population having 6 observable values (corresponding to the number of dots showing on each side of the die).

(a) Calculate the range of this population.

(b) We define the *sample range* to be the difference between the highest and the lowest values obtained from the two tosses. The sampling distribution of the sample range is provided in the table below. Calculate the expected value of the sample range.

(c) Comparing your answers to (a) and (b), do you conclude that the sample range is an unbiased estimator of the population range? Explain.

Possible Sample Range	Elementary Events	Probability
0	(1,1) (2,2) (3,3) (4,4) (5,5) (6,6)	6/36
1	(1,2) (2,1) (2,3) (3,2) (3,4) (4,3) (4,5) (5,4) (5,6) (6,5)	10/36
2	(1,3) (3,1) (2,4) (4,2) (3,5) (5,3) (4,6) (6,4)	8/36
3	(1,4) (4,1) (2,5) (5,2) (3,6) (6,3)	6/36
4	(1,5) (5,1) (2,6) (6,2)	4/36
5	(1,6) (6,1)	2/36
		36/36

7-3 INTERVAL ESTIMATES OF THE MEAN USING A LARGE SAMPLE

The interval estimate is the preferred form for reporting the results of sampling studies because it serves several purposes: (1) it provides an estimated range of values for the unknown parameter; (2) it acknowledges the presence of uncertainty as to the parameter's actual value; and (3) it may very simply show the degree of precision achieved by the results obtained. Thus, the interval estimate has a distinct advantage over the point estimate; by itself, the latter partially accomplishes only the first purpose. Indeed, it is not uncommon to mistake the point estimate for a precise expression of the parameter value itself. Whenever the evidence is obtained by a sample, perfect results are most improbable.

An interval estimate, being a range of possible values, tacitly illustrates the presence of uncertainty. It is ordinarily accompanied by a statement indicating the degree of uncertainty, expressed as a likelihood that the parameter being estimated actually does lie within the stated interval.

In this section, we will discuss the suitability of an interval estimate and show how it is constructed.

Confidence Interval Estimate of the Mean When σ Is Known

To illustrate how to establish suitable goals for obtaining the interval estimate and to show how one is obtained, we will begin with the problem of estimating a population mean when the standard deviation of the population is known.

Form of the Interval Estimate

We wish to establish an interval containing μ that will be in the form

$$a \leqslant \mu \leqslant b \tag{7-2}$$

We will use the sample mean, our estimator, to accomplish this. Ordinarily, the interval end points a and b are chosen in such a way that the interval is centered at the value of $\bar{X}$ actually obtained. Also, we wish to attach to the interval a measure of likelihood, such as .95 or .99, which will vary with the situation and which we will denote by C. Since we know that the sampling distribution of $\bar{X}$ is normal, for any value of C a particular normal deviate value z may be found so that

$$C = P[\mu - z\sigma_{\bar{X}} \leqslant \bar{X} \leqslant \mu + z\sigma_{\bar{X}}] \tag{7-3}$$

The inequality in expression (7-3) may be transformed into the order in expression (7-2) by rearranging the terms inside the brackets so that we obtain

$$C = P[\bar{X} - z\sigma_{\bar{X}} \leqslant \mu \leqslant \bar{X} + z\sigma_{\bar{X}}] \tag{7-4}$$

Although this places μ inside inequalities, μ is constant (and of unknown value) and cannot change. The lower and upper limits for μ are subject to chance variation, since they involve the random variable $\bar{X}$.

The event inside the brackets in expression (7-4) corresponds to the interval estimate obtained using the actual observed sample mean $\bar{X}$. Using $\sigma/\sqrt{n}$ in place of $\sigma_{\bar{X}}$, we may use the following expression to calculate the ***interval estimate for the population mean***

$$\bar{X} - z\frac{\sigma}{\sqrt{n}} \leqslant \mu \leqslant \bar{X} + z\frac{\sigma}{\sqrt{n}} \tag{7-5}$$

This interval is centered at $\bar{X}$, and the end points are partly determined by the value chosen for C, which in turn establishes z (the value of the normal deviate). This interval is sometimes expressed more compactly as $\mu = \bar{X} \pm z\sigma/\sqrt{n}$.

Meaning of an Interval Estimate

Keep in mind that the interval in expression (7-5) may not actually contain μ. This is because the end points depend upon the observed value for $\bar{X}$, which may be quite different from the value of μ due to sampling error. *Before* sample results are obtained, there is a probability C that μ will lie within $\bar{X} \pm z\sigma_{\bar{X}}$. But the interval in expression (7-5) is determined by a known value $\bar{X}$ calculated *after* taking the sample, so that we cannot attach a probability measure to it. Once we have a particular value for the sample mean, the interval estimated from it is certain to either contain or not contain μ. But we cannot know which will be true. Figure 7-1 shows the conceptual implications of the interval estimate when $C = .90$.

Confidence Level

Because we cannot attach a probability value to the truthfulness of our interval estimate, instead we employ a related term, the *confidence level*, to which we assign the following

DEFINITION The ***confidence level***, denoted by C, is the proportion of interval estimates—obtained from many repeated samples (of the same size) taken from the same population—that will contain the actual value of the parameter being estimated. Although a single sample is ordinarily taken, the confidence level is usually expressed as a percentage.

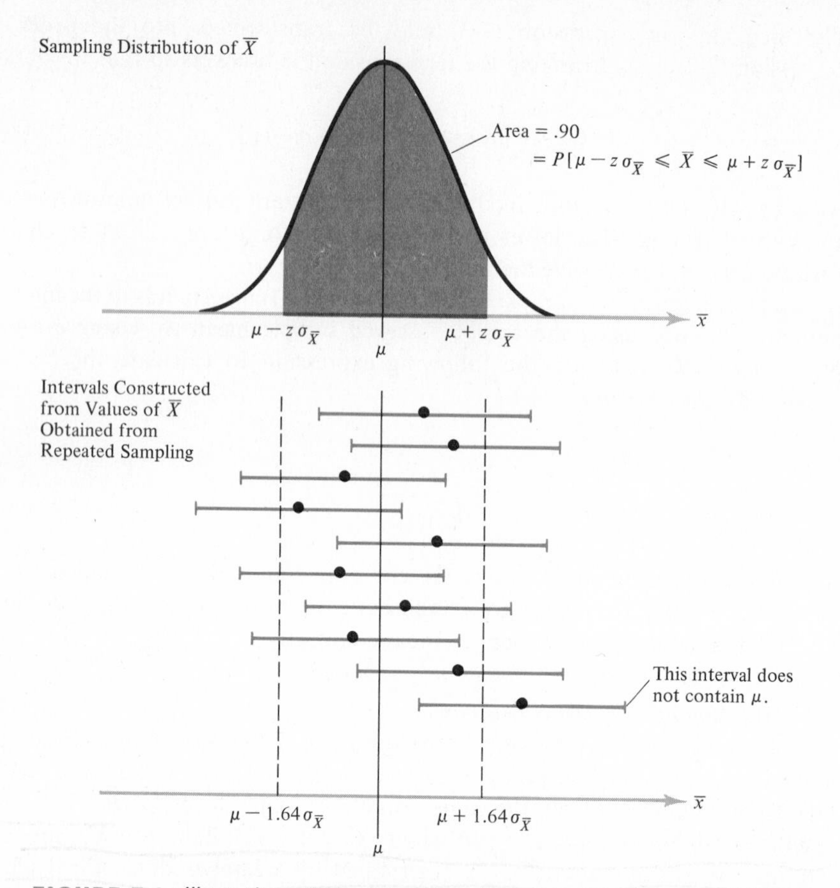

FIGURE 7-1 Illustration of the interval estimate concept. Intervals have been obtained for ten values of $\overline{X}$ calculated for different samples taken from the same population. Notice that nine of the intervals contain μ. Were many repeated samples to be taken from a population, the percentage of intervals obtained which contained μ would be approximately the same as the .90 probability that $\overline{X}$ lies within $\mu \pm 1.64\sigma_X$.

Because in practice just one sample is obtained, so that only a single value of $\overline{X}$ is available, just one interval is constructed from expression (7-5). This is referred to as a *confidence interval*. The level of confidence determines the value of z to be used. The larger the value of C, and hence z, the wider the resulting confidence interval. The value of z required for the construction of an interval corresponding to the desired confidence level may be found from Appendix Table D, which gives the areas under the normal curve. It is the normal deviate for which the area of the portion to the right of the mean is $C/2$. Thus, for a confidence level of .95 (or 95 percent), $C = .95$ and $C/2 = .475$, so that the required z is found from Appendix Table D to be 1.96. To simplify our task, normal deviates for some of the more commonly used confidence levels are provided in Table 7-2.

We will now illustrate how a confidence interval estimate may be obtained. Suppose that a sample of size $n = 100$ families is selected from the population of

TABLE 7-2 Normal Deviates for Common Confidence Levels and Reliablity Probabilities

Reliability or Confidence Level C	*Area Between Mean and z*	*Normal Deviate* [a] z
.80	.40	1.28
.90	.45	1.64
.95	.475	1.96
.99	.495	2.57
.995	.4975	2.81
.998	.4990	3.08

[a] These are the closest normal deviates for the given areas. Ties occur at C=.90, where either z=1.64 or z=1.65 could be used, and at C=.99, where either z=2.57 or z=2.58 would suffice. Throughout the book, we will use 1.64 and 2.57.

a large city and that the sample mean income is $\overline{X} = \$15{,}549.63$. We have *prior knowledge* that the standard deviation for the population of family incomes is $\sigma = \$5{,}000$. We wish to determine a 99-percent confidence interval estimate of the true population mean μ.

From Table 7-2, we determine the required normal deviate when $C = .99$ to be $z = 2.57$. Substituting the appropriate values into expression (7-5), we have as our 99-percent confidence interval estimate

$$\$15{,}549.63 - 2.57\frac{\$5{,}000}{\sqrt{100}} \leqslant \mu \leqslant \$15{,}549.63 + 2.57\frac{\$5{,}000}{\sqrt{100}}$$

or

$$\$14{,}264.63 \leqslant \mu \leqslant \$16{,}834.63$$

We *cannot* say that the stated interval contains μ with a .99 probability. (It either does or does not, but we don't know which case applies.) The proper interpretation of this 99-percent confidence interval is: ***If we were to repeat the above procedure often, each time selecting a different sample from the same population, then, on the average, 99 out of every 100 similar intervals obtained would contain μ, while one would not.***

Confidence Interval Estimate of the Mean When σ Is Unknown

The end points of a confidence interval estimate of μ described above depend upon the value for the standard error of $\overline{X}$. This, in turn, depends upon the population standard deviation σ, which is usually unknown and which, like μ, ordinarily must be estimated from the sample data. As we saw in Section 7-2, we may use the actual sample standard deviation s as the estimator of σ, so that the standard error of $\overline{X}$, $\sigma/\sqrt{n}$, may be estimated by $s/\sqrt{n}$. Thus, we may modify expression (7-4) to obtain the probability expression

$$C = P\left[\overline{X} - z\frac{s}{\sqrt{n}} \leqslant \mu \leqslant \overline{X} + z\frac{s}{\sqrt{n}}\right]$$

In order to evaluate this probability, we note that it is the same as

$$P\left[-z \leqslant \frac{\overline{X} - \mu}{s/\sqrt{n}} \leqslant z\right]$$

so that it is necessary for us to determine the sampling distribution of the statistic

$$\frac{\overline{X} - \mu}{s/\sqrt{n}}$$

This expression is what we would obtain from $(\overline{X} - \mu)/\sigma_{\overline{X}}$ (which the central limit theorem says is approximately normally distributed) by replacing $\sigma_{\overline{X}}$ with $s/\sqrt{n}$. Multiplying the numerator and denominator by $\sqrt{n}$ and canceling terms, we will use the equivalent expression for the above statistic:

$$\frac{(\overline{X} - \mu)\sqrt{n}}{s}$$

As we will see in Section 7-5, there is a special sampling distribution for the above statistic. However, for large values of n, it is approximately normally distributed for nearly all populations. For *large samples*, we may therefore use the following ***confidence interval estimate of μ when σ is unknown***

$$\overline{X} - z\frac{s}{\sqrt{n}} \leqslant \mu \leqslant \overline{X} + z\frac{s}{\sqrt{n}} \qquad (7\text{-}6)$$

In effect, the sampling error in using $s/\sqrt{n}$ to estimate $\sigma_{\overline{X}}$ is ignored. Large values of s will tend to stretch the interval beyond the desired width, and small values of s will have the opposite effect. In practice, ignoring the variability in s is not serious, since, on the average, about as many intervals obtained from repeated samples will contain μ as would be the case if σ were known.

Expression (7-6) only applies when the population itself is large. When the mean of a small population is to be estimated by a sample selected without replacement, then the standard error of $\bar{X}$ includes the finite population correction factor, so that the ***confidence interval estimate of μ for small populations is***

$$\bar{X} - z\frac{s}{\sqrt{n}}\sqrt{\frac{N-n}{N-1}} \leqslant \mu \leqslant \bar{X} + z\frac{s}{\sqrt{n}}\sqrt{\frac{N-n}{N-1}} \tag{7-7}$$

Example 7-3 A certain city maintains several public swimming pools. Since the local water supply is quite hard, special water-softening tanks are required to reduce the frequent buildup in water alkalinity. A test has been made of three different types of chemical agents in order to estimate the population mean alkalinity drop achieved in 10,000-gallon quantities of water of varying hardness as they pass through pool separation tanks. In each case, the same quantity of chemical has been applied. The following reductions in alkalinity (in parts per million) have been obtained for the three samples, each of size $n = 100$

	Chemical	
A	*B*	*C*
$\bar{X}$ = 51.3 ppm	$\bar{X}$ = 44.7 ppm	$\bar{X}$ = 43.4 ppm
s = 6.8	s = 7.3	s = 5.9

A confidence interval may be constructed for each chemical agent. The desired level of confidence is 99 percent, so $z = 2.57$. For chemical A, the following confidence interval is calculated using expression (7-6):

$$51.3 - 2.57\frac{6.8}{\sqrt{100}} \leqslant \mu \leqslant 51.3 + 2.57\frac{6.8}{\sqrt{100}}$$

or

$$\mu = 51.3 \pm 1.75$$

so that

$$49.55 \leqslant \mu \leqslant 53.05 \text{ ppm for chemical A}$$

Analogously, the following confidence intervals may be calculated for the other two water softeners:

$$\mu = 44.7 \pm 1.88 \quad \text{or} \quad 42.82 \leqslant \mu \leqslant 46.58 \text{ ppm for chemical B}$$

$$\mu = 43.4 \pm 1.52 \quad \text{or} \quad 41.88 \leqslant \mu \leqslant 44.92 \text{ ppm for chemical C}$$

Note that there is so little difference between the results for chemicals B and C that the confidence intervals overlap. Of the three softeners, chemical A seems to provide the greatest reduction in alkalinity. By using chemical A, the city *might* (remember, we do not know the value of μ) be able to achieve the greatest mean reduction in water alkalinity in its swimming pools.

Features Desired in a Confidence Interval

What features are desirable in a confidence interval estimate? Two considerations seem to be involved. One is the level of confidence itself, which expresses the degree of *credibility* that may be attached to the results. The other is the *precision* of the estimate, which is gauged by the width of the interval itself. But credibility and precision are competing ends in themselves. For a fixed sample size, a reduction in the interval width, which causes greater precision, can be achieved only at the expense of reducing z, which is tantamount to using a lower confidence level. Conversely, greater confidence can be obtained only at the expense of precision. Thus, a confidence interval may be very precise but not at all credible, or vice versa. For example, if z were .5 in Example 7-3, then the confidence interval for chemical A would become

$$\mu = 51.3 \pm .5(6.8)/\sqrt{100} = 51.3 \pm .34$$

or

$$50.96 \leqslant \mu \leqslant 51.64 \text{ ppm}$$

which is more precise than before, but the confidence level would be only 38 percent. Likewise, increasing the confidence level to 99.8 percent so that $z = 3.08$ would yield the less precise interval estimate

$$\mu = 51.3 \pm 3.08(6.8)/\sqrt{100} = 51.3 \pm 2.09$$

or

$$49.21 \leqslant \mu \leqslant 53.39 \text{ ppm}$$

The only way to increase both confidence and precision is to collect a larger sample in the first place. For example, suppose that $n = 1{,}000$ sample batches of water were softened with chemical A, and that the same sample results were obtained. With $C = .998$, the confidence interval would be

$$\mu = 51.3 \pm 3.08(6.8)/\sqrt{1{,}000} = 51.3 \pm .66$$

or

$$50.64 \leqslant \mu \leqslant 51.96 \text{ ppm}$$

a *more precise* interval with a *greater confidence level* than the one originally obtained.

EXERCISES

7-5 The board of trustees of a college is evaluating the performance of its president in order to decide whether or not to retain him. One trustee proposes that each trustee rate the quality of the president's performance on a scale from 1 to 5 in five areas:(1) endowment growth; (2) personnel relations; (3) operating efficiency achieved; (4) increased faculty research potential; and (5) successful new curriculum development. The ratings will be summarized in terms of frequency distributions, with the mean rating calculated for each category and submitted to the trustees at their next meeting, when a vote on retention will be taken. The board agrees to the procedure.

The chairman is concerned with the average ratings. He is worried about their precision and wonders what level of confidence should be attached to the results. A trustee who happens to be an expert in statistics diplomatically informs the chairman that these considerations are irrelevant. Explain the basis for the expert's stand.

7-6 Construct a 95-percent confidence interval for the means of large populations, given the following sample results:
(a) With $n = 100$, $\bar{X} = 100.53$ minutes, and $s = 25.3$ minutes.
(b) With $n = 200$, $\bar{X} = 69.2$ inches, and $s = 1.08$ inches.
(c) With $n = 350$, $\bar{X} = \$12.00$, and $s = \$7.00$.

7-7 Repeat Exercise 7-6, assuming in each case that a relatively small population of size $N = 1{,}000$ was used.

7-8 A psychometrist wishes to estimate the mean IQ of the eighth graders in her school district. Using a sample of $n = 100$ students, she found $\bar{X} = 105.6$ and $s = 14.7$. Assuming that the population is large, construct a 95-percent confidence interval estimate for the true mean.

7-9 A hospital administrator desires to improve the level of emergency-room service by increasing support personnel. To justify the increase, he must estimate the mean waiting time experienced by patients before being attended by a physician. From a random sample of $n = 100$ previously recorded emergencies, the sample mean waiting time was established at 70.3 minutes, with a standard deviation of 28.2 minutes. Construct a 99-percent confidence interval that the administrator might use to estimate the actual mean waiting time.

7-10 A tire manufacturer has obtained the following sample results for the tread life of $n = 50$ radial tires tested:

$$\bar{X} = 52{,}346 \text{ miles}$$

$$s = 2{,}911 \text{ miles}$$

(a) Construct a 99-percent confidence interval estimate of the mean tread life for all such tires manufactured.
(b) What is the interpretation of the interval you found in part (a)?
(c) Suppose that an estimate having a precision of ± 500 miles is desired. Find the corresponding confidence level.

7-11 In attempting to analyze the causes of the astronomical rise in the cost of medical care over the past decade, a government agency needs to estimate the mean fee for various operations. Suppose that a random sample of 250 splenectomies yields the following results:

$$\bar{X} = \$374.00$$

$$s = \$\ 56.25$$

(a) Construct a 99-percent confidence interval estimate of the current mean splenectomy fee.
(b) Suppose that ten years ago, a similar study based upon 150 operations

yielded the following results at a 95-percent level of confidence

$$\$130.00 \leqslant \mu \leqslant \$150.00$$

Find the values of $\overline{X}$ and s used to construct this confidence interval.

(c) Using the value of $\overline{X}$ obtained in (b) and the one given in (a) as point estimates of the mean fees in two years, estimate the percentage increase in splenectomy fees over the ten-year period.

7-4 INTERVAL ESTIMATES OF THE POPULATION PROPORTION

The population proportion π may be estimated by means of an interval, in much the same way as μ. Just as we used $\overline{X}$ to estimate μ, the sample proportion P may be used as the estimator of π. To do this, we take advantage of the normal approximation to the binomial sampling distribution of P (which, as we saw in Chapter 6, is allowable only for certain sample sizes).

We will use P to estimate π by an interval in the form

$$P - z\sigma_P \leqslant \pi \leqslant P + z\sigma_P \tag{7-8}$$

Again, we are faced with the difficulty arising from not knowing σ_P (which depends upon π, the value being estimated). This, too, must be determined from the sample results. For this purpose, we may use P in place of π in

$$\sigma_P = \sqrt{\frac{\pi(1-\pi)}{n}}$$

obtaining the point estimate

$$\sqrt{\frac{P(1-P)}{n}}$$

Substituting this into expression (7-8), we obtain the form of our ***confidence interval for estimating the population proportion***

$$P - z\sqrt{\frac{P(1-P)}{n}} \leqslant \pi \leqslant P + z\sqrt{\frac{P(1-P)}{n}} \tag{7-9}$$

Expression (7-9) applies for large populations or when sampling with replacement.

Example 7-4 A marketing research firm was retained by a food processor to determine what proportion of food buyers favored the quality of its canned corn over the similar product of one of its major competitors. A panel of $n = 100$ persons was randomly selected. Each panelist was given three cans of brand X and three cans of brand Y with the labels removed. The panelists were asked to use only this corn over a period of two months. Each can was numbered in the sequence in which it was to be used and were assigned so that brand X and brand Y would be alternated. Half of the group started with brand X; the other half with brand Y.

The testers were not told how many brands were involved. They were asked to rate each can of corn on a scale of 1 to 5 for each of four factors: tenderness, sweetness, consistency, and color. The scores were aggregated when all the data had been collected. This was done by brand, so that a comparison could be made—the brand receiving the highest score by a tester would be preferred. The test results are as follows

Number preferring brand X:	59
Number preferring brand Y:	37
Number of ties:	4
Total	100

A 99-percent confidence interval is desired in estimating π, the proportion of the buying population which favors brand X. Thus, $C = .99$, so that the normal deviate obtained from Table 7-2 is $z = 2.57$. The sample proportion of buyers preferring brand X is

$$P = 59/100 = .59$$

We may use expression (7-9) to determine the 99-percent confidence interval

$$.59 - 2.57\sqrt{\frac{.59(1-.59)}{100}} \leqslant \pi \leqslant .59 + 2.57\sqrt{\frac{.59(1-.59)}{100}}$$

or

$$\pi = .59 \pm .13$$

or

$$.46 \leqslant \pi \leqslant .72$$

Should the sample be taken from a small population without replacement, the finite population correction factor is incorporated. In such cases, we use a slightly different ***confidence interval for the population proportion for small populations***

$$P - z\sqrt{\frac{P(1-P)}{n}}\sqrt{\frac{N-n}{N-1}} \leqslant \pi \leqslant P + z\sqrt{\frac{P(1-P)}{n}}\sqrt{\frac{N-n}{N-1}} \qquad (7\text{-}10)$$

EXERCISES

7-12 Find the 99-percent confidence intervals for π, the proportion of undersize boards passing out of a saw mill's planing machine under the following assumptions:

(a) The number of boards measured is $n = 500$, and $P = .25$.
(b) $n = 1{,}000$ and $P = .1$.
(c) $n = 300$ and $P = .5$.

7-13 A poll has been taken to estimate the President's current popularity. Each person in a random sample of $n = 1{,}000$ voters was asked to agree with one of the following statements: (1) The President is doing a good job; (2) The President is doing poorly; (3) It is not possible to say. A proportion of .59 chose statement (1). Assuming that the actual number of voters is very large, construct a 95-percent confidence interval for the population proportion of voters who will choose statement (1).

7-14 Find the 95-percent confidence interval for the proportion of defective items in shipments of parts, when:
(a) $N = 10{,}000$, $n = 1{,}000$, and $P = .2$.
(b) $N = 5{,}000$, $n = 2{,}000$, and $P = .4$.
(c) $N = 8{,}000$, $n = 500$, and $P = .01$.

7-15 The highway-patrol director in a certain state has ordered a crackdown on drunken drivers. To see if his safety campaign is working, the director has ordered a sampling study to estimate the proportion of all fatal traffic accidents caused by drinking. In a random sample of $n = 100$ accidents, 42 percent were attributable to alcohol. Assuming that the accident population is large, construct a 95-percent confidence interval for the population proportion.

7-16 A printer has negotiated a contract with a telephone company to print directories. Because a disgruntled customer—stuck with a misspelled name or listed with the number of a mortuary—must suffer for a whole year (and will complain bitterly), the telephone company is anxious to avoid errors. Its contract therefore states that the printer must forfeit $100 for every page containing errors caused by the printing operation.

Counting the errors in a completed directory is an onerous chore. The majority are found by customers' complaints and most of these turn out not to be the printer's fault. Thus, a sample will be taken in order to determine what penalty to charge the printer. The printer has agreed to these terms. It is so difficult to find errors that many persons must independently search for them. A company accountant selects a sample of 50 pages from a 600-page directory and has 20 people in succession review each page for errors. An estimate will be made of the proportion π of pages containing errors.
(a) Construct a .998 confidence interval estimate of π, if $P = .37$.
(b) The contract spells out that the estimate of π must be a point estimate. What are the losses to the printer if the estimate is too large by .02?
(c) Calculate the probability that the estimator P would be .37 or over if the actual error rate is only $\pi = .35$.

7-5 INTERVAL ESTIMATES OF μ WHEN SMALL SAMPLES ARE USED

We have indicated that when a small sample is taken and σ is unknown, there are special problems associated with finding an appropriate sampling distribution for the statistic $(\bar{X} - \mu)\sqrt{n}/s$. For small sample sizes n, this statistic is not even approximately normally distributed, and we therefore cannot obtain the value for z for our desired level of confidence from the table of areas under the normal curve. We introduce a new probability distribution to serve this purpose.

The Student t Distribution

Shortly after the turn of the century, the statistician W.S. Gosset first studied this problem. Because his employer, a brewery, forbade him to publish, he chose the nom de plume "Student." Gosset derived a probability distribution, now referred to as the Student t distribution, for a random variable t. The sampling distribution of $(\bar{X}-\mu)\sqrt{n}/s$ is the t distribution. Whenever the population being sampled is normally distributed

$$t = \frac{(\bar{X}-\mu)\sqrt{n}}{s} \tag{7-11}$$

In practice, however, the t distribution may be used for samples taken from any population not highly skewed.

The Student t distribution, like the normal distribution, has a relative frequency curve that is bell-shaped and symmetrical, as shown in Figure 7-2.

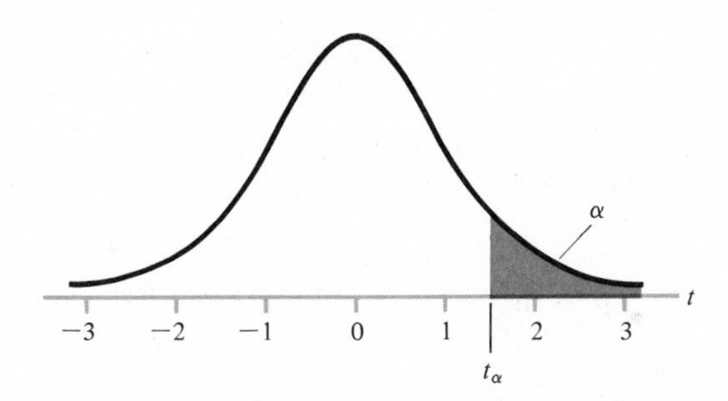

FIGURE 7-2 The Student t distribution.

The single parameter that determines the shape of its curve is called the *number of degrees of freedom*. When t represents $(\bar{X}-\mu)\sqrt{n}/s$, the number of degrees of freedom is $n-1$ (because for a fixed value of $\bar{X}$, there are only $n-1$ "free choices" for the values of the n observations used in calculating $\bar{X}$ and s).

To find the probability that t exceeds some value, it is necessary to use the table of areas under the Student t distribution in Appendix Table E on page 565. There is a separate distribution, and hence a separate set of values, for each number of degrees of freedom; these are shown in separate rows in the table. The values in the body of the table are the points t_α corresponding to the respective upper-tail area α (column) and degrees of freedom (row). The reason for this strange layout (compared to the normal curve table) is that we will be concerned with finding the point that corresponds to a given upper-tail area.

As with the normal curve, there is no need to have separate table entries for the left-hand portion of the t curve. Thus, in order to find t_α so that

$$P[t_\alpha \leqslant t] = \alpha$$

we read down the column headed by the probability value α, stop at the row corresponding to the number of degrees of freedom, and read the desired t_α value.*

For example, suppose that we wish to find $t_{.01}$ (the value for which the probability is .01 that t is greater than or equal to that value) and that the number of degrees of freedom is 20. From Appendix Table E, we obtain $t_{.01} = 2.528$. Thus

$$P[2.528 \leqslant t] = .01$$

A common guideline is that the Student t distribution need only be used when the sample size is less than 30; for larger samples, the normal distribution is ordinarily used. In actual fact, $(\bar{X} - \mu)\sqrt{n}/s$ has the t distribution for samples from normal populations, regardless of the magnitude of n. But for large values of n, the curves for the t distribution and the standard normal curve are very nearly the same. The choice of $n = 30$ as the demarcation point is quite arbitrary, but it is traditional. [As a practical matter, R.A. Fisher, whose table is universally copied (and is used in this book), jumps from 30 to 40 degrees of freedom, so that values in between are unavailable.]

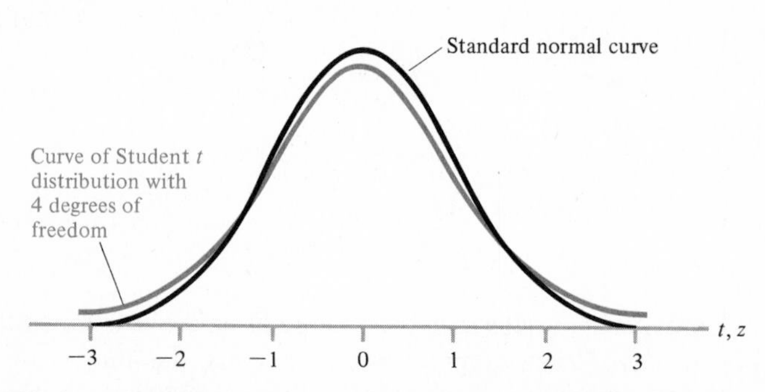

FIGURE 7-3 Comparison of the relative frequency curves of the standard normal distribution and the Student t distribution.

Figure 7-3 shows the curves for the standard normal and t distributions when the number of degrees of freedom is 4. Note that the normal curve lies below the t curve for random variable values lying in the tails. This reflects the fact that the t distribution assigns higher probabilities to extreme-valued outcomes

* The reader is cautioned that many other books place the value of the sum of both the lower- and the upper-tail areas at the heads of the columns.

than the normal distribution does, which is caused by the extra element of uncertainty arising from not knowing σ.

Importance of the t Distribution

The availability of the Student t distribution for making interval estimates has proved advantageous in many applications of statistics where there are often severe limitations upon the size of the samples obtained. The reasons may be economic, since the cost of experimentation with large samples could become prohibitive. Sometimes a large sample cannot be collected quickly enough to provide the answers required. This is the case when observations must be obtained sequentially from a manufacturing process having a low rate of output—large commercial jet aircraft production, for example. Or the sample observations must be made of rare phenomena, such as nuclear reactor accidents. Experiments in medicine, education, and psychology, where people are observed, are often limited to small samples.

Constructing the Confidence Interval

We may now use t_α in the same manner that we used z with large samples—to construct ***the confidence interval for estimating the population mean***

$$\bar{X} - t_\alpha \frac{s}{\sqrt{n}} \leqslant \mu \leqslant \bar{X} + t_\alpha \frac{s}{\sqrt{n}} \tag{7-12}$$

We must find the particular value of t_α that provides the same probability that $\bar{X}$ lies within $\mu \pm t_\alpha s/\sqrt{n}$ as the desired level of confidence C:

$$\begin{aligned} C &= P\left[\mu - t_\alpha \frac{s}{\sqrt{n}} \leqslant \bar{X} \leqslant \mu + t_\alpha \frac{s}{\sqrt{n}}\right] \\ &= P\left[-t_\alpha \leqslant \frac{(\bar{X}-\mu)\sqrt{n}}{s} \leqslant t_\alpha\right] \\ &= P[-t_\alpha \leqslant t \leqslant t_\alpha] \end{aligned}$$

The required value of t_α may be obtained from Appendix Table E for an upper-tail area of $\alpha = (1-C)/2$.

To show how this works, let us find a 90-percent confidence interval for sample results obtained from a survey. Since $C = .90$, we have $\alpha = (1-.90)/2 = .05$. Substituting this value into expression (7-12) gives us

$$\bar{X} - t_{.05} \frac{s}{\sqrt{n}} \leqslant \mu \leqslant \bar{X} + t_{.05} \frac{s}{\sqrt{n}}$$

To carry this further, suppose that $n = 25$ and the results obtained gave $\bar{X} = 10.0$

and $s = 5.0$. Then the degrees of freedom would be $n-1 = 24$, and for an $\alpha = .05$ area under the Student t distribution, $t_{.05} = 1.711$. The 90-percent interval is constructed by substituting these values into expression (7-12)

$$10 - \frac{1.711(5.0)}{\sqrt{25}} \leqslant \mu \leqslant 10 + \frac{1.711(5.0)}{\sqrt{25}}$$

so that

$$\mu = 10.0 \pm 1.7$$

or

$$8.3 \leqslant \mu \leqslant 11.7$$

Example 7-5 A psychologist is investigating the relationships between dreams and late-night television viewing. Dreaming time is roughly gauged by the length of periods when rapid-eye-movement (rem) occurs during sleep. These rem periods can be recorded by special equipment. The rem durations for a random sample of ten volunteers who have watched "Frankenstein" with Boris Karloff are provided in the table below. These data are used to estimate the population mean rem duration for all persons who watch "Frankenstein" prior to sleep. Similar data will be obtained for different kinds of late-night television programing.

Rem Duration (minutes) X	X^2
55	3,025
131	17,161
194	37,636
48	2,304
173	29,929
109	11,881
76	5,776
97	9,409
114	12,996
133	17,689
1,130	147,806

$$\bar{X} = \frac{\sum X}{n} = \frac{1{,}130}{10} = 113 \text{ minutes}$$

$$s = \sqrt{\frac{\sum X^2 - n\bar{X}^2}{n-1}} = \sqrt{\frac{147{,}806 - 10(113)^2}{9}} = 47.28 \text{ minutes}$$

The psychologist desires a 95-percent confidence interval estimate for the population mean. The number of degrees of freedom is $10-1=9$. Here, $\alpha = (1-.95)/2 = .025$, so that we obtain $t_{.025} = 2.262$. From expression (7-12)

$$113 - \frac{2.262(47.28)}{\sqrt{10}} \leqslant \mu \leqslant 113 + \frac{2.262(47.28)}{\sqrt{10}}$$

so that

$$\mu = 113 \pm 33.8$$

and

$$79.2 \leqslant \mu \leqslant 146.8 \text{ minutes}$$

Note that the confidence interval is quite wide, which is to be expected with such a large sample standard deviation when such a small sample must be employed.

EXERCISES

7-17 Determine the value t_α corresponding to each upper-tailed area of the Student t distribution specified below:

	(a)	(b)	(c)	(d)	(e)
α:	.05	.01	.025	.01	.005
Degrees of Freedom:	10	13	21	120	30

7-18 Construct 95-percent confidence intervals for estimates of the means of populations yielding the following sample results:
(a) $n = 6$, $\bar{X} = \$8.00$, $s = \$2.50$.
(b) $n = 15$, $\bar{X} = 15.03$ minutes, $s = .52$ minutes.
(c) $n = 25$, $\bar{X} = 27.30$ pounds, $s = 2.56$ pounds.

7-19 Construct 99-percent confidence intervals for (a), (b), and (c) in Exercise 7-18.

7-20 A biological researcher wishes to estimate the longevity of wild rats raised under laboratory conditions. A random sample of $n = 25$ second-generation offspring was reared from birth and monitored until all of the rats were deceased. Lifetime statistics of $\bar{X} = 31.2$ months and $s = 5.4$ months were obtained. Construct a 95-percent confidence interval estimate of the population mean longevity.

7-21 A medical researcher has a contract with a state prison that provides volunteers to test new drugs. He recently tested a drug to be marketed for the treatment of an exotic virus. Each patient was given injections of the live virus until he was infected and was then treated with the drug. Because of the obvious dangers, only 15 volunteers could be obtained. One of the parameters of interest is the mean recovery time. The following recovery times (in days) were obtained:

7	12	17
18	13	6
9	5	11
8	9	12
32	8	18

(a) In preparing his case for presentation to the Food and Drug Administration, a 95-percent confidence interval estimate of the mean recovery time had to be obtained. Find the values of $\bar{X}$ and s, and then calculate the confidence interval used.
(b) What do you think of the appropriateness of ascribing the above result to the population as a whole?

7-22 Suppose that the psychologist in Example 7-5 used a sample of $n=17$ volunteers who watched Walt Disney's "Fantasia" and obtained the following rem-duration results:

$$\bar{X} = 90 \text{ minutes}$$
$$s = 35.1 \text{ minutes}$$

Construct a 95-percent confidence interval estimate for the population mean rem duration.

7-23 An electronics firm currently trains its chassis assemblers on the job. It has been proposed that new employees be provided with one week's training. In order to test the quality of the training program, the output rate of ten trainees is to be compared to that of ten new hires receiving on-the-job training. Each girl from the trainee group is matched with one from the other group, according to scores received on an aptitude test. After training, each girl is given several chassis to wire, and the time to complete the task is recorded. For each of the $n=10$ pairs of girls, the time of the trainee is subtracted from that of the matching girl not in the training program. The following results for the differences are obtained:

$$\bar{X} = 3 \text{ minutes}$$
$$s = 2 \text{ minutes}$$

(a) Calculate a 95-percent confidence interval estimate of the mean time difference.

(b) If there is really no difference between the two training methods, then assuming that the population standard deviation is $\sigma=2$, find the probability that the sample mean would be 3 or more.

7-24 A large casualty insurance company is revising its rate schedules. A staff actuary wishes to estimate the average size of claims resulting from fire damage in apartment complexes having between 10 and 20 units. He will use the current year's claim-settlement experience as a sample. There were 19 claim settlements for buildings in this category. The average claim size was $73,249, with a standard deviation of $37,246. Construct a 90-percent confidence interval estimate of the mean claim size.

7-25 To compare the IQ scores of professional persons, a sociologist administered the Stanford-Binet test to random samples of 16 doctors and 25 lawyers. The following results were obtained:

Doctors	Lawyers
$\bar{X}$ = 121.4	$\bar{X}$ = 128.6
s = 12.0	s = 15.0

(a) Construct a 95-percent confidence interval for the mean IQ of doctors.

(b) Construct a 95-percent confidence interval for the mean IQ of lawyers.

(c) Do the sample data indicate that lawyers have higher IQs than doctors? Explain your answer.

7-6 DETERMINING THE REQUIRED SAMPLE SIZE

A statistic that is consistent becomes more reliable when a larger sample is used. Therefore, the reliability of an estimate made from such a statistic may be controlled by choosing a sample of appropriate size. But what makes a sample size appropriate? That depends upon (1) the disadvantages of erroneous estimates, and (2) the costs of obtaining the sample. As we have indicated, larger samples are more reliable, but they are also more costly. The costs of obtaining the sample must therefore be balanced against any potential damage from error, as reliability and economy are competing ends.

Error and Reliability

The risks of error are dependent upon two things: the chances of making them and the penalties they cause. In a sense, implicit costs are incurred when an estimate is erroneous if it leads to consequences that may be damaging. For example, consider an estimate of a politician's popularity, as expressed by the proportion of voters that prefer him. If the campaign manager estimates a proportion of preference for his candidate that is different from the actual one, then campaign planning may turn out to be far from optimal. In extreme cases, a hopeless campaign might be prolonged to the extent of committing political suicide (for a poor finish could preclude the politician from seeking office in the future), or a very attractive candidate might become too discouraged to continue the battle.

In many statistical applications, it is very difficult to place a monetary value on the results of an erroneous estimate. Also, the seriousness of an error will vary. Being off by half an inch in estimating the mean height of a group of persons would not be as serious as a six-month error in the estimate of expected lifespan, for longevity statistics can have a terrific influence on a nation's health planning. Nevertheless, all errors do have some implicit cost. Fortunately, when estimating from samples, the likelihood of major errors is smaller than the likelihood of minor errors. A more reliable sample will yield even smaller chances of major error. In selecting the sample size, the decision-maker must assess his preferences toward the risks involved by considering the chances of making the errors and thus incurring their implicit costs. Ideally, he should then select a sample size that achieves the most desirable balance between the chances of making errors, their costs, and the costs of sampling.

Example 7-6 Lake Airlines, a small regional carrier, derives a substantial amount of business as a feeder to the larger airlines. Passengers purchasing a ticket for a cross-country trip are issued a single ticket for the entire journey at the point of embarkation. When Lake sells these tickets, it is entitled to keep only its portion of the revenue and must forfeit shares to the other airlines involved. The reverse occurs for passengers originating with other airlines who use Lake for a single leg. In these cases, Lake must collect its share of the ticket revenue from the other airline. At the end of each month, the airlines balance their accounts based upon a detailed enumeration of the interairline tickets expired. This has proved to be a very costly and time-consuming process.

A consulting statistician has suggested that the airlines balance their accounts by means of random samples of collected tickets, arguing that only a small portion of the tickets must be analyzed, and so that clerical costs could be considerably reduced. There would be a certain amount of risk, however, for the samples obtained could provide erroneous figures.

Using a sample has drawbacks. Because of chance sampling error, Lake's claims against other airlines will be understated in some months, and overstated in other months. Viewing the long-run average or expected revenue losses as the cost of sampling error, the statistician has computed (by a procedure too detailed

Sample Size *n*	*Cost of Collecting Sample*	*Sampling Error Costs*	*Total Cost of Sampling*
100	$ 100	$40,222	$40,322
1,000	1,000	13,059	14,059
5,000	5,000	5,666	10,666
5,200	5,200	5,454	10,654
6,000	6,000	5,202	11,202
10,000	10,000	4,028	14,028

to discuss here) the cost for various sample sizes. Assuming that each sample ticket processed costs Lake $1, the costs of sampling have also been calculated. The results are provided in the table above. The sample size yielding the minimum cost to Lake is about $n=5{,}200$.

Figure 7-4 illustrates the concepts involved in finding the optimal sample size, which minimizes the total cost of sampling. The costs of collecting the sample data increase with the value of n. But larger samples are more reliable, so that the risks of loss from chance sampling error decline. The total cost of sampling—the sum of collection costs and error costs—will achieve a minimum value for some optimal n. This is the sample size that should be used.

Because of the difficulties associated with finding the costs of sampling error, the above procedure is not usually used to determine the required sample size in traditional applications. Instead, we may focus upon a single number that delineates insignificant errors from decidedly undesirable ones. This is called the *tolerable-error level.* In determining this level, we acknowledge that all error is

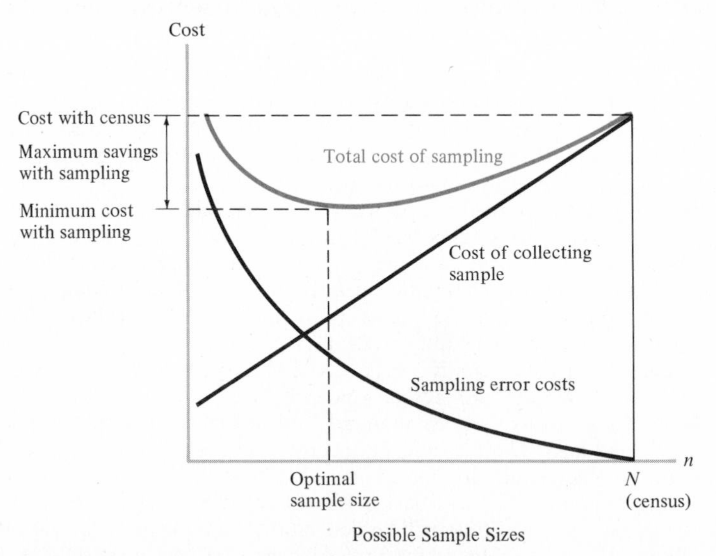

FIGURE 7-4 Relationship among costs in sampling.

undesirable, but that potential error must be accepted as the price for using a sample instead of a census. There are no guarantees that the tolerable error will not be exceeded. However, *large errors may be controlled by keeping the chances of their occurrence small.* Furthermore, as we have seen, reducing the chance of error increases reliability. Thus, we may speak of reliability in terms of the probability that the estimate will differ from the parameter's true value by no more than the tolerable error.

Estimating μ by $\bar{X}$

In order to develop a procedure for relating reliability to sample size, we will denote the tolerable error, which is the maximum amount by which the estimate ought to be above or below the parameter, by the letter e, a number in the same units as the parameter being estimated. As applied in traditional statistics, we define reliability in terms of e.

DEFINITION The ***reliability*** associated with using $\bar{X}$ to estimate μ is the probability that $\bar{X}$ differs from μ by no more than the tolerable error level e:

$$\text{Reliability} = P[\mu - e \leqslant \bar{X} \leqslant \mu + e] \tag{7-13}$$

When the sample is sufficiently large, we have seen that $\bar{X}$ tends to be normally distributed. Estimation reliability may therefore be described in terms of the normal curve. Figure 7-5 illustrates this when the reliability is equal to .95. Here, the colored area provides the probability that $\bar{X}$ will lie within a distance of amount e from μ, either above or below.

For a fixed e, only one normal curve has a shape that provides the desired area. Stated differently, since the shape of a normal curve is specified by its standard deviation, there is only one value of $\sigma_{\bar{X}}$ for which the desired area holds.

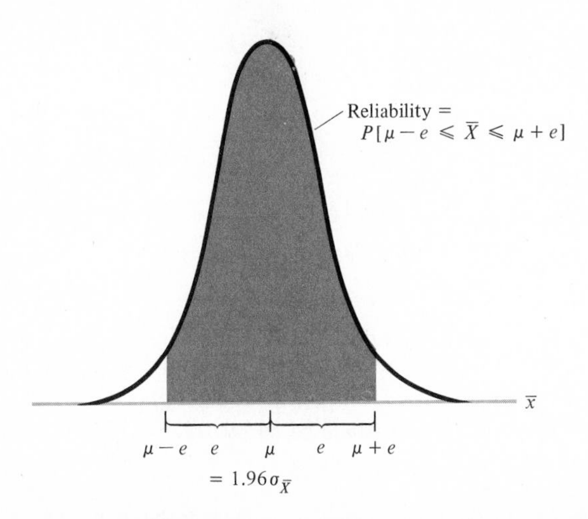

FIGURE 7-5 Relationship among reliability, tolerable error, and the sampling distribution of $\bar{X}$, when reliability (colored area) = .95.

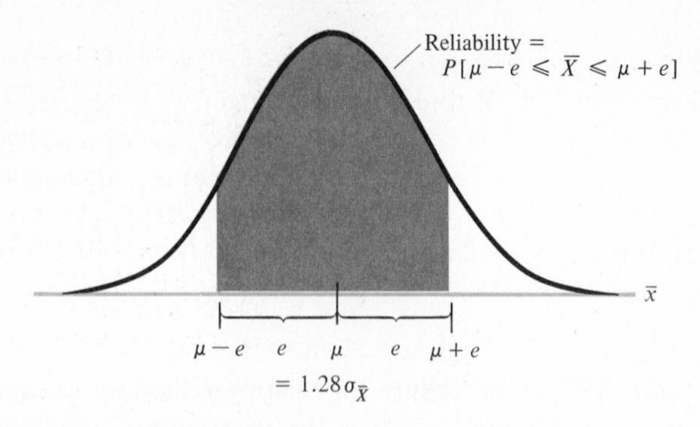

FIGURE 7-6 Relationship among reliability, tolerable error, and the sampling distribution of $\bar{X}$, when reliability (colored area) = .80.

This may be demonstrated with another case, shown in Figure 7-6. Here the reliability has been established at .80 for the same estimating situation. Note that this curve is flatter than the one in Figure 7-5 for the higher level of reliability. This indicates more variability, so that the curve in Figure 7-6 must have a larger $\sigma_{\bar{X}}$ value.

These two examples demonstrate that there is a relation between reliability and the standard deviation of $\bar{X}$. For a specified reliability and tolerable error e, there is a unique corresponding normal curve. Since the shape of this curve is dictated by $\sigma_{\bar{X}}$, the level of reliability itself is determined by the value of $\sigma_{\bar{X}}$. The desired level is ordinarily set in advance of sampling, so that a value for $\sigma_{\bar{X}}$ must be established in order to meet this goal. We must therefore determine the relationship between the desired reliability and the necessary value of the standard deviation of $\bar{X}$.

By rearranging the terms in expression (7-13) and dividing by $\sigma_{\bar{X}}$, we may express the reliability probability as

$$\text{Reliability} = P\left[\frac{-e}{\sigma_{\bar{X}}} \leqslant \frac{\bar{X} - \mu}{\sigma_{\bar{X}}} \leqslant \frac{e}{\sigma_{\bar{X}}}\right] \tag{7-14}$$

Recall that $(\bar{X} - \mu)/\sigma_{\bar{X}}$ is equal to the standard normal random variable, so that we may equivalently state that

$$\text{Reliability} = P\left[\frac{-e}{\sigma_{\bar{X}}} \leqslant Z \leqslant \frac{e}{\sigma_{\bar{X}}}\right] \tag{7-15}$$

Because the normal curve is symmetrical, half of the above probability lies between zero and $e/\sigma_{\bar{X}}$; thus, we may find the appropriate normal deviate for any reliability level. For .95, for example, the area between the mean and $z = 1.96$ is $.95/2 = .4750$. In this case, $\sigma_{\bar{X}}$ must be equal to that unique value permitting the following relationship to hold:

$$1.96 = \frac{e}{\sigma_{\bar{X}}}$$

Likewise, for a reliability of .80, the corresponding normal deviate would be $z = 1.28$, and the following must hold:

$$1.28 = \frac{e}{\sigma_{\bar{X}}}$$

Since each level of reliability corresponds to a unique area under the normal curve, choosing the reliability is tantamount to selecting a value for the normal deviate z. In a sense, a value for z can express the desired reliability level, so that we may summarize the relationship reliability, tolerable error, and $\sigma_{\bar{X}}$ by means of

$$z = \frac{e}{\sigma_{\bar{X}}} \qquad (7\text{-}16)$$

Keeping e fixed, since the value for z is established by the specified reliability, only $\sigma_{\bar{X}}$ can be adjusted to make the above hold. But if we are sampling from a population, its standard deviation σ is fixed, so that $\sigma_{\bar{X}}$ may be changed only by adjusting the size of the sample. This is so because, as we have seen, $\sigma_{\bar{X}}$ depends upon n (and also upon N, in the case of a small population), so that estimation reliability also depends upon the sample size, and vice versa.

Determining Required Sample Size in Estimating μ by $\bar{X}$

Independent Sample Observations

Sample observations will be independent either when they are obtained with replacement or when the population is large. In such cases, the standard deviation of $\bar{X}$ is found from

$$\sigma_{\bar{X}} = \frac{\sigma}{\sqrt{n}}$$

Since the relationship among reliability, tolerable-error level, and sample size is summarized by expression (7-16), we may substitute the above for $\sigma_{\bar{X}}$, obtaining

$$z = \frac{e}{\sigma/\sqrt{n}} = \frac{e\sqrt{n}}{\sigma}$$

Squaring both sides of the above expression, then dividing both sides by e^2, and finally multiplying both sides by σ^2, we obtain *the required sample size for sampling with replacement or when the population is large*

$$n = \frac{z^2\sigma^2}{e^2} \qquad (7\text{-}17)$$

The above expression tells how large n must be in order to obtain the desired reliability probability that $\bar{X}$ does not differ from μ by more than the tolerable error. This reliability probability fixes a value for z. The level of the tolerable error e establishes an arbitrary frame of reference for gauging reliability. Both z and e are under the statistician's control. The other factor, the population variance σ^2, remains fixed and is not subject to choice. The effect of each of these three factors is described below.

Three Influences on Sample Size

1. *The required sample size is directly proportional to the population variance.*

Thus, a population that is quite disperse will require a larger sample size than one that exhibits little variability. This tells us that for a constant level of reliability and the same tolerable error, a larger sample size will be necessary to obtain an estimate of the mean for a highly variable population.

Example 7-7 The problem of estimating the mean distance traveled by college professors during the previous summer vacation may be compared to the problem of estimating the mean number of miles that Cadillacs can travel on a tank of gasoline. The mileage of the professors will be quite varied: some opted to spelunk in Tasmania, while others traveled no further than the local golf course. We can also expect some variability in total gasoline mileage, depending upon the routes traveled, the amount of traffic, who is driving, the brand of gasoline used, and so on. The assumed standard deviation of $\sigma = 500$ miles for the population of distances traveled by professors is 50 times the value of $\sigma = 10$ miles we shall assume for the driving distance of Cadillacs without refueling.

If we desire estimates that differ from the true means by 5 miles or more with a reliability of .99, then, in either case

$$e = 5 \text{ miles}$$
$$z = 2.57$$

From expression (7-17), the sample sizes would be

$$n = \frac{(2.57)^2(500)^2}{5^2} = 66{,}049 \text{ (for professors)}$$

$$n = \frac{(2.57)^2(10)^2}{5^2} = 26.42, \text{ or } 27 \text{ (for Cadillacs)}$$

Note that we must always use the next largest whole number for the sample size. We see that about 2,500 (which is the square of 50) times as many professors as Cadillacs would be needed in the sample in order to provide the same reliability for a tolerable error of 5 miles.

This does not imply that estimating the mean length of professorial odysseys would actually require such a large sample. It is unreasonable to assume that the tolerable error should be as small as the one which might be required by General Motors for its Cadillacs. Also, there may be little need for so low a level of reliability, as the consequences of being off by a few hundred miles for professors may not be particularly harmful.

2. *The required sample size is inversely proportional to the square of the tolerable error.*

This tells us that decreasing or increasing the tolerable error by a factor results in increasing or decreasing the required sample size by that factor squared.

Thus, if we reduce e by half, then we need four times as large a sample—holding reliability constant. Likewise, if we increase e ten times, the sample size will be one-hundredth as large. Therefore, if we let $e = 50$, ten times the tolerable error used for professors in Example 7-7, then the sample size will be one-hundredth as large as before:

$$n = \frac{(2.57)^2 (500)^2}{50^2} = 660.49, \text{ or } 661$$

The cost of sampling, as reflected by the sample size, is highly sensitive to the magnitude of the tolerable error. As used in statistics, the tolerable error has roughly the same effect on cost as it does on precision machining. It is an order of magnitude more costly to have machine tolerances of one ten-thousandth of an inch versus one-thousandth of an inch.

3. *The required sample size is directly proportional to the square of the reliability-level normal deviate z.*

Thus, doubling z will cause n to be four times as large. We cannot say that doubling z will ordinarily raise reliability by a factor of 2, however, as z does not have a linear relationship to normal curve areas. Table 7-2 on page 257 provides normal deviates for the more commonly used levels of reliability. For example, decreasing z from 2.57 to 1.28 (a reduction in z by about half, so that n would be about one-fourth as large) lowers the reliability from .99 to .80—not quite a 20-percent drop.

The choices of e and z are interrelated, since the reliability usually depends upon the selected tolerable error. Since the tolerable error serves merely as a convenient cutoff point between serious and insignificant errors, its choice may be inseparable from the selection of the reliability level.

Should the sample size be fixed at some level—due to a budgetary limitation or some other practical constraint, like the availability of volunteers to test the effects of a new drug—then the tolerable error can be adjusted upward or downward only by a compensating increase or decrease in the level of reliability. Expression (7-17) shows that by raising or lowering e, z must be correspondingly changed when n and σ are fixed.

Sampling Without Replacement from a Small Population

When the sample is to be taken from a small population of size N, the expression for the standard error of $\bar{X}$ is

$$\sigma_{\bar{X}} = \frac{\sigma}{\sqrt{n}} \sqrt{\frac{N-n}{N-1}}$$

Substituting the terms on the right-hand side into expression (7-16) for $\sigma_{\bar{X}}$ and solving for n, we obtain ***the required sample size when sampling without replacement from small populations***

$$n = \frac{Nz^2\sigma^2}{(N-1)e^2 + z^2\sigma^2} \tag{7-18}$$

Example 7-8 A university research center wishes to estimate the mean monthly housing rental paid by minority families with one or more children. A sample is to be selected from the residents of a medium-sized city, where $N = 5{,}000$ such families live. The tolerable error has been set at $e = \$1.00$. Previous studies indicate that the standard deviation should be about $\sigma = \$20.00$. A sample size n must be determined so that the probability that the sample mean differs from the population mean by more than the tolerable error is only .95.

The required normal deviate is found from Table 7-2 to be $z = 1.96$. Substituting these values into expression (7-18), we obtain

$$n = \frac{5{,}000(1.96)^2(20)^2}{4{,}999(1)^2 + (1.96)^2(20)^2} = 1{,}175.5, \text{ or } 1{,}176$$

Determining Required Sample Size for Estimating π by P

The same sort of analysis that we applied to the sample mean may also be applied to find the number of observations required to estimate a population proportion from the value of its sample counterpart. The roles of tolerable error and reliability, as they affect n, are completely analogous. Tolerable error e represents the maximum allowable deviation between π and its estimator P, and e is expressed as a *decimal fraction*. Reliability is the probability that P differs from π by no more than e.

Using the normal approximation, the variability in the sampling distribution of P is summarized by σ_P. Analogously with using $\bar{X}$ to estimate μ, we have the requirement that σ_P be of appropriate size so that

$$z = \frac{e}{\sigma_P} \tag{7-19}$$

Since σ_P depends upon n (and upon the population size N as well, when sampling without replacement from a small population), the desired size for σ_P is obtained through the correct choice of n.

Independent Sample Observations

When the sample is obtained from a continuous process, or when N is large, so that the sample outcomes are independent, we have established the standard deviation of P to be

$$\sigma_P = \sqrt{\frac{\pi(1-\pi)}{n}}$$

Note that σ_P, and hence the sampling distribution of P, depend upon π. But it is π that we are trying to estimate. Thus, we cannot know σ_P until π is known. In practice, we must pick a value for π that seems reasonable and use it to establish σ_P. When the sample values are actually obtained, the point estimate made of π may turn out to be smaller or larger than the value used to select σ_P. Unlike estimating with $\bar{X}$, we cannot, therefore, be sure that our standard deviation of

P, and hence n itself, is of an appropriate size to ensure satisfying the prescribed level of reliability (although when using $\bar{X}$, the choice of n depends upon σ, which is not ordinarily known either).

Substituting the expression for σ_P into expression (7-19) and solving the resulting expressions for n, we obtain ***the required sample size when sampling with replacement or from a large population***

$$n = \frac{z^2\pi(1-\pi)}{e^2} \tag{7-20}$$

Note that this depends upon π, which must be *guessed* before we take the sample to estimate it. We do not use the actual value for π in expression (7-20), because it remains unknown.

Expression (7-20) shows that n is inversely proportional to the square of the tolerable error e and directly proportional to the square of the reliability factor z—just as it is with $\bar{X}$. In place of the population variance, we have the product $\pi(1-\pi)$. Thus, n depends upon π. When the product $\pi(1-\pi)$ is large, then the resulting n must also be large. This term is largest when $\pi = .5$:

$$\pi(1-\pi) = .5(1-.5) = .25$$

It can also be quite small, as when $\pi = .001$,

$$\pi(1-\pi) = .001(.999) = .000999$$

(As indicated in Table 6-5 on page 238, the normal approximation holds for $\pi = .001$ only when $n \geqslant 5{,}000$.)

Example 7-9 A gas company has noticed that there is a continual discrepancy between the cubic feet leaving its storage tanks and the actual consumption reported by company meter readers. Some of the discrepancy may be caused by inaccurate meters and minor leaks, but much is believed to be due to the lackadaisical behavior of the meter readers. Therefore, the proportion of misread meters is to be estimated for each reader. A sample of homes from each meter reader's route is selected for audit. A company supervisor, whose presence is unknown, will follow each reader on a portion of his route and reread the meters. Later, the figures will be compared. Initially, it will be assumed that 10 percent of the readings are in error in order to choose the sample size (not all readers will have the same error rate, however). The true proportion of erroneous readings is to be estimated within a tolerable error of .02 with a reliability probability of .90.

We have $e = .02$, $z = 1.64$, and a *guess* of .10 for π. Substituting these values into expression (7-20), we obtain the desired sample size for each route:

$$n = \frac{(1.64)^2(.10)(1-.10)}{(.02)^2} = 605.16, \text{ or } 606$$

Interpreting Sample Results

As noted, whether or not the sample results actually achieved provide the desired risk protection depends upon how good the original guess of π is. Suppose that the sample results for one meter reader's route indicate that he erroneously read 105 of his 623 meters; for him

$$P = \frac{105}{623} = .17$$

We must use this value as an estimator of π.

Substituting .17 in place of π in expression (7-20), we see that the sample size that gives the desired reliability should have been approximately

$$n = \frac{(1.64)^2(.17)(1-.17)}{(.02)^2} = 948.8, \text{ or } 949$$

which is larger than the sample size obtained when we assumed $\pi = .10$. The present sample is too small to adequately protect against the risks of an erroneous estimate. (Note that we still do not know the precise value of π and that we never will, but .17 is a better estimate than the value of .10 used initially.)

On the other hand, it is possible to use a sample size larger than the one actually required. For example, another meter reader may have misread only 32 of his meters, so that

$$P = \frac{32}{623} = .051$$

Substituting this value into expression (7-20) in place of π, we obtain

$$n = \frac{(1.64)^2(.051)(1-.051)}{(.02)^2} = 325.4, \text{ or } 326$$

so that, in this case, more observations than necessary would have been obtained.

Samples Taken Without Replacement from a Small Population

The procedure for determining n must be slightly modified when the sample is taken without replacement from a finite population of size N. Recall that the standard error of P must then be found from

$$\sigma_P = \sqrt{\frac{\pi(1-\pi)}{n}}\sqrt{\frac{N-n}{N-1}}$$

Substituting this into expression (7-19) and solving the resulting equation for n, using the level of reliability, population size, and tolerable error, and our initial guess at the value of π, we obtain the following expression for ***the required sample size when sampling without replacement from small populations***

$$n = \frac{Nz^2\pi(1-\pi)}{(N-1)e^2 + z^2\pi(1-\pi)} \qquad (7\text{-}21)$$

Reliability vs. Confidence and Tolerable Error vs. Precision

It may prove helpful at this point to clarify the key concepts that have been introduced in this chapter. An obvious question is: What is the difference between reliability and confidence? Another might be: How do we distinguish between tolerable error and precision? In both cases, are we not really dealing with two terms which are synonymous?

In one sense, confidence and reliability mean the same thing. But the distinction between these concepts is that they apply in *different stages* of a sampling study. The notion of reliability is mainly a concern in the *planning stage*, where an appropriate sample size is found to satisfy the statistical analyst's goals. Greater reliability may only be achieved either by increasing the sample size, and hence the data-collection costs, or by reducing the tolerable-error level. In selecting a reliability level, the statistician seeks to find a satisfactory balance between risks of error and cost. The concept of confidence applies in the last phase of a sampling study—the *data analysis and conclusions stage*. At this final point in an investigation, the statistician is concerned with communicating his results, so that his overriding concern is with *credibility*. Although often of the same value, the reliability probability need not be equal to the reported confidence level. Indeed, there are instances where the sample size n is not under the statistician's control, so that reliability considerations do not even enter into his planning.

We have already encountered the notion of precision in constructing confidence intervals. As the limits of the interval estimate become narrower, we can say that the resulting estimate becomes more precise. Like confidence, precision applies in the final stage of a sampling study. At this point in time, all sample data have been collected and precision can be improved only by a compensating reduction in confidence, or vice versa. The tolerable error level, on the other hand, is merely a convenient point of demarcation which separates serious errors from those that may be acceptable; it is used only in the planning stage and, with reliability, serves primarily to guide the statistician in selecting an appropriate sample size. Only when the reliability probability coincidentally equals the confidence level, and the planning guess as to σ or π is on target, will tolerable error be of equal magnitude to reported precision.

EXERCISES

7-26 The mean time required by automobile assemblers to hang a car door is to be estimated. Assuming a standard deviation of 10 seconds, determine the required sample size under the following given conditions.

(a) The desired reliability probability of being in error by no more than 1 second (in either direction) is .99.

(b) The desired reliability probability is .95 for a tolerable error of 1 second.

(c) A reliability of .99 is desired, with a tolerable error of 2 seconds. How does the sample size obtained compare with your answer to (a)?

(d) Same as (a), but suppose that $\sigma = 20$ seconds instead. By how much does the sample size increase or decrease?

7-27 The cost of obtaining the sample in Exercise 7-26 is \$1.50 per observation.

(a) For a tolerable error of 1 second, find the reliability that can be achieved by an expenditure of \$300.

(b) For a reliability probability of .95, find the tolerable error that must correspond to an expenditure of \$300.

7-28 A quality control manager has a flexible sampling policy. Because certain parts have different quality requirements, he must have a separate policy for each type of part. For example, a pressure seal used on some assemblies must be able to withstand a maximum load of 5,000 pounds per square inch (psi) before bursting. If the mean maximum load of a sample of seals taken from a shipment is less than 5,000 psi, then the entire shipment must be rejected.

The sampling policy for this item requires a sample to be large enough that the probability that the sample mean differs from the population mean by 10 psi or more (in either direction) is no greater than .05. Historically, it has been established that the standard deviation for bursting pressures of this seal is 100 psi.

(a) Find the required sample size for a shipment of $N = 10{,}000$ seals. Then do the same for a shipment of $N = 5{,}000$ seals. How much larger is the sample size for the larger lot? Why shouldn't it be twice as large?

(b) If the true mean maximum bursting pressure for the seals in a shipment is actually $\mu = 5{,}025$ psi, then use your sample size for $N = 10{,}000$ in (a) to determine the probability that the shipment will be rejected.

7-29 The purchasing officer for a government hospital that orders supplies from many vendors has the following policy: The quality of each shipment is to be estimated by means of a sample that will always include 10 percent of the items, regardless of the size of the shipment. Do you think this is a good policy? Discuss your answer.

7-30 The engineer responsible for the design of a communications satellite wishes to determine how many extra batteries to include in the power supply so that its useful life will be likely to exceed minimum specifications. A random sample of a particular type of battery is selected for testing. The engineer wishes to find the sample size that will provide an estimate of the true proportion π of batteries whose performance will be satisfactory—to an accuracy of $e = .05$ with a .95 reliability. We may assume that the population of batteries is very large.

(a) Find the necessary sample size, with an initial guess that $\pi = .5$ [which, from expression (7-20), would yield the largest possible sample size needed to satisfy the requirements].

(b) Suppose that after testing the number of batteries found in (a), only 250 prove satisfactory. What is the point estimate of π? If this were the true value of π, how many more batteries were tested unnecessarily?

REVIEW EXERCISES

7-31 The following salaries were determined from a sample of ten university presidents:

\$35,000	\$67,500	\$51,500	\$53,000	\$38,000
\$42,000	\$29,500	\$31,500	\$46,000	\$37,500

(a) Find the value of an efficient, unbiased, and consistent estimator of the mean for all the salaries of university presidents.

(b) Find the value of an unbiased and consistent estimator of the variance for all the salaries of university presidents.

(c) Find the value of an unbiased estimator of the proportion of all university presidents' salaries above $40,000.

7-32 An estimate is to be made of the mean number of square feet of scrap and wasted wood lost per log in the manufacture of one-half-inch exterior plywood. For this purpose, a random sample of 100 logs of various sizes from several mills was chosen. Each log's volume was carefully measured and the theoretical number of square feet of plywood was computed. This value was compared to the actual quantity achieved at the end of production, and the wood loss was calculated. When the study data were assembled, the sample mean and the standard deviation were computed as $\bar{X} = 1{,}246 \text{ ft}^2$ and $s = 114.6 \text{ ft}^2$, respectively. Construct a 95-percent confidence interval estimate for the mean wood loss.

7-33 A psychologist has designed a new aptitude test to be used by life insurance companies to screen applicants for beginning actuarial positions. In order to estimate the mean score achieved by all future applicants who will eventually take the test, the psychologist has administered it to nine persons. The sample mean and the standard deviation have been calculated as $\bar{X} = 83.7$ and $s = 12.9$ points, respectively. Determine the confidence interval estimates for the mean for each of the following levels of confidence:

(a) 99 percent
(b) 95 percent
(c) 90 percent
(d) 99.5 percent

7-34 The respective proportions of viewers watching each of the earliest prime-time television programs of one network on four successive nights are to be estimated within a tolerance of $\pm .01$ in such a manner that there is a 95-percent chance that this precision will be achieved (or reliability = .95). In doing this, the sample proportion P will be used. Determine the required sample sizes when the assumed population proportion for: (a) Tuesday is .3; (b) Wednesday is .6; (c) Thursday is .5; (d) Friday is .1.

7-35 A random sample of $n = 100$ widths for two-by-fours was selected from a shipment of $N = 500$ boards. The results show that $\bar{X} = 3.5$ inches and $s = .1$ inches. Construct a 95-percent confidence interval estimate of the boards in the entire shipment.

7-36 The following IQ scores have been obtained for a random sample of persons taken from a large population:

70	110	110
110	100	80
120	110	90

Construct a 95-percent confidence interval estimate of the population mean IQ.

7-37 The proportion of children born with defects is to be estimated, with a tolerable error of .1 and a reliability level of .99. A very conservative guess is that 20 percent of all children born have defects.

(a) Find the necessary sample size.

(b) Suppose that 13 defective children were found, using the sample size in (a), construct a 95-percent confidence interval estimate for the population proportion of children with defects.

7-38 A pharmaceutical house wants to estimate the mean number of milligrams of drug its machinery inserts into each capsule in four different sizes of pills. In each case, a standard deviation of 1 milligram is assumed to hold and a reliability level of 95 percent is desired. However, the amount of error tolerated varies with the pill size. Find the required sample size for: (a) tiny pills, .05 milligram error; (b) small pills, .1 milligram error; (c) medium pills, .2 milligram error; (d) large pills, .5 milligram error.

8 Hypothesis Testing

They are entirely fortuitous you say? Come! Come! Do you really mean that? . . . When the four dice produce the venus-throw, you may talk of accident: But suppose you made a hundred casts and the venus-throw appeared a hundred times; could you call that accidental?

Quintus, in Cicero's *De Divinatione*, Book I

In Chapter 7, we discussed how sample information may be used to estimate a population parameter. In this chapter, we will introduce another basic type of inference where a sample may be used in deciding between two complementary courses of action. Many important decisions made in the face of uncertainty involve a choice between just two alternatives. For example, in revising school curriculum, a superintendent must make a series of decisions: whether or not to try a new learning concept on an experimental basis, then to adopt it, and finally to accommodate it by dropping an existing program. At each decision-making stage, sample data may be used to facilitate his choice. Similarly, a medical researcher is often faced with deciding whether a new treatment is an improvement over an existing one. Often sample data are the only basis for making such a choice. Every purchasing agent, whether in business, government, public health, or education, must determine if the items he receives are of adequate quality. Many of these decisions are based upon sample outcomes and fall into the broad category of acceptance sampling, discussed in Chapter 6.

The statistical methods used in making these types of decisions are based upon two complementary assumptions regarding the true nature of a population. A new high-school curriculum will either raise average college entrance examination scores by 30 points, or it will not. A new treatment will reduce a patient's recovery time, or it will not. A shipment of supplies will have 10 percent or fewer defective items, or it will have a greater percentage of defective items. These assumptions are traditionally referred to as hypotheses. The observation of the sample is viewed as a test. Thus, the procedures for making decisions such as those above fall into an area of statistical inference called *hypothesis testing*. Because of their involvement in the decision-making process, hypothesis-testing procedures provide more dynamic forms of statistical inference than the more passive estimating techniques we have previously encountered.

In this chapter, we will introduce the basic concepts underlying hypothesis testing as they relate to decisions based upon population parameters such as the mean and the proportion. In later chapters, hypothesis testing will be presented in a broader context.

8-1 BASIC CONCEPTS OF HYPOTHESIS TESTING

We will develop the many concepts that comprise hypothesis testing by means of an extensive example involving the decision of whether to approve the manufacture of a new drug.

Vita Synthetics, a pharmaceutical manufacturer, has just developed a dietary supplement believed capable of significantly reducing heart disease resulting from a lifetime of poor eating habits.* The product introduces a chemical into the patient's blood stream that dissolves entrenched arterial deposits of cholesterol. One of the major causes of coronary ailments is a hardening of the arteries caused by the accumulation of cholesterol; by reducing its deleterious effects, Vita's product would effectively remove cholesterol from the roll of major killers.

Before the product can be approved, it must be extensively tested in order to assess its effectiveness in reducing blood cholesterol levels in human beings. The only evidence obtained to date has been from some limited testing of rhesus monkeys, but the Food and Drug Administration will not allow Vita's product to be sold without strong evidence that it will work on humans. For the test, a sample of 100 middle-aged men with no previous history of heart disease is selected. Each man will have his cholesterol level measured prior to the experiment and will then be given a standard daily dosage of the supplement; any other changes in lifestyle will be discouraged. At the end of the period, each man's cholesterol level is to be measured again.

Results of earlier studies indicate that human cholesterol levels fluctuate naturally over time. They also indicate that in about 10 percent of cases of high cholesterol levels, a major reduction occurs naturally and unaided. The government drug administrator has therefore determined that the supplement will be judged effective enough to be approved if 20 percent of all patients treated

* This situation is completely hypothetical, and any resemblance to an actual product is coincidental.

experience a major reduction in their cholesterol levels over a three-month span. Thus, there are two results to consider: an *ineffective supplement* yielding a proportion of major reductions equal to .10 for the population of treated persons, and an *effective supplement* providing a proportion of major reductions equal to .20.

Factors Influencing the Decision

On the basis of the experimental results, the administrator has to decide between two courses of action: (1) to approve the supplement, or (2) to disapprove it and to require further developmental research. There are some risks involved due to the uncertainty that will remain with respect to the supplement's true effectiveness even after the evidence has been collected. This residual uncertainty arises from the fact that a sample whose results may by chance deviate from the population must be used in the test. The decision really depends upon the answers to several questions, including:

1. What costs potentially ensue from incorrect decisions? *A decision is incorrect if the action taken is contrary to what would have been done were there no uncertainty.* In our example, an incorrect decision would result in (a) approving an ineffective supplement, or (b) not approving an effective supplement. It is important to understand that incorrect decisions of this kind would never occur were it not for the uncertainties arising from the sample. The costs resulting from incorrect decisions may be expressed in terms of their financial impact or their social damage.
2. What levels of risk are acceptable? Answering this question essentially requires that the government official look at the costs and assess his preferences with respect to the various probabilities of incurring such costs. It will be shown that the decision-maker has some freedom, although it is limited, with which to balance or to reduce the risks associated with making an incorrect decision. The necessarily subjective evaluations must be his.
3. When does the sample evidence favor a particular assumption or hypothesis regarding the population? A rule must be developed for the decision-maker, so that he will be able to decide whether the sample evidence confirms one of the assumptions—that the supplement either is or is not effective. His decision should be consistent with his attitudes toward the risks involved.

Testing the Value of a Population Proportion

Formulating the Hypotheses

So that we can make some generalizations that will be useful in analyzing other testing situations, certain statistical concepts must be introduced. It is common to refer to the possible assumed states of the effectiveness of Vita's dietary supplement in terms of hypotheses. Since the drug administrator is considering only two possible states—ineffective versus effective—the hypotheses are

Supplement is ineffective

Supplement is effective

We let π represent the proportion of middle-aged men in the population of those who would take the supplement (were it made available) who would experience a major reduction in cholesterol level. It must be emphasized that this is a *target population* that does not exist until (and will not exist unless) the supplement is widely available. However, it is still possible to draw a sample from the "imaginary" population, as the men in the sample will be treated in the same way. The first hypothesis is customarily referred to as the *null hypothesis* (here, "null" represents no change from the natural variation in cholesterol level). In terms of the population parameter π, we will use the following expression for the ***null hypothesis***

$$H_0 : \pi = \pi_0 = .10$$

since it has been decided that a 10-percent major reduction rate corresponds to what would occur without the supplement. The second hypothesis is called the *alternative hypothesis*. We analogously express the ***alternative hypothesis***

$$H_1 : \pi = \pi_1 = .20$$

since .20 is the rate of major cholesterol reduction at which the administrator would be willing to approve the supplement. The symbols H_0 and H_1 represent the null and alternative hypotheses, respectively. π_0 is the value of the parameter π when the null hypothesis H_0 is true, and π_1 is the value of π when the alternative hypothesis H_1 is true. The hypotheses will always be formulated so that H_0 and H_1 are opposites—when one is true the other is false.

At this point, it must be emphasized that although we are considering just two values of π (.10 and .20), *this does not imply that one or the other must be the true value*. Other values, such as .15 or .50, are conceivable. But we wish only to consider these two values at this time. These are two meaningful extremes which the health official can focus upon in evaluating his decision. In Section 8-2, we will find out how to cope with these additional possibilities.

Making the Decision

The decision to be made is the selection of the most appropriate course of action consistent with the sample evidence and the decision-maker's evaluations of risk. The actions may be expressed in terms of the hypotheses formulated above. The action "disapprove the supplement" corresponds to a belief that the null hypothesis is true (or that the supplement is ineffective); thus, it may also be expressed in an equivalent form: "accept the null hypothesis." Analogously, the action "approve the supplement" may be expressed as "reject the null hypothesis" (or equivalently as "accept the alternative hypothesis").

The sample evidence must be expressed in a manner that will meaningfully support one of these hypotheses. The health official should want to approve the drug if it works for a high percentage of the sample patients, as this would seem to deny H_0 (that the supplement is truly ineffective). Likewise, it seems that the official ought to disapprove the supplement if it is not very effective during the sample test, when the evidence would tend to confirm H_0. The sample information may be conveniently summarized in terms of the proportion P of men in the sample for which there is a major reduction in cholesterol level. We refer to P as the *test statistic*. Our decision-maker will accept H_0 for small values of P and reject H_0 for large values of P. The question is, where should he draw the line?

Selecting the point of demarcation for P will occur when the drug administrator decides what action he will eventually take. For now, we will suppose that a value of $P = .15$ is chosen. The test statistic is then used to formulate a decision rule. A *decision rule* translates sample evidence into the basis for making a choice of action. The value .15 is commonly referred to as the *critical value* or *acceptance number*. Once the critical value is established, we have a decision rule. In this instance, we may express this as:

DECISION RULE

Accept H_0 (disapprove the supplement) if $P \leqslant .15$

Reject H_0 (approve the supplement) if $P > .15$

The decision rule identifies two regions for the possible value of P. These can be illustrated as follows:

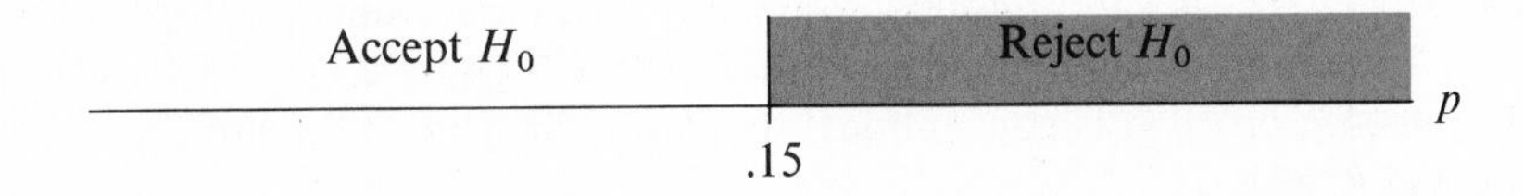

Having established the decision rule, the ultimate action to be taken by the government will be determined by the observed sample results after the test is conducted, the number of major reductions in cholesterol level are observed, and the value of P is calculated. This value of P will fall within either the acceptance or the rejection region, and H_0 will be accepted or rejected accordingly. When H_0 is rejected, the results are said to be *statistically significant*. Here, we use the word "significant" in a special sense; all hypothesis-testing results are important, but only when the null hypothesis is rejected is the test statistically significant.

TABLE 8-1 Hypothesis-Testing Decision Outcome Table

	Possible States of the Population Considered	
	Null Hypothesis True	***Null Hypothesis False (Alternative True)***
Alternative Course of Action	$\pi = \pi_0 = .10$ (ineffective supplement)	$\pi = \pi_1 = .20$ (effective supplement)
Accept Null Hypothesis *resulting* $P \leq .15$ (disapprove)	***Correct Decision*** Probability $1-\alpha$ (disapprove ineffective supplement)	***Incorrect Decision*** Probability β *Type II error* (disapprove effective supplement)
Reject Null Hypothesis (Accept Alternative) *resulting* $P > .15$ (approve)	***Incorrect Decision*** Probability α Significance level α *Type I error* (approve ineffective supplement)	***Correct Decision*** Probability $1-\beta$ (approve effective supplement)

The structure of the government's decision is shown in Table 8-1. Two courses of action are considered: to approve or to disapprove the supplement. There are also two population states: either the supplement is effective or it is not. For each combination of an action and a supplement quality, the outcome represents either a correct or an incorrect decision. Two possible outcomes are (1) to accept a true null hypothesis and disapprove an ineffective supplement, and (2) to reject a false null hypothesis and approve an effective supplement. Both of these would be correct decisions. Because of chance sampling error, however, the decision rule may also result in two other, undesirable outcomes, where in each case an incorrect decision or an error is made. The incorrect decisions are referred to as Type I and Type II errors. The *Type I error is to reject H_0 when it is true* (to approve an ineffective supplement), and the *Type II error is to accept H_0 when it is false* (to disapprove an effective supplement).

To help to evaluate the decision rule, we will find the probability of committing each error. It is conventional to denote the probabilities of these errors as α and β, where

$$\alpha = P[\text{Type I error}] = P[\text{reject } H_0 \mid H_0 \text{ true}]$$

$$\beta = P[\text{Type II error}] = P[\text{accept } H_0 \mid H_0 \text{ false}]$$

The value α is sometimes called the *significance level* of the test.

Finding the Error Probabilities

How good was the choice of .15 as the critical value? Recall that the decision-maker should be aware of the risks of incorrect decisions in making his decision and hence in establishing the decision rule. His attitude toward the risks depends upon both the costs resulting from incorrect decisions and the respective probabilities that these incorrect decisions will occur. The classical approach to hypothesis testing focuses on the error probabilities, providing for their adjustment so that acceptable weight is given to the costs. The error probabilities are found by analyzing the sampling distributions of the test statistic when each of the hypotheses is true.

Consider first the case in which the null hypothesis is true. The curve in the top portion of Figure 8-1 represents the normal distribution being used to approximate the binomial sampling distribution for P. Its mean and standard deviation are

$$\pi = \pi_0 = .10$$

$$\sigma_P = \sqrt{\frac{\pi_0(1-\pi_0)}{n}} = \sqrt{\frac{.1(1-.1)}{100}} = .03$$

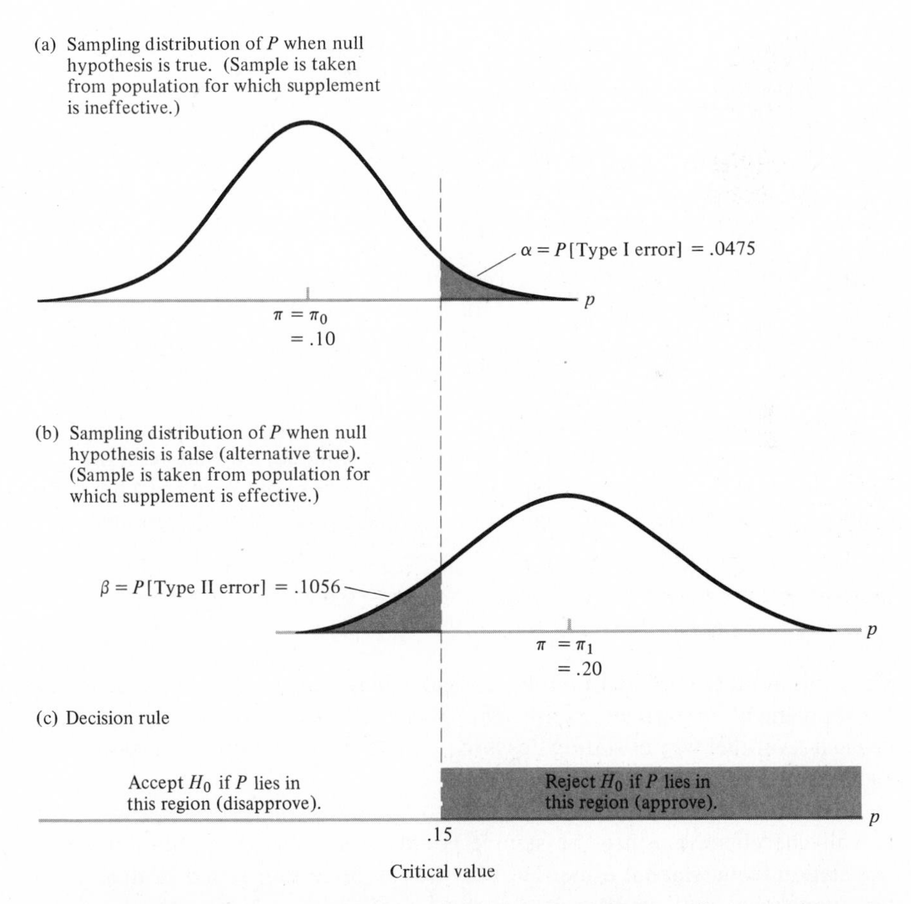

FIGURE 8-1 Sampling distributions of P for hypothesis-testing example.

The probability of making an incorrect decision of Type I (shown as the colored area to the right of .15 on the top curve) may be calculated:

$$\begin{aligned}\alpha &= P[.15 < P \mid H_0 \text{ true } (\pi = \pi_0 = .10)] \\ &= P\left[\frac{.15-.10}{.03} < \frac{P-\pi_0}{\sigma_P}\right] = P[1.67 < Z] \\ &= .5000 - .4525 = .0475\end{aligned}$$

When the null hypothesis is false (when the alternative is true), a different sampling distribution for P must be used, as is shown in Figure 8-1(b). This is because *the sample is selected from a different population* (namely, that for which the diet supplement is effective) and hence has a different value of π. The mean and the standard deviation of this sampling distribution are

$$\pi = \pi_1 = .20$$

$$\sigma_P = \sqrt{\frac{\pi_1(1-\pi_1)}{n}} = \sqrt{\frac{.2(1-.2)}{100}} = .04$$

(Note that σ_P changes value with π.) The probability of a Type II error (shown as the colored area to the left of .15 on the bottom curve) is thus determined to be

$$\begin{aligned}\beta &= P[P \leqslant .15 \mid H_0 \text{ false } (\pi = \pi_1 = .20)] \\ &= P\left[\frac{P-\pi_1}{\sigma_P} \leqslant \frac{.15-.20}{.04}\right] = P[Z \leqslant -1.25] \\ &= .5000 - .3944 = .1056\end{aligned}$$

Interpreting the Sample Results

Once a suitable decision rule has been formulated, all that remains is to await the results of the sample. Suppose that 23 men turn out to have significantly reduced cholesterol levels; the value of the test statistic is then determined to be

$$P = \frac{23}{100} = .23$$

Since this number is greater than the critical value .15, the evidence indicates that the supplement is effective; *H_0 would be rejected and the decision to approve would be made*. Another way of stating this is to say that the results would be statistically significant.

Even with P so large, there is still no guarantee that the supplement will actually be effective, since the sample could be a poor representation of the population from which it came. The only way to be certain would be to approve the supplement and monitor its progress in all heart patients. However, the probability of obtaining a value for P this large or greater is infinitesimal if H_0 is true and the supplement is actually ineffective. The normal deviate would be $z = (.23 - .10)/.03 = 4.33$.

EXERCISES

8-1 Find the error probabilities α and β for testing the hypotheses

$$H_0: \pi = \pi_0 = .1$$
$$H_1: \pi = \pi_1 = .2$$

when the sample size is $n = 100$ and the decision rule is

$$\textit{Accept } H_0 \text{ if } P \leqslant .16$$
$$\textit{Reject } H_0 \text{ if } P > .16$$

8-2 A quality control manager in an electronics assembly plant takes random samples of size $n = 100$ from the items in shipments received from outside suppliers. His decision rule is: If the proportion P of defective items is less than or equal to .07, then accept the shipment; otherwise, reject it. Assuming that $\pi_0 = .05$ and $\pi_1 = .10$, find the error probabilities α and β. (Assume that the size of the shipment is large, so that the binomial distribution—approximated by the normal—applies.)

8-3 Indicate whether the following statements are true or false. If false, explain why.

(a) The Type II error is the same as accepting the alternative hypothesis when it is false.
(b) Either a Type I or Type II error must occur.
(c) The significance level is the probability of rejecting the null hypothesis when it is true.

8-4 For each of the following situations, indicate the Type I and Type II errors and the correct decisions.

(a) H_0: New system is no better than the old one.
  (1) Adopt new system when new one is better.
  (2) Retain old system when new one is better.
  (3) Retain old system when new one is not better.
  (4) Adopt new system when new one is not better.
(b) H_0: New product is satisfactory.
  (1) Introduce new product when unsatisfactory.
  (2) Do not introduce new product when unsatisfactory.
  (3) Do not introduce new product when satisfactory.
  (4) Introduce new product when satisfactory.
(c) H_0: Batch of transistors is of good quality.
  (1) Reject good quality batch.
  (2) Accept good quality batch.
  (3) Reject poor quality batch.
  (4) Accept poor quality batch.

8-5 For each of the following hypothesis-testing situations, (1) state the Type I and Type II errors in nonstatistical terms (for example, approving a poor drug or disapproving a good drug); (2) identify the relevant population parameter (for example, proportion defective); (3) reexpress the hypotheses in terms of the parameter, using appropriate symbols; and (4) identify an appropriate test statistic.

(a) The null hypothesis is that a new diagnostic procedure is no improvement over the existing procedure, which is 90 percent satisfactory. The alternative hypothesis is that the new procedure has a higher quality where it proves to be 95 percent satisfactory. If there is an improvement in the percentage satisfactory, the new procedure will be adopted; otherwise, no change will be made.

(b) A union is negotiating with management for a four-day week with nine-hour working days—all at a slightly higher pay than at present. The union has argued that its members' output will actually increase by 10 percent using the new schedule, due to greater efficiency and higher morale. Management agrees to test the plan for one month in a small plant in which daily output has been a steady 100 units per man. Choosing as its null hypothesis that no change will occur, management will sign the contract if there is adequate evidence that the union claims are true; otherwise, some hard negotiating sessions are in store.

(c) The results of a public-opinion survey are being prepared for publication. The objective of the statistical study is to determine whether a stronger consumer protection law would be approved by a higher proportion of voters (.60) than the present law was approved by (.50). Taking as his null hypothesis that the modified law is not desired by more persons, with the reverse alternative hypothesis, the author must formulate a conclusion for his report.

8-6 A university introduces a new degree program.

(a) What null and alternative hypotheses are being tested if the Type I error is to incorrectly conclude that the program will succeed?

(b) What hypotheses are being tested if the Type II error is to incorrectly conclude that the program will succeed?

8-2 FORMULATING HYPOTHESES AND CHOOSING A DECISION RULE

Considerations in Choosing the Decision Rule

The decision rule tells us how to translate a sample result into action. The test statistic P served the purpose in the preceding example of the dietary supplement, being compared to a critical value. Because of sampling error, the test statistic may be unusually large or small, thus leading to an erroneous decision. As we have seen, the probabilities of such errors result from the choice of a decision rule. In Chapter 7, we saw that the chances of sampling error may be reduced if a larger sample is used. Thus, the sample size n will also affect the attained probabilities of error; n must be chosen so that the ensuing error probabilities will be of magnitudes consistent with the decision-maker's attitudes toward the risks involved. It is therefore necessary to consider both the sample size and the critical value in selecting an appropriate decision rule.

Effect of Changing the Critical Value

Recall that a decision rule establishes a point of demarcation, called the critical value, which tells us when to accept or to reject the null hypothesis, depending upon where the test statistic falls. We will denote the critical value for the test statistic P by the symbol P^*. In the example of the dietary supplement, we established a critical value of $P^* = .15$. We will now consider two other decision rules: one with $P^* = .17$, and another with $P^* = .13$. With these different rules, how will α and β be changed?

The two situations are portrayed in Figure 8-2. It may be helpful to compare them to the results for the original decision rule in Figure 8-1. In raising the

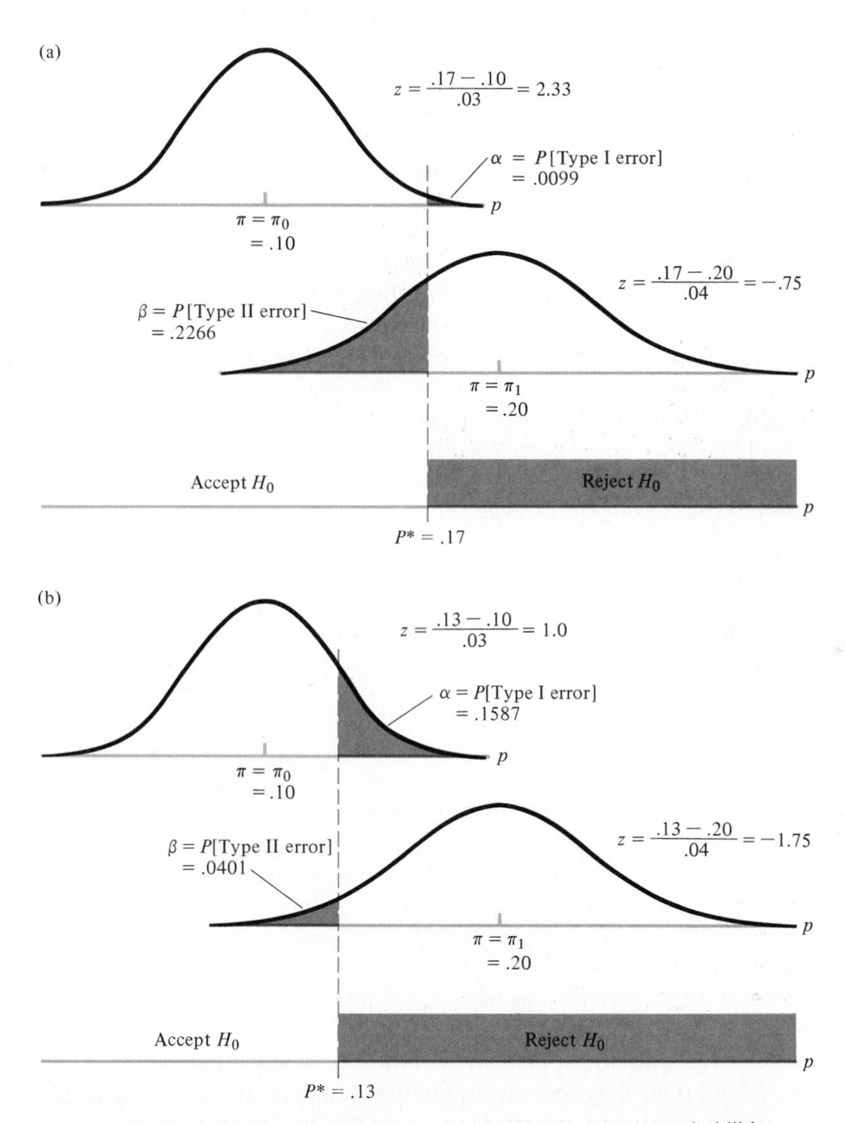

FIGURE 8-2 The effect of changing critical value upon error probabilities.

critical value from $P^* = .15$ to $P^* = .17$ [Figure 8-2(a)], we "stack the deck" in favor of accepting and thereby reduce the chance of rejecting H_0 when it is true. The new Type I error probability is smaller, $\alpha = .0099$ compared to the previous figure of .0475. Thus, the new decision rule with $P^* = .17$ decreases the chance of approving an ineffective supplement. The price paid in reducing α is to increase the chance that an untypical sample outcome will result in accepting H_0 when it is false. For by increasing P^*, we also raise the Type II error probability from $\beta = .1056$ to $\beta = .2266$, so that the chance of disapproving an effective supplement is more than doubled. The opposite effect is achieved by reducing the critical value from $P^* = .15$ to $P^* = .13$ [Figure 8-2(b)]; in this case, α increases while β decreases.

These illustrations demonstrate that for a fixed sample size *the chance of one error can only be reduced at the expense of increasing the probability of the other error.* The decision rule must be chosen so that an acceptable balance is achieved between the chances of the two errors.

Effect of Increasing the Sample Size

The only way to decrease both types of error probabilities is to increase the sample size. This will make the sampling distribution curves like those in Figures 8-1 and 8-2 become more peaked and have shorter tails, as their standard deviations will decrease with larger values of n (see Chapter 6, page 220). The net effect will be smaller error probabilites. For example, if the sample size were increased (so that $n = 200$), then the error probabilities, using a critical value of .15, would be determined as follows:

When Null Hypothesis Is True

$$\pi = \pi_0 = .10$$

$$\sigma_P = \sqrt{\frac{\pi_0(1-\pi_0)}{n}} = .021$$

$$\begin{aligned}\alpha &= P[.15 < P \mid \text{null hypothesis true}] \\ &= P\left[\frac{.15-.10}{.021} < \frac{P-\pi_0}{\sigma_P}\right] \\ &= P[2.38 < Z] \\ &= .5000 - .4913 = .0087\end{aligned}$$

When Null Hypothesis Is False

$$\pi = \pi_1 = .20$$

$$\sigma_P = \sqrt{\frac{\pi_1(1-\pi_1)}{n}} = .028$$

$$\begin{aligned}\beta &= P[P \leqslant .15 \mid \text{alternative true}] \\ &= P\left[\frac{P-\pi_1}{\sigma_P} \leqslant \frac{.15-.20}{.028}\right] \\ &= P[Z \leqslant -1.79] \\ &= .5000 - .4633 = .0367\end{aligned}$$

Figure 8-3 compares these error probabilities to those when the sample size is smaller.

Increasing the sample size reduces the chances that both kinds of incorrect decisions will be made when the same decision rule is applied. This will always be true. There is usually a practical limit as to how large the sample may be, for the cost of sampling may become prohibitive. In the testing of a drug, constraints may be enforced by government regulations limiting the allowable sample size.

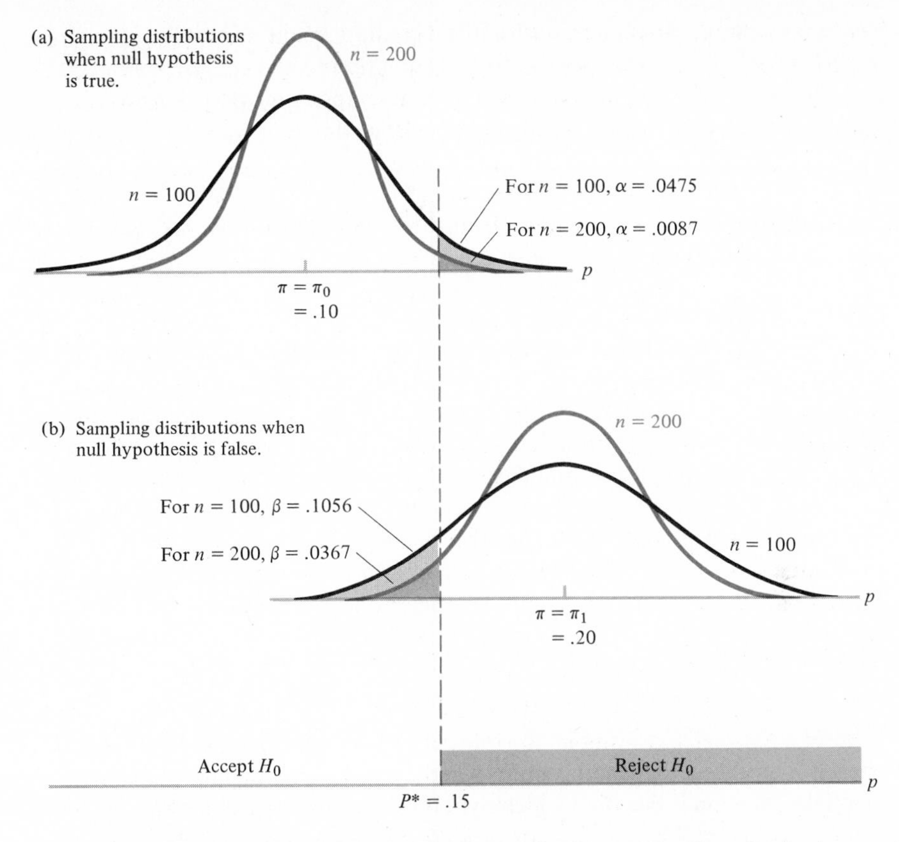

FIGURE 8-3 The effect of increased sample size on sampling distributions and error probabilities.

Usually, cost is the main consideration. Such a study would have limited funding, and doubling the sample size might not be practicable.

Determining Acceptable Error Probabilities

The basic question we will consider here is: Under what circumstances would the decision-maker find another decision rule better? The answer is that the better decision rule is the one that yields the more acceptable levels of risk, which can only be determined after considering the *costs* of incorrect decisions versus the *costs* of taking the sample.

Approving an ineffective supplement could have deleterious effects upon the long-run success of the pharmaceutical firm. The fact that the supplement did not work well would soon become known, probably generating considerable ill will, which could affect the credibility of the company's claims for its entire product line. In any case, the costs of setting up production and promotion would probably never be recovered. Having "burned his fingers" once, the president might be reluctant to reenter the diet-supplement market for some time, thereby reducing the likelihood of an early solution to the cholesterol problem. Also, the public, becoming somewhat skeptical about all new pharmaceutical developments, would need even stronger evidence that new products work.

On the other hand, failure to approve an effective supplement could result in considerable opportunity losses for the firm. There would also be the social cost of the thousands of premature deaths that could have been prevented if an anticholesterol supplement had been available.

Balancing the Risks

The decision maker must choose α and β himself, and his choices must be based upon his own *subjective evaluations* as to how the risks of incorrect decisions should be balanced. Both the Type I and Type II errors are undesirable. A decision rule must be chosen that will provide a lower probability of the more serious error. Only the decision-maker can judge whether society's loss of a beneficial drug is more or less serious than approving an ineffective one. Only he can establish the degree to which one error is more abhorrent than the other. He should therefore be wary of setting α and β at arbitrary or traditional levels.

Formulating Hypotheses

The null and alternative hypotheses are opposites, so that when one is true the other is false. This allows us to consider just two actions and two outcomes for each hypothesis, depending upon whether H_0 is true or false. One area of potential confusion is deciding which hypothesis to label H_0. In earlier statistical applications, the null hypothesis corresponded to the assumption that no change occurs, which accounts for the adjective "null." For example, in testing a drug, the usual hypothesis would be that it yields no improvement over nature. The H_0 used in the dietary supplement illustration corresponds to this designation. In modern applications, other considerations determine which is the null hypothesis, and H_0 may represent a change rather than no change. For instance, there would have been no material difference in our discussions in Section 8-1 if the H_0 and H_1 designations had been reversed so that the null hypothesis was $\pi = .20$ instead of $\pi = .10$. Of course, the other labels would also be different, so that the Type I and Type II errors would also be reversed and so would α and β.

Another often used guideline is that the null hypothesis ought to be the one that the decision-maker wishes to disprove. *A common practice is to designate as H_0 that hypothesis for which rejecting when true is the most serious error*, so that the Type I error would be worse than the Type II error. As we saw in the dietary supplement illustration, the decision rule was chosen so that this happened to be the case.

Simple and Composite Hypotheses

Deciding whether or not to approve the new dietary supplement in our example was purposefully kept simple to ease the development of fundamental hypothesis-testing concepts. Recall that the null hypothesis corresponded to no change over nature, so that the proportion of major cholesterol reductions achieved by the dietary supplement would then be $\pi = .10$. Likewise, the government considered the supplement to be "fully effective" when $\pi = .20$. Thus, the hypotheses were formulated as:

$$H_0 : \pi = \pi_0 = .10 \qquad \text{(ineffective supplement)}$$
$$H_1 : \pi = \pi_1 = .20 \qquad \text{(effective supplement)}$$

These may be referred to as *simple hypotheses.* In fact, $\pi = \pi_1 = .20$ is only one of many alternative hypotheses that the drug administrator could have considered. Realistically, we may suppose that π might be equal to $.11, .12, \ldots, .19, .20$, or some greater value. Indeed, there may be no reason why π might not assume any value above .10 (from being imperceptibly effective to perfectly so), in which case we might have $\pi = 1.0$.

Because the government would be interested in the entire range of possibilites for π, the following hypotheses are more realistic:

$$H_0 : \pi = \pi_0 (=.10)$$
$$H_1 : \pi > \pi_0$$

We refer to such an expression of H_1 as a *composite hypothesis*, because it includes all possible alternative values for π. Problem situations where the decision outcomes are determined by a population parameter will ordinarily involve a composite hypothesis. Sometimes the null hypothesis is also composite. For example, if the dietary supplement were to actually increase some patients' cholesterol, then π might fall below the natural level and we might use $H_0 : \pi \leqslant \pi_0$ (=.10). (We will assume that this cannot happen for this particular supplement.)

Sometimes a hypothesis can be *two-sided.* For example, a manufacturer may wish to test the output of a production process. As his null hypothesis, he may assume that the mean weight μ of his items is the desired output, so he uses $H_0 : \mu = \mu_0$. Because, on the average, the items produced may be over- or underweight, the alternative hypothesis would take the form $H_1 : \mu \neq \mu_0$; that is, H_0 is violated either when $\mu < \mu_0$ or when $\mu > \mu_0$. Tests with two-sided hypotheses will be discussed in Section 8-4.

EXERCISES

8-7 Assume that each of the following situations involve the hypotheses

$$H_0 : \pi = \pi_0 = .1$$
$$H_1 : \pi = \pi_1 = .2$$

In each case, determine the values of the error probabilities.
(a) $n = 50$, $P^* = .15$; (b) $n = 50$, $P^* = .18$; (c) $n = 200$, $P^* = .18$.

8-8 An opinion survey has taken a random sample of $n = 100$ persons from a population where the proportion favoring a certain presidential candidate is to be tested. The candidate desires to cancel his campaign if the proportion preferring him is $\pi_0 = .30$, his null hypothesis. But he would continue the campaign if he knew that this proportion was $\pi_1 = .40$, his alternative hypothesis. Unfortunately, he can only know the sample result, and has he established a decision rule with a critical value of $P^* = .35$.

(a) What are the values of α and β?

(b) Will α increase or decrease if the decision rule is changed to a critical value of $P^* = .36$? What will the new value of α be? What will be the new value for β? Did β increase or decrease?

8-3 LARGE-SAMPLE TESTS USING MEAN OR PROPORTION

With the basic concepts of hypothesis testing firmly in hand, we are ready to describe the usual procedure for such investigations. Ordinarily, the sampling study begins with the formulation of hypotheses. After some introspection on the part of the decision-maker, a Type I error probability α (significance level) is determined. Everything that follows is largely mechanical: The sample data are collected, and a critical value corresponding to the desired value of α is found. Comparing the computed value of the test statistic to this critical value, the decision rule indicates the appropriate action to take.

Although we initially considered a qualitative population where the sample proportion was used as the test statistic for making inferences about π, hypotheses regarding quantitative populations usually involve the mean. In such cases, the sample mean serves as the test statistic. The procedure for using $\bar{X}$ to test for μ differs only in minor ways from tests involving P.

Testing the Population Mean

To illustrate testing the population mean, we will consider an investigation of the influence of childrens' reading achievements upon their IQ scores. Since the popular IQ tests largely reflect verbal skills, it is widely believed that poor readers test lower in IQ than children whose reading is satisfactory. A reading coordinator in a large public-school system is conducting a study to determine whether this is true for the fifth-grade pupils in her district. The population of interest in her study is the IQ scores of children whose reading skills are at least one grade level lower.

Historically, fifth-grade pupils in her school district have established a mean IQ score of 105. Since she regards as the most serious error wrongly concluding that low reading scores yield lower IQs, the coordinator chooses to test with the following hypotheses:

$$H_0: \mu \geqslant \mu_0\ (=105) \quad \text{(IQs are just as high)}$$

$$H_1: \mu < \mu_0 \quad \text{(IQs are lower)}$$

Here, μ represents the population mean IQ for poor readers, and μ_0 represents the mean IQ level of all fifth graders without regard to reading skill. In this type of investigation, the null hypothesis is usually a composite one, which, in this case, gives poor readers the "benefit of the doubt" by allowing for the contingency that as a group they may have greater than average measurable intelligence.

Because of expenses, the coordinator has to rely upon sample data. For a random sample of $n = 100$ poor readers, she is able to determine current IQ scores from school records. As her test statistic, she uses the sample mean IQ $\bar{X}$, which she compares to a critical value $\bar{X}^*$. Since a low sample mean IQ score would tend to refute H_0 that poor readers are at least as intelligent as satisfactory ones, a value for $\bar{X}$ lower than $\bar{X}^*$ should lead to rejecting this null hypothesis. Because she desires a 5-percent chance for the Type I error of incorrectly rejecting her null hypothesis, the coordinator chooses a value for $\bar{X}^*$ so that the colored area in the lower tail of the sampling distribution shown in Figure 8-4 is equal to $\alpha = .05$.

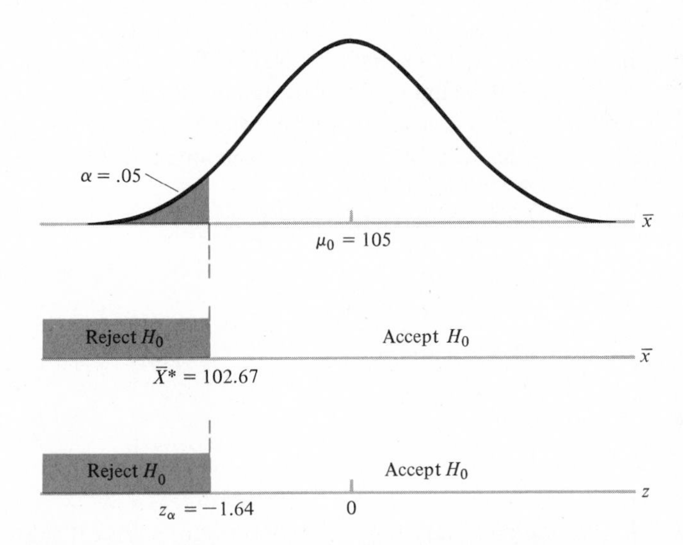

FIGURE 8-4 Sampling distribution of $\bar{X}$, with decision rule and Type I error probability.

Although here H_0 allows μ to be any IQ value ≥ 105, in setting α, we focus on the smallest of these—placing μ at 105. (In testing with any composite null hypothesis, we set μ_0 or π_0 at the value found by ignoring the "$<$" or "$>$" portions, so that the value for α always reflects the worst case.)

Thus, we use the normal curve with a mean of $\mu_0 = 105$ to represent the sampling distribution of $\bar{X}$. The standard deviation for this curve is $\sigma_{\bar{X}} = \sigma/\sqrt{n}$. Even though the population standard deviation has an unknown value, we may still use the normal curve because the sample size is large. The coordinator estimates the value of σ from the sample data by the sample standard deviation

$s = 14.2$. Thus $\sigma_{\bar{X}}$ is estimated by

$$\frac{s}{\sqrt{n}} = \frac{14.2}{\sqrt{100}} = 1.42$$

In employing the normal curve, any value for $\bar{X}$ corresponds to a particular normal deviate by the relation

$$z = \frac{\bar{X} - \mu_0}{\sigma_{\bar{X}}}$$

The critical value of the test statistic $\bar{X}^*$ provides a very important normal deviate value, for which the lower-tailed area under the standard normal curve is equal to α; we denote this *critical normal deviate* as z_α, and it may be read from Appendix Table D (in reverse). For $\alpha = .05$, we may find z_α (which we can also denote as $z_{.05}$) by finding the area under the normal curve that is closest to $.5000 - .05 = .4500$ (which, in this case, is either .4495 or .4505). We arbitrarily choose the smaller of these. Using the .4495 figure, the corresponding normal deviate happens to be $z_{.05} = -1.64$. Here, $z_{.05}$ is negative because the area is a lower-tailed one. For convenience, the values of z_α for commonly encountered significance levels are provided in Table 8-2 on page 304.

We can relate a critical normal deviate to $\bar{X}^*$ by

$$z_\alpha = \frac{\bar{X}^* - \mu_0}{\sigma_{\bar{X}}}$$

Solving the above for $\bar{X}^*$, we obtain

$$\bar{X}^* = \mu_0 + z_\alpha \sigma_{\bar{X}}$$

Using $s/\sqrt{n} = 1.42$ as the estimate for $\sigma_{\bar{X}}$, the coordinator's critical value is

$$\bar{X}^* = 105 - 1.64(1.42) = 102.67$$

This may be expressed in terms of the following decision rule:

Accept H_0 (conclude IQs are just as high) if $\bar{X} \geq 102.67$

Reject H_0 (conclude IQs are lower) if $\bar{X} < 102.67$

After compiling her data, the reading coordinator computes $\bar{X} = 101.5$ for the sample mean IQ of the poor readers in her sample. Since this value is less than $\bar{X}^* = 102.67$, she must *reject* the null hypothesis and conclude that poor fifth-grade readers indeed test lower in intelligence than satisfactory readers. Expressed another way, the test results are statistically significant at a level of $\alpha = .05$.

As an alternative to the above decision rule, the same choice can be made using normal deviate values. The observed sample mean can be transformed into a particular point on the standard-deviation scale, using $s/\sqrt{n} = 1.42$ as the estimate of $\sigma_{\bar{X}}$:

$$z = \frac{\bar{X} - \mu_0}{\sigma_{\bar{X}}} = \frac{101.5 - 105}{1.42} = -2.46$$

The value $z = -2.46$ gauges how *rare* it would be under the null hypothesis to obtain a sample mean as small or smaller than the $\bar{X}$ value actually observed. Assuming that 105 is the true population mean IQ for the poor readers and that the population standard deviation is 14.2, then the area under the normal curve between $\mu_0 = 105$ and $\bar{X} = 101.5$ is found from Appendix Table D to be .4931, so that the probability of getting a result at least as small is $.5 - .4931 = .0069$. Since this probability is smaller than the desired Type I error probability of $\alpha = .05$, the sample result is rare enough under the null hypothesis to justify rejecting it. (Remember that the actual value of σ is not known but has been estimated by $s = 14.2$. Unless 14.2 happens to be the true value, .0069 is only an *approximate* probability.)

The testing approach outlined here can be shortened by comparing the computed value for z to the critical normal deviate, so that H_0 is accepted if $z \geqslant z_\alpha$ and rejected if $z < z_\alpha$. Since $z = -2.47$ is smaller than $z_{.05} = -1.64$, H_0 must be rejected.

In this illustration, the decision rule was actually formulated after the sample data were collected. Such a procedure is correct because once μ_0 is determined, the sample size is prescribed, and the desired significance level is decided, everything that follows is automatic. But it would be improper to change μ_0, n, or α *after* evaluating the sample results. As a practical matter, the decision rule can be established either before or after the data are collected. One advantage of establishing a rule before sampling is that it is then possible to see what Type II error probabilities β result from the initial rule and to make any modifications that may be necessary to provide a better balance between α and β. But, again, the decision rule should never be altered after analyzing the data. Unless σ is known in advance when testing with $\bar{X}$, the decision rule can be found only after the sample statistics have been computed.

Summary of Procedure

It is convenient to categorize hypothesis-testing situations in terms of what portion of the test statistic's sampling distribution corresponds to α. In the preceding illustration, α represents the lower-tail area under the normal curve for

$\overline{X}$; this situation is therefore a *lower-tailed test*. In the dietary-supplement example earlier, the hypotheses were designated so that an *upper-tailed test* resulted. In general, when H_0 and H_1 correspond to the smaller and the larger

TABLE 8-2 Critical Normal Deviate Values for Common Significance Levels

Note: For lower-tailed areas, z_α is negative.

α	z_α
.20	.84
.10	1.28
.05	1.64
.025	1.96
.01	2.33
.005	2.57
.001	3.08

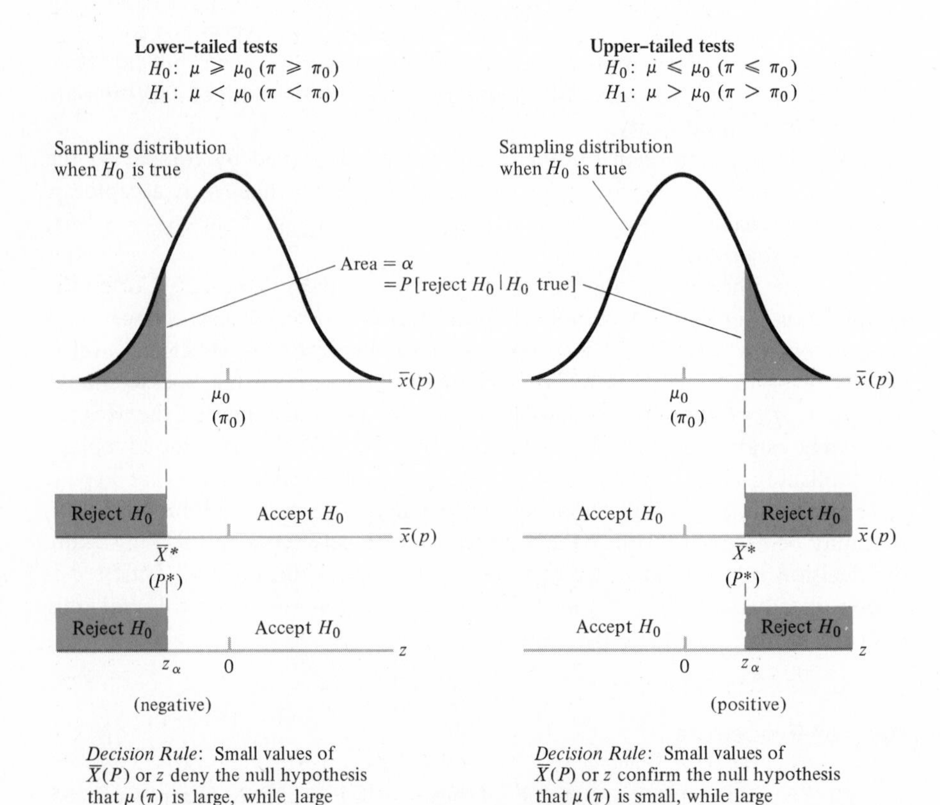

FIGURE 8-5 Upper-tailed and lower-tailed testing situations.

population parameter values, respectively, an upper-tailed test will result. A lower-tailed test results when H_0 values are larger than H_1 values. Upper- and lower-tailed tests are summarized in Figure 8-5.

The general approach to one-sided hypothesis-testing situations may be summarized by the following steps:

1. Determine a null hypothesis for which α may be determined from the sampling distribution of the desired statistic and express the alternative hypothesis.
2. For the prescribed significance level α, determine the value of z_α from Table 8-2 on page 304. For an upper-tailed test, z_α is positive, and for a lower-tailed test z_α is negative.

   Either use: ***(a) In tests where a critical value for $\bar{X}$ or P is desired.***

   (1) In the case of a one-sided test using P, ***the critical value is***

$$P^* = \pi_0 + z_\alpha \sqrt{\frac{\pi_0(1-\pi_0)}{n}} \tag{8-1}$$

   In case of sampling without replacement from small populations, the last term would be multiplied by the finite population correction factor $\sqrt{(N-n)/(N-1)}$.

   (2) When a one-sided test uses $\bar{X}$, ***the critical value is***

$$\bar{X}^* = \mu_0 + z_\alpha \frac{s}{\sqrt{n}} \tag{8-2}$$

   If the population size N is small, the last term would be multiplied by $\sqrt{(N-n)/(N-1)}$.

   Or use: ***(b) In tests where only a critical normal deviate is required.***

   The value of z is found from the following:

   (1) For tests involving P:

$$z = \frac{P-\pi_0}{\sqrt{\pi_0(1-\pi_0)/n}} \tag{8-3}$$

   (2) For tests involving $\bar{X}$:

$$z = \frac{(\bar{X}-\mu_0)\sqrt{n}}{s} \tag{8-4}$$

   NOTE: If the decision rule is constructed before the sample data are collected, then s would not be available. In those cases, we would use an assumed value for σ in place of s in expression (8-2).

3. Formulate the decision rule:

	(1)	(2)
	Using $\bar{X}$ or P	*Using z*
(a) Upper-tailed tests	*Accept H_0* if $\bar{X} \leqslant \bar{X}^*$, $P \leqslant P^*$ *Reject H_0* if $\bar{X} > \bar{X}^*$, $P > P^*$	*Accept H_0* if $z \leqslant z_\alpha$ *Reject H_0* if $z > z_\alpha$
(b) Lower-tailed tests	*Accept H_0* if $\bar{X} \geqslant \bar{X}^*$, $P \geqslant P^*$ *Reject H_0* if $\bar{X} < \bar{X}^*$, $P < P^*$	*Accept H_0* if $z \geqslant z_\alpha$ *Reject H_0* if $z < z_\alpha$

4. Apply the decision rule to the sample data obtained and accept or reject the null hypothesis.

Illustrations of Procedure

To continue with our first example, where government approval of a dietary supplement is being considered, suppose that the drug administrator orders a sample size $n = 100$ and wishes to find a critical value P^* for the test statistic P when $\pi_0 = .10$, using a smaller significance level than before, $\alpha = .01$. Since large values of P will refute H_0, this is an *upper-tailed test*, and the critical normal deviate z_α corresponds to that point where the upper-tailed area under the normal curve is equal to $\alpha = .01$. Thus, from Table 8-2, we find $z_\alpha = z_{.01} = 2.33$, and a positive value is used.

Substituting the appropriate values given above into expression (8-1), we find the critical value

$$P^* = .10 + 2.33\sqrt{\frac{.10(1-.10)}{100}} = .17$$

(This is larger than the value $P^* = .15$ used before, for here α has been set at a lower level than the .0475 achieved at the outset.)

Suppose that 18 percent of the patients experienced a major cholesterol reduction, so that $P = .18$. Since this value exceeds $P^* = .17$, H_0 must be *rejected.*

Viewed another way, when $P = .18$, expression (8-3) provides the normal deviate

$$z = \frac{.18-.10}{\sqrt{.10(1-.10)/100}} = 2.67$$

Because $z = 2.67$ exceeds $z_{.01} = 2.33$, this indicates that were H_0 to be true, the sample outcome would be rarer than the smallest permitted Type I error. Again we see that H_0 must be rejected.

Example 8-1 Have you ever wondered why most people are right-handed? Right-handers outnumber lefties by 19 to 1, regardless of culture, race, or nationality. One researcher has proposed the following theory to partially justify this phenomenon:

1. Handedness would be strictly a "coin-flipping" proposition, except for environmental influences biasing the development of the right hand.
2. Mothers' hearts tend to be on their left sides.
3. Babies feel secure when they sense heart-beat rhythms and are happier when placed close to their mothers' hearts. Unconsciously accommodating this behavior, babies are held predominantly on the mothers' left side *supported by her left arm.*
4. This leaves only the mother's right arm free for fondling the baby. As the mother's right hand must therefore be used for most hand-to-hand contact, it is easier for her to have contact with the baby's *right hand* because it matches her free hand. Therefore, the baby begins its life using its right hand more than its left.
5. In effect, heart placement helps to determine handedness. If the human heart were on the right side instead, the above theory indicates that more babies than otherwise would end up being left-handed.

To test her theory, the researcher collected a random sample of $n = 100$ new mothers. The bias toward right-handedness was eliminated for their babies. Each mother was asked to wear a tiny amplifier that makes the heart beat loudest on the mother's right side, in effect repositioning the apparent heart location. If the above theory applies, this should favor the motor development of the babies' left hands. Each sample mother was told to wear the device when holding her baby. (The purposes of the study device and the experiment were not known to the mothers to avoid biasing their behavior.) At the end of the experiment, the babies were tested to determine their dominant hand.

As her null hypothesis, the researcher assumed that the proportion π of babies who become left-handed when reared under these experimental conditions will be no more than for the general population:

$$H_0 : \pi \leqslant \pi_0 \ (=.05) \text{ (theory untrue)}$$

The alternative is that experimental conditions will yield a higher proportion of left-handed babies:

$$H_1 : \pi > \pi_0 \text{ (theory true)}$$

Since large values of the sample proportion P of left-handed babies will refute the null hypothesis, this is an *upper-tailed test.* The critical normal deviate will therefore be positive. Using an $\alpha = .001$ significance level for the probability of incorrectly rejecting the null hypothsis when it is true, from Table 8-2, we find that $z_\alpha = z_{.001} = 3.08$.

The critical value for the sample proportion is

$$P^* = .05 + 3.08\sqrt{\frac{.05(1-.05)}{100}} = .117$$

The decision rule is

$$\textit{Accept } H_0 \text{ (discard theory) if } P \leqslant .117$$

$$\textit{Reject } H_0 \text{ (publish theory) if } P > .117$$

At the end of the experiment, the researcher found that 15 babies were left-handed, so that $P = 15/100 = .15$. Since this is greater than $P^* = .117$, she must *reject* H_0 that the proportion of left-handed babies reared with a left-hand bias is no larger than for the general population. She will therefore publish her findings and promulgate her theory.

To see just how rare this result would be were H_0 true, the normal deviate may be calculated from expression (8-3):

$$z = \frac{.15 - .05}{\sqrt{.05(1 - .05)/100}} = 4.59$$

Such a result is quite inconsistent with the null hypothesis rejected above.

Example 8-2 Over the past 100 years, there has been a pronounced trend in the United States for successive generations to grow taller than their parents. This tendency has been attributed to improved health care and nutrition. One researcher believes that this trend has ended. As his reasons, he cites the degraded environment, diminished exercise by children, and diets dominated by packaged convenience foods that lack sound nutritional value. To test his theory, the researcher has collected a random sample of heights for $n = 100$ 20-year-old men. As his null hypothesis, he assumes that the past trend is unchanged, so that the mean height of young men is assumed to be at least as high now as it was 20 years before, when it has been established at $\mu_0 = 70$ inches. Thus

$$H_0 : \mu \geqslant 70 \text{ inches (growth trend continues)}$$

$$H_1 : \mu < 70 \text{ inches (growth trend is over)}$$

As his sample results, the researcher obtains $\bar{X} = 69.75$ inches, with $s = 2.71$ inches. He chooses $\alpha = .05$ as his significance level. From Table 8-2, $z_{.05} = -1.64$ which is negative because the test is *lower-tailed* (that is, small values of $\bar{X}$ in the lower tail of the normal curve refute H_0). The researcher's critical value for the sample mean is

$$\bar{X}^* = 70 - (1.64)\frac{2.71}{\sqrt{100}} = 69.56 \text{ inches}$$

and his decision rule is:

Accept H_0 (conclude growth trend continues) if $\bar{X} \geqslant 69.56$

Reject H_0 (conclude growth trend is over) if $\bar{X} < 69.56$

In terms of normal deviates, the equivalent decision rule, using -1.64 for z_α, is

Accept H_0 if $z \geqslant -1.64$

Reject H_0 if $z < -1.64$

The normal deviate for the actual sample results is

$$z = \frac{(69.75 - 70)\sqrt{100}}{2.71} = -.92$$

Since $\bar{X} = 69.75$ inches is larger than the critical value of 69.56 inches, and $z = -.92$ exceeds -1.64, either decision rule leads to the same conclusion—that the null hypothesis must be *accepted*. Another way of saying this is the test results are not significant at the 5-percent level. The researcher cannot conclude that the present generation is shorter in height than the preceding one. The sample data do not support his theory.

Type II Error Considerations

Example 8-2 raises an important issue in hypothesis testing. In accepting the null hypothesis that the present trend in successively taller generations is continuing, the researcher has used a rule that protects him with a probability of .05 from committing the Type I error of wrongly concluding that the growth trend is over when it has not ended. *But he may still have made an incorrect decision*: He may have committed the Type II error of accepting the null hypothesis when it is not true. In other words, maybe the growth trend has ended and the true mean height for the present generation is actually some smaller value such as $\mu = 69.5$ inches. If this figure happens to be the true mean, then the probability of incorrectly accepting H_0 would be substantial. The traditional procedure outlined in this chapter always guarantees the desired level for α. But there is the ever-present danger of ignoring the Type II error probability β. As a practical matter, we obtain smaller βs by using a large sample size. A more detailed discussion of this point will appear in Section 8-6.

EXERCISES

8-9 In testing the hypotheses for a large population

$$H_0: \pi \leqslant \pi_0 (=.5) \qquad \text{and} \qquad H_1: \pi > \pi_0$$

the desired Type I error probability is $\alpha = .10$. A sample size of $n = 100$ is to be used.

(a) Is the test lower- or upper-tailed?

(b) Find the value of the critical normal deviate z_α.

(c) Using the sample proportion P as the test statistic, find the critical value P^* and formulate the decision rule.

(d) If the sample proportion obtained is .55, would the decision rule obtained in part (c) lead to accepting or rejecting H_0?

8-10 In testing the hypotheses

$$H_0: \mu \geqslant \mu_0 (=30) \qquad \text{and} \qquad H_1: \mu < \mu_0$$

the desired Type I error probability is $\alpha = .05$. A sample size of $n = 100$ is to be used. Assume that $s = 10$ and that the population is large.

(a) Is the test upper- or lower-tailed?

(b) Using the sample mean $\bar{X}$ as the test statistic, find the critical value $\bar{X}^*$ and formulate the decision rule.

(c) If the sample mean obtained is 33, should H_0 be accepted or rejected?

(d) Suppose that the population is small with $N = 500$. Determine $\bar{X}^*$ and the decision rule that would apply. In this instance, should H_0 be accepted or rejected?

8-11 For each of the following testing situations, find the value of the critical normal deviate z_α. Then, for each sample result obtained, calculate the normal deviate value z and indicate whether the null hypothesis should be accepted or rejected. Finally, for each outcome, determine the approximate probability of obtaining a value of the test statistic as rare as or rarer than the one obtained by finding the tail area under the normal curve corresponding to the computed z. (Assume that the respective populations are large.)

(a) $H_0: \mu \leqslant 105$; $H_1: \mu > 105$; $\alpha = .05$; $n = 100$; $\bar{X} = 108$; and $s = 15$.

(b) $H_0:\pi \geqslant .5$; $H_1:\pi < .5$; $\alpha = .01$; $n = 100$; and $P = .49$.
(c) $H_0:\mu \geqslant .8$; $H_1:\mu < .8$; $\alpha = .10$; $n = 36$; $\bar{X} = .6$; and $s = .6$.
(d) $H_0:\pi \leqslant .10$; $\pi > .10$; $\alpha = .001$; $n = 900$; and $P = .11$.

8-12 A psychologist believes that age has an influence on a person's IQ. A random sample of 100 middle-aged persons who had been tested at age 16 were tested again. Subtracting their earlier scores from the new scores, a mean difference of $\bar{X} = 5$ points and a standard deviation of $s = 8$ points were determined. Using $\alpha = .01$ as her significance level, the psychologist wishes to test the null hypothesis $H_0:\mu \leqslant 0$ that no improvement in IQ occurs with age.
(a) Is this test upper- or lower-tailed? Find the critical normal deviate.
(b) Calculate the normal deviate for the test results. Does this value indicate that H_0 must be accepted or rejected?

8-13 A congressman wishes to test the null hypothesis that at least 50 percent of the voters recognize his name. From a random sample of $n = 25$ persons, only 10 could identify him as a U.S. Representative. Using an $\alpha = .05$ significance level:
(a) Is this test upper- or lower-tailed? Find the critical normal deviate.
(b) Calculate the normal deviate for the test results. Does this value indicate that the null hypothesis must be accepted or rejected?

8-14 A random sample of 100 rolls of paper has been selected from a large shipment by the superintendent of a printing plant. Their lengths have been measured, and the results show that the average length is $\bar{X} = 531$ feet, with a standard deviation of $s = 52$ feet. The null hypothesis that $\mu \geqslant 525$ is to be tested against the alternative that $\mu < 525$.
(a) Assuming that $\alpha = .01$, determine the critical normal deviate. Use this to find the critical value $\bar{X}^*$.
(b) Should the null hypothesis be rejected or accepted? Why?

8-15 A city purchasing department checks samples from each shipment of supplies and returns poor-quality shipments. Policy states that when sampling is used for assessing the quality of a shipment, the probability of erroneously disqualifying a "good" shipment actually having 5 percent or fewer defective should be $\alpha = .05$.
(a) Express the null and alternative hypotheses in terms of the proportion defective π. Is the test lower- or upper-tailed?
(b) Suppose a sample of $n = 100$ items is selected for testing. If the critical value is $P^* = .07$, what is the value for the Type I error probability? (You may assume that N is large.)
(c) Determine the critical value P^* necessary to satisfy the desired α, when the sample size is $n = 100$.

8-16 In testing the null hypothesis that $\mu \leqslant 100$ for the IQs of juvenile delinquents, you may presume that the population standard deviation is $\sigma = 16$. You cannot decide whether to use a sample of $n = 64$ or $n = 100$, but you will use the smallest n and corresponding decision rule that provides a β probability of at most .10 for accepting H_0 when μ is actually 105. In all cases, use $\alpha = .05$.
(a) Find the critical value $\bar{X}^*$ and the decision rule when $n = 64$.
(b) Find the critical value $\bar{X}^*$ and the decision rule when $n = 100$.
(c) Calculate the value of β at $\mu = 105$ when $n = 64$ and when $n = 100$. Which sample size should be used?

8-17 A subsystem will be incorporated into a communications satellite if its mean time between failures (MTBF) obtained from a sample of units subjected to environmental testing is greater than or equal to $\bar{X}^* = 500$ hours. This action is equivalent to accepting the null hypothesis that the actual MTBF is

$\mu_0 = 550$ hours or more. Assuming that the standard deviation of failure times for the units is $\sigma = 200$ hours and that N is large:

(a) Find the significance level α of the above test, assuming that $n = 100$.

(b) Determine the critical value and formulate the decision rule if it is required that $\alpha = .05$. Use a sample size of $n = 100$.

8-4 TWO-SIDED HYPOTHESIS TESTS USING THE MEAN

As we have seen in the preceding section, the position of the alternative hypothesis with respect to the null hypothesis may differ with the situation. When an alternative places the population parameter values to one side of the null hypothesis, it is a *one-sided alternative* and it results in a lower- or upper-tailed test. We will now discuss the situation that arises when the alternative lies to either side of the null hypothesis, so that the test has a *two-sided alternative.*

An important application of hypothesis testing occurs in monitoring quality in production operations. Since quality control is an important branch of applied statistics in which two-sided tests are frequently used, we can illustrate two-sided procedures for testing the mean with the following detailed example.

Illustration of the Testing Procedure

A food company advertises that there are 105 chunks of beef in every 15-ounce can of its chili. It has spent a great deal of money creating the image that it offers a high-quality product; this has been predicated upon a policy of including generous portions of expensive ingredients in its canned foods. The company wants to be assured that its policies are being met in chili production. Therefore, periodic checks of the quality of production are made; these consist of taking random samples of 15-ounce chili cans and measuring the quantity of beef in each sample.

Management cannot, of course, ensure that 105 chunks of beef will go into every can, unless each can is carefully filled individually with precisely the desired amount of beef. This is not practicable, because the chili is cooked in large kettles and the ingredients are added as the recipe requires. Each kettle contains several hundred pounds of chili, and the actual number of chunks of beef distributed to each can will vary. Management can control only the mean number of chunks per can. How close the mean actually approaches the desired goal will depend upon the precision used in mixing all of the ingredients during preparation.

Management's goal is therefore to see that the mean number of beef chunks is very close to the advertised 105. If a sample batch has a lower mean, then some remedial action must be taken in the kitchen. On the other hand, since beef is by far the most expensive chili ingredient, a mean higher than the desired 105 will also require remedial action. Management wishes to avoid extreme values in the mean beef quantity, both high and low.

Assuming that the standard deviation will be constant under either hypothesis, the following hypotheses regarding the mean number of beef chunks per can are formulated:

$$H_0 : \mu = \mu_0 \; (=105) \text{ (no remedial action is required)}$$

$$H_1 : \mu \neq \mu_0 \qquad \text{(remedial action is required)}$$

The alternative hypothesis states that the mean quantity of beef is *different* from μ_0; the direction of difference is unimportant. This is an example of a *two-sided alternative*.

The decision rule will have the form:

Accept H_0 (do not make remedies) if $\bar{X}_1^* \leq \bar{X} \leq \bar{X}_2^*$

Reject H_0 (make remedies) either if $\bar{X} < \bar{X}_1^*$ *or* if $\bar{X} > \bar{X}_2^*$

Note that there are two critical values for the test statistic. This is because there are two bases for rejection: insufficient beef and excessive beef.

The decision rule is chosen to be symmetrical about μ_0. Figure 8-6 shows

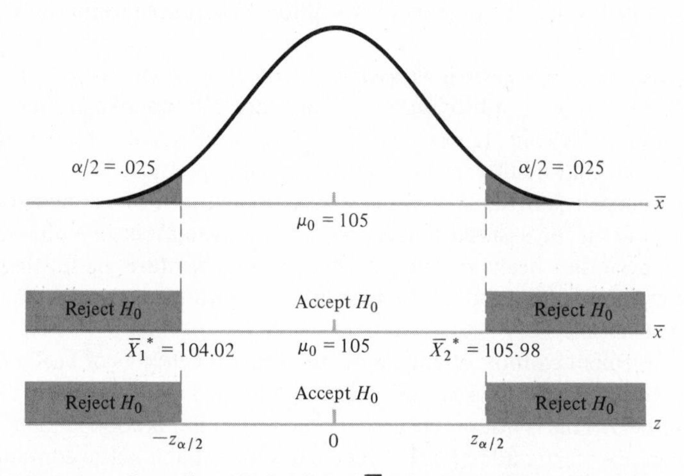

FIGURE 8-6 Sampling distribution of $\bar{X}$, when the null hypothesis is true.

the sampling distribution when the null hypothesis is true, as well as the critical values of the decision rule obtained below. Since the sum of the areas in both tails represents the probability of rejection, this test is called a *two-sided test*.

In two-sided tests there are two Type I error situations; in this example, a Type I error is to take unnecessary remedial action for either overfilling or underfilling. The significance level α is usually evenly divided between these two errors. There are also two Type II error situations; here, a Type II error results from not taking remedial action either when there is overfilling or when there is underfilling.

Finding Critical Values for the Two-Sided Test

Suppose that a sample of size $n = 100$ is taken and that the sample standard deviation is $s = 5$ chunks per can. The test statistic will be the sample mean chunks found per can $\bar{X}$. Management wants to find a decision rule that will lead to rejecting a true null hypothesis (Type I error) with a probability of .05. The significance level is therefore $\alpha = .05$.

The following provide ***the critical values for a two-sided test of the mean***

$$\bar{X}_1^* = \mu_0 - z_{\alpha/2}\frac{s}{\sqrt{n}}$$

$$\bar{X}_2^* = \mu_0 + z_{\alpha/2}\frac{s}{\sqrt{n}} \qquad (8\text{-}5)$$

The critical normal deviate $z_{\alpha/2}$ is a *positive quantity* representing the upper-tailed area under the normal curve that corresponds to half of the Type I error probability. Since $z_{\alpha/2}$ expresses distance in standard deviation units, the lower critical value $\bar{X}_1^*$ is found by subtracting $z_{\alpha/2}$ times $s/\sqrt{n}$ (the estimate for $\sigma_{\bar{X}}$) from μ_0; the upper critical value $\bar{X}_2^*$ is found by adding $z_{\alpha/2}$ times $s/\sqrt{n}$ to μ_0. As with one-sided tests, we may construct the decision rule in advance of sampling, provided that σ is known, by using σ in place of s in expression (8-5).

This procedure applies when the population is large. When the sample is taken from a small population, the last term in both equations in expression (8-5) must be multiplied by the finite population correction factor $\sqrt{(N-n)/(N-1)}$.

Returning to our example, Table 8-2 (page 304) provides the critical normal deviate $z_{\alpha/2} = z_{.025} = 1.96$. Since the population of chili cans is large, we ignore the effect of the population size N in using expression (8-5) to find the critical values for $\bar{X}$:

$$\bar{X}_1^* = 105 - 1.96\frac{5}{\sqrt{100}} = 104.02$$

$$\bar{X}_2^* = 105 + 1.96\frac{5}{\sqrt{100}} = 105.98$$

The decision rule may be expressed in terms of these values:

Accept H_0 (do not make remedies) if $104.02 \leqslant \bar{X} \leqslant 105.98$

Reject H_0 (make remedies) either if $\bar{X} < 104.02$ *or* if $\bar{X} > 105.98$

Suppose that the sample mean achieved is $\bar{X} = 106.3$. Since this does not lie between the critical values $\bar{X}_1^* = 104.02$ and $\bar{X}_2^* = 105.98$, *management must reject* H_0 *and take remedial action.* If a different result had been obtained, such as $\bar{X} = 104.5$, then H_0 would have been accepted, as in that case $\bar{X}$ would fall between 104.02 and 105.98.

As with the tests discussed previously, the computation of $\overline{X}_1^*$ and $\overline{X}_2^*$ can be by-passed, and the normal deviate z for the results obtained can be calculated instead. Should z fall within the limits $-z_{\alpha/2}$ to $z_{\alpha/2}$, H_0 would be accepted; otherwise H_0 would be rejected.

With our example, the normal deviate is

$$z = \frac{(\overline{X}-\mu_0)\sqrt{n}}{s} = \frac{(106.3-105)\sqrt{100}}{5} = 2.6$$

This value is greater than $z_{\alpha/2} = 1.96$, so as before, H_0 must be rejected.

Explanation of Hypothesis Testing in Terms of Confidence Intervals

Two-sided hypothesis-testing procedures may be described in terms of the confidence-interval estimate. The *acceptance region* of this type of test has the form:

$$\overline{X}_1^* \leqslant \overline{X} \leqslant \overline{X}_2^*$$

which can be equivalently expressed as

$$\mu_0 - z_{\alpha/2}\frac{s}{\sqrt{n}} \leqslant \overline{X} \leqslant \mu_0 + z_{\alpha/2}\frac{s}{\sqrt{n}}$$

Each two-tailed decision rule must be established so that there is a probability $1-\alpha$ of accepting H_0 when the null hypothesis that $\mu = \mu_0$ is true. Thus, when $\alpha = .05$, so that $1-\alpha = .95$, then 95 percent of all samples taken from a population where μ_0 is indeed the true mean will result in accepting this null hypothesis.

Rearranging the above inequality, and replacing the symbol μ_0 with μ, an equivalent expression provides the ***confidence interval***

$$\overline{X} - z_{\alpha/2}\frac{s}{\sqrt{n}} \leqslant \mu \leqslant \overline{X} + z_{\alpha/2}\frac{s}{\sqrt{n}} \tag{8-6}$$

We note that when $\mu = \mu_0$, 95 percent of all samples yield values for $\overline{X}$ and s that provide intervals, calculated from expression (8-6), which contain the value μ_0. But 5 percent of the samples do not contain μ_0. Thus, when $\alpha = .05$, expression (8-6) provides a 95-percent confidence-interval estimate of μ. Finding whether or not H_0 falls into the acceptance region produces the same result that finding whether or not μ_0 falls into the above confidence interval does.

We may then approach the two-sided hypothesis-testing procedure by first constructing a confidence interval. *If μ_0 is a value lying inside the confidence* *interval constructed from the sample results, H_0 must be accepted. But if these limits fall completely above or below μ_0, H_0 must be rejected.* Only 5 percent of these intervals can ever be so far off that they will not bracket μ_0 when it is the true mean; that is, only 5 percent of the time this procedure is employed will the Type I error be committed of rejecting H_0 when it is true.

In the chili illustration, $\bar{X} = 106.3$ beef chunks, $s = 5$, and $n = 100$. Using these data to construct a 95-percent confidence interval, we have

$$106.3 - 1.96\frac{5}{\sqrt{100}} \leqslant \mu \leqslant 106.3 + 1.96\frac{5}{\sqrt{100}}$$

or

$$105.32 \leqslant \mu \leqslant 107.28 \text{ beef chunks}$$

Since the above interval does not contain the value $\mu_0 = 105$, H_0 must be *rejected.* This is the same conclusion we reached earlier in this section.

EXERCISES

8-18 In testing the hypotheses $H_0 : \mu = \mu_0 (=0)$ and $H_1 : \mu \neq \mu_0$, answer the following, assuming that $\sigma = 10$ and that a sample of size $n = 100$ is taken from a large population.

(a) Assuming that $\alpha = .05$ and that a symmetrical two-tailed test is made, find $z_{\alpha/2}$ and the critical values $\bar{X}_1^*$ and $\bar{X}_2^*$, and then formulate the decision rule.

(b) Using the decision obtained in (a), calculate the probability of accepting the null hypothesis when $\mu = 1.0$.

(c) Suppose that the sample results show $\bar{X} = 1.2$ and $s = 9.5$. Calculate the resulting normal deviate. Should H_0 be accepted or rejected?

8-19 For each of the following two-sided testing situations, construct a 95-percent confidence interval for μ and use it to indicate whether H_0 must be accepted or rejected.

(a) $\mu_0 = 10$; $\bar{X} = 10.1$; $s = .3$; $n = 100$; and N is large.

(b) $\mu_0 = .72$; $\bar{X} = .705$; $s = .13$; $n = 169$; and $N = 800$.

(c) $\mu_0 = 0$; $\bar{X} = -.7$; $s = 2.5$; $n = 625$; and N is large.

8-20 A researcher believes that marijuana has a latent effect upon the duration of dreams, but he does not know whether dreaming is increased or decreased. The night's sleep of 36 volunteers is monitored 24 hours after each has experienced a marijuana high. Previous research indicates that, under ordinary conditions, the population mean dream duration is 1.5 hours. Taking that figure as his null hypothesis, he tests this at the .01 significance level against the two-sided alternative. With $\bar{X} = 1.8$ hours and $s = .6$ hours, what conclusion should the researcher make?

8-21 The output of a chemical process is checked periodically in order to determine the level of impurities in the final product. Too many impurities will cause the product to be reclassified; too few are an indication that an expensive catalytic ingredient is being consumed in undesirably large quantities. The process has been designed so that the total amount of impurities desired is $\mu = .05$ grams per liter, with a standard deviation of $\sigma = .02$ grams.

A sample of one-liter vials is taken at random times throughout each four-hour period and the level of impurities is measured. The process should be stopped if there are either too many impurities (the tanks are then purged) or if too few impurities (the control valves are then readjusted). Criteria are to be established so that the chief engineer can determine when the process should be stopped.

(a) Formulate the hypotheses for the situation.

(b) Suppose the chief engineer desires a Type I error probability of $\alpha = .01$ for stopping the process when the mean level of impurities is $\mu_0 = .05$ grams per liter. Find the critical value $\bar{X}^*$ when $n = 100$ liters are used in the sample. Then formulate the decision rule.

(c) Using the above decision rule, find the probability of erroneously stopping the process when the mean level of impurities shifts to (1) $\mu = .043$; (2) $\mu = .045$; (3) $\mu = .052$; (4) $\mu = .055$. What do you notice about your answers to (2) and (4)? Explain this.

8-5 TESTS OF THE MEAN USING SMALL SAMPLES

Often a statistical decision must be made using data from a small sample. In tests involving people—especially in medical research, education, and the behavioral sciences—sample data are frequently skimpy and sample sizes of less than 30 are quite common. Ordinarily, the population standard deviation σ is of unknown value and must be estimated by its sample counterpart s. As we saw in Chapter 7, the normal curve does not perform satisfactorily as the sampling distribution for $\bar{X}$ under these circumstances. In its place, we therefore employ the Student t distribution.

The Student t statistic is found in the same manner as the normal deviate z—by subtracting the population mean from the observed value of the sample mean—except that we divide this difference by the *estimated* standard error for $\bar{X}$, $s/\sqrt{n}$, instead of the actual value of $\sigma_{\bar{X}}$, $\sigma/\sqrt{n}$, which remains unknown because the value of the population standard deviation is unavailable. (If the value of σ is established in advance of sampling, we may use the procedures outlined in Sections 8-3 and 8-4, and the Student t distribution will not apply.) Under the null hypothesis, μ is equal to an assumed value, μ_0, so that the following expression is used to compute the ***Student t test statistic***

$$t = \frac{(\bar{X} - \mu_0)\sqrt{n}}{s} \tag{8-7}$$

Procedure Using t Statistic

1. Calculate t from expression (8-7).
2. From Appendix Table E, determine t_α for the desired significance level α. This will be positive for upper-tailed tests and negative for lower-tailed

ones. For a two-sided test, find $t_{\alpha/2}$ corresponding to an upper-tail area of $\alpha/2$. The number of degrees of freedom is $n-1$.

3. Apply the decision rule:

   (a) For an upper-tailed test:

$$\textit{Accept } H_0 \text{ if } t \leqslant t_\alpha$$

$$\textit{Reject } H_0 \text{ if } t > t_\alpha$$

   (b) For a lower-tailed test:

$$\textit{Accept } H_0 \text{ if } t \geqslant t_\alpha$$

$$\textit{Reject } H_0 \text{ if } t < t_\alpha$$

   (c) For a two-sided test:

$$\textit{Accept } H_0 \quad \text{if } -t_{\alpha/2} \leqslant t \leqslant t_{\alpha/2}$$

$$\textit{Reject } H_0 \text{ either if } t < -t_{\alpha/2} \textit{ or } \text{if } t > t_{\alpha/2}$$

We may illustrate this procedure by means of the following example involving a medical experiment with artificial blood.

Example 8-3 A team of medical researchers is seeking a chemical that may be used as artificial blood. A blood substitute would provide many advantages to patients undergoing major surgery, which is severely complicated by the need for whole blood, and might also serve in flushing out and cleansing diseased body organs like the liver or the kidneys.

The research team is currently experimenting with rats to determine what deleterious effects result from a complete substitution of blood with a chemical solution:

> The scientists insert a plastic tube into a white rat's jugular vein and slowly drain away all its blood. Simultaneously, they pump a milky-white artificial blood into the rat's circulatory system. The substitute is a man-made chemical concoction containing fluorocarbons, or compounds of fluorine and carbon that are more commonly used in aerosol propellants, refrigerants, and fire-extinguishing agents.
>
> Shortly after the blood exchange has taken place, the rat begins to move about its cage, washing itself or perhaps taking a drink of water. It not only survives, but it thrives, illustrating for the first time anywhere that animals can live without any element of natural blood in their bodies.*

In attempting to decide whether or not the same chemical agent should be used in future experiments, a sample of $n = 25$ "middle-aged" rats has been selected. Each rat's blood has been totally replaced by the same fluorocarbon solution. Over a period of several days, this artifical blood has been gradually replaced internally by new blood regenerated naturally. The null hypothesis is

* Jeffery A. Perlman, "Artificial Blood, Long a Goal of Scientists, Sustains Lab Animals," *The Wall Street Journal* (January 29, 1974), 1. Although the experiments themselves are real, the decision situation and the data presented in this example are strictly hypothetical.

that the mean remaining life span of injected rats is as long as that of their untreated brethren of the same age:

$$H_0 : \mu \geqslant \mu_0 \; (=4 \text{ months}) \quad \text{(artificial blood not harmful)}$$

$$H_1 : \mu < \mu_0 \quad \text{(artificial blood harmful)}$$

The one-sided alternative provided above has been chosen, because the scientists are concerned with changing chemicals only if the fluorocarbon solution is harmful in the long run. Since the Type I error, rejecting H_0 when it is true, would unnecessarily slow down further research while new chemical substitutes are evaluated, a fairly low significance level of $\alpha = .01$ has been established.

After all the injected rates had died of natural causes, the following terminal life span data were obtained:

$$\bar{X} = 3.8 \text{ months}$$

$$s = .5 \text{ month}$$

Applying step (1) of the t-statistic procedure, the value of t is calculated from expression (8-7):

$$t = \frac{(3.8-4.0)\sqrt{25}}{.5} = -2.0$$

Next we apply step (2). The critical value for an $\alpha = .01$ tail area under the t distribution having $n-1 = 25-1 = 24$ degrees of freedom is found from Appendix Table E to be

$$t_\alpha = t_{.01} = -2.492$$

which is negative because the test is lower-tailed. (Small values of $\bar{X}$ will tend to refute H_0 that the mean life span of treated rats is not reduced from the natural level.)

Applying step (3), the scientists' decision rule may be expressed as

Accept H_0 (test fluorocarbon further) if $t \geqslant -2.492$

Reject H_0 (seek a new blood substitute) if $t < -2.492$

Because the resulting value $t = -2.0$ calculated for the actual results exceeds the critical value $t_{.01} = -2.492$, step (3) indicates that H_0 must be *accepted* and the test results are not significant at level .01. Further investigations will be made using the fluorocarbon solution as artificial blood.

An Assumption of the t Test

The Student t distribution is based upon the assumption that the underlying population frequency distribution is represented by the normal curve. As a practical matter, this requirement poses no difficulties, as long as the population frequency distribution is unimodal and fairly symmetrical. Thus, the t test is fairly insensitive to this requirement. Statisticians would say that the t test is, to some degree, *robust* with respect to the non-normality of the population.

To avoid potential difficulties that may arise when the normality assumption does not hold, certain nonparametric tests may be employed. (These will be discussed in Chapter 14.)

EXERCISES

8-22 For each of the following hypothesis-testing situations, calculate the value of the test statistic t. Indicate whether the test is lower-tailed, upper-tailed, or two-sided. Then determine the critical value and formulate the decision rule for each. Indicate whether each H_0 should be accepted or rejected.

(a) $\bar{X} = 13$; $s = 2$; $n = 9$; and $\alpha = .01$.
$H_0: \mu \geq 15$ and $H_1: \mu < 15$

(b) $\bar{X} = 170$; $s = 30$; $n = 25$; and $\alpha = .05$.
$H_0: \mu \leq 160$ and $H_1: \mu > 160$

(c) $\bar{X} = 102$; $s = 24$; $n = 16$; and $\alpha = .10$.
$H_0: \mu = 100$ and $H_1: \mu \neq 100$

(d) $\bar{X} = -2.8$; $s = 2.0$; $n = 25$; and $\alpha = .01$.
$H_0: \mu = 0$ and $H_1: \mu \neq 0$

8-23 A structural engineer is testing the strength of a newly designed steel beam required in cantilever construction. As his null hypothesis, he assumes that the mean strength will be at most as great as the 100,000 pounds per square inch (psi) for traditional beams. A sample of $n = 9$ new beams has been tested; the results show $\bar{X} = 105{,}000$ psi, with $s = 10{,}000$ psi. At the $\alpha = .05$ significance level, should the engineer accept or reject his null hypothesis?

8-24 A statistics professor believes his 25 students are a representative random sample of all beginning statistics students he has taught at the university. At the end of the term, he gives all of his students a standardized test. Over the years, they have established a mean score of 75 points on this test.

The professor wishes to assess the impact of spending twice as much time as before on probability theory, taking some time away from that devoted to all other "core" statistical material. Should he find that this results in a score improvement on his standardized test, he will adopt the new policy in the future. As his null hypothesis, he assumes that there will be no score improvement, so that $H_0: \mu \leq 75$ points. The professor assigns an $\alpha = .01$ probability of making the Type I error of adopting the new policy (rejecting H_0) when it does not actually improve test scores.

The sample results show $\bar{X} = 86$ points, with $s = 10$ points. Should the new policy be adopted?

8-25 A cannery inspector desires that the mean weight of ingredients in 16-ounce cans be precisely that figure. If the mean weight falls above or below this figure, some remedial action must be taken. In deciding whether or not to order corrections, he has established an $\alpha = .05$ significance level for a sampling test, so that the Type I error (taking unnecessary action) is avoided 5 percent of the time.

Each hour, a sample of $n = 25$ cans is taken and each can is opened and the contents are weighed.

(a) What is the decision rule for the value of the t statistic?

(b) Suppose that $\bar{X} = 15.9$ ounces and $s = .4$ ounces. Should the inspector take remedial action?

(c) Suppose that $\bar{X} = 16.2$ ounces and $s = .3$ ounces. Should he order corrections?

8-6 SELECTING THE TEST

Some Important Questions

We have now covered the basic concepts of hypothesis testing as it is applied to means and proportions. It is now time to consider some fundamental questions.

1. *What other test statistics might we use?* Many test statistics other than $\bar{X}$ and P are applicable to hypothesis-testing situations. In the preceding section, we presented one of these based upon the Student t distribution. Usually, the nature of the test determines the type of test statistic to be used. In later chapters, a host of test statistics will be introduced which may be used to compare two or more population parameters, to determine whether or not two population characteristics are independent, and to determine whether or not a sample is random. Even in a single application, such as finding if a new medical procedure results in recovery improvement, several test statistics might apply; in addition to testing with $\bar{X}$, a number of so-called nonparametric tests, (to be described in Chapter 14) may be used.
2. *How can we determine which particular test statistics will work best?* Either $\bar{X}$ or P might be used to obtain the estimate of a population mean in some applications. In certain testing situations, it is also true that a decision can be made using either the sample mean or the sample proportion as the test statistic. Often, we may choose from several alternative procedures. Perhaps the most important factor in making this choice is the relative efficiency of each test.
3. *How might we evaluate a particular decision rule in terms of the protection it provides against both kinds of incorrect decisions: the Type I and Type II errors?* Both types of decision errors should be avoided. But we have seen that when the choice is based upon sample data, some probability for each—α and β—must be tolerated. Ordinarily, only α, the probability of the more serious Type I error, is explicitly considered in selecting a decision rule. But the statistician may revise his initial rule upon finding it woefully inadequate in protecting against the Type II error.
4. *What assumptions underlie the proposed testing procedure, and how do they affect its applicability?* Later in this book, tests will be discussed which are based upon certain key assumptions. In the previous section, for example, the t test was described; this test assumes that the population itself is normally distributed. Just how critical are such assumptions? At one extreme, they may so severely limit a test's applicability that it may be totally inappropriate for many decision-making situations.

On the other hand, an assumption may prove to be relatively unimportant. A test that works pretty much as it is intended to even when a basic requirement is not exactly met is said to be *robust* with respect to a violation of that assumption.

The Power Curve

When a composite alternative hypothesis is used, there is no single Type II error probability value for β. Recall that β is the probability of accepting H_0

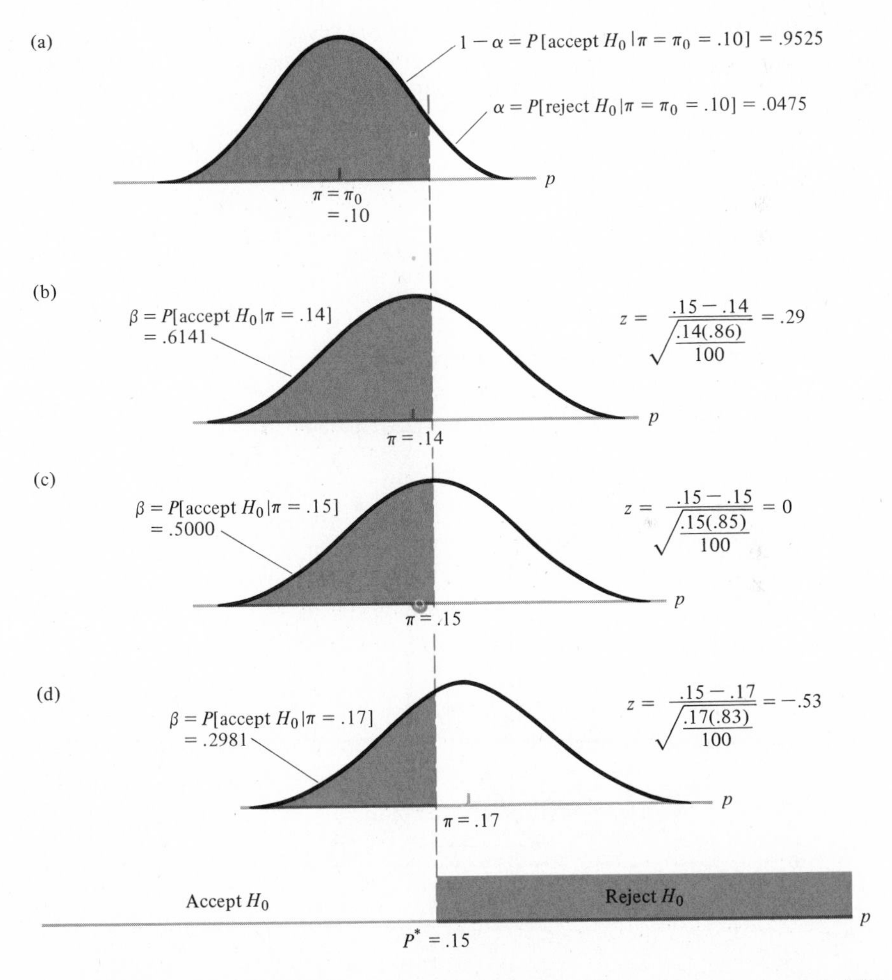

FIGURE 8-7 The effect of possible intermediate values of π upon the probability of accepting H_0. (a) is the sampling distribution of P when the null hypothesis is true; (b), (c), and (d) are the sampling distributions of P for intermediate values of π.

when H_0 is false (and H_1 is true). If we use the composite H_1 that $\pi > .10$ in the dietary-supplement illustration, we can compute β only for a particular value of π greater than .10, such as .14, .15, or .17. Figure 8-7 illustrates the values of β obtained for these values of π using the decision rule, with critical value $P^* = .15$ and sample size $n = 100$.

To evaluate the choices of critical value and sample size, a decision-maker may assess the effects of the entire range of possible values of π by obtaining the corresponding βs. To this end, a graph like the one in Figure 8-8 may be constructed. Such a graph is called the *power curve* of the test. The height of that point on the curve directly above each possible parameter value provides the probability that H_0 will be rejected when that value happens to be true. When π is equal to π_0, this curve provides the Type I error probability α. For larger values of π, where the dietary supplement improves on nature and H_0 is false, so that rejecting H_0 (approving the supplement's release) is a correct decision, the height of the power curve provides values equal to $1-\beta$. The Type II error probabilities β of accepting H_0 when it is false are 1 minus these; thus, for any value of π where H_0 is false, we can determine the value of β by subtracting the

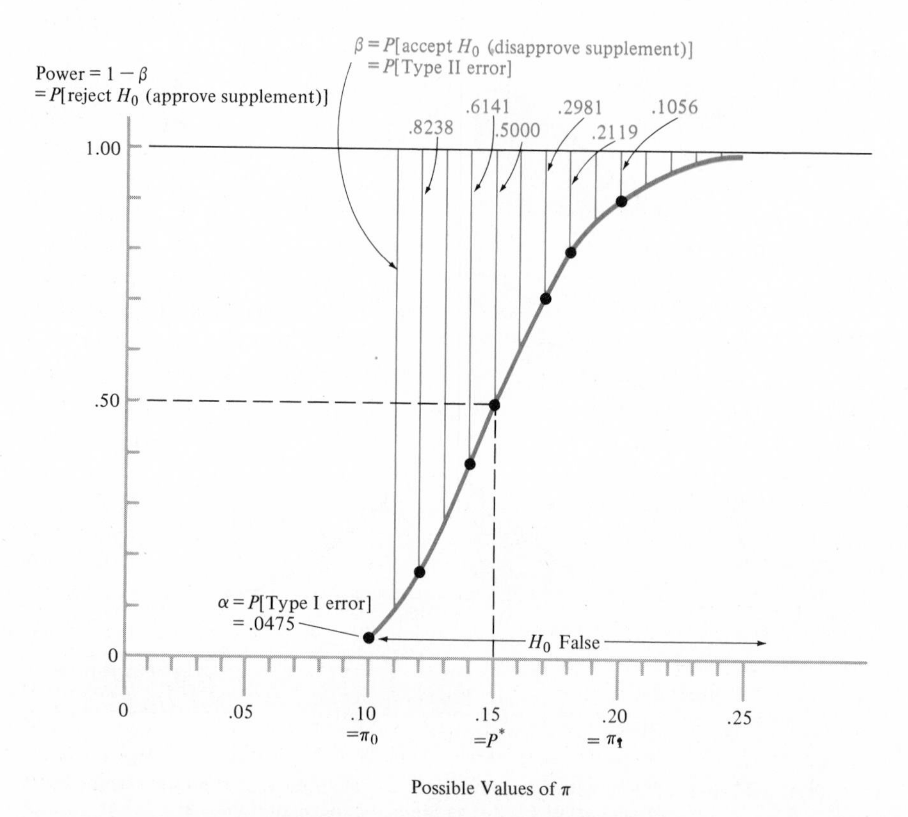

FIGURE 8-8 Power curve for dietary-supplement approval decision, when $P^* = .15$ and $n = 100$.

height of the curve from 1. Since the maximum height of the power curve must be 1, the β values are provided by the distances above the curve to a horizontal line 1 unit tall.

Note that in the power curve in Figure 8-8, as the possible values of π become larger, the curve grows taller and the β distances grow smaller. This reflects the corresponding higher levels of the anticholesterol supplement's effectiveness and the reduced chances of making the Type II error (accepting H_0 that the dietary supplement is not effective). Also notice that the power curve in Figure 8-8 is shaped like the letter S. All power curves for upper-tailed tests will be S-shaped; curves for lower-tailed tests will always have the shape of an inverted S, because the probability of rejecting H_0 becomes smaller for larger population-parameter values. Finally, notice that when π happens to be equal to P^*, the probability of rejecting H_0 is .50. Here there is a 50–50 chance that H_0 will be either accepted or rejected. This is true of any decision rule when the population parameter happens to equal the critical value of the test statistic.

Using the Power Curve

Recall that a decision rule is usually found for a fixed sample size n by first setting α and then determining the corresponding critical value. By studying the power curve, the decision-maker may find that the β values are higher than desired. In a previous section, we established that there are two ways to lower an error probability. One is to revise the critical value, which may be a remedy if the two kinds of risks, α and β, are not suitably balanced by the decision rule. The drug administrator could reduce the levels for β (that is, the chances of disapproving supplements of varying degrees of effectiveness) by lowering P^*, thereby shifting the power curve leftward. But recall that this can be done only at the expense of increasing the probability α of rejecting a true H_0 (in other words. of approving an ineffective supplement). If the decision-maker finds any reduction in α intolerable, he can lower the β values only by raising the sample size n. This is his second remedy. In such a case, more resources must be devoted to collecting additional sample observations. The resultant test would thereby become more discriminating, and the new power curve would have a different shape.

Evaluating the Test's Capability to Discriminate

The power curve for the dietary-supplement decision (Figure 8-8) shows that there are some quite large probabilities for rejecting H_0 and approving the supplement when it is only a minor improvement over nature. For example, the probability of approving the supplement when the actual rate of major cholesterol reduction is only $\pi = .12$ becomes $1-\beta = 1-.8238 = .1762$. Were π to be as low as .12, it would be unlikely that the government would want to approve the supplement. On the other hand, the power curve shows that if the actual rate were to be $\pi = .18$, which is quite close to the target .20 identified at the outset, then the probability of accepting H_0 and disapproving would be $\beta = .2119$. For $\pi = .18$, the level of effectiveness achieved might still be high enough so that approving the supplement would be desirable. Thus, we see that if π is slightly above .10 or slightly below .20, there is about a 20-percent chance of a decision leading to an undesired course of action. We may therefore conclude that *this statistical test provides less than desirable protection against making the wrong choice when the major reduction deviates only slightly from the originally hypothesized values of* π. One way out of this dilemma is to make the test more discriminating. This can usually be achieved by paying the price to increase the sample size.

To see how a larger sample would affect the drug administrator's discriminating capability, consider what happens when n is raised to 1,000 but the critical value is kept at $P^* = .15$. Figure 8-9 compares the resulting power curve with the previous curve in Figure 8-8. Note that the slope of the new curve (in color) is very steep for intermediate values of π. β is a high .9982 when $\pi = .12$, so that the probability of approving a barely effective supplement is $1-\beta = .0018$, a considerable reduction. Also, β is a low .0069 when $\pi = .18$, so that the probability of disapproving a fairly good supplement is substantially lowered.

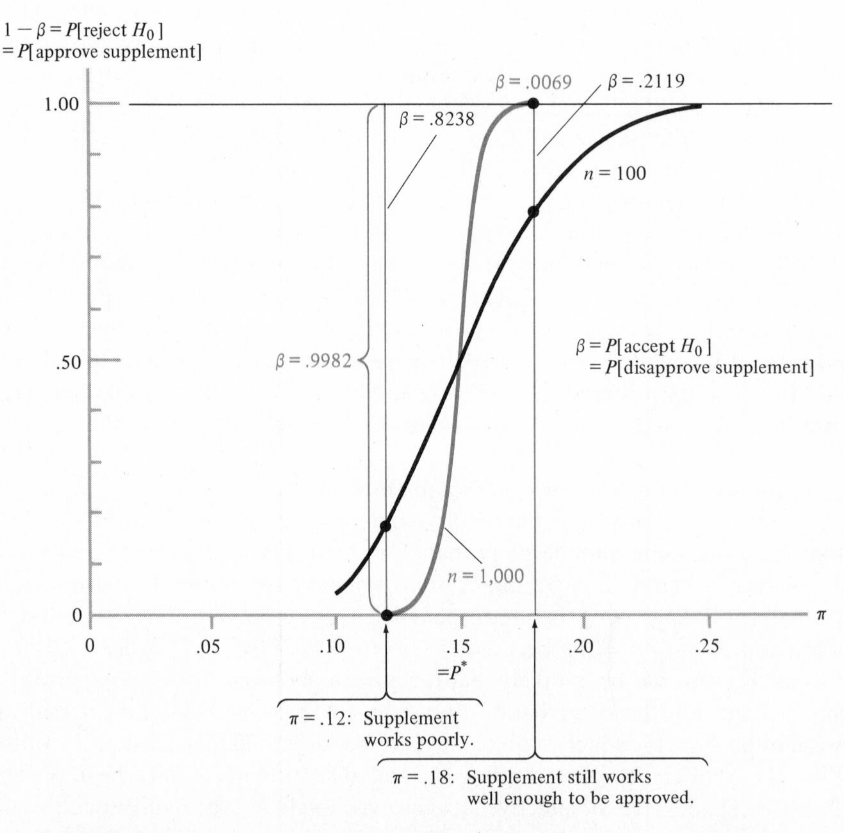

FIGURE 8-9 Comparison of power curves, when sample size is increased.

Efficiency and Power

A very important question considered in advanced applications of hypothesis-testing theory concerns the choice of the test statistic. In many decision situations, there may be a number of statistical tests that can be performed. For instance, in the dietary-supplement example, instead of finding the percentage of patients with improved conditions, the statistician could measure the actual amount of change in cholesterol levels for each patient in the experiment. The drug administrator could then use the mean change in cholesterol level $\bar{X}$ as the test statistic in making the decision and μ_0 as the mean change in the cholesterol levels occurring naturally in untreated patients. Generally, $\bar{X}$ provides more information than P, so $\bar{X}$ would be a more efficient test statistic than P. That is, for the same sample size and Type I error probability α, the decision rule obtained for $\bar{X}$ should provide smaller Type II error probabilities β (and hence higher probabilities of rejecting H_0 when it is false). The $\bar{X}$ test in this case is said to be more *powerful* than the P test.

REVIEW EXERCISES

8-26 For each of the following situations, find the Type I error probability α that corresponds to the stated decision rule. In all cases, use a sample size of $n = 100$ and assume that the population size is large.

(a) *Accept* H_0 if $P \leqslant .15$ *Reject* H_0 if $P > .15$ when $\pi_0 = .1$.

(b) *Accept* H_0 if $\bar{X} \geqslant 12$ *Reject* H_0 if $\bar{X} < 12$ when $\mu_0 = 14$ and $\sigma = 18$.

(c) *Accept* H_0 if $\bar{X} \leqslant 100$ *Reject* H_0 if $\bar{X} > 100$ when $\mu_0 = 97$ and $\sigma = 20$.

(d) *Accept* H_0 if $P \geqslant .42$ *Reject* H_0 if $P < .42$ when $\pi_0 = .5$.

8-27 Each of the following hypothesis-testing situations contains a sample size of $n = 100$ taken from a large population, and an $\alpha = .05$ significance level is desired. In each case: (1) give the value of μ_0; (2) indicate whether the test is lower-tailed, upper-tailed, or two-sided; (3) find the critical value(s) for $\bar{X}$; (4) calculate the normal deviate for the sample results; and (5) state whether H_0 should be accepted or rejected.

(a) $H_0: \mu \geqslant 25$; $H_1: \mu < 25$; $\bar{X} = 23.0$; and $s = 4.7$.

(b) $H_0: \mu \leqslant 14.7$; $H_1: \mu > 14.7$; $\bar{X} = 15.3$; and $s = 1.5$.

(c) $H_0: \mu = 168$; $H_1: \mu \neq 168$; $\bar{X} = 169$; and $s = .92$.

8-28 In the manufacture of a certain brand of paper, the desired thickness is .06 millimeters. If the mean value of the individual sheets is above or below this value, then the machinery must be adjusted. Because the amount of production time lost during adjustment is considerable, unnecessary adjustments should occur no more than 5 percent of the time. Historically, it has been established that the standard deviation in sheet thickness is $\sigma = .003$ millimeters. Each day a random sample of 100 sheets of paper is selected, and based upon the sample mean thickness, a decision of whether or not to adjust the equipment is made.

(a) State H_0 and H_1 in equation form. Is the test lower-tailed, upper-tailed, or two-sided?
(b) Does adjusting the machine correspond to accepting or rejecting H_0?
(c) Does not adjusting the machine correspond to accepting or rejecting H_0?
(d) Calculate the critical value(s) for the test statistic $\bar{X}$ and then formulate the decision rule.
(e) Calculate the Type II error probability, using the decision rule from part (d) and assuming that (1) the mean thickness is actually .061 millimeters, and (2) the mean thickness is actually .0597 millimeters.

8-29 A presidential candidate plans to campaign in only those primaries where he is preferred by at least 20 percent of the voters in his party. A random sample of 100 voter preferences is to be obtained from each state. In each case, the null hypothesis will be that the state meets the above criterion.
(a) Identify the appropriate parameter for the test. Formulate the null and alternative hypotheses in terms of specific values for this parameter.
(b) What test statistic should be used? Is the test upper- or lower-tailed?
(c) If an $\alpha = .05$ Type I error probability is desired, find the critical value for the test statistic and then formulate the decision rule.
(d) Based upon your answers to (c), indicate whether the null hypothesis should be accepted or rejected in each of the following states. Also indicate whether or not the candidate will enter the campaign in each of these states.

State	*Number in Sample Preferring Candidate*
New Hampshire	12
Florida	23
Wisconsin	15
Massachusetts	10

8-30 To assess the time required to cure a disease using an experimental drug, the drug is administered to a sample of $n = 25$ persons. The medical team is basing its decision of whether or not to continue its research upon whether or not the drug is deemed effective. An $\alpha = .01$ Type I error probability is desired for not continuing the research when the drug indeed yields a mean advantage in cure time of at least two days over present treatment time.
(a) State H_0 and H_1 in equation form. Is the test lower-tailed, upper-tailed, or two-sided?
(b) What test statistic should be used here? Determine the critical value for this test statistic and express the decision rule.
(c) Suppose that the sample results for the cure-time advantage are $\bar{X} = 1.5$ and $s = .5$ day. Should the null hypothesis be accepted or rejected? Does this mean that the drug experiment must be continued or terminated?

8-31 A shoe manufacturer must decide whether or not to tighten down on the sizing department's work. Due to ordinary variability, a shoe does not fall precisely into a size, like $8\frac{1}{2}$ EEE. Finished shoes must be individually measured and then classified to the nearest appropriate size. In order to determine whether or not to take corrective action, the plant manager orders a sample size of 100 randomly selected shoes to be meticulously resized and the proportion that were originally correctly sized determined. As his null hypothesis, he assumes that the true proportion of all shoes incorrectly sized will be $\leq .05$, the desirable level. He must construct a decision rule that will reject this hypothesis when it is true with a probability of .05.

(a) Is this test lower- or upper-tailed? Determine the critical value P^* and express the decision rule.

(b) Calculate the probability of making the Type II error, assuming that the true proportion of missized shoes is .10.

8-32 A civil service psychologist for a large city wishes to give his standard screening examination to some recent job applicants in order to decide whether or not job seekers score more highly than applicants did ten years ago. If they do, he will use a new examination in the future. The test is to be administered to a random sample of $n = 25$ persons. In analyzing the results, the psychologist desires only a 1-percent chance of incorrectly changing procedures when the actual mean screening examination score is ≤ 86, the historical mean figure ten years ago. Assuming that the mean score achieved by the sample group is 88, with a standard deviation of 10 points, should the present screening examination be kept or changed?

9 Regression and Correlation Analysis

"What is Fate?" Nasrudin was asked by a scholar.

"An endless succession of intertwined events, each influencing the other."

"That is hardly a satisfactory answer. I believe in cause and effect."

"Very well," said the Mulla, "look at that." He pointed to a procession passing in the street.

"That man is being taken to be hanged. Is that because someone gave him a silver piece and enabled him to buy the knife with which he committed the murder; or because someone saw him do it; or because nobody stopped him?"

A very important area of statistics involves predicting the value of a variable. A college admissions director is concerned with predicting the success of an applicant, which may be expressed in terms of his or her grade-point average (GPA) after matriculation. A chemical engineer may wish to predict the level of impurities in his final product. Many economists make predictions of gross national product (GNP). In each case, knowledge of one factor may be used to better predict another factor. The admissions director may use high-school grades as the basis for predicting college success. The chemical engineer may use process temperatures or concentrations of ingredient chemicals to forecast impurities. The economist may use current interest rates, unemployment levels, and government spending in making his GNP prognostications.

In this chapter we will discuss the techniques that fit into a broad category called *regression and correlation analysis.* Regression and correlation analysis comprises a body of statistical methodology that investigates the relation between variables. In Chapter 7, we initially investigated one form of statistical inference by using the sample mean to estimate the value of the population mean. In Chapter 8, a second kind of inference was described—one that uses a sample mean to test a hypothesis about a population mean. In this chapter, a third kind of statistical inference is to be introduced. Here, we will be concerned with measuring the *association* between two or more variables.

Regression analysis tells us how one variable is related to another. It provides an equation wherein the known value of one or more variables may be used to estimate the unknown value of the remaining variable. For instance, regression analysis can be used by a medical researcher to estimate a laboratory animal's longevity (a variable of unknown value) resulting from the caloric content of its daily diet (a variable of known value). This information can lead to further research on how diet affects human life spans. Similarly, an economist may use regression analysis to show how one variable, such as percentage unemployment, can be used to predict the percentage inflation rate. The resulting mathematical relationship provides a graphical display (called the Phillips curve). As another example, the future sales potential for a fried-chicken franchise outlet at a candidate site can be estimated by measuring the traffic density along the access road to the site and then substituting this value into a mathematical equation. By including other relevant factors, such as the distance to the closest competitor or the number of apartment units within walking distance, a finer estimate of sales potential might be obtained. This example illustrates that more than one variable can be used to estimate an unknown variable. When several variables are used to make a prediction, the technique is called *multiple regression.* (This will be discussed in Chapter 10.)

Correlation analysis tells us the degree to which two variables are related. It is useful in expressing the efficiency achieved by using one variable to estimate the value of another. Correlation analysis can also identify which factors of a multiple-characteristic population are highly related, either directly or by a common connection to another variable.

Regression analysis takes its name from studies made by Sir Francis Galton around the turn of this century. Galton compared the heights of persons to the heights of their parents. His major conclusion was that the offspring of unusually tall persons tend to be shorter than their parents, while children of unusually short parents tend to be taller. In a sense, the successive generations of offspring from tall persons "regress" downward toward the mean height of the population, while the reverse is true of the offspring from short families.* Since one variable (the height of the parent) was used to predict another (the height of the child), the original term"regression" came to be applied to more general analyses involving the prediction of one variable by another. Aside from the context of prediction, the term "regression" as used in this chapter has little relationship to Galton's original notion of regressing toward the mean.

9-1 REGRESSION ANALYSIS

The primary goal of regression analysis is to obtain predictions of one variable using the known values of another. These predictions are made by means

* But the distribution of heights for the total population continues to have the same variability from generation to generation. This is because the more prevalent parents of near-average height produce more tall offspring than do the relatively rare tall parents.

of an equation such as $Y = a + bX$, which provides the estimate of an unknown variable Y when the value of another variable X is known. Such an expression is referred to as a *regression equation*. Knowing the regression equation, a prediction of Y may be readily obtained from a given X. Unlike the results from ordinary mathematical equations (such as $A = b \times h$ for the area of a rectangle, or interest $= i \times P \times t$, where i is the rate of interest, P is the principal, and t is the time), we cannot be certain about the value of Y obtained from the regression equation. This is due to inherent statistical variability. Thus, predictions made from the regression equation are subject to error and are only *estimates* of the true values.

To set the stage, we will consider the problem of the admissions director at a certain college. In predicting the college grades of potential students, regression analysis can establish a relationship between the college grades of present students and predictive factors (such as high-school grades, Scholastic Aptitude Test (SAT) scores, and extracurricular activities). For the present, we will consider just one predictive factor, high-school grades, since they are reputed to be one of the most reliable indicators of future college success. The relationship determined will be based upon sample data obtained for present students at the college.

TABLE 9-1 Sample Observations of High-School and College Grade-Point Averages. (Based upon 4 points for "A," 3 for "B," and so on.)

Student	*High-School GPA X*	*College GPA Y*
1. Anita Juarez	4.0	3.8
2. Harry Brown	3.7	2.7
3. Virgil Kiser	2.2	2.3
4. JoAnn Hampton	3.8	3.2
5. Warren Schultz	3.8	3.5
6. Niki Scott	2.8	2.4
7. Robert Beaman	3.0	2.6
8. Charles Jackson	3.4	3.0
9. Sally Adamson	3.3	2.7
10. Trevor Wallacek	3.0	2.8

Regression analysis begins with a set of data comprised of pairs of observed values (one number for each variable). Table 9-1 shows the observations of high-school and college GPA for a sample of graduating college seniors. These data will be used to predict college grades for entering freshmen by means of a regression equation. Because of sampling error, the regression equation obtained may not be truly representative of the actual relationship between the variables in the population as a whole. In order to reduce the chances of a large sampling error, a sample size considerably greater than 10 ought to be used. We have chosen such a small number for our sample here merely to make calculations easier.

The Scatter Diagram

A first step in regression analysis is to plot the value pairs as points on a graph, as we have done in Figure 9-1 for the data in Table 9-1. The horizontal axis corresponds to the values of the high-school GPA, a variable we will denote by X. The vertical scale represents the values of the other variable, college GPA, which we designate as Y. A point is found for each student. For example, the grade-point averages of Harry Brown are $X=3.7$ for high school and $Y=2.7$ for college. These GPAs are represented on the graph by the point ($X=3.7$, $Y=2.7$). The ten points obtained in Figure 9-1 are spread in an irregular pattern. For this reason, such a plot is referred to as a *scatter diagram*. It is customary

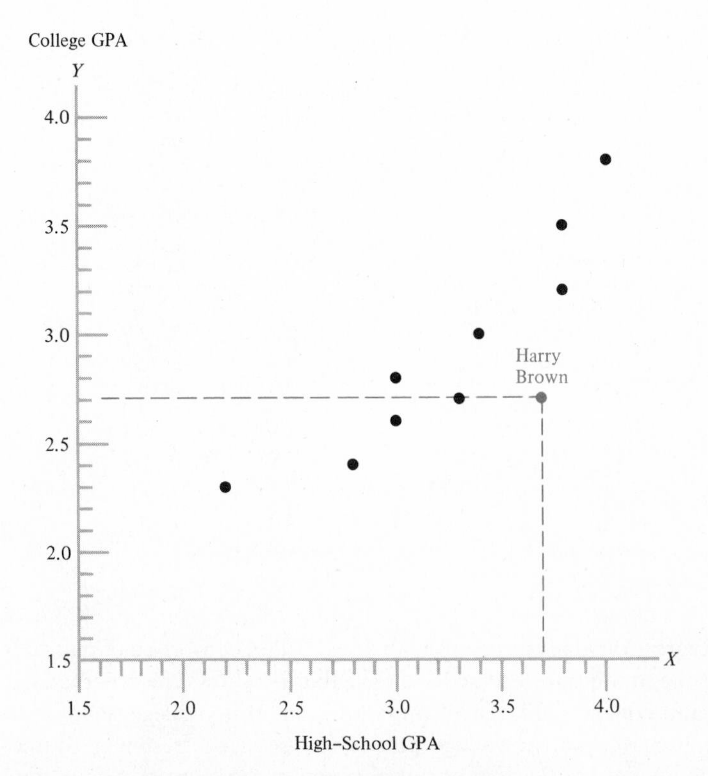

FIGURE 9-1 Scatter diagram for student high-school and college grade-point averages (GPAs). (Values from Table 9-1)

to refer to the variable whose value is known as the *independent variable*; its possible values are represented by the X axis of the scatter diagram. The variable being predicted is called the *dependent variable*; the values on the Y axis of the scatter diagram represent the possible magnitudes of the dependent variable. Thus, high-school GPA X is the independent variable, because it can be readily determined for any potential student; college GPA Y is the dependent variable, because it is in part predictable from high-school GPA X.

These designations follow from simple algebra, where the X axis represents the independent variable and the Y axis provides values for the dependent variable by means of a function or an equation. Thus, Y is a function of X. The dependence of Y upon X does not necessarily mean that Y is *caused* by X. The type of relationship found by regression analysis is a statistical one. (We can find a statistical relationship expressing family household expenditures as a dependent variable that is a function of an independent variable, family disposable income. But having money to spend does not mean that appliances will be purchased. A purchase is a voluntary decision that is merely allowed to occur because money has been made available; a purchase need not be made just because there is sufficient income.)

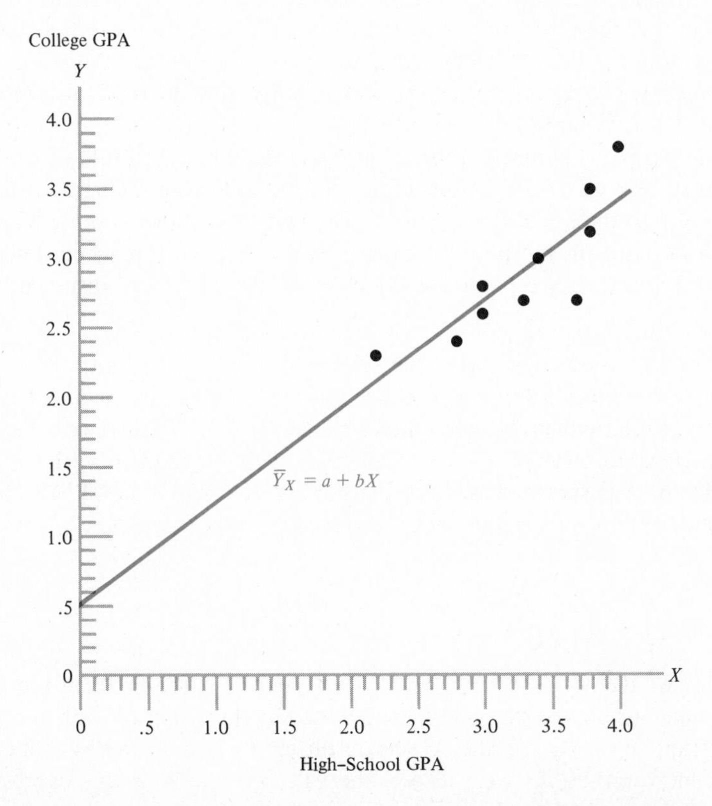

FIGURE 9-2 Fitting a regression line to the student GPA data in Table 9-1.

The Data and the Regression Equation

The next step in regression analysis is to find a suitable function to use for the regression equation, which will provide the predicted value of Y for a given value of X. The clue to finding an appropriate regression equation is the general pattern presented by the points on the scatter diagram. A cursory examination of the student-grade data in our example indicates that a straight line, like the one shown in Figure 9-2, might be a meaningful summary of the information provided by the sample. This line seems to "fit" the rough scatter pattern of the data points.

Regression Line

A linear relationship between the variables X and Y is conceptually the simplest. The general equation for a straight line is $Y = a+bX$. The constant a is the value of Y obtained when $X = 0$, so that $Y = a+b(0) = a$. This is the value for Y at which the line intersects the Y axis, so that a may be referred to as the Y *intercept*. The constant b is the *slope* of the line. It represents the change in Y due to a one-unit increase in the value of X. Figure 9-3 shows the line for the equation $Y = 3+2X$. Here, the Y intercept is $a = 3$ and the slope is $b = 2$. Y increases by two units for every one-unit change in X. To review how to find the value of Y for X, suppose we wish to find the Y corresponding to $X = 5$. Substituting 5 for X in the above expression, we obtain $Y = 3+2(5) = 13$. The same value may be read directly from Figure 9-3 by following the vertical black line from $X = 5$ to the line relating Y to X. The vertical distance represents the value of Y. Any point on the line, for example ($X = 5$, $Y = 13$), can be described by the horizontal distance from the Y axis (5) and the vertical distance from the X axis (13).

The line used to describe the relationship between Y and X is generally obtained from sample data and is called the *estimated regression line*. It expresses the average relationship between the variables X and Y. The estimated regression line provides an estimate of the mean level of the dependent variable Y, when the value of X is specified. We use the symbol $\overline{Y}_X$ ("Y-bar sub X") to represent the values obtained from the linear ***estimated regression equation***

$$\overline{Y}_X = a + bX \tag{9-1}$$

With $\overline{Y}_X$, we distinguish estimates of the dependent variable from the observed data points, which, for simplicity, are denoted by the symbol Y. When expression (9-1) is applied to a specific X, the resulting $\overline{Y}_X$ is a *predicted* value for the dependent variable. As we will see, the values a and b in this expression are found from sample data. They are referred to as *estimated regression coefficients*.

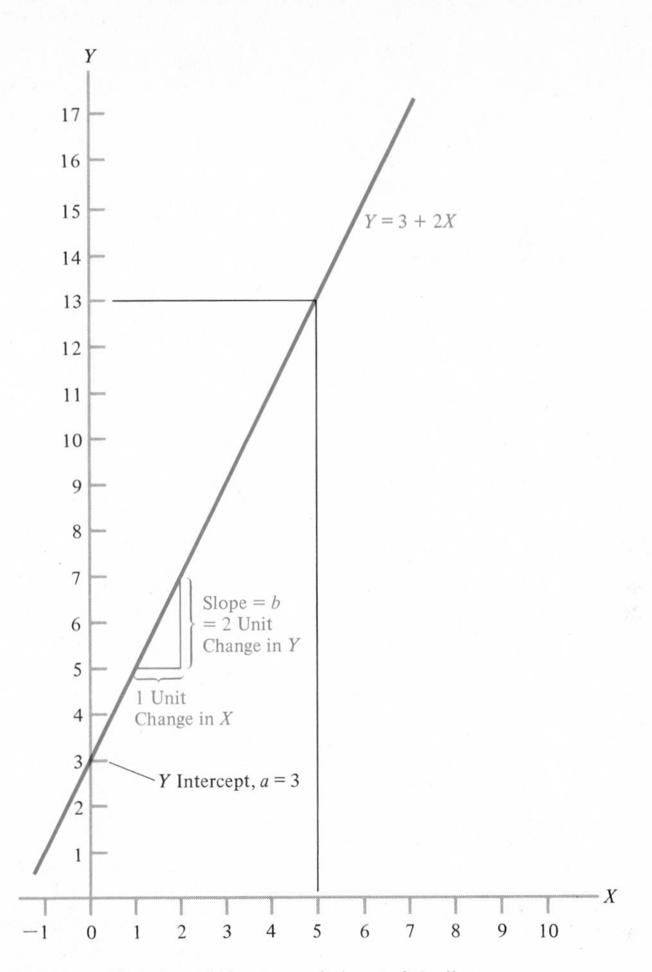

FIGURE 9-3 Slope and Y intercept for a straight line.

Regression Curves

It must be emphasized that a straight line is not always an appropriate function relating Y to X. The scatter diagrams in Figure 9-4 show cases where various types of nonlinear functions more closely fit the data. Notice that all the functions are curves. Such relationships between X and Y are called *curvilinear*.

Figure 9-4(a) shows the relationship between crop yield Y and quantity X of fertilizer applied. The curve fits data obtained from test plots to which various amounts of fertilizer have been applied. As the amount of fertilizer increases to a certain point, it proves beneficial in increasing the harvest. But beyond this point, the benefits become negative, because additional fertilizer burns plant roots and causes the crop yield to decline. The greatest increases in Y occur for small values of X, with Y increasing at a decreasing rate. The peak value of Y may be referred to as the "point of negative returns," after which Y decreases at an

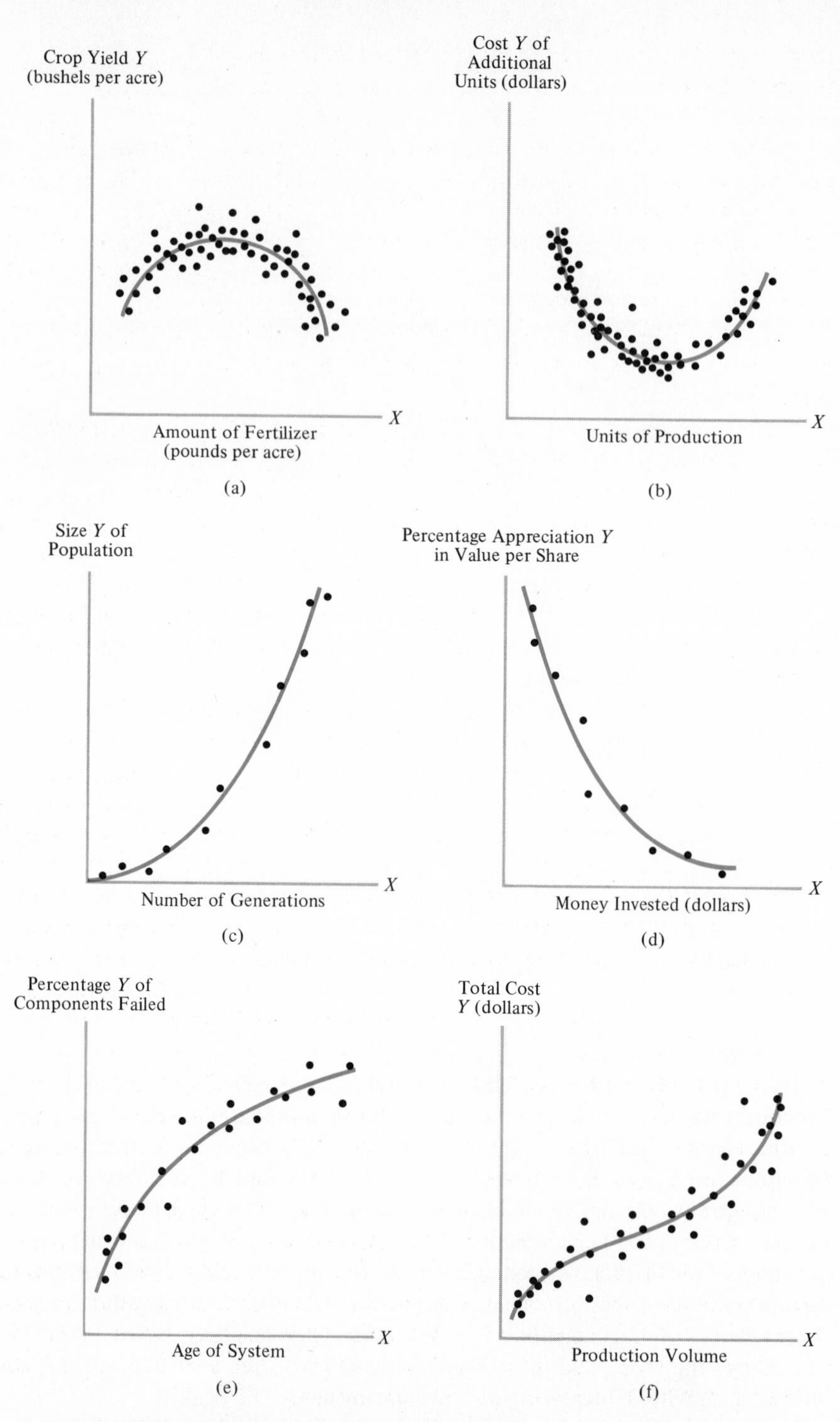

FIGURE 9-4 Examples of curvilinear relationships found for scatter diagrams.

increasing rate. This curve expressing Y as a function of X has the shape of an inverted U. One regression equation that could be used here is the form $Y = a + bX - X^2$.

This is contrasted to the U-shaped curve in Figure 9-4(b), where incremental or marginal cost Y of production is plotted against volume X. The regression equation has the form $Y = a + bX + X^2$. Data points have been obtained for the various levels of activity experienced in a plant. For low levels of plant activity, all factors of production are employed less efficiently. But as volume increases, costs of additional units decline as inputs become more efficiently employed. Increases in efficiency become less pronounced, until a point is reached where extra production can be handled, but less efficiently than before, and marginal costs begin to rise.

In Figure 9-4(c), the size Y of a population of *Drosophila* (fruit flies) is shown as a function of the number of generations X since the colony was established. The data points represent observations made after successive hatchings. There are no natural checks upon the population growth, since the flies are reared in an artificial environment. The curve obtained is an exponential or geometric growth curve, so that Y is related to X by an expression like $Y = a^X$ or $Y = X^c$. Plots of the world's human population growth during the past several centuries have the same basic shape. With exponential growth, Y increases at an increasing rate for larger values of X.

This is in contrast to the negative exponential curve in Figure 9-4(d), where the percentage appreciation Y in the value of each share of a mutual fund is plotted against the total money invested by the fund. The points represent several successful mutual funds at various stages of growth. The appropriate regression equation would be in the form $Y = ae^{-bX}$. Negative exponential curves correspond to values of Y that decrease at a decreasing rate as X becomes large. A rationale for such a result is that a small mutual fund can be very selective in choosing its portfolio, so that it has the opportunity to buy small-company stocks that may appreciate greatly. But as the fund grows larger, it cannot buy as heavily into the rather limited number of small, growing companies and must invest more money in the stocks of larger, less promising firms.

Figure 9-4(e) shows a logarithmic curve relating the percentage Y of original components of a particular kind that have failed to the age X of the system. The points were obtained from the histories of several systems. Some components survive almost indefinitely, although most of the original ones fail early. Here, the values for Y increase at a decreasing rate as X becomes large. We may express Y in terms of X by the equation $Y = a + b \log(1 + X)$.

In Figure 9-4(f), the total costs Y of production experienced by a plant over a particular time period are plotted against the associated levels of activity X. These points are fitted by a curve that shows total cost increasing at a decreasing rate as greater volumes of production are achieved. Beyond a point of diminishing returns for X, the larger volume reduces incremental efficiency and the curve shows the values of Y increasing at an increasing rate.

Throughout the remainder of this chapter, only linear relationships will be considered. The procedure for determining which particular line best fits the data is called *linear regression analysis.* One reason for our emphasis on linear equations is that they are easier to explain. The methods of analysis are also simpler and may be directly extended to curvilinear reltaionships. Straight lines are useful for describing a great many phenomena, so they are among the most common regression relationships.

Some Characteristics of the Regression Line

Some important general properties of the regression line and its fit to the data are illustrated in Figure 9-5. We will first consider the manner in which Y is related to X. There are two basic kinds of regression lines. If the values of the dependent variable Y increase for larger values of the independent variable X, then Y is *directly related* to X, as is shown in Figure 9-5(a). Here, the slope of the line is positive, so that b is greater than zero; this is because Y will increase as X becomes larger. Figure 9-2 shows that college and high-school GPAs are directly related variables. Other examples of directly related variables are age and salary (during employment years), weight and daily caloric intake, and the number of passengers and the quantity of luggage on commercial aircraft flights. In Figure 9-5(b), the slope of the regression line is negative, with the value of b less than zero. Here, Y becomes smaller for larger values of X, so the variables X and Y are *inversely related.* Examples of inverse relationships include remaining tire tread and miles driven, crop damage by insects and the quantity of insecticide applied, and the typical economic demand curve where demand decreases as price increases.

The degree to which two variables are related is reflected by the amount of scatter of the data points about the regression line. In Figure 9-5(c), the data points all lie on the regression line. This unusual "perfect fit" indicates that X and Y are *perfectly correlated.* In Figure 9-5(d), the data points are widely scattered about the regression line, so that we may say that the data indicate that X and Y are weakly correlated. Contrast this scatter diagram with the one in Figure 9-5(a), where the points cluster more closely about the line. Predictions of Y tend to be more accurate when there is less scatter, for then the sample results are less varied in relation to the regression line. As we saw in Chapter 7, small sample variability indicates a smaller standard error in the estimate; the sampling error is less pronounced, and more reliable estimates are obtained. When there is less scatter, the degree of correlation is higher. This suggests that correlation analysis can be used to qualify the accuracy of estimates made from the regression line. When X and Y are *uncorrelated,* as in Figure 9-5(e), the regression line is horizontal, indicating that values of Y have sizes independent of the value of X. This means that X is a worthless predictor of Y, so that the poorest predictions of Y are obtained by regression when Y is uncorrelated with X.

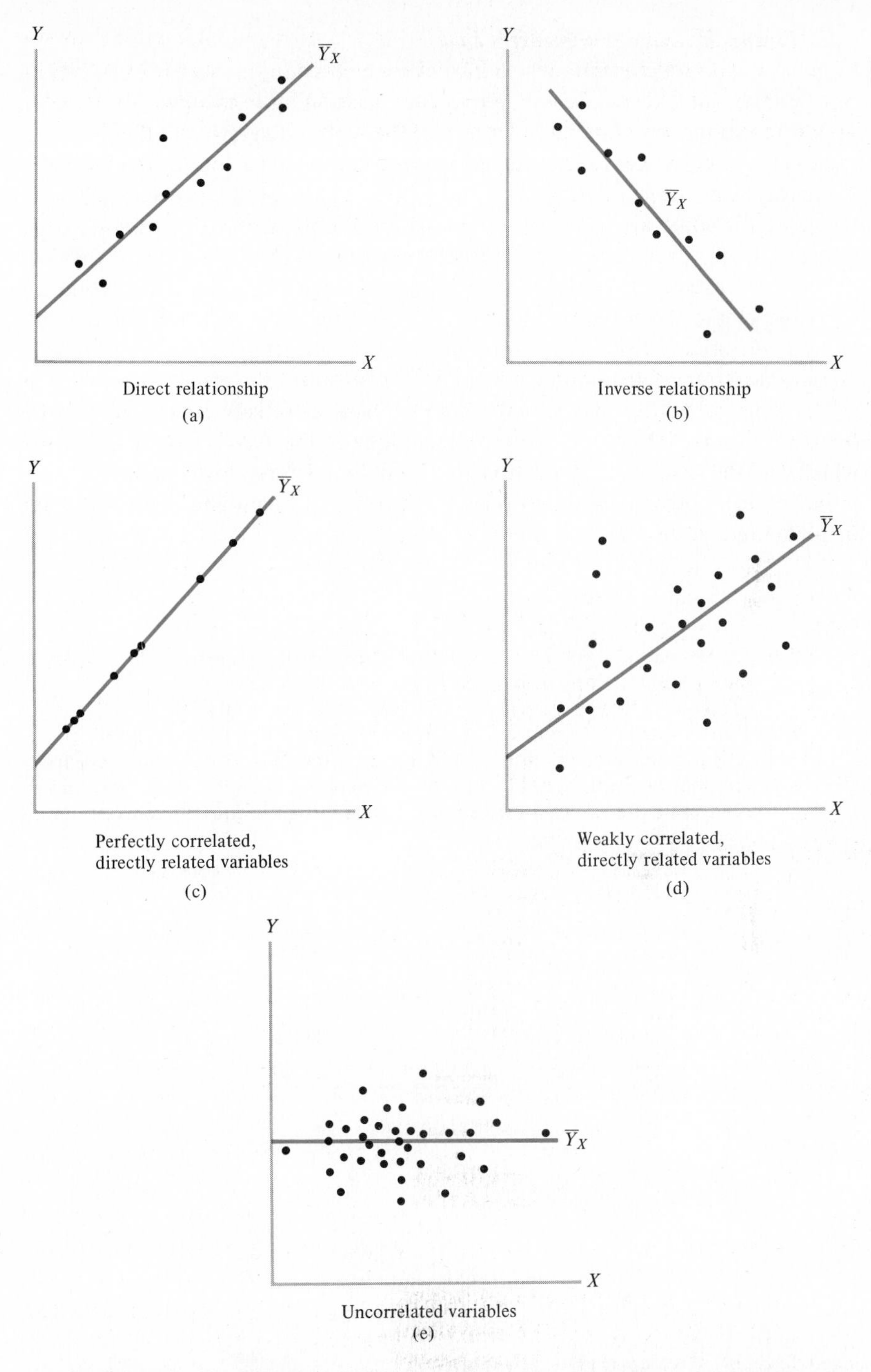

FIGURE 9-5 Properties of the regression line and possible relationships between variables.

Fitting the Data by a Straight Line

How do we determine which particular regression line to use? One simple procedure would be to use our judgment, positioning a straight edge until it appears to summarize the linear pattern of the scatter diagram and then drawing the line. The equation of this regression line can then be found by reading its Y intercept and slope directly from the graph. Although this procedure may be adequate for some applications, estimates for Y obtained in this way are often crude. A major drawback to using freehand methods in fitting a line is that two persons will usually draw different lines for the same data. Freehand fitting can therefore create unnecessary controversy regarding the conclusions of the data analysis. Another serious objection is that statistical methodology cannot then be employed to qualify (by confidence limits, for instance) the errors of estimation.

These difficulties may be overcome by using a statistical method to fit the line to the data. The most common technique is the *method of least squares*, which will be discussed in Section 9-2. This procedure, from many different points of view, provides the best possible fit to a set of data and thereby the best possible predictions.

EXERCISES

9-1 On a sheet of graph paper, construct the following regression lines and indicate whether X and Y are directly or inversely related:
(a) $\bar{Y}_X = -5+2X$ (b) $\bar{Y}_X = 5+3X$ (c) $\bar{Y}_X = 20-2X$

9-2 Plot a scatter diagram for the following data on a piece of graph paper. Then, using a ruler, draw a line through the data points that appears to summarize the underlying relationship between X and Y. From your graph, determine the values for a and b, and then write the equation corresponding to your regression line.

Number of Pages X	*Hours of Typing Time* Y	*Number of Pages* X	*Hours of Typing Time* Y
10	20	100	50
20	10	40	40
20	30	60	40
30	30	110	30
50	20	120	50

9-3 Plot a scatter diagram for each of the following sets of data. Sketch the shape of the regression curve that best seems to fit each relationship.

(a)		(b)		(c)	
Number of Men X	*Output per Man-Hour* Y	*Number of Units* X	*Total Cost* Y	*Minutes Between Rest Periods* X	*Pounds Lifted per Minute* Y
5	10	600	$11,000	5.5	350
8	4	50	3,100	9.6	230
1	3	470	10,200	2.4	540
1	2	910	15,700	4.4	390
1	7	160	6,300	.5	910
8	8	950	19,500	7.9	220
7	10	690	13,900	2.0	680
10	2	90	1,800	3.3	590
2	5	310	8,800	13.1	90
3	8	1,000	25,700	4.2	520

9-4 Consider two variables: IQ (a purported measure of intelligence), and college GPA. For each of the following studies, indicate which of these two variables would be independent and which would be dependent.

(a) A college admissions director wishes to predict the potential college grades for all applicants who have an IQ of 120.

(b) A sociologist wishes to find out how "smart" his colleagues are by using their undergraduate college grades.

(c) An educator wishes to find a way to predict college success using IQ scores.

9-2 METHOD OF LEAST SQUARES

Introduction

Least squares regression is a technique for fitting a regression equation to the observed data. The least squares criterion has a great many desirable properties that make it the most commonly used tool in regression analysis. Although much of this chapter is restricted to linear relationships between two variables, the procedures we will describe here can also be extended to a variety of situations where a curvilinear fit is desired. We will begin by describing how the method of least squares is applied to our previous GPA data, which are again plotted in Figure 9-6. The least squares criterion requires that a line be chosen to fit our data so that the *sum of the squares of the vertical deviations separating the points from the line will be a minimum.* The deviations are represented by the lengths of vertical line segments that connect the points to the estimated regression line in the scatter diagram.

Rationale for Least Squares

To explain how this procedure may be interpreted, we investigate Harry Brown, who earned a high-school GPA of $X = 3.7$. Our data show that his college GPA is $Y = 2.7$. This value is represented in Figure 9-6 by the vertical distance along the thin line from the X axis at $X = 3.7$ to the corresponding Y data point (2.7). A predicted or estimated college GPA for a new freshman with identical high-school grades may be obtained by following this vertical distance all the way up to the regression line (the height of the thin vertical line plus the length of the colored vertical segment, or a GPA of $\overline{Y}_X = \overline{Y}_{3.7} = 3.192$). The difference between Harry Brown's observed college GPA, $Y = 2.7$, and the predicted value for Y is the difference $Y - \overline{Y}_X = Y - \overline{Y}_{3.7} = 2.7 - 3.192 = -.492$. The number .492 is the length of the colored vertical line segment connecting the point to the regression line. Because the observed value of Y lies below the predicted value, a negative deviation is obtained; if the observed Y lay above the line, the deviation would be positive. The vertical deviation represents the amount of *error associated with using the regression line to predict* a new student's college GPA. We want to find values for a and b that will minimize the sum of the squares of

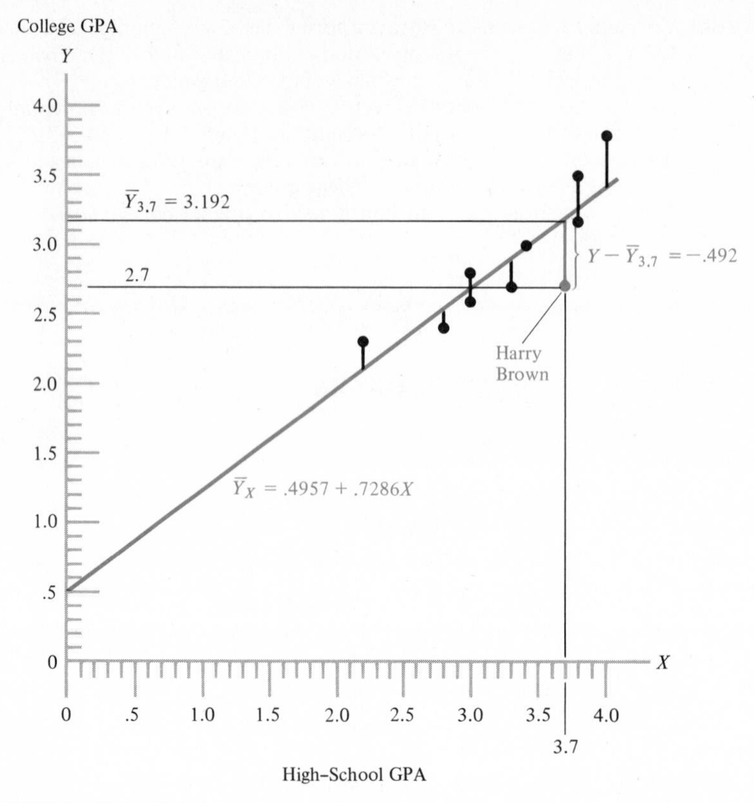

FIGURE 9-6 Fitting a regression line to the student GPA data, using the method of least squares.

these vertical deviations (or prediction errors):

$$\sum(Y-\overline{Y}_X)^2 \tag{9-2}$$

One reason for minimizing the sum of the squared vertical deviations is that some of the differences $Y-\overline{Y}_X$ are negative while others are positive. For any set of data, a great many lines can be drawn for which the sum of the unsquared deviations would be zero, but most of these lines would poorly fit the data.

Finding the Regression Equation

Substituting $a+bX$ for $\overline{Y}_X$ into expression (9-2), the sum to be minimized becomes

$$\sum(Y-a-bX)^2$$

which is a function having two unknowns, a and b. Mathematically, it may be shown that the required values must simultaneoulsy satisfy the following expressions, referred to as the *normal equations**:

$$\sum Y = na + b\sum X \tag{9-3}$$

$$\sum XY = a\sum X + b\sum X^2 \tag{9-4}$$

Solving the above algebraically, we can obtain the following final expression for b:

$$b = \frac{n\sum XY - \sum X\sum Y}{n\sum X^2 - (\sum X)^2} \tag{9-5}$$

The equation for a may then be obtained by substituting the value for b into expression (9-3) and then solving for a:

$$a = \frac{1}{n}(\sum Y - b\sum X) \tag{9-6}$$

The above expressions may be further simplified using the mean values $\overline{X} = (\sum X)/n, \overline{Y} = (\sum Y)/n$ in order to compute the ***estimated regression coefficients***

$$b = \frac{\sum XY - n\overline{X}\overline{Y}}{\sum X^2 - n\overline{X}^2} \tag{9-7}$$

$$a = \overline{Y} - b\overline{X} \tag{9-8}$$

The advantage of calculating a and b from expressions (9-7) and (9-8) is that every step involves computations with values of moderate size. Although this may increase the danger of rounding errors, such errors ordinarily prove to be negligible.

Illustration of the Method

We are now ready to find the regression equation for the admissions director's regression line obtained from the $n = 10$ observations. In order to evaluate the expressions for a and b, we must perform a set of intermediate calculations (shown in Table 9-2). To find b, we must calculate $\overline{X}$, $\overline{Y}$, $\sum X^2$, and $\sum XY$. Columns of values for X, Y, X^2, and XY are used for this purpose. An extra column for the squares of the dependent variable observations, Y^2, is computed for use in later discussions of the regression line. Using the

* The word "normal" as used here has nothing to do with the normal curve. Rather, *normal equations* receive their name from a mathematical property of linear algebra.

TABLE 9-2 Intermediate Calculations for Obtaining the Estimated Regression Line.

(1) Student	(2) High-School GPA X	(3) College GPA Y	(4) XY	(5) X^2	(6) Y^2
1	4.0	3.8	15.20	16.00	14.44
2	3.7	2.7	9.99	13.69	7.29
3	2.2	2.3	5.06	4.84	5.29
4	3.8	3.2	12.16	14.44	10.24
5	3.8	3.5	13.30	14.44	12.25
6	2.8	2.4	6.72	7.84	5.76
7	3.0	2.6	7.80	9.00	6.76
8	3.4	3.0	10.20	11.56	9.00
9	3.3	2.7	8.91	10.89	7.29
10	3.0	2.8	8.40	9.00	7.84
	33.0	29.0	97.74	111.70	86.16

$$\bar{X} = \frac{\sum X}{n} = \frac{33.0}{10} = 3.3 \qquad \bar{Y} = \frac{\sum Y}{n} = \frac{29.0}{10} = 2.9$$

$$\sum XY = 97.74 \qquad \sum X^2 = 111.70 \qquad \sum Y^2 = 86.16$$

intermediate values obtained, we can first find the value for b by employing expression (9-7). (We find b first because it is used to calculate a.)

$$b = \frac{\sum XY - n\bar{X}\bar{Y}}{\sum X^2 - n\bar{X}^2} = \frac{97.74 - 10(3.3)(2.9)}{111.70 - 10(3.3)^2} = \frac{2.04}{2.80} = .7286$$

Substituting $b = .7286$ into expression (9-8), we obtain

$$\begin{aligned} a &= 2.900 - .7286(3.3) \\ &= 2.900 - 2.4043 = .4957 \end{aligned}$$

Thus we have determined the following equation for the estimated regression line graphed in Figure 9-6:

$$\bar{Y}_X = .4957 + .7286X$$

We may now use the above regression equation to predict the college GPA $\bar{Y}_X$ for a student whose high-school GPA X is known. For instance, when $X = 3.7$, we have

$$\bar{Y}_{3.7} = .4957 + .7286(3.7) = 3.192$$

Thus, the prediction is that any new freshman having a 3.7 high-school GPA will have a college GPA of $\bar{Y}_{3.7} = 3.192$.

A Check for Computational Accuracy

The least squares regression line has two important features. One is that it goes through the point $(\bar{X}, \bar{Y})$ corresponding to the mean of the observations of X and Y. The other is that the sum of the deviations of the Ys from the regression line is zero. That is

$$\sum (Y - \bar{Y}_X) = 0 \qquad (9\text{-}9)$$

Thus, the positive and negative deviations about the regression line cancel, so that the least squares line goes through the center of the data scatter. This can be a useful check to determine if any miscalculations were made in finding a and b. In Table 9-3, the deviations are calculated for the student data. (The sum of the deviations is slightly less than zero, $-.002$, due to rounding errors.)

TABLE 9-3 Computation of Student GPA Regression Line Deviations as a Check for Consistency.

X	Y	$\bar{Y}_X = a + bX$ $= .4957 + .7286X$	$Y - \bar{Y}_X$
4.0	3.8	3.410	.390
3.7	2.7	3.192	−.492
2.2	2.3	2.099	.201
3.8	3.2	3.264	−.064
3.8	3.5	3.264	.236
2.8	2.4	2.536	−.136
3.0	2.6	2.682	−.082
3.4	3.0	2.973	.027
3.3	2.7	2.900	−.200
3.0	2.8	2.682	.118
33.0	29.0		$\sum (Y - \bar{Y}_X) = -0.002$

Meaning and Use of the Regression Line

Once the regression equation has been obtained, predictions or estimates of the dependent variable may be made. The admissions director now has a basis for screening out applicants who do not have a reasonable chance of succeeding in his college. The value of the slope, $b = .7286$, tells us that for each complete grade-point rise in a student's high-school GPA there will be a predicted college GPA rise of .7286 of a grade point, indicating that grades must be reduced to about 73 percent of their high-school level before predictions of college performance can be made from them. This is consistent with the widely accepted fact that college is harder than high school, so that most students' grades drop somewhat in college. In one sense, we might say that a student is only about 73 percent as capable of achieving the same level of college grades that he earned in high school (that is, his capability of obtaining grades is reduced by 27 percent).

Because b is positive, superior performance in high school still results in greater college success, but the regression line shows that a further adjustment must still be made to the lowered college grades.

This second adjustment is where the Y intercept applies. The value $a = .4957$ indicates that after reducing the high-school grades we must adjust all of them upwardly by this uniform amount. This adjustment corrects the distortion that would be incurred if every student's GPA were reduced by a flat 27 percent. The data indicate that people who were "C" students in high school will continue to be "C" students in college (when they are accepted). Their grades change less in the "pressure-cooker" college environment than do those of the "A" students in high-school, who, in large measure, become "B" students in college. Upwardly adjusting all grade points by .4957 produces a net college GPA reduction that actually hits harder, in terms of points lost, the better the student did in high school.

Viewed differently, we might say that college cuts everybody down to size (since the slope of the regression line is less than unity), but that college professors then give an equal bonus of nearly half a grade point (the Y intercept).

Knowing that a student earns a 3.7 GPA in high school, $\bar{Y}_{3.7} = 3.192$ may be used as a point estimate for that student's college GPA. But the proper interpretation of the 3.192 GPA is that *on the average* all new students having a 3.7 high-school GPA will earn a 3.192 GPA in college. The point estimate is an average because students differ by a host of factors, including the high schools they graduated from, their levels of motivation, and their study habits. College GPA will vary from student to student, even among those who appear to have done equally well in high school. This leads us to the next important consideration of regression analysis.

Measuring Variability of Results

The fundamental expression of variability available from the sample data is a measure of the spread or scatter about the estimated regression line. As we have noted, estimates made from the regression line will be more precise when the data are less scattered. Hence, we may investigate the degree of scatter to determine an expression for the error involved in making estimates through regression. This suggests employing a measure that fits naturally into the scheme of least squares regression. Recall that we obtained the Y intercept a and the slope b of the regression line by minimizing the squared deviations about the regression line, $\sum(Y-\bar{Y}_X)^2$. As we saw in selecting the fundamental measure of a population's variability, the variance is the average of the squared deviations from the mean. This suggests that we can similarly select as our measure of variability the mean of the squared deviations about the regression line for a sample of size n:

$$\frac{\sum(Y-\bar{Y}_X)^2}{n}$$

Standard Error of the Estimate

The square root of the mean squared deviations is referred to as the *standard error of the estimate* about the regression line. This suggests that it may be used to estimate the true variability in Y. Because we will use it for this purpose, for convenience we will modify the above expression before taking its square root, using as our standard error of the estimate

$$s_{Y \cdot X} = \sqrt{\frac{\sum(Y-\overline{Y}_X)^2}{n-2}} \tag{9-10}$$

Here, we use the letter s in accordance with our convention to indicate that the calculations have been made from sample data. The subscript $Y \cdot X$ indicates that the deviations are about the regression line, which provides values of Y for given levels of X. We divide the sum of the squared deviations by $n-2$, which will make $s^2_{Y \cdot X}$ an unbiased estimator of the true variance of the Y values about the regression line. We subtract 2 from n because 2 degrees of freedom are lost since the values of a and b making up the expression for $\overline{Y}_X$ have been calculated from the same data.

The standard error of the estimate resembles the *sample standard deviation* calculated for the individual Ys, which we designate as s_Y:

$$s_Y = \sqrt{\frac{\sum(Y-\overline{Y})^2}{n-1}} \tag{9-11}$$

Here, X does not appear in the subscript because s_Y makes no reference to the values for X. The sample standard deviation is the square root of the mean of the squared deviations about the center $\overline{Y}$ of the sample: $(Y-\overline{Y})^2$. Thus, s_Y represents the *total variability* in Y. Ordinarily, the deviations about $\overline{Y}$ are larger than their counterparts about the estimated regression line, so that s_Y will be larger than $s_{Y \cdot X}$. Figure 9-7 illustrates this concept. Note that s_Y and $s_{Y \cdot X}$ summarize the dispersions of separate sample frequency distributions.

Although expression (9-10) serves to define $s_{Y \cdot X}$, in practice it is cumbersome to use in calculating the standard error, because the $\overline{Y}_X$ values must first be calculated from the estimated regression equation. The following mathematically equivalent expression is often used instead for ***the standard error of the estimate***

$$s_{Y \cdot X} = \sqrt{\frac{\sum Y^2 - a\sum Y - b\sum XY}{n-2}} \tag{9-12}$$

This is simpler to use than expressions appearing in some other textbooks because it allows us to take advantage of the calculations of a and b that we already made when determining the regression equation. We will use expression (9-12) to calculate $s_{Y \cdot X}$ for our college admissions illustration. Using the values

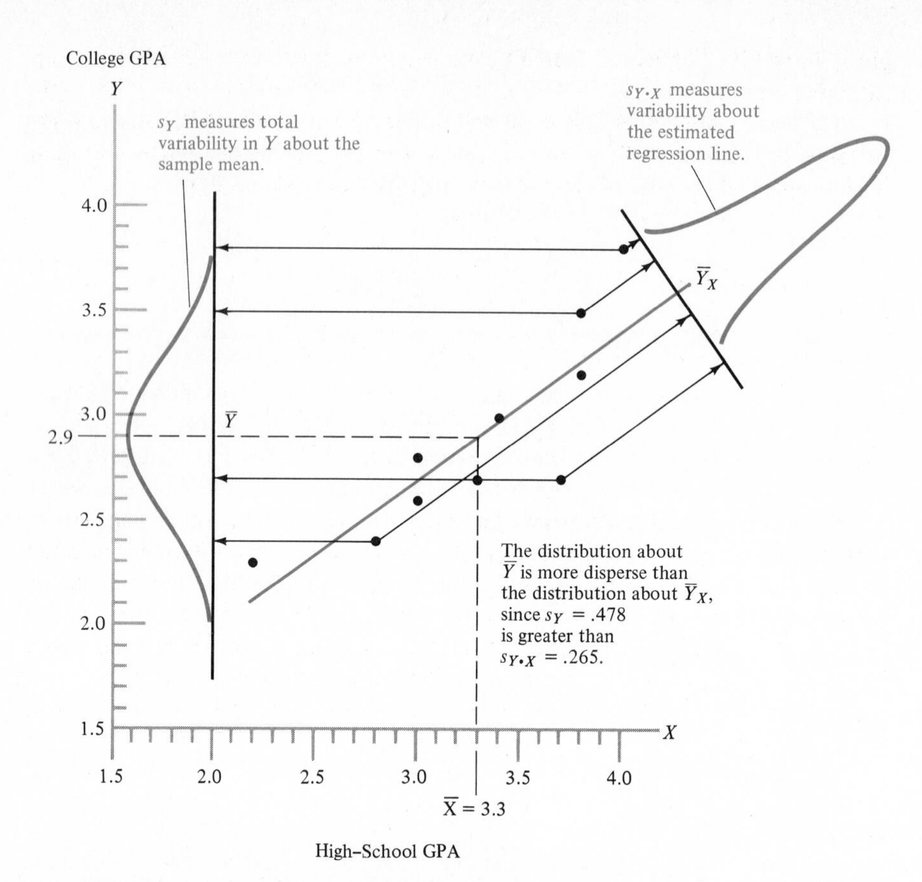

FIGURE 9-7 Illustration of the difference between total variability and variability about the estimated regression line.

obtained previously for a and b and taking the intermediate calculations from Table 9-2, we have

$$s_{Y \cdot X} = \sqrt{\frac{86.16 - .496(29.0) - .7286(97.74)}{10-2}}$$

$$= .265$$

Variability and Knowledge of X

The standard error of the estimate directly expresses the degree of scatter in the data. In the extreme case, when there is no scatter, all the Y observations fall on the regression line, as shown in Figure 9-5(c), and the vertical deviations are all zero, making $s_{Y \cdot X} = 0$. This presents strong evidence that the true variability of the Ys about the regression line is zero, so that the regression line provides perfect predictions. But when the degree of scatter is high, making $s_{Y \cdot X}$ large, the values of the dependent variable are quite disperse, so that for a given X, the predictions of Y made from the regression line are subject to considerable sampling error.

When knowledge of X is ignored, so that a regression line is not available, we have seen that the total variability in Y is summarized by the sample standard deviation s_Y, which may be more conveniently calculated from

$$s_Y = \sqrt{\frac{\sum Y^2 - n\bar{Y}^2}{n-1}} \tag{9-13}$$

Using the student GPA data from Table 9-2, we find

$$s_Y = \sqrt{\frac{86.16 - 10(2.9)^2}{10-1}} = .478$$

This value of s_Y is larger than the value found earlier for $s_{Y \cdot X}$, reflecting the fact that the variability about the regression line is smaller than the total variation in Y. This is indicated in Figure 9-7, where the underlying frequency curve for deviations about the estimated regression line is more compact than the one that might be constructed for the Ys without a knowledge of X. Thus, a prediction interval calculated using s_Y would be wider. A general conclusion is that predictions will tend to be more reliable and accurate when X is considered than when X is not included in the data.

EXERCISES

9-5 A statistician for the Civil Aeronautics Broad has selected a random sample of ten freight invoices in order to determine an equation that relates destination distance to freight charges for a standard-sized crate. He obtains the following results:

Distance (in hundreds of miles) X	*Charge (to nearest dollar)* Y
14	68
23	105
9	40
17	79
10	81
22	95
5	31
12	72
6	45
16	93

(a) Plot a scatter diagram for the above data.
(b) Using the method of least squares, determine the equation for the estimated regression line.
(c) Check your calculations by computing $\sum(Y - \bar{Y}_X)$. Allowing for rounding errors, this should equal zero. If it does not, find your error. Then plot the regression line on the scatter diagram.

9-6 For each of the following data, determine the estimated regression equation $\bar{Y}_X = a + bX$.

(a) $\bar{X} = 10$; $\bar{Y} = 20$; $\sum XY = 3{,}000$; $\sum X^2 = 2{,}000$; $n = 10$.
(b) $\bar{X} = 10$; $\bar{Y} = 20$; $\sum XY = 1{,}000$; $\sum X^2 = 2{,}000$; $n = 10$.
(c) $\bar{X} = 50$; $\bar{Y} = 10$; $\sum XY = 30{,}000$; $\sum X^2 = 135{,}000$; $n = 50$.

9-7 A personnel manager wishes to evaluate various employee aptitude tests in order to find one which will predict productivity. A sample of $n = 100$ electronics assembly workers has been selected. Each person is administered a test currently being evaluated, and his aptitude score X is determined. The production foreman has previously evaluated the performance of each worker by means of a productivity index Y. The following intermediate calculations have been obtained:

$$\sum X = 5{,}000 \qquad \sum Y = 600 \qquad \sum XY = 50{,}000$$

$$\sum X^2 = 350{,}000 \qquad \sum Y^2 = 8{,}600$$

Suppose the test is adopted. Determine the regression equation $\bar{Y}_X = a + bX$ and plot it on graph paper.

9-8 A government economist wishes to establish the relationship between annual family income X and savings Y. A sample of $n = 100$ families has been randomly chosen from various income levels between \$5,000 and \$20,000. A thorough investigation of these families has been made, and the following intermediate calculations have been obtained (X and Y are measured in thousands of dollars):

$$\sum X = \$1{,}239 \qquad \sum Y = \$79$$

$$\sum XY = 1{,}613 \qquad \sum X^2 = 17{,}322 \qquad \sum Y^2 = 293$$

(a) Determine the equation for the estimated regression line.
(b) State the meaning of the slope b and Y intercept a.
(c) Calculate $s_{Y \cdot X}$ and s_Y. Does a comparison between these indicate that the regression line may be a useful tool for predicting family savings? Why or why not?

9-9 A stereo-cartridge manufacturer has conducted a regression analysis to estimate the average cartridge lifetime (in hours) at various record tracking forces X (in grams). The following regression equation has been obtained for a sample of $n = 100$ cartridges that were played until worn out at various tracking forces: $\bar{Y}_X = 1{,}300 - 200X$. The standard error of the estimate for cartridge lifetimes about this line is $s_{Y \cdot X} = 100$ hours.

(a) Plot the regression line on graph paper.
(b) State the meaning of the slope of the regression line.
(c) Calculate $\bar{Y}_X$ when $X = 1$, $X = 2$, and $X = 3$ grams.

9-3 ASSUMPTIONS AND PROPERTIES OF LINEAR REGRESSION ANALYSIS

The introduction to regression in the preceding section is largely a mechanical process of fitting a line to the data. In this section, we will provide the assumptions of a theoretical model for regression analysis. Our purpose is to lay the groundwork for measuring the error associated with using the regression line in making estimates.

Assumptions of Linear Regression Analysis

Suppose that in our grade-point-average illustration we consider each possible college GPA level Y earned by *all* students, past and future, whose high-school GPAs are at a specified level X. For this fixed X, the values of Y represent a population and they will fluctuate and cluster about a central value. Similarly, for any other high-school GPA X, there will be an analogous population of Y values. Since the means of these populations depend upon the respective values for X, we may represent them symbolically by $\mu_{Y \cdot X}$, where, as before, the subscript $Y \cdot X$ signifies that the values of Y are for a given value of X.

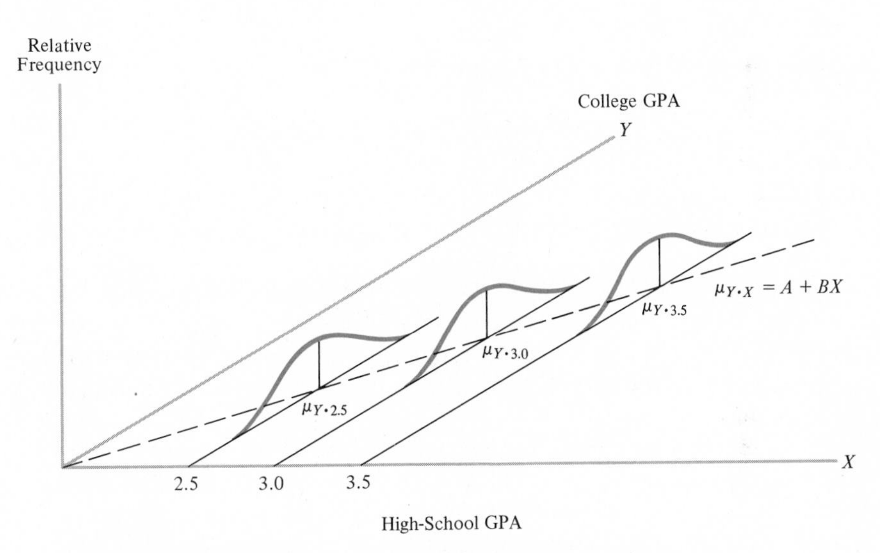

FIGURE 9-8 Populations for Y at various levels given for X.

Figure 9-8 portrays several populations for Y, showing how they fit into the context of linear regression. This graph is three dimensional, with an extra axis perpendicular to the XY plane. This vertical axis represents the relative frequency for Y at a specified level X. The curves are drawn with their centers at a distance $\mu_{Y \cdot X}$ from the X axis. Thus, we may refer to $\mu_{Y \cdot X}$ as the *conditional mean* of Y given X. There will be a different frequency curve for each X. Here, we show these curves for all students having high-school GPAs of $X = 2.5$, $X = 3.0$, and $X = 3.5$. The respective conditional means are denoted by $\mu_{Y \cdot 2.5}$, $\mu_{Y \cdot 3.0}$, and $\mu_{Y \cdot 3.5}$.

The Assumptions

Linear regression analysis makes four theoretical assumptions regarding the populations for Y:

1. All populations have the same standard deviation, denoted by $\sigma_{Y \cdot X}$, no matter what the value of X is.
2. The means $\mu_{Y \cdot X}$ all lie on the same straight line, having the equation

$$\mu_{Y \cdot X} = A + BX \tag{9-14}$$

which is the expression for the *true regression line*.
3. Successive sample observations are independent.
4. The value of X is known in advance.

Sometimes an additional assumption is that the population for Y be normally distributed. But many regression analysis results do not require such a strong and sometimes unrealistic condition.

Estimating the True Regression Equation

We have seen how the method of least squares may be used to derive the estimated regression equation $\overline{Y}_X = a + bX$. We will now investigate how this equation is related to the true regression equation $\mu_{Y \cdot X} = A + BX$. The values A and B, which we will call the *true regression coefficients*, are generally unknown. The reason for our choice of the notation $\overline{Y}_X$ is now apparent: $\overline{Y}_X$ will be used to estimate $\mu_{Y \cdot X}$, in the same way we used $\overline{X}$ to estimate μ in Chapter 7. The two regression equations differ only in the values for the Y intercept and slope. We may consider a, which is calculated from the sample data by expression (9-8), to be a point estimate of A. Likewise, from expression (9-7), we consider b the estimate of B.

The values calculated for a and b depend upon the sample observations obtained. The equation $\overline{Y}_X = .4957 + .7286X$ relating college and high-school grades resulted from the particular GPAs of the ten students chosen for the sample. Had different students been chosen, the regression equation obtained from them would most likely differ—perhaps considerably so—from the one we found.

When a and b are calculated by the method of least squares, they are *unbiased estimators* of the true coefficients A and B. This means that if the experiment of collecting samples is repeated a large number of times, and the regression line is found by the least squares method each time, then the average value of the Y intercepts a would tend to be close to the true Y intercept A. Likewise, on the average, the values of b would be close to their target, B. Unbiasedness was one of several properties mentioned in Chapter 7 as desirable to any estimator. One reason for the universal choice of the least squares criterion for fitting a regression line to the data is that this method exhibits so many of these desirable features. It has been established that of all unbiased estimators for linear regression coefficients, the least squares criterion provides the estimators of smallest variance, making the method of least squares the *most efficient* of all conceivable estimators. Being the most efficient, a and b minimize chance sampling error, so that estimates made from the regression line $\bar{Y}_X = a + bX$ are the most reliable ones for a fixed sample size. The values for a and b obtained by the least squares method are also *consistent* estimators of A and B.* Recall from Chapter 7 that a consistent estimator becomes progressively closer to the target parameter with increasing sample size. This can be attributed to the sampling distributions of a and b, whose variances decrease with n.

9-4 STATISTICAL INFERENCES USING THE REGRESSION LINE

A variety of inferences may be made using the estimated regression line. These fall into two broad categories: (1) predictions of the dependent variable, and (2) inferences regarding the regression coefficients A and B. Since predictions are made more often, we will discuss them first.

Prediction Intervals in Regression Analysis

The major goal of regression analysis is to predict Y from the regression line at given levels of X. This may be done in two ways. One involves predicting the value of the conditional mean $\mu_{Y \cdot X}$. However, sometimes it is useful to make a second kind of prediction—one for an individual Y value rather than for a mean.

For example, the college admissions director might want to predict the mean college GPA level that will be achieved by all students who earned straight "B" averages in high school. In this case, $X = 3.0$, and the best point estimate for $\mu_{Y \cdot X}$ will be the fitted Y value from the regression line, which we denote by $\bar{Y}_X$. Here, $\bar{Y}_X = \bar{Y}_{3.0}$, and

$$\begin{aligned}\bar{Y}_{3.0} &= a + b(3.0)\\ &= .4957 + .7286(3.0)\\ &= 2.68\end{aligned}$$

This same value may be used to estimate the college GPA of an entering freshman. To distinguish a mean value from an *individual value*, both of which can only be estimated from the sample, we use the special symbol Y_I.

* When the Ys are normally distributed, the least squares estimators fall into a broad class referred to as maximum likelihood estimators (MLE). In addition to being consistent, the MLE are most efficient and are normally distributed.

Either kind of estimate will involve sampling error, which can be acknowledged and expressed in terms of confidence intervals. Because of the special nature of regression analysis, the numbers obtained are usually referred to as *prediction intervals*.

Prediction Interval for Conditional Mean

A prediction interval is found in a manner similar to the way the confidence intervals we encountered in Chapter 7 were derived. There, we used

$$\mu = \bar{X} \pm z\frac{s}{\sqrt{n}} \qquad \text{for large samples}$$

and

$$\mu = \bar{X} \pm t_{\alpha}\frac{s}{\sqrt{n}} \qquad \text{for small samples}$$

where $s/\sqrt{n}$ served as the estimator of the standard error of $\bar{X}$, $\sigma_{\bar{X}} = \sigma/\sqrt{n}$.

In regression analysis, Y (not X) is the variable being estimated, and in the present notational context, intervals of analogous form are required to estimate a conditional mean for Y:

$$\mu_{Y \cdot X} = \bar{Y}_X \pm (z \text{ or } t_{\alpha}) \text{ estimated } \sigma_{\bar{Y}_X}$$

The standard error for $\bar{Y}_X$, which we denote by $\sigma_{\bar{Y}_X}$, represents the amount of variability in possible $\bar{Y}_X$ values at the particular level for X that a prediction is desired. In the context of the college GPA prediction problem, a somewhat different line $\bar{Y}_X = a + bX$ (such as $\bar{Y}_X = .62 + .83X$ or $\bar{Y}_X = .41 + .69X$) might have fitted the least squares procedure if some other random sample of ten students had been selected. Thus, for some other sample $\bar{Y}_{3.0}$, the point estimate of college GPA when $X = 3.0$, might have computed to ad ifferent value than 2.68. For every level of X, the potential set of $\bar{Y}_X$ values would have a distribution with a standard deviation of $\sigma_{\bar{Y}_X}$.

There are two components of the variability in $\bar{Y}_X$:

$$\sigma^2_{\bar{Y}_X} = \begin{matrix}\text{Variability in the} \\ \text{mean of } Y\text{s}\end{matrix} + \begin{matrix}\text{Variability caused by the} \\ \text{distance of } X \text{ from } \bar{X}\end{matrix}$$

The first source is analogous to the variability in the sample mean, which as we saw in Chapter 7, depends upon the potential standard deviation and the sample size. The second source of variation is associated with the distance that X lies from $\bar{X}$. Figure 9-9 shows why this is so. Here, several estimated regression lines have been plotted, each representing different samples of students taken

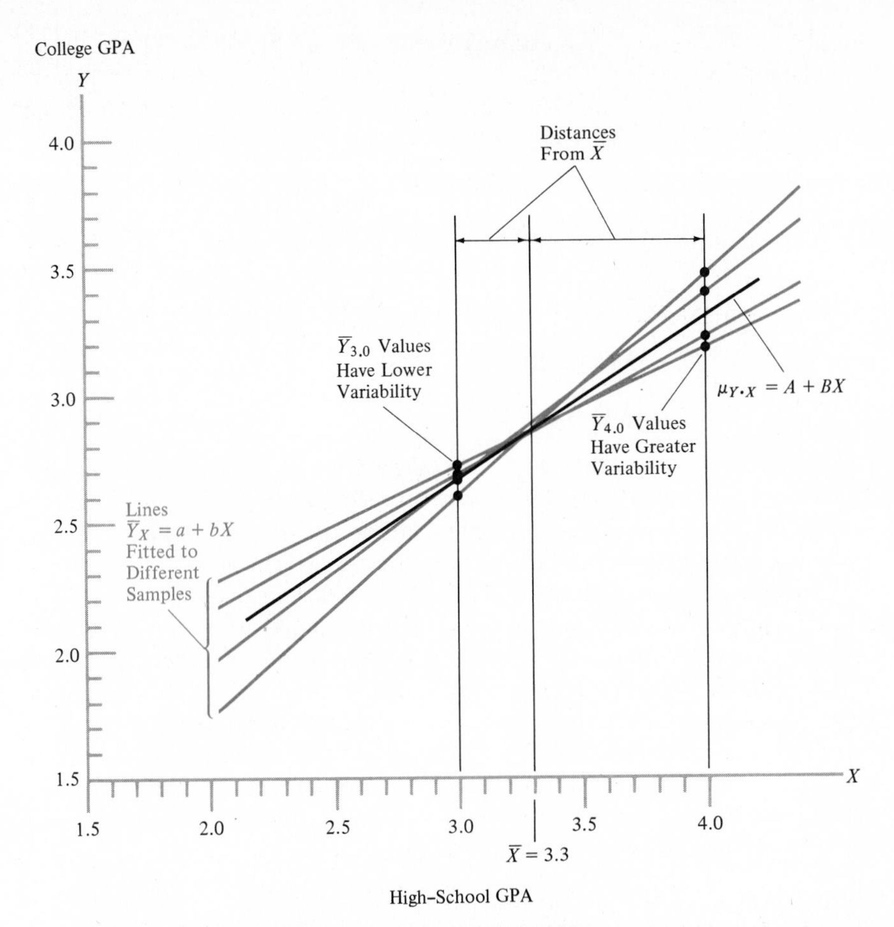

FIGURE 9-9 Illustration of how variability in $\overline{Y}_X$ increases for larger distances separating X from $\overline{X}$.

from the same population. Note that these lines tend to diverge and that their separations become greater as the distance between X and $\overline{X}$ increases. Therefore, the values of $\overline{Y}_X$ become more varied the farther they are from $\overline{X}$.

Small Sample Predictions

With small samples (generally, when $n < 30$), the following expression is used to calculate the ***prediction interval for the conditional mean***

$$\mu_{Y \cdot X} = \overline{Y}_X \pm t_\alpha s_{Y \cdot X} \sqrt{\frac{1}{n} + \frac{(X - \overline{X})^2}{\sum X^2 - n\overline{X}^2}} \tag{9-15}$$

where t_α is found from Appendix Table E for $n-2$ degrees of freedom.* Representing $\mu_{Y \cdot X}$ when $X = 3.0$ as $\mu_{Y \cdot 3.0}$, we may construct the 95-percent prediction interval for the conditional mean college GPA using $10-2=8$ degrees of freedom, so that $\alpha = (1-.95)/2 = .025$, and $t_{.025} = 2.306$:

$$\mu_{Y \cdot 3.0} = \bar{Y}_{3.0} \pm t_{.025} s_{Y \cdot X} \sqrt{\frac{1}{n} + \frac{(3.0-\bar{X})^2}{\sum X^2 - n\bar{X}^2}}$$

$$= 2.68 \pm 2.306(.265) \sqrt{\frac{1}{10} + \frac{(3.0-3.3)^2}{111.70-10(3.3)^2}}$$

$$= 2.68 \pm .22$$

$$2.46 \leqslant \mu_{Y \cdot 3.0} \leqslant 2.90$$

We would therefore conclude that the college GPA for straight "B" students in high school is *on the average* somewhere between 2.46 and 2.90. Our confidence that this statement is correct rests upon the procedure used, which provides similar intervals containing the true mean about 95 percent of the time.

We may calculate 95-percent confidence intervals for the other values of X, thus obtaining prediction limits for $\mu_{Y \cdot X}$ over the entire range of X. In Figure 9-10, this has been done for the student GPA data. Notice that the width of the confidence band is dependent upon the distance of the X values from the mean. As we already noted, this is because the slope error is magnified; hence $\sigma_{\bar{Y}_X}$ increases as the distance separating X from $\bar{X}$ increases. Since one term is zero for $X = \bar{X}$, the narrowest portion of the band occurs here. Thus, we may conclude that estimates of $\mu_{Y \cdot X}$ made from the regression line are most precise for X values near their mean, with precision decreasing for values farther from $\bar{X}$.

* To show how this is obtained, we reexpress $\bar{Y}_X = a + bX = (\bar{Y} - b\bar{X}) + bX = \bar{Y} + (X - \bar{X})b$. Then, since $\bar{Y}$ and b may be shown independent, the addibility of variance provides

$$\sigma^2_{\bar{Y}_X} = \sigma^2_{\bar{Y}} + \sigma^2[(X-\bar{X})b]$$

and because $(X-\bar{X})$ is a constant, we can square it and obtain

$$\sigma^2_{\bar{Y}_X} = \sigma^2_{\bar{Y}} + (X-\bar{X})^2\sigma^2_b$$

Since we have assumed that the respective Y populations have the same variance regardless of X, then $\sigma^2_{\bar{Y}} = \sigma^2_{Y \cdot X}/n$. We may express b in a form equivalent to expression (9-7)

$$b = \frac{\sum(X-\bar{X})Y}{\sum(X-\bar{X})^2}$$

The terms involving X are constants and each Y observation is independent and has a variance of $\sigma^2_{Y \cdot X}$, so applying the addibility of variance and squaring the constants gives us

$$\sigma^2_b = \frac{\sum(X-\bar{X})^2}{[\sum(X-\bar{X})^2]^2}\sigma^2_{Y \cdot X} = \frac{1}{\sum(X-\bar{X})^2}\sigma^2_{Y \cdot X} = \frac{1}{\sum X^2 - n\bar{X}^2}\sigma^2_{Y \cdot X}$$

Thus

$$\sigma^2_{\bar{Y}_X} = \sigma^2_{Y \cdot X}\left(\frac{1}{n} + \frac{(X-\bar{X})^2}{\sum X^2 - n\bar{X}^2}\right)$$

The estimated standard error for $\bar{Y}_X$ follows, using $s_{Y \cdot X}$ in place of $\sigma_{Y \cdot X}$.

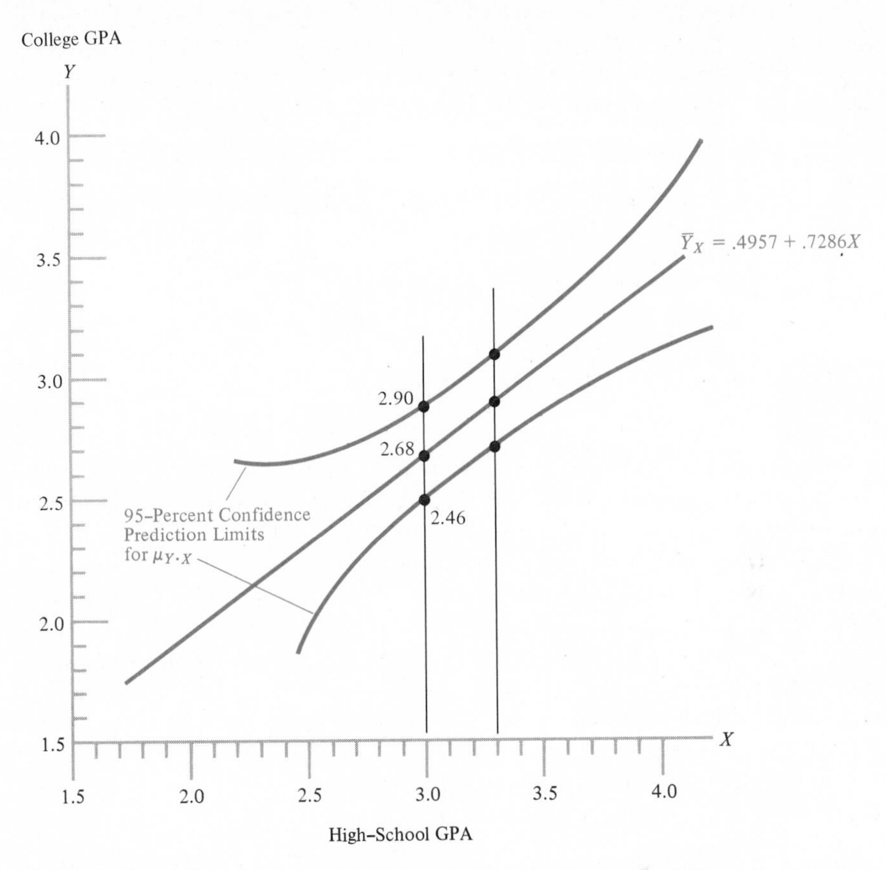

FIGURE 9-10 Confidence limits for predictions of the mean college GPA.

Large Sample Predictions

When n is large (30 or more), the normal curve applies, in which case the normal deviate z replaces t_α in expression (9-15). Further simplification is ordinarily made for large samples, for then the fraction involving X values becomes very small. Thus, we may instead use an abbreviated but approximate expression for the ***prediction interval for the conditional mean using large samples***

$$\mu_{Y \cdot X} = \overline{Y}_X \pm z \frac{s_{Y \cdot X}}{\sqrt{n}} \qquad \text{(9-16)}$$

Note that expression (9-16) resembles the one we used to construct a confidence interval for the population mean. Here, $\overline{Y}_X$ is analogous to the sample mean, while $s_{Y \cdot X}$ replaces s.

Prediction Interval for Individual Value of Y Given X

Predicting an individual value of Y given X is similar to predicting the mean. If our admissions director wished to predict the college GPA of the next entering freshman with a high-school GPA of $X = 3.0$, then the same point estimate, $\overline{Y}_{3.0} = 2.68$, would be made from the regression equation. The following expression provides ***the prediction interval for an individual value of Y when using small samples***

$$Y_I = \overline{Y}_X \pm t_\alpha s_{Y \cdot X} \sqrt{\frac{1}{n} + \frac{(X - \overline{X})^2}{\sum X^2 - n\overline{X}^2} + 1} \tag{9-17}$$

Expression (9-17) is the same as expression (9-15) for $\mu_{Y \cdot X}$, except that an extra 1 has been added to the radical. This reflects the fact that when $\overline{Y}_X$ is used to estimate Y_I, a third source of variability is present—the dispersion of individual Y values about the regression line. (Even if the true regression line were known, the Ys at a particular level for X would have a variance of $\sigma^2_{Y.X}$*).

We may now construct a 95-percent confidence interval for Y_I, using the student data when $X = 3.0$. Substituting the appropriate values into expression (9-17), we predict Y_I by

$$Y_I = \overline{Y}_{3.0} \pm t_{.025} s_{Y \cdot X} \sqrt{\frac{1}{n} + \frac{(3.0 - \overline{X})^2}{\sum X^2 - n\overline{X}^2} + 1}$$

$$= 2.68 \pm 2.306(.265) \sqrt{\frac{1}{10} + \frac{(3.0 - 3.3)^2}{111.70 - 10(3.3)^2} + 1}$$

$$= 2.68 \pm .65$$

or

$$2.03 \leqslant Y_I \leqslant 3.33$$

Note that this interval is considerably wider than the interval obtained previously for $\mu_{Y \cdot 3.0}$. This is to be expected, because Y_I is the estimate for the college GPA of *a particular student*, not a mean, and the greater width is attributable to the added variability that would be present even if the true regression line were available in making the prediction.

* So that we must add $\sigma^2_{Y \cdot X}$ to $\sigma^2_{\overline{Y}X}$:

$$\sigma^2_{\overline{Y}X} = \sigma^2_{Y \cdot X}\left(\frac{1}{n} + \frac{(X - \overline{X})^2}{\sum X^2 - n\overline{X}^2}\right) + \sigma^2_{Y \cdot X}$$

Using $s_{Y \cdot X}$ in place of $\sigma_{Y \cdot X}$ we obtain the estimated standard error for the estimator of Y_I.

For *large samples*, the normal curve may be assumed, so that an appropriate normal deviate value z replaces t_α. Ordinarily, we can use the following abbreviated equation for computing ***the prediction interval of an individual value of Y when using large samples***

$$Y_I = \overline{Y}_X \pm z s_{Y \cdot X} \tag{9-18}$$

Dangers of Extrapolation

In establishing a regression equation, a set of observations is used that covers a limited range of values for the independent variable X. Caution must be exercised when making predictions of the dependent variable Y whenever X falls outside this range. Such predictions are called *extrapolations*.

In the student grade-point-average illustration, the regression line was computed for ten students whose high-school GPAs ranged from 2.2 to 4.0. It would be inappropriate to predict the college grades of a high-school student whose GPA was lower than 2.2, since such a poor prospect would probably never be considered for college admission, except under extenuating circumstances.

Regression analysis is limited only to the range of actual observations. These observations—and not qualitative reasoning—are used to quantify the relationship between X and Y. Qualitative reasoning is useful initially in selecting the form of the regression equation (linear versus curvilinear) and later in interpreting the results, but it cannot be used in place of actual observation. We are not ruling out extrapolation here, but are merely indicating its potential pitfalls. If there are no data available beyond the range of required predictions, then extrapolation may be the only suitable alternative. Keep in mind that *extrapolation assigns values using a relationship that has been measured for circumstances differing from those for the prediction.*

Circumstances may differ for reasons other than extrapolation. Should the underlying populations change over time—as would be the case for a college admissions director who uses the above regression results to screen out poor high-school students—then the regression line would not represent the student body in subsequent years. Often, a regression analysis is short-lived in its applicability because underlying relationships can change over time.

Inferences Regarding Slope of Regression Line*

Second in importance to prediction intervals are inferences regarding slope B of the true regression line. This is especially true in statistical applications where the underlying relationship between X and Y is more important than predicting Y for a particular level of X. For example, much economic theory

* Much of the material in this section relies on background material in Chapter 8 and may be skipped with no loss of continuity.

relies upon regression analysis to substantiate hypothetical models requiring supply and demand curves; the *coefficients* of these demand and supply equations—rather than a predicted quantity for a given price—are our main interest. Similarly, a metallurgist might use regression analysis to develop a mathematical relationship between alloy concentrations and strength properties; he might be more concerned with how much extra shearing force is needed to break a metal for each unit increase in alloy material than in predicting a particular force (that is, he might wish to estimate B rather than $\mu_{Y \cdot X}$).

Confidence Interval Estimate of B

An unbiased estimate of slope B of the true regression line may be obtained from its sample counterpart b. The following equation* is used to construct a ***confidence interval estimate for B***

$$B = b \pm t_{\alpha} \frac{s_{Y \cdot X}}{\sqrt{\sum X^2 - n\bar{X}^2}} \tag{9-19}$$

Here, t_{α} is the value of t found from Appendix Table E for which the upper-tailed area under the t curve is α—an amount equal to half of 1 minus the confidence level. The degrees of freedom are $n-2$ (rather than $n-1$ used in earlier chapters) because $s_{Y \cdot X}$ depends upon the two values, a and b, calculated from the sample results.

We may use our student GPA illustration to apply this procedure. Suppose that a 95-percent confidence interval is desired of the true regression coefficient B. Using $\alpha = (1-.95)/2 = .025$ and $10-2 = 8$ degrees of freedom, we have, from Table E, $t_{\alpha} = t_{.025} = 2.306$. Using the intermediate calculations from Table 9-2 and the values previously determined for b and $s_{Y \cdot X}$, we have as our 95-percent confidence interval:

$$\begin{aligned} B &= .7286 \pm 2.306(.265)/\sqrt{111.70 - 10(3.3)^2} \\ &= .7286 \pm .3652 \\ .3634 &\leqslant B \leqslant 1.0938 \end{aligned}$$

This means that we are 95 percent confident that the true value of B—the mean number of points in the students' college GPAs attributable to each point in

* Here we use the principal that

$$B = b \pm t_{\alpha} \text{ estimated } \sigma_b$$

It was established in the footnote on page 356 that

$$\sigma_b = \frac{\sigma_{Y \cdot X}}{\sqrt{\sum X^2 - n\bar{X}^2}}$$

Using $s_{Y \cdot X}$ in place of $\sigma_{Y \cdot X}$, we obtain an estimate of the standard error for b.

their high-school GPAs—lies between .3634 and 1.0938 points. If we repeated this procedure with 100 different samples of 10 students, about 95 percent of the time we would construct such an interval containing the true value of B.

This interval estimate of B is not very precise and would probably be of little use to an admissions director. As with the confidence interval for the population mean discussed in Chapter 7, precision can be increased by using a larger sample. It is advisable that a sample considerably larger than the one used in this illustration be taken.

In practice, when n is large, the Student t distribution closely fits the normal distribution. In these cases, the normal deviate z may replace t_α in expression (9-19) to determine the confidence interval for B.

Testing Hypotheses about B

We may extend tests of hypotheses to inferences concerning B. Ordinarily, the fact of greatest importance in testing for the value of B is whether it equals zero. Figure 9-11 illustrates a regression line having zero slope. Note that no matter what the value of X, $\mu_{Y \cdot X}$ remains at A parallel to the X axis. Thus, if $B = 0$, then, since the population distributions for Y have the same mean and variance, we may usually conclude that the Ys are identically distributed for all values of X. This means that there is no statistical relationship between X and Y; we will bring up this point again later in context with our discussion of correlation. Thus, if $B = 0$, regression analysis will be of no value in making predictions of Y.

In making the two-sided test, our hypotheses are

$$H_0 : B = 0$$

$$H_1 : B \neq 0$$

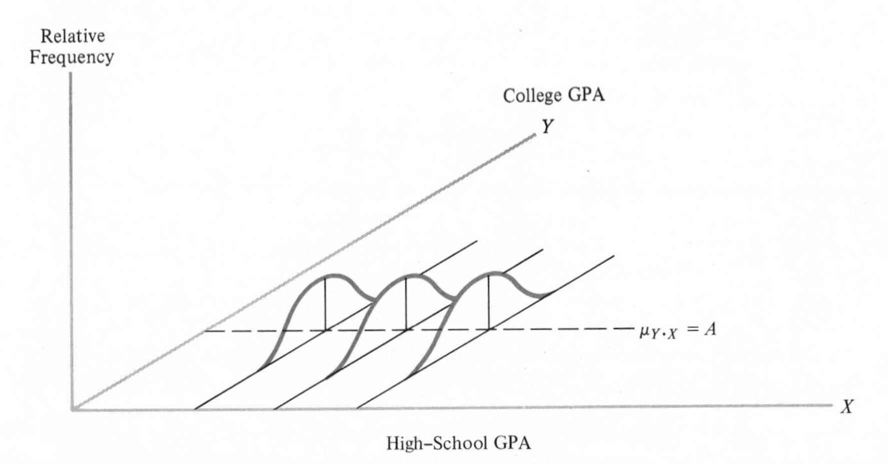

FIGURE 9-11 Illustration of a true regression line having zero slope, so that X and Y are uncorrelated.

If we choose .05 (1 minus the previous confidence level) as our significance level, then we only need to find out if our confidence interval contains the point $B = 0$. If it does not, we may reject the null hypothesis. Because the lower limit of the 95-percent confidence interval calculated previously is .3634, a number greater than $B = 0$, we may reject the null hypothesis at a .05 significance level, concluding that high-school GPA does affect college GPA.

It may be appropriate instead to employ a one-sided upper-tailed test with $H_1 : B > 0$. This would be better if Y varies directly with X, as seems natural in the case of college and high-school GPAs. The first step is to calculate the t statistic, using zero for B under the assumption of H_0 that $B \leq 0$:

$$t = \frac{b}{\dfrac{s_{Y \cdot X}}{\sqrt{\sum X^2 - n\bar{X}^2}}} \tag{9-20}$$

The value obtained is then compared to the critical value t_α, found from Table E, that corresponds to the prescribed significance level α. If t is smaller than t_α, the null hypothesis is accepted. The reverse is true for a lower-tailed test, where the alternative is that $B < 0$ (the case when Y bears an inverse relationship to X.)

We may illustrate the one-sided test with the student-grade results. For example, if we use $\alpha = .005$, then from Table E and using $10-2 = 8$ degrees of freedom, we obtain the critical value $t_{.005} = 3.355$. Applying expression (9-20), we have

$$t = \frac{.7286}{\dfrac{.265}{\sqrt{111.70 - 10(3.3)^2}}} = 4.601$$

Since $t = 4.601$ exceeds the critical value $t_{.005} = 3.355$, the null hypothesis is *rejected* at a significance level of $\alpha = .005$, which indicates that the slope of the true regression line is greater than zero.

EXERCISES

9-10 The estimated regression line providing the weight (in grams) of a laboratory rat fed a daily diet of X calories is $\bar{Y}_X = 500 + 5X$. This result was obtained for a sample of $n = 100$ animals for whom $s_{Y \cdot X} = 10$ grams. Calculate 95-percent prediction intervals for $\mu_{Y \cdot X}$ and Y_I when $X = 150$ calories.

9-11 The relationship between the total weight Y (in pounds) of luggage stored in an aircraft's baggage compartment and the number of passengers X on the flight manifest is $\bar{Y}_X = 250 + 27X$. This will be used by superintendents at airports to determine how much additional freight can safely be stored on a flight, after taking into consideration the fuel load and the weight of the passengers themselves. The data were obtained from a sample of $n = 25$ flights. Other results are: $s_{Y \cdot X} = 100$ pounds, $s_Y = 300$ pounds, $\sum X^2 =$

64,000, and $\bar{X} = 50$. Construct a 95-percent confidence interval estimate for both $\mu_{Y \cdot X}$ and Y_I when: (a) $X = 50$; (b) $X = 75$; (c) $X = 100$.

9-12 From a sample of 400 families, a city planner has found the number of square feet in each home X and the size of each family Y. He has established the following estimated regression line in order to predict the family size for given square-footage levels: $\bar{Y}_X = .5 + .002X$. The values $s_Y = .30$ and $s_{Y \cdot X} = .10$ have been obtained. When the home has 1,000 square feet, construct 95-percent prediction intervals for (a) the mean family size, and (b) the number of persons in a particular family.

9-13 The credit manager of a department store has determined the following regression equation for a customer credit-rating index X and the proportion of customers Y who eventually incur bad debts: $\bar{Y}_X = .09 - .002X$. The index values range from zero to 40. A sample of $n = 25$ is taken, and other calculations show that $\sum X^2 = 23{,}000$, $s_{Y \cdot X} = .02$, and $\bar{X} = 30$.

(a) Determine the 90-percent confidence prediction interval for the mean proportion of customers with ratings of $X = 20$ who will incur bad debts.

(b) The proper interpretation for Y_I is in this case the *probability* that a particular individual will incur a bad debt. Determine the 90-percent confidence interval for Y_I when $X = 20$.

9-14 Referring to the information provided in Exercise 9-11:

(a) Construct a 99-percent confidence interval estimate slope B of the true regression line.

(b) In testing H_0 that $B = 0$ against the two-sided alternative that $B \neq 0$, should the null hypothesis be accepted or rejected at the .01 significance level?

(c) In testing H_0 that $B \leqslant 0$ against the one-sided alternative that $B > 0$, should the null hypothesis be accepted or rejected at the .01 significance level?

9-15 Referring to the information provided in Exercise 9-13:

(a) Construct a 95-percent confidence interval estimate of slope B of the true regression line.

(b) In testing H_0 that $B = 0$ against the two-sided alternative that $B \neq 0$, should the null hypothesis be accepted or rejected at the .05 significance level?

(c) In testing H_0 that $B \geqslant 0$ against the one-sided alternative that $B < 0$, should the null hypothesis be accepted or rejected at the .05 significance level?

9-5 CORRELATION ANALYSIS

The goal of correlation analysis is to measure the *degree* to which two variables are related. Regression analysis provides an equation by which one variable's value may be estimated from another's. Correlation analysis shows how closely two variables can move together by means of a single number calculated from the same data.

Correlation analysis is very useful as an auxiliary tool in regression analysis, because it can be used to describe how well the regression line explains the variation in the values of the dependent variable. It is used instead of regression when the only question is how strongly two variables are related. One such application is in isolating statistically related characteristics of a population in order to explain their differences. For example, a pharmacologist may be interested in identifying those chemicals that can be formulated into a drug to alleviate various symptoms of a particular disease, such as anemia, pain, and poor appetite. A highly positive correlation between dosages X of a specific chemical and appetite Y (as measured by the quantity of food consumed) may make the chemical a good candidate for inclusion in the final drug.

The central focus of correlation analysis is to find a suitable index indicating how strongly X and Y are related. It is convenient to initially treat correlation as an adjunct to regression analysis, so this index will first be explained by using the regression line. Later, a parallel explanation will be given that does not require prior knowledge of the estimated regression line.

Measuring Degree of Association

As we saw in Section 9-1, the degree to which X and Y are related may be explained in terms of the magnitude of data scatter about the regression line. One extreme case occurs when scatter is so great that the regression line has a zero slope and is parallel to the X axis, as shown in Figure 9-12(a). Here, levels of Y have no relationship to the value of X. We say that the degree of correlation is zero, since knowledge of X cannot add to the accuracy of predictions of Y. Figure 9-12(b) illustrates the opposite extreme. Here, a perfect fit between Y and X observations is achieved because all of the data points happen to lie on the same line. Since there is no scatter about the regression line, the data indicate that Y will change by some predetermined amount for each increment in X, showing the strongest possible relationship between X and Y. We can say that the degree of correlation is perfect, so that knowledge of X allows perfect predictions to be made.

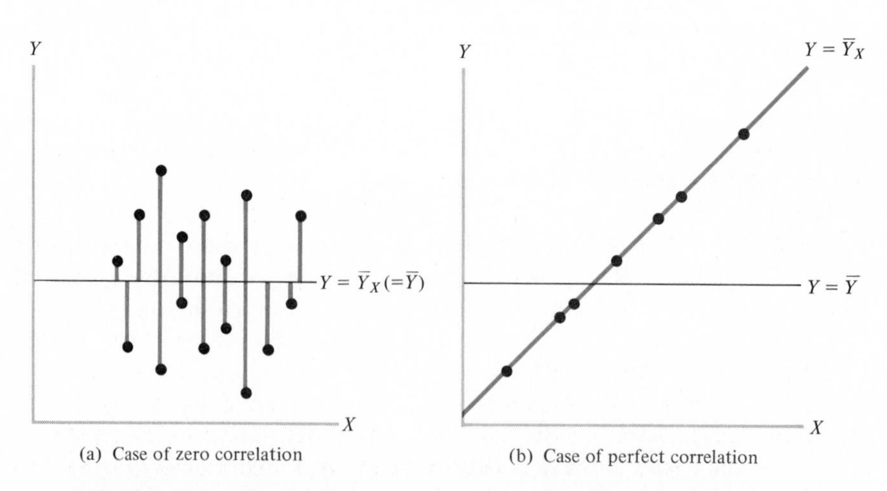

(a) Case of zero correlation

(b) Case of perfect correlation

FIGURE 9-12 Two extreme cases illustrating degrees of correlation.

We will now develop two indices to summarize the strength of association. The more important is the *coefficient of determination*, which expresses the relative reduction in the variation of Y that can be attributed to a knowledge of X and its relationship to Y by way of the regression line. From this, another useful index, the *correlation coefficient*, may then be obtained.

Sample Coefficient of Determination

In Section 9-2, the scatter of data about the estimated regression line was summarized by the standard error of the estimate $s_{Y \cdot X}$. Its square, $s^2_{X \cdot Y}$, is the *mean* of the squared vertical deviations of the data points around the regression line. It is now convenient to summarize the scatter with the *sum* of the squared deviations about $\bar{Y}_X$:

$$\sum (Y - \bar{Y}_X)^2$$

This may be compared to the scatter of the sample observations about their mean, represented by

$$\sum (Y - \bar{Y})^2$$

which is the sum of the squared vertical deviations around the horizontal line $Y = \bar{Y}$. The terms in this sum of squares are easily recognizable as the numerators of $s^2_{Y \cdot X}$ and s^2_Y, respectively.

These sums of squares will be used to construct the indices measuring strength of association. When we compared $s_{Y \cdot X}$ to s_Y previously, we saw that the predictive power of knowing X is indicated by their relative sizes. Using the corresponding sums of squares, we may construct the *sample coefficient of determination* to express how strongly X is associated with Y:

$$r^2 = 1 - \frac{\sum (Y - \bar{Y}_X)^2}{\sum (Y - \bar{Y})^2} \qquad \text{(9-21)}$$

When X and Y have a zero correlation, as they do in Figure 9-12(a), the regression line has a zero slope and $\bar{Y}_X = \bar{Y}$, in which case the deviations about $\bar{Y}_X$ are the same as those about $\bar{Y}$. This makes the numerator the same as the denominator in expression (9-21), so that the fraction must be equal to 1; thus, $r^2 = 1 - 1 = 0$. If X and Y are perfectly correlated, as in Figure 9-12(b), then $\sum (Y - \bar{Y}_X)^2 = 0$, so that $r^2 = 1 - 0 = 1$. The value obtained for r^2 must lie somewhere between 0 and 1.

Population Coefficient of Determination

Recall that data observations are taken from a sample and may present a distorted image of the true association between X and Y. The sample coefficient of determination may indicate either a stronger or a weaker relationship than actually exists. The true measure of association is obtained from the *population coefficient of determination*, defined by

$$\rho^2 = \frac{\sigma_Y^2 - \sigma_{Y \cdot X}^2}{\sigma_Y^2} = 1 - \frac{\sigma_{Y \cdot X}^2}{\sigma_Y^2} \tag{9-22}$$

The symbol ρ is the Greek letter *rho*. There, $\sigma_{Y \cdot X}^2$ is the variance of the various population distributions for Y and σ_Y^2 is the variance of the combined population of Ys. The former expresses the variation in Y about the true regression line $\mu_{Y \cdot X} = A + BX$, while the latter indicates the variability of Y about its true mean without regard to X. The difference between the variances, $\sigma_Y^2 - \sigma_{Y \cdot X}^2$, in the numerator of the first expression for ρ^2 represents the net reduction in the variance in Y that is due to knowing X and the true regression line. Dividing this difference by σ_Y^2 gives us ρ^2. This may be interpreted as the proportional reduction in the variance of Y due to regression.

The expressions for ρ^2 and r^2 are similar. Each is 1 minus the ratio of variability about a regression line to the variability about a mean value for Y. ρ^2 is defined in terms of true parameter values, while r^2 is based upon sample values that are estimators of their population counterparts. Thus, we may consider the sample coefficient of determination r^2 as a point estimator of its population counterpart ρ^2.*

Calculating the Coefficient of Determination

In actual practice, calculating r^2 from expression (9-21) can be quite cumbersome. Instead, we may use the estimated regression coefficients and the intermediate values obtained in finding these coefficients to calculate r^2 from the following mathematically equivalent expression for the ***sample coefficient of determination***

$$r^2 = \frac{a \sum Y + b \sum XY - n\bar{Y}^2}{\sum Y^2 - n\bar{Y}^2} \tag{9-23}$$

This equation uses values previously obtained from regression analysis.

* Some texts define the sample coefficient of determination as

$$r_a^2 = 1 - \frac{s_{Y \cdot X}^2}{s_Y^2}$$

which can be obtained directly from expression (9-22), using the unbiased point estimators $s_{Y \cdot X}^2$ and s_Y^2 for the variances $\sigma_{Y \cdot X}^2$ and σ_Y^2. r_a^2 is referred to as the *adjusted coefficient of determination* and is related to r^2 by

$$r_a^2 = 1 - (1 - r^2)\left(\frac{n-1}{n-2}\right)$$

For large values of n, r^2 is nearly identical to r_a^2. We do not use r_a^2 because of difficulties with the divisors $(n-2)$ and $(n-1)$ used to calculate $s_{Y \cdot X}^2$ and s_Y^2. An expression equivalent to (9-21) is

$$r^2 = 1 - \frac{s_{Y \cdot X}^2}{s_Y^2}\left(\frac{n-2}{n-1}\right)$$

In our student GPA illustration, we may calculate r^2 from expression (9-23), using $a = .4957$, $b = .7286$, and the intermediate calculations from Table 9-2:

$$r^2 = \frac{.4957(29) + .7286(97.74) - 10(2.9)^2}{86.16 - 10(2.9)^2}$$

$$= \frac{1.4973}{2.06} = .723$$

Interpretation of Coefficient of Determination

How should we interpret our finding that $r^2 = .723$? Our initial motivation in comparing variation about $\overline{Y}_X$ to variation about $\overline{Y}$ was to show how knowledge of X can reduce errors in predicting Y. Next, we will establish that r^2 can be deduced from an explanation of variations in Y.

Figure 9-13 shows the regression line calculated from the previous sample of student grade-point averages. Suppose that a point estimate is required of the next entering freshman's college GPA. If we do not bother to use the regression equation—in effect, if we ignore X—then our estimate of that student's college GPA is $\overline{Y} = 2.9$. If he happened to have had an "A" average in high school, so

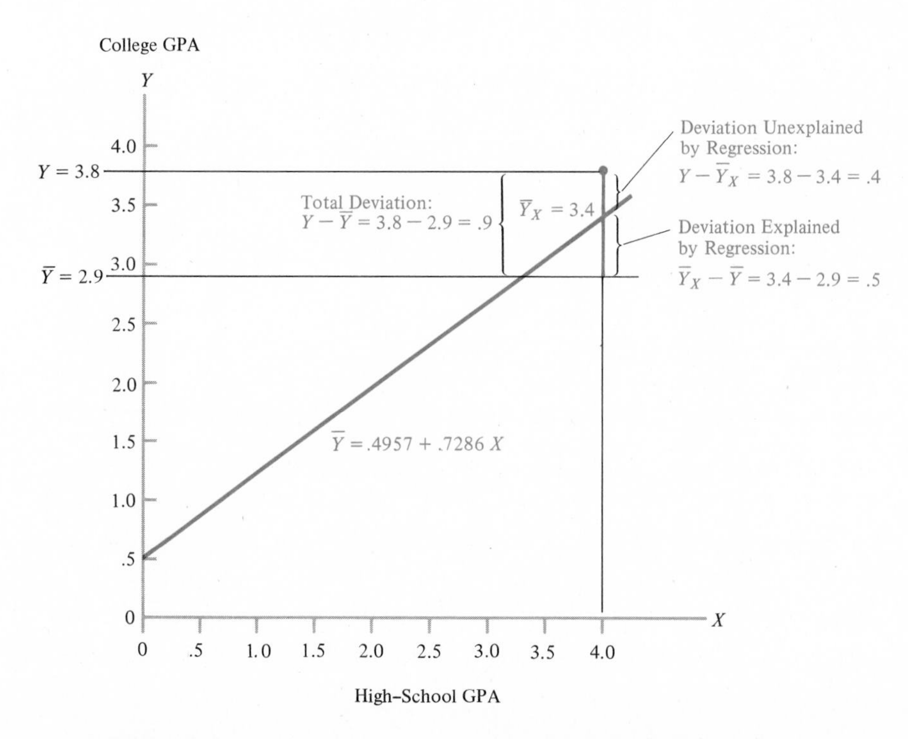

FIGURE 9-13 Illustration of total, explained and, unexplained variations in Y.

that $X = 4.0$, this would result in considerable error. We represent this error by the *total deviation* of the estimate from the observed college GPA of $Y = 3.8$ actually obtained for a student in the sample: $Y - \bar{Y} = 3.8 - 2.9 = .9$. If we use the regression line calculated from the sample data instead, then our best estimate is $\bar{Y}_{4.0} = 3.4$ (rounded), which lowers the error so that the actual deviation from the estimate is $Y - \bar{Y}_X = 3.8 - 3.4 = .4$. By using the regression line, our error has been reduced by $\bar{Y}_X - \bar{Y} = 3.4 - 2.9 = .5$. Thus, a large part of the total error may be attributed to using the regression line, so that $\bar{Y}_X - \bar{Y} = .5$ is the *explained deviation.* The remaining portion of the error is due to unidentifiable causes, so $Y - \bar{Y}_X = .4$ is an *unexplained deviation.* The total deviation of the observed Y may then be expressed as

$$(Y - \bar{Y}) = (\bar{Y}_X - \bar{Y}) + (Y - \bar{Y}_X) \tag{9-24}$$

which we can state in words:

$$\text{Total deviation} = \text{Explained deviation} + \text{Unexplained deviation}$$

The above relationship suggests that the *total variation* in Y has two components, which may be stated as

$$\text{Total variation} = \text{Explained variation} + \text{Unexplained variation}$$

The total variation expresses the amount of vertical scatter by the data points about their mean $\bar{Y}$. This may be measured by $\sum(Y - \bar{Y})^2$. Likewise, we use $\sum(Y - \bar{Y}_X)^2$ to express the unexplained variation—the magnitude of scatter about the estimated regression line. Thus, the explained variation may be expressed as the difference:

$$\text{Explained variation} = \sum(Y - \bar{Y})^2 - \sum(Y - \bar{Y}_X)^2$$

If we determine the ratio of the explained variation to the total variation, we obtain

$$\frac{\text{Explained variation}}{\text{Total variation}} = \frac{\sum(Y - \bar{Y})^2 - \sum(Y - \bar{Y}_X)^2}{\sum(Y - \bar{Y})^2} = 1 - \frac{\sum(Y - \bar{Y}_X)^2}{\sum(Y - \bar{Y})^2}$$

The right-hand side of the above equation is the sample coefficient of determination, which is also expression (9-21). Thus, we have

$$r^2 = \frac{\text{Explained variation}}{\text{Total variation}} = 1 - \frac{\sum(Y - \bar{Y}_X)^2}{\sum(Y - \bar{Y})^2} \tag{9-25}$$

This provides us with the following interpretation of r^2: *The sample coefficient of determination is the proportion of the total variation in Y explained by the regression line.* In our illustration, we calculated $r^2 = .723$. This signifies that 72.3 percent of the total variation or scatter of average college grades Y about their mean can be explained by the relationship between this variable and the corresponding average high-school grades X, as estimated by the regression line for X and Y.

Because the unexplained variation can never exceed the total variation, their ratio is at most 1, so again we see that the greatest possible value of r^2 is 1. Likewise, when the explained variation is zero, so that knowledge of the regression line cannot reduce prediction errors, the value for r^2 must be zero. This corresponds to the example previously discussed where the regression line was horizontal and had both a slope and a correlation of zero.

Correlation Coefficient

Although the rationale for using the coefficient of determination to express the degree of relationship between X and Y is well justified, statisticians sometimes use another value calculated from the sample data. This is called the *sample correlation coefficient*, which is the square root of the coefficient of determination, calculated from

$$r = \sqrt{r^2} \tag{9-26}$$

The *population correlation coefficient* ρ is defined analogously,

$$\rho = \sqrt{\rho^2} \tag{9-27}$$

We may consider r to be a point estimator of ρ.

Since the square root of any number may be positive or negative, the correlation coefficient may be a more useful expression of the strength of association. The symbol r, positive or negative, can be used to signify the *direction* of the relationship between X and Y. Thus, when Y varies *directly* with X, r is *positive*; when Y bears an *inverse* relationship to X, r is *negative*.

For the student GPA data, the sample correlation coefficient is

$$r = \sqrt{r^2} = \sqrt{.723} = .850$$

We choose a positive sign for r, since the data indicate that college GPAs should be greater for students with better high-school grades, so that X and Y are directly related.

Figure 9-14 shows the values of the correlation coefficient r calculated for various sets of data. The values of $r = \sqrt{r^2}$ may range from -1 to 1. Negative

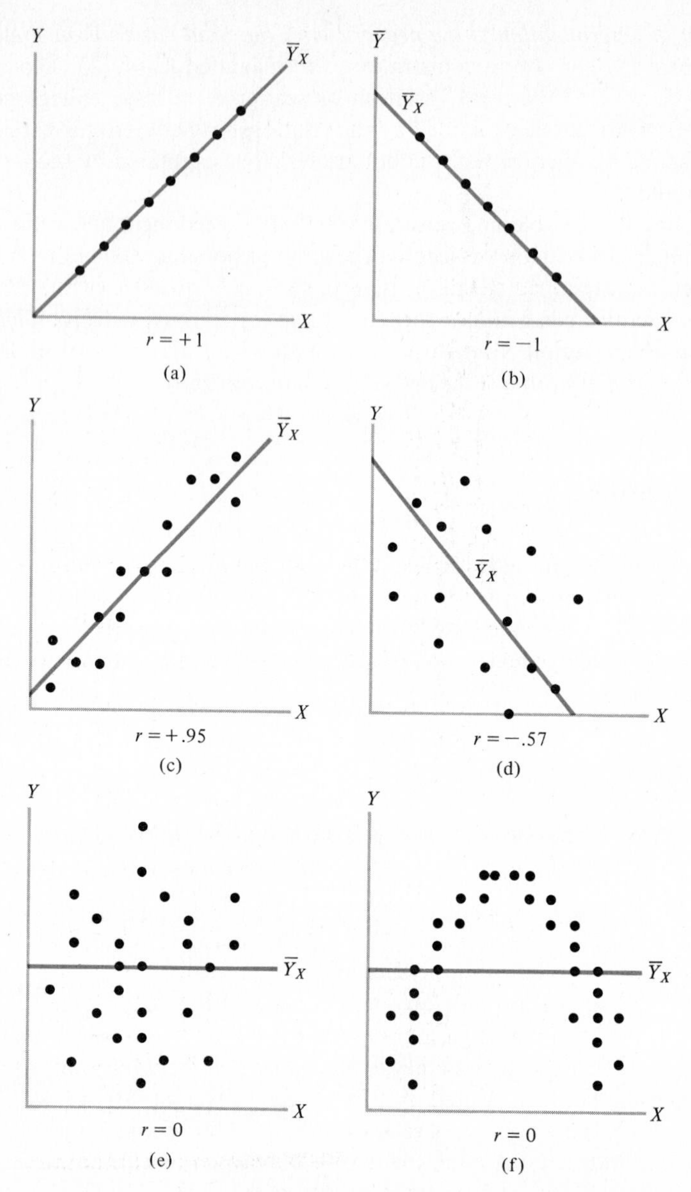

FIGURE 9-14 Scatter diagrams for sample results having various degrees of correlation.

values for r are obtained for the data sets in diagrams (b) and (d), where Y varies inversely with X. The slope of the estimated regression lines fitted to each set of these data will be negative, because Y decreases with increasing X. In diagrams (a) and (c), the correlation coefficients are positive, because the data in each case indicate a direct relationship between X and Y; the slope of the estimated regression lines will therefore be positive. Thus, *the signs for the correlation coefficient and the slope of the regression line must agree.*

Just like the coefficient of determination, the value of the correlation coefficient approaches zero as the degree of scatter becomes greater. Diagrams (a) and (c) in Figure 9-14 illustrate this point: In (a), the data are perfectly correlated, so that $r = 1$; in (c), considerable scatter indicates a smaller degree of correlation, so that $r = .95$. Likewise, for the perfectly correlated data in diagram (b), $r = -1$, while the less correlated, more scattered data in (d) yield $r = -.57$. As with r^2, $r = 0$ signifies a zero correlation. Diagrams (e) and (f) represent two instances where $r = 0$.

In (e), there is no apparent relationship between X and Y. But diagram (f) also shows $r = 0$, indicating no statistical relationship between X and Y; yet clearly, the data there appear to have a well-pronounced curvilinear relationship. Thus, the conclusion of no relationship is fallacious in this case, indicating that our *correlation coefficient must be restricted to instances where the underlying relationship between X and Y is believed to be linear.* Different procedures are required for calculating the strength of association for data having a curvilinear relationship.

Comparison of r and r^2

As we have already seen, the sample coefficient of determination r^2 shows the proportion of the total variation in Y that is explained by the regression line. As an adjunct to regression analysis, r^2 is a more useful measure of association. Since r is a decimal fraction, it will always be larger (in absolute value) than r^2. Thus, $r = .70$ may give us the false impression that there is a substantial explanation of the variation in Y through regression. The reduction in total variation amounts to only $r^2 = (.70)^2 = .49$, which is less than $\frac{1}{2}$. Statisticians usually prefer to use the coefficient of determination r^2 to explain the strength of association between X and Y when employing regression analysis.

When regression analysis is not used, however, the correlation coefficient r may be more meaningful. In such a case, r would be calculated from a different expression (which is to be explained next) that automatically provides the proper sign indicating whether a direct or an inverse relationship exists.

Correlation Explained Without Regression

The value of the correlation coefficient may be computed without performing regression analysis first. In many statistical applications, only the correlation coefficient is desired. For example, in determining if a chemical's concentration X improves patient recovery time Y, a medical researcher may only be interested in finding out whether or not raising the drug's dosage reduces symptoms; that is, he would be concerned with the *degree* to which X and Y are related, which he could determine from the correlation coefficient. The researcher may not be at all interested in *predicting* the number of days it will take a patient given a 50 cc dosage to recover, which is what he would learn from a regression line.

We may directly calculate from the following expression* for ***the sample correlation coefficient***

$$r = \frac{\sum XY - n\bar{X}\bar{Y}}{\sqrt{(\sum X^2 - n\bar{X}^2)(\sum Y^2 - n\bar{Y}^2)}} \qquad (9\text{-}28)$$

For example, we may use the intermediate calculations found earlier for high-school GPA X and college GPA Y:

$$n = 10 \qquad \bar{X} = 3.3 \qquad \bar{Y} = 2.9$$

$$\sum XY = 97.74 \qquad \sum X^2 = 111.70 \qquad \sum Y^2 = 86.16$$

Substituting these values into expression (9-28), the sample correlation coefficient is

$$r = \frac{97.74 - 10(3.3)2.9}{\sqrt{[111.70 - 10(3.3)^2][86.16 - 10(2.9)^2]}} = .850$$

which is the same value that we found earlier by using expression (9-23) to obtain r^2 and then taking the positive square root.

To explain the foregoing procedure for calculating r, we can express the sample correlation coefficient in an equivalent form. The numerator may be rewritten equivalently as $\sum(X-\bar{X})(Y-\bar{Y})$; also, the two terms in the denominator are the same as the sample variances of the X and Y values without the $n-1$ divisors. Thus

$$r = \frac{\sum(X-\bar{X})(Y-\bar{Y})}{\sqrt{[(n-1)s_X^2][(n-1)s_Y^2]}} = \frac{\sum(X-\bar{X})(Y-\bar{Y})}{(n-1)s_X s_Y}$$

Here, we use s_X^2 to represent the sample variance of the Xs in order to distinguish it from s_Y^2, which summarizes the total variability in the sample Ys. The above

* This equation may be obtained from expression (9-23) for r^2. First substitute $\bar{Y} - b\bar{X}$ for a. Then the right-hand side of expression (9-7) on page 343 may be substituted for b. After canceling and collecting terms in the equation that results, taking the square root of both sides provides the above expression for r.

expression may then be further transformed into the following equivalent expression for r:

$$r = \frac{\sum \left(\frac{X-\overline{X}}{s_X}\right)\left(\frac{Y-\overline{Y}}{s_Y}\right)}{n-1} \tag{9-29}$$

This relationship suggests that the sample correlation coefficient may be explained in terms of the deviations $X-\overline{X}$ and $Y-\overline{Y}$.

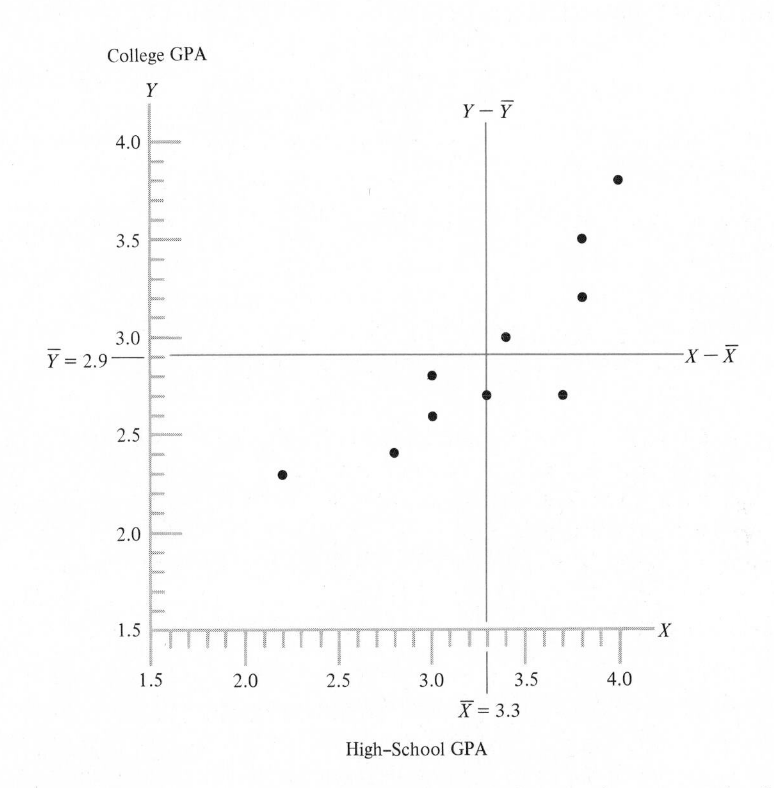

FIGURE 9-15 Illustration explaining correlation without using a regression line.

Figure 9-15 shows how these deviations can be plotted on the original scatter diagram by shifting the horizontal axis upward and the vertical axis to the right. The new axes are relabeled $X-\overline{X}$ and $Y-\overline{Y}$, with their origin at the points where $X=\overline{X}$ and $Y=\overline{Y}$. This graph shows how college and high-school

GPAs move together in terms of the deviation products, the sum of which is denoted by

$$\sum(X-\overline{X})(Y-\overline{Y})$$

If Y increases with X, then most of the deviations will fall in the northeast and southwest quadrants. Thus, the sum of the deviation products will be positive, reflecting a direct relationship between X and Y. But if the points are predominantly in the northwest and southeast quadrants, indicating that Y decreases for increasing X, this sum will be negative. If the points are scattered nearly evenly throughout all four quadrants, indicating poor responses of Y to X, the sum of the deviation products will be close to zero. Thus, we may conclude that the deviation-product sum will carry the same sign as the correlation coefficient and will have a value of zero when X and Y demonstrate no statistical relationship.

By itself, the deviation-product sum cannot serve as a suitable index of correlation, because it is affected by the units employed. (For example, a different value would result if GPAs were based upon some other point assignment than 4 for an "A," 3 for a "B," and so on). To compensate for this, each deviation may be standardized by dividing by the respective sample standard deviation and obtaining $(X-\overline{X})/s_X$ and $(Y-\overline{Y})/s_Y$. Expression (9-29) shows that the resultant standardized deviation-product sum actually underlies the r computation. To account for the influence of the sample size (for instance, the standardized deviation-product sum would be about 10 times as large if the sample size were 10 times as large), this is divided by $n-1$. The resulting index for r is therefore an average of standardized deviation products.

Correlation and Casuality

The correlation coefficient measures only the strength of *association* between two variables. This is a statistical relationship, and a large positive or negative value for r does not indicate whether a high value of one variable may cause the other to be large. Examples of nonsense or *spurious* correlations abound. We may illustrate such a correlation by finding the value for r that measures the association between the number of sperm whales Y caught from 1964 to 1968 and *Standard and Poor's Price Index* for 500 stocks X. The intermediate calculations are provided in Table 9-4. We find $r=-.89$, indicating a highly negative correlation between X and Y. How do we interpret this result? Clearly, there is no apparent logical connection between whaling and the New York Stock Exchange prices for common stocks. The values of X and Y have simply moved in opposite directions by approximately the same relative amounts for the five years considered. Statistically, large values of X have occurred with small values of Y, and vice versa. Obviously, r does not measure *causation* here, for we cannot say that the increasing stock prices have caused a declining catch of sperm whales, nor the reverse.

TABLE 9-4 Calculation of Correlation Coefficient for the Sperm-Whale Catch and Common Stock Price Index.

Year	*Number of Sperm Whales Caught (thousands)* X	*Standard and Poor's Common Stock Price Index* Y	X^2	Y^2	XY
1964	29	81	841	6,561	2,349
1965	25	88	625	7,744	2,200
1966	27	85	729	7,225	2,295
1967	26	92	676	8,464	2,392
1968	24	99	576	9,801	2,376
Totals	131	445	3,447	39,795	11,612

$\sum X = 131$ $\bar{X} = 26.2$ $\sum Y = 445$ $\bar{Y} = 89$

$\sum X^2 = 3{,}447$ $\sum Y^2 = 39{,}795$ $\sum XY = 11{,}612$

$$r = \frac{\sum XY - n\bar{X}\bar{Y}}{\sqrt{(\sum X^2 - n\bar{X}^2)(\sum Y^2 - n\bar{Y}^2)}} = \frac{11{,}612 - 5(26.2)(89)}{\sqrt{[3{,}447 - 5(26.2)^2][39{,}795 - 5(89)^2]}} = -.89$$

SOURCES: *Yearbook of Fishery Statistics*, Food and Agricultural Organization, United Nations, and *Economic Report of the President, February 1970.*

Statistical Inferences About ρ*

The population correlation coefficient ρ is the square root of the population coefficient of determination. The latter is defined by 1 minus the ratio of the variance of Y about the true regression line to the variance of Y about its mean. In order to make probability statements regarding ρ, *both* X and Y must be treated as random variables. Correlation theory requires this to be so before we can qualify inferences about ρ. When using r as the estimator, one additional restrictive—and often unrealistic—assumption is ordinarily made to avoid mathematical difficulties in finding the sampling distribution of r. This is that X and Y have a particular joint probability distribution called the bivariate normal distribution. Discussion of this distribution is beyond the scope of this book. For this reason, we will not describe the sampling distribution of r, and we therefore cannot discuss how to construct confidence interval estimates of ρ.

The far more common inference desired is a test to decide whether $\rho = 0$; that is, whether there is any statistical relationship at all between X and Y. For this purpose, we may substitute the test regarding the slope of the regression line described in Section 9-4. Recall that a slope of zero for the regression line

* This section relies on background material in Chapter 8 and may be skipped with no loss of continuity.

indicates a zero correlation. Thus, a test of whether $B = 0$ may be used to reject or to accept the null hypothesis $\rho = 0$. *A test rejecting the null hypothesis that $B = 0$ will also reject the assumption that $\rho = 0$.*

EXERCISES

9-16 Determine the value of the sample correlation coefficient for each of the following situations:

(a) $\bar{Y}_X = 10 + 2X$; $r^2 = .974$ (c) $\bar{Y}_X = 1.1 + 5.5X$; $r^2 = .950$

(b) $\bar{Y}_X = .3 - .1X$; $r^2 = .640$ (d) $\bar{Y}_X = 20 - 2X$; $r^2 = .810$

9-17 For each of the following situations, indicate whether a correlation analysis, a regression analysis, or both would be appropriate. In each case, give the reasons for your choice.

(a) In order to choose advertising media, an agency account executive is investigating the relationship between a woman's age and her annual expenditures on a client firm's cosmetics.

(b) A trucker needs to establish a decision rule that will enable him to determine when to inspect or to replace his tires, based upon the number of miles driven.

(c) A government agency wishes to identify which field offices of various sizes (in numbers of employees) are out of alignment with the prevailing pattern of working days lost due to sick leave.

(d) A research firm conducts attitude surveys in two stages. The first stage identifies coincident factors, such as age and income. The second stage is more detailed, involving separate study to predict values of one variable using the known values of other variables associated with it in the initial stage.

9-18 The following data represent disposable personal income and personal consumption expenditures (in billions of dollars) for the United States during the five-year period from 1964 through 1968. (Source: *Economic Report of the President, February 1970.*)

Year	*Disposable Personal Income* X	*Personal Consumption Expenditures* Y
1964	401	438
1965	433	473
1966	512	466
1967	547	492
1968	590	537

(a) Use the method of least squares to find the estimated linear regression equation which provides consumption expenditure predictions for specified levels of disposable income.

(b) Is the Y intercept negative or positive? Why do you think this is so?

(c) State verbally the meaning of the slope of the regression line.

(d) Using your intermediate calculations from (a) and the values found for the estimated regression coefficients, calculate the coefficient of determination using expression (9-23).

9-19 In helping to evaluate her district's effectiveness in teaching reading, a superintendent is studying the following results obtained for a sample of

$n = 10$ fifth-grade students. The data provide the reading readiness scores for each student at the beginning of two successive school years.

Student	*Prior Score* *X*	*Current Score* *Y*
Grace Brown	90	83
Patrick Gray	75	72
Lisa White	80	84
Homer Black	65	76
John Green	85	77
Linda Jones	90	82
Carl Smith	95	95
Freddy Tyler	75	68
Lisa Adams	70	78
Karen Johnson	60	55

(a) Use the method of least squares to determine the equation for the estimated regression line.
(b) Calculate the sample coefficient of determination using expression (9-23).
(c) What percentage of the total variation in Y is explained by the estimated regression line?

9-20 Referring to the information contained in Exercise 9-18 and to your answers to that exercise:
(a) What percentage of the total variation in Y is explained by the regression line?
(b) What is the value of the correlation coefficient?
(c) Calculate the correlation coefficient another way, using expression (9-28).

9-21 A statistics instructor wants to know if there is a correlation between his students' homework point totals and their average examination scores. A random sample of five of his students has produced the following results:

Homework Point Total *X*	*Average Examination Score* *Y*
140	90
80	80
90	60
150	80
110	70

Calculate the correlation coefficient for the above data.

REVIEW EXERCISES

9-22 For each of the following results, calculate (1) the sample coefficient of determination, and (2) the sample correlation coefficient.
(a) $\overline{Y}_X = 10 + .5X$; $\sum Y = 200$; $\sum XY = 5{,}000$; $\sum Y^2 = 5{,}250$; $n = 10$.
(b) $\overline{Y}_X = 15 - .5X$; $\sum Y = 500$; $\sum XY = 8{,}000$; $\sum Y^2 = 4{,}500$; $n = 100$.

9-23 A statistician has obtained the following data on the mean litter size of mice Y from mothers of various ages X:

Age (months) X	Mean Litter Size Y	Age (months) X	Mean Litter Size Y
12	10	2	14
7	9	9	12
4	13	5	10
9	8	8	12
7	13	4	11

(a) Find the estimated regression equation that may be used to predict mean litter size from a mother mouse's age.
(b) Using your results from part (a), find (1) the sample coefficient of determination, and (2) the correlation coefficient.
(c) What percentage of the variation in mean litter size is explained by the regression line?

9-24 A linear least squares relationship is used to determine how the grade-point average Y (GPA is measured on a scale where 4.0 is "straight 'A'") of a particular student in a certain university relates to his or her average weekly study time X (in hours per week). A random sample of 100 students was selected, and the following intermediate calculations were obtained:

$$\sum X = 3{,}000 \qquad \sum Y = 260 \qquad \sum XY = 8{,}050$$

$$\sum X^2 = 92{,}500 \qquad \sum Y^2 = 775$$

(a) Find the estimated regression equation.
(b) State in words the meaning of the slope of your estimated regression line.

9-25 An economist has established that personal income X may be used to predict personal savings Y by the relationship $\bar{Y}_X = 24.0 + .06X$ (billions of dollars). For each of the following levels of personal income, calculate the predicted value for personal savings:
(a) $X = \$300$ billion (b) $X = \$500$ billion (c) $X = \$700$ billion

9-26 A California rancher has kept records over the past $n = 10$ years of the amount of rainfall X (in inches) in his county and of the number of alfalfa bales Y he has had to buy to supplement grazing grass for his herd until he has been able to sell his excess cattle. The following estimated regression line has been obtained: $\bar{Y}_X = 20{,}000 - 500X$. The rancher has calculated $s_Y = 1{,}000$ bales, $s_{Y \cdot X} = 500$ bales, and $\sum X^2 = 2{,}500$. The mean rainfall for the time period considered has been $\bar{X} = 15$ inches. In order to arrange bank financing, the rancher wishes to predict how much alfalfa he must buy for the remainder of the current year. Since the dry season has arrived, he knows how much rain has fallen. Construct 95-percent prediction intervals for the required number of bales if this year's rainfall is (a) 15 inches; (b) 20 inches; (c) 10 inches.

9-27 An agronomist believes that over a limited range of fertilizer-application levels X (in gallons per acre), a prediction of crop yield Y (in bushels per acre) may be obtained using linear least squares regression. Using a sample of $n = 25$ plots, he has established that $\bar{Y}_X = 50 + .05X$, with $s_Y = 8$ bushels per acre, $s_{Y \cdot X} = 2$ bushels per acre, $\bar{X} = 210$ gallons, and $\sum X^2 = 1{,}105{,}000$.
(a) Construct a 95-percent prediction interval for the conditional mean bushels per acre $\mu_{Y \cdot X}$ for an application of 200 gallons of fertilizer per acre.
(b) Construct a 95-percent prediction interval for the yield of a particular one-acre plot where 200 gallons of fertilizer will be applied.

9-28 Suppose that the estimated regression line providing total ingredient cost for chemical batches of size X (in thousands of liters) is $\bar{Y}_X = \$30{,}000 + \$5{,}000X$. This result was obtained from a sample of $n = 100$ production runs, for which $s_{Y \cdot X} = \$400$.

(a) Construct the 95-percent prediction interval for the conditional mean batch cost $\mu_{Y \cdot X}$ when $X = 10$ thousand liters.

(b) Construct the 99-percent prediction interval for the total cost of the next 10–thousand-liter batch.

10 Multiple Regression and Correlation

A man saw Nasrudin searching for something on the ground. "What have you lost, Mulla?" he asked.

"My key," said the Mulla.

So the man went down on his knees, too, and they both looked for it.

After a time, the other man asked: "Where exactly did you drop it?"

"In my own house."

"Then why are you looking here?"

"There is more light here than inside my own house."

The techniques discussed in Chapter 9 for the analysis of the relationship between two variables are referred to as *simple* regression and correlation analysis. There we learned how to construct a regression equation to make predictions of a dependent variable when the value of a single independent variable is known. We have already noted that such predictions may be too imprecise for practical application, because a substantial amount of the variation in Y cannot be explained by X. In this chapter, these techniques will be expanded to include *multiple regression and correlation analysis*, which involves *several* independent predictor variables. The total variation in Y can then be explained by two or more variables, which in many situations, allows a more precise prediction to be made than is possible in simple regression analysis.

The essential advantage in using two or more independent variables is that the available information is put to greater use. For example, a regression line expressing a student's college GPA in terms of his high-school grades should yield a poorer prediction than an equation that also considers his admissions test scores and extracurricular activities.

10-1 LINEAR MULTIPLE REGRESSION INVOLVING THREE VARIABLES

Linear multiple regression analysis extends simple linear regression analysis by considering two or more independent variables. In the case of two independent variables, denoted as X_1 and X_2, we use ***the estimated multiple regression equation***

$$Y_C = a + b_1 X_1 + b_2 X_2 \tag{10-1}$$

Here, Y_C, with the subscript C for "computed," denotes values for Y calculated from the estimated regression equation. This is analogous to the term $\overline{Y}_X$ we used before, but Y_C is used in dealing with several independent variables because it is too cumbersome to place X_1 and X_2 in subscripts. With two independent variables and one dependent variable, a total of three variables must be considered. The sample data will consist of three values for each sample unit observed, so that a scatter pattern portraying these observations will be three dimensional.

Regression in Three Dimensions

To explain how sample data can be portrayed in three dimensions, we will make an analogy between the walls and the floor of a room. Letting a corner of the room represent the situation when all three variables have zero value, the data points may be represented by marbles suspended in space at various distances from the floor and the two walls. A marble's height above the floor can represent the value of Y for that observation. Its distance along the wall on the right may then represent the observed value for X_1, while its distance in the wall on the left may express the value for X_2. A pictorial representation of one three-dimensional scatter is shown in Figure 10-1 for a hypothetical set of data.

The regression equation (10-1) corresponds to a *plane*,* which must be slanted in the way that provides the best fit to the sample data. We refer to the three-dimensional surface so obtained as the *regression plane*. The choice of this plane is analogous to determining how to position a pane of glass through the suspended marbles so that its inclination approximates that presented by the pattern of scatter. Of course, we want the same regression plane to be obtained by everyone, so we will adapt the method of least squares to three dimensions for this purpose.

* Although Y is related to X_1 and X_2 by a plane instead of a line, we still say that the relationship is linear. The three-dimensional extension of a two-dimensional line is a plane. Although a line can also exist in three dimensions, it is defined as the intersection of two planes. Thus, a three-dimensional line is like a point in two dimensions, which can be defined by the intersection of two lines.

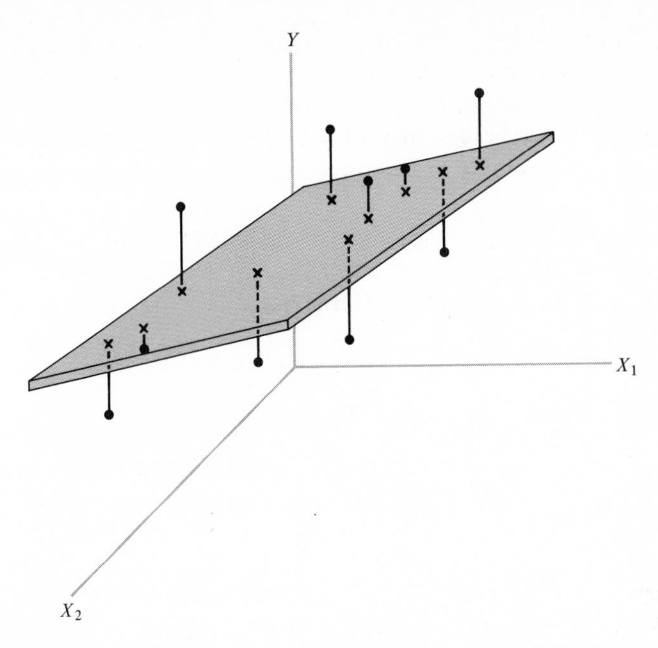

FIGURE 10-1 Regression plane for multiple regression, using three variables.

The constants a_1, b_1, and b_2 in the equation of the regression plane, $Y_C = a + b_1 X_1 + b_2 X_2$, are called the estimated regression coefficients. As with $\overline{Y}_X = a + bX$, a is the value of Y where the regression plane cuts the Y axis, so that we still refer to a as the Y intercept. However, the interpretations of b_1 and b_2 are somewhat different than in simple regression. The constant b_1 expresses the net change in Y for a one-unit increase in X_1, holding X_2 fixed at a constant value. We may view b_1 as the slope of the edges obtained by slicing the regression plane with cuts made parallel to the X_1 axis at a distance X_2 from the origin. Likewise, b_2 is the net change in X_2, holding X_1 fixed. Because either b_1 or b_2 alone only partially shows the total movement in Y in response to increases in the respective independent variables, b_1 and b_2 are referred to as *estimated partial regression coefficients*.

For example, suppose that average family food expenditures per week Y (in dollars) are related to weekly take-home pay, denoted by X_1, and family size X_2 by the equation $Y_C = 15 + .20X_1 + 5X_2$. Here, $a = 15$, so that the regression plane cuts the Y axis at $Y = \$15$. The partial regression coefficient $b_1 = .20$ signifies that a family's average food expenditures will increase by \$.20 for each additional dollar in take-home pay, regardless of family size. The constant $b_2 = 5$ indicates that each additional family member will raise the food bill by \$5, regardless of the family's level of income. Thus, an $X_2 = 4$ person family taking home $X_1 = \$200$ would have average weekly food expenses of

$$Y_C = 15 + .20(200) + 5(4) = \$75$$

Because of individual differences in eating habits, physiology, and general life style for the families in the sample used to obtain the above regression equation, we cannot assume that a particular family necessarily spends $75 per week for food. As we have seen with simple regression, we must account for both the inherent variability in Y and the variability in Y due to sampling error.

10-2 MULTIPLE REGRESSION USING LEAST SQUARES

Mathematically, the method of least squares involves inclining the regression plane in such a way that the sum of the squared vertical deviations from its surface to the data points is minimized. The dots in Figure 10-1 represent the observed data points, and the vertical deviations of these points from the plane are shown as line segments. The crosses indicate the corresponding points on the regression plane having identical values for X_1 and X_2. The height of a cross above the $X_1 X_2$ plane (or the floor in our analogy) represents the value of Y_C computed from expression (10-1). The difference between the observed and computed heights, $Y - Y_C$, will be positive for dots lying above the regression plane and negative for those lying below, so that the method of least squares minimizes

$$\sum (Y - Y_C)^2 \qquad (10\text{-}2)$$

As with simple regression, the coefficients of the estimated regression plane may be determined by solving a set of three equations in three unknowns. As before, these are referred to as the ***normal equations***

$$\begin{aligned} \sum Y &= na + b_1 \sum X_1 + b_2 \sum X_2 \\ \sum X_1 Y &= a \sum X_1 + b_1 \sum X_1^2 + b_2 \sum X_1 X_2 \\ \sum X_2 Y &= a \sum X_2 + b_1 \sum X_1 X_2 + b_2 \sum X_2^2 \end{aligned} \qquad (10\text{-}3)$$

These equations may be obtained from the equation for a plane, $Y = a + b_1 X_1 + b_2 X_2$. The first normal equation is found by summing each term of the regression equation and using the fact that $\sum a = na$. The second normal equation is obtained by multiplying every term in the regression equation by X_1 and then summing the resultant products. The third equation is found in a similar manner by multiplying every term by X_2 and then summing.*

We will not introduce separate expressions for the regression coefficients here, since the resulting equations would be extremely involved and cumbersome to use. Thus, we find a_1, b_1, and b_2 by first calculating the required sums from the

* This procedure is a mnemonic aid. Calculus was used to actually derive the equations in expression (10-3).

data for the various combinations of Y, X_1, and X_2 and then substituting these sums into the normal equations, which are solved simultaneously. To show how this works, we will expand the college grade-point-average illustration we used in Chapter 9.

An Illustration

Suppose that our admissions director has a record of the verbal SAT scores for each student in the original sample. In the past, these scores have been used as a supplement to high-school grades in making college admissions decisions—the belief being that certain students who do poorly in high school because of immaturity or personal problems and yet score highly on the SAT often do well in college. Also, some sterling high-school performers fizzle out in college because of a lack of aptitude, and a low SAT score often confirms this. By performing multiple regression analysis using two independent variables, it is believed that a better prediction of college success can be made. Using college GPA as the dependent variable Y, we now have high-school GPA, denoted by X_1, and verbal SAT score X_2 as the independent variables. The sample results are summarized in Table 10-1.

The further intermediate calculations necessary to find the regression

TABLE 10-1 Sample Observations of College Grade-Point Averages, Using High-School GPA and Verbal SAT Score as Independent Variables.

Student	*College GPA* Y	*High-School GPA* X_1	*Verbal SAT Score* X_2	X_2Y	X_1X_2	X_2^2
1	3.8	4.0	750	2,850	3,000	562,500
2	2.7	3.7	380	1,026	1,406	144,400
3	2.3	2.2	580	1,334	1,276	336,400
4	3.2	3.8	510	1,632	1,938	260,100
5	3.5	3.8	620	2,170	2,356	384,400
6	2.4	2.8	440	1,056	1,232	193,600
7	2.6	3.0	540	1,404	1,620	291,600
8	3.0	3.4	650	1,950	2,210	422,500
9	2.7	3.3	480	1,296	1,584	230,400
10	2.8	3.0	550	1,540	1,650	302,500
	29.0	33.0	5,500	16,258	18,272	3,128,400

$\sum Y = 29.0$ $\quad \sum X_1 = 33.0$ $\quad \sum X_2 = 5{,}500$

$\bar{Y} = 2.9$ $\quad \bar{X}_1 = 3.3$ $\quad \bar{X}_2 = 550$

$\sum X_1Y = 97.74$ $\quad \sum X_2Y = 16{,}258$ $\quad \sum X_1X_2 = 18{,}272$

$\sum X_1^2 = 111.70$ $\quad \sum X_2^2 = 3{,}128{,}400$ $\quad \sum Y^2 = 86.16$

equation are also shown in Table 10-1. The column totals and the relevant totals found earlier in Table 9-2 are also provided. Substituting the appropriate sum values into expression (10-3) gives us the following normal equations:

$$
\begin{aligned}
29 &= 10a + 33b_1 + 5{,}500b_2 \\
97.74 &= 33a + 111.70b_1 + 18{,}272b_2 \\
16{,}258 &= 5{,}500a + 18{,}272b_1 + 3{,}128{,}400b_2
\end{aligned}
$$

Solving these equations simultaneously for the unknowns a, b_1, and b_2, we obtain

$$
\begin{aligned}
a &= -.412 \\
b_1 &= .631 \\
b_2 &= .00223
\end{aligned}
$$

The simultaneous solution of three normal equations by hand can be quite a chore. This task, as we will see later in this chapter, can be enormously simplified by using a computer.

The above values for a, b_1, and b_2 provide the estimated multiple regression equation

$$Y_C = -.412 + .631X_1 + .00223X_2$$

This equation may then be used to forecast the college GPA of any entering freshman. For example, for a student with a high-school GPA of $X_1 = 3.0$ and verbal SAT score of $X_2 = 600$, the regression plane provides the estimated college GPA

$$Y_C = -.412 + .631(3.0) + .00223(600) = 2.819$$

Interpreting the Results

The value $b_1 = .631$ means that a student's college GPA can be predicted to be .631 points higher for each additional point in his high-school GPA, regardless of how well that student did on the SAT. Likewise, each additional SAT score point will result in an estimated college-GPA rise of $b_2 = .00223$ grade points, so that 100 extra points on the SAT will raise a student's GPA by .223 grade points (again, regardless of that student's high-school grade level). The result $a = -.412$ is the Y intercept of the estimated regression plane. This value signifies that every GPA prediction based only upon X_1 and X_2 must be further adjusted downward by .412 points to fully reflect the relation between the dependent and independent variables. In a sense, the high-school GPA and the SAT score contributions by themselves inflate the college GPA forecast by about .412 points.

Advantages of Multiple Regression

Two interesting questions may be posed regarding the use of multiple rather than simple regression analysis: Is the simultaneous analysis of two independent variables through multiple regression any improvement over analysis employing two separate simple regressions? How can we show that the accuracy of predictions is improved by the use of multiple regression analysis? We will begin to answer these questions by continuing with our student GPA illustration.

Both of these questions may be partly answered through correlation analysis. We have already computed the sample correlation coefficient for high-school and college GPAs to be .850. Since a similar figure may be obtained for any pair of the three variables discussed, we can denote the above sample correlation coefficient by the double-subscripted symbol r_{Y1}, where the term $Y1$ indicates that the term measures the strength of association between the variables Y and X_1. Thus, we have $r_{Y1} = .850$. A simple correlation analysis for Y and X_2 provides another sample correlation coefficient, denoted by the symbol r_{Y2}, with a value of .667. [As a review exercise, you may verify this number by using the verbal SAT score as the single dependent variable X and applying expression (9-28).]

The above correlation coefficients indicate that a stronger association exists between college GPA Y and high-school GPA X_1 than exists between Y and verbal SAT score X_2. Thus, a simple regression analysis for Y and X_1 should provide more reliable and precise predictions of college GPA than would result from using Y and X_2 alone. However, when X_1 and X_2 are considered *together* as independent variables, a multiple regression analysis should "outperform" either simple regression analysis. As we will see, the multiple regression equation reduces the variation in values of Y that is left unexplained by either two-variable analyses. The following example resolves an interesting paradox involving two independent variables.

The Cows and Acres Problem

To illustrate, suppose that Y represents dairy-farm profits. Using the sample data, a multiple regression could be performed on two independent variables: X_1 = number of cows, and X_2 = number of acres on the farm.* For certain kinds of data (say, $r_{Y1} = .9$) profit will be highly positively correlated with number of cows, which is what we would expect. However, for the same farms, there may be no correlation between profit and number of acres, so that $r_{Y2} = 0$, which is contrary to what we may believe (especially today, when big farms seem to be absorbing the less profitable, smaller farms). Even more

* This example is a simplification of the one contained in Mordecai Ezekial and Karl A. Fox, *Methods of Correlation and Regression Analysis, Third Edition* (New York: John Wiley & Sons, Inc., 1959).

perplexing, the multiple regression equation could show that dairy profits increase if either number of cows or number of acres increases (that is, b_1 and b_2 are both positive). But why?

The reason is that our hypothetical dairy-farm sample exhibits a sizable *negative correlation* between number of cows and number of acres (so that the simple correlation between the independent variables X_1 and X_2 could be expressed by a number like $r_{12} = -.7$). This is because the sample contains a mixture of dairies where feed grain is grown on the premises (so that many acres are needed) and dairies that purchase all feed grain (so that relatively few acres are needed). Thus, a dairy can be profitable either with very few acres or with many acres. The multiple regression equation still reflects that, in either case, more acres will accommodate a larger herd and hence will provide greater profits.

This dairy-farm example illustrates that separate simple correlations between profit and the individual independent variables should not be relied upon to eliminate predictor variables. Only after the interactions between the independent variables have been determined may we safely discard one of them. The multiple regression analysis itself can provide this information.

Next we will see how to measure the improvement over simple regression that may be achieved through multiple regression.

Standard Error of the Estimate

We have seen that the errors in making predictions from the regression line are smaller when the scatter of the data is less pronounced. The degrees of such variation in Y can be expressed by the standard error of the estimate for values of Y. The same is true in multiple regression. The standard error of the estimate for values of Y about the regression plane is defined by

$$S_{Y \cdot 12} = \sqrt{\frac{\sum (Y_C - Y)^2}{n-3}} \qquad (10\text{-}4)$$

We use a different subscript notation, $Y \cdot 12$, to show that two independent variables X_1 and X_2 are being used to predict Y. As in simple regression, the standard error of the estimate is found by taking the square root of the mean squared deviations of observed Y values from the estimated regression plane. The divisor $n-3$ is chosen because 3 degrees of freedom are lost in estimating the regression coefficients. This makes $S_{Y \cdot 12}^2$ an unbiased estimator of the variance of Y about the true regression plane. In practice, it is simpler to use the following expression to calculate ***the standard error of the estimate***

$$S_{Y \cdot 12} = \sqrt{\frac{\sum Y^2 - n\bar{Y}^2 - b_1(\sum X_1 Y - n\bar{X}_1 \bar{Y}) - b_2(\sum X_2 Y - n\bar{X}_2 \bar{Y})}{n-3}} \qquad (10\text{-}5)$$

which does not require the calculation of Y_C for every data point.

Using the data in Table 10-1 and the partial regression coefficients found previously, we calculate $S_{Y \cdot 12}$ for the student GPA data as

$$S_{Y \cdot 12} = \sqrt{\frac{86.16 - 10(2.9)^2 - .631[97.74 - 10(3.3)(2.9)] - .00223[16{,}258 - 10(550)(2.9)]}{10 - 3}}$$

$$= .111$$

This value may be compared to the earlier standard error found in Chapter 9 for the simple regression analysis using high-school and college GPAs only. In keeping with our revised subscript notation, that may be denoted by $s_{Y \cdot X_1}$. We found the standard error of the estimate about the regression *line* to be $s_{Y \cdot X_1} = .265$. Comparing this with the standard error of the estimate about the regression *plane*, $S_{Y \cdot 12} = .111$, we see that multiple regression provides for less variation in college GPA Y. As we noted earlier, the standard error expresses the amount of variation in Y that is left unexplained by regression analysis. Since $S_{Y \cdot 12}$ is smaller than $s_{Y \cdot X_1}$, our regression plane (incorporating both high-school GPA and verbal SAT score as *two* independent variables) explains more of the variation in Y than the regression line does (where high-school GPA serves as the *single* independent variable or, in other words, where there is more unexplained variation). This indicates that inclusion of verbal SAT score data in the analysis will indeed provide improved forecasts of college GPA. Before we see how such predictions can be made from multiple regression results, we must clarify some theoretical points.

Assumptions of Multiple Regression

The assumptions of linear multiple regression are similar to those of simple regression. The least squares method provides a plane that is an estimate of

$$\mu_{Y \cdot 12} = A + B_1 X_1 + B_2 X_2 \qquad (10\text{-}6)$$

where $\mu_{Y \cdot 12}$ denotes the conditional mean of Y, given X_1 and X_2. The true regression coefficients are A, B_1, and B_2. Their values remain unknown and are estimated by a, b_1, and b_2, respectively. For each combination of Y, X_1, and X_2, there is a corresponding point on the true regression plane. The height of this point is $\mu_{Y \cdot 12}$, the mean of the corresponding populations for Y. As with simple regression, each such distribution is assumed to have the same variance $\sigma^2_{Y \cdot 12}$.

It is possible to extend the theoretical developments of simple regression to multiple regression in order to make inferences of A, B_1, and B_2. For instance, a confidence interval can be obtained for B_1 using estimator b_1, just as we did for B with b. Prediction intervals for $\mu_{Y \cdot 12}$ and individual values for Y can also be constructed using the estimated regression plane.

Prediction Intervals in Multiple Regression

Prediction Interval to Estimate Mean Value of Y

The procedures for constructing prediction intervals to qualify estimates made from the regression plane are similar to those in simple regression. The following equation* provides *the prediction interval estimate for* $\mu_{Y \cdot 12}$

$$\mu_{Y \cdot 12} = Y_C \pm t_\alpha \frac{S_{Y \cdot 12}}{\sqrt{n}} \tag{10-7}$$

where Y_C is the value for Y computed from the regression plane at the given levels for X_1 and X_2, and t_α is the critical value of the t statistic for an upper-tailed area of α (half of 1 minus the confidence level) when the degrees of freedom are $n-3$.

We may illustrate this with our earlier sample student data. Suppose that $\mu_{Y \cdot 12}$ is to be estimated by a 95-percent confidence interval, so that $\alpha = .025$. This will be done for all entering freshmen with a "B" average in high school who score 600 on the SAT. Thus, $X_1 = 3.0$ and $X_2 = 600$. We have previously found that $Y_C = 2.819$, and from Appendix Table E, the critical value for $n-3 = 10-3 = 7$ degrees of freedom is $t_\alpha = t_{.025} = 2.365$. Substituting these values, with $S_{Y \cdot 12} = .111$, into expression (10-7), we obtain the prediction interval

$$\mu_{Y \cdot 12} = 2.819 \pm \frac{2.365(.111)}{\sqrt{10}}$$

$$= 2.819 \pm .083$$

$$2.736 \leqslant \mu_{Y \cdot 12} \leqslant 2.902$$

Thus, we are 95 percent confident that the true mean college GPA for all such students lies between 2.736 and 2.902. Note that this interval is narrower than the one we obtained earlier (page 356) from simple regression using only high-school grades for prediction.

As with all sampling situations, a larger sample size would provide an even more precise estimate of mean college GPA. *When the sample size is large (30 or more), the normal distribution can be used,* so that t_α is replaced in expression (10-7) by the normal deviate z for an upper-tailed normal curve area of α.

Prediction Interval to Estimate Individual Y

An individual value of Y is denoted here by Y_I. The following expression

* This equation is an approximation of a more complicated one, sometimes used in other textbooks, which provides nearly the same values for most sample sizes n.

provides ***the prediction interval estimate for Y_I***

$$Y_I = Y_C \pm t_\alpha S_{Y \cdot 12} \sqrt{\frac{n+1}{n}} \tag{10-8}$$

We may continue with the college grade prediction example. Again, using a *particular* student with a high-school GPA of $X_1 = 3.0$ and a SAT score of $X_2 = 600$, we still have $Y_C = 2.819$. Substituting these values, with $t_{.025} = 2.365$ and $S_{Y \cdot 12} = .111$, as before, we may compute the 95-percent prediction interval from expression (10-8):

$$Y_I = 2.819 \pm 2.365(.111) \sqrt{\frac{10+1}{10}}$$

$$= 2.819 \pm .275$$

$$2.544 \leqslant Y_I \leqslant 3.094$$

Because the prediction pertains to an individual's GPA and not to a mean GPA, the above interval is wider than the one previously found.

When n is large, expression (10-8) is modified by replacing t_α with the appropriate normal deviate z.

EXERCISES

10-1 An economist wishes to predict the incomes of restaurants over two years old. This prediction will be made from a regression equation using total floor space and number of employees. For a sample of $n = 5$ restaurants, the following data have been obtained:

Income (thousands of dollars) Y	*Floor Space (thousands of square feet)* X_1	*Number of Employees* X_2
20	10	15
15	5	8
10	10	12
5	3	7
10	2	10

(a) Calculate the estimated regression equation for the above data.

(b) State verbally the meaning of the partial regression coefficients b_1 and b_2.

10-2 Suppose that a college admissions director has used high-school GPA X_1 and IQ score X_2 to predict college GPA Y. Using the regression equation

$$Y_C = .5 + .8X_1 + .003X_2$$

calculate the predicted college GPA for each of the following students:

	(a)	(b)	(c)	(d)
High-School GPA	2.9	3.0	2.7	3.5
IQ Score	123	118	105	136

10-3 The editor of a statistics journal wishes to use the intermediate sample data calculations provided below to determine a regression equation that predicts the total typing hours Y for article drafts. As independent variables, she uses number of words in the draft X_1 (expressed in tens of thousands) and an index X_2 for level of difficulty on a scale from 1 (least difficult) to 5 (most difficult).

$$n = 25 \qquad \sum Y = 200 \qquad \sum X_1 = 100 \qquad \sum X_2 = 75$$
$$\sum X_1 Y = 1{,}000 \qquad \sum X_2 Y = 800 \qquad \sum X_1^2 = 600$$
$$\sum X_2^2 = 325 \qquad \sum Y^2 = 3{,}800 \qquad \sum X_1 X_2 = 200$$

(a) Determine the equation for the regression plane $Y_C = a + b_1 X_1 + b_2 X_2$.

(b) Explain the meaning of values obtained for b_1 and b_2.

10-4 A record manufacturer uses special machines to press a recording's grooves onto blank disks from a die. Each die lasts for about 1,000 pressings. Because of the time constraints involved in the record business, it is sometimes necessary to use several pressing machines simultaneously. Since each machine requires an expensive die disk and many production runs are completed before the useful lifetime of each die disk has been achieved, this raises production costs. For $n = 100$ production runs, the total manufacturing cost Y (in thousands of dollars) has been determined for the number of pressings made X_1 (in thousands) and the number of die disks required X_2. The following results have been calculated:

$$Y_C = 1.082 + 1.2X_1 + .553X_2$$
$$\bar{Y} = 32.0 \qquad \bar{X}_1 = 23.0 \qquad \bar{X}_2 = 6.0$$
$$\sum Y^2 = 104{,}000 \qquad \sum X_1 Y = 75{,}000 \qquad \sum X_2 Y = 15{,}000$$

(a) Calculate the estimated cost Y_C for each of the following production runs:

	Number of Pressings (thousands)	*Number of Die Disks*
(1)	15	5
(2)	20	3
(3)	15	4
(4)	100	10

(b) Calculate the standard error of the estimate for values of Y about the regression plane.

(c) For each production-run situation in (a), construct the 95-percent confidence interval for the *mean* production costs.

(d) Same as (c), but construct intervals for costs estimates for *individual* production runs.

10-5 Referring to the information contained in Exercise 10-3 and to your answers to that exercise:

(a) Determine the equation for the estimated regression line $\bar{Y}_{X_1} = a + bX_1$, using only a single independent variable X_1.

(b) Calculate $s_{Y \cdot X_1}$, using X_1 instead of X in expression (9-12) on page 347, and $S_{Y \cdot 12}$, using expression (10-5). Do you think the inclusion of the level of difficulty variable X_2 will result in better predictions than those obtained using X_1 alone? Why?

(c) Suppose that a particular article containing 50,000 words is rated at a difficulty level of 3. Determine the prediction interval for the typing hours required at a 95-percent confidence level. Do the same for the mean typing time of all articles of the same length and difficulty.

10-3 REGRESSION WITH MANY VARIABLES: COMPUTER APPLICATIONS

Multiple regression may use more than two independent variables. For example, in predicting college GPA, we might also use the number of extracurricular, high-school activities—especially since this information is considered important in making college admissions decisions. Inclusion of this third independent variable, denoted by X_3, would provide the regression equation

$$Y_C = a + b_1 X_1 + b_2 X_2 + b_3 X_3 \tag{10-9}$$

In solving for the regression coefficients, additional product sums involving X_3 must be computed and four normal equations must be solved simultaneously. *Ordinarily, the necessary calculations are made with a digital computer rather than by hand*, so a detailed discussion of the computations will not be provided in this book.

As in multiple regression with X_1 and X_2, by including X_3 we may compute the standard error of the estimate about the regression *hyperplane*. (Including Y, there will now be four variables, so the regression surface cannot be graphed in only three dimensions. Expression (10-9) yields a four-dimensional plane, which mathematicians call a "hyperplane.") The standard error, denoted by the symbol $S_{Y \cdot 123}$, is calculated in the same way we calculate $S_{Y \cdot 12}$, but with an extra term involving b_3 in the numerator; the denominator will be $n-4 = 10-4 = 6$. The number of degrees of freedom will also be $n-4$, and we may calculate the prediction intervals as we did before with this reduced value.

When more predictor variables are desired, we can expand the regression analysis by including X_4, X_5, and so on. When there are m total variables (one dependent and $m-1$ independent), the number of degrees of freedom will be $n-m$, which is used as the denominator for computing the standard error and in finding t_α when constructing prediction intervals.

To illustrate multiple regression with three independent variables, a computer application will be discussed next.

Using the Computer in Multiple Regression Analysis

By hand, the computations necessary to perform a regression analysis are at best tedious, even with the assistance of a hand-operated calculator. Although hand calculation may be feasible when the number of observations and the number of independent variables are small, such tedium can be alleviated by using a digital computer. The computer saves not only time and energy but also prevents the inevitable cascade of errors which occurs whenever one error, such as mispunching a calculator button, is committed. Also, a computer generally provides greater levels of accuracy because of its superior capability to handle a large number of significant figures.

It is not necessary for the user to prepare his own computer program to perform regression and correlation analysis. Library programs already written for this purpose are widely available. These "canned" programs vary considerably in their format and in the type of output data they provide, but they generally include a determination of the regression coefficients (a, b_1, b_2, and so on), standard errors, and correlation coefficients. Such programs can be quite elaborate, and may include lower-dimensional regression equations for various variable combinations, values for the t statistic (used in constructing confidence intervals for regression coefficients and in obtaining prediction intervals), and coefficients of partial determination (discussed in Section 10-4). The most complex programs provide inference-making computations too sophisticated to include here.

Here we will describe how to use one particular program: the Multiple Linear Regression Routine of the International Timesharing Corporation. Although this program is of the time-sharing type, similar programs are available that operate in a batch-processing mode. In those cases, the input data and operating parameters must be punched onto cards rather than fed directly through a teletype terminal.

To illustrate the use of the computer, we will expand our college GPA illustration to include a third independent variable—the number of extracurricular high-school activities. To do this, we will augment our previous sample data by including the ten observations for X_3 provided in Table 10-2.

Based on the data in Table 10-2, Figure 10-2 shows the log of terminal communications, how the data are entered, and the output of the regression analysis. The computer results indicate that college GPA predictions may be

TABLE 10-2 Sample Observations of College Grade-Point Averages, Using High-School GPA, Verbal SAT Score, and Number of Extra-curricular Activities as Independent Variables.

Student	*College GPA* Y	*High-School GPA* X_1	*Verbal SAT Score* X_2	*Number of Activities* X_3
1	3.8	4.0	750	7
2	2.7	3.7	380	6
3	2.3	2.2	580	5
4	3.2	3.8	510	6
5	3.5	3.8	620	3
6	2.4	2.8	440	2
7	2.6	3.0	540	8
8	3.0	3.4	650	2
9	2.7	3.3	480	7
10	2.8	3.0	550	4

```
I T S  STATISTICAL PACKAGE

WANT TO SEE STAT INDEX? NO
WANT ADVANCED USER STATUS(YES OR NO)?  NO
TYPE THE CODE OF THE ROUTINE YOU WISH TO USE  MU
TYPE NO. OF VARIABLES AND OBSERVATIONS  4.,10
DATA ORDERED BY OBSERVATION OR BY VARIABLE(OBS OR VAR)?  OBS
ENTER DATA FILE NAME OR 'TERMINAL' --  TERMINAL
INPUT DATA IN FORM XX.X,XX.X,XX.X,...XX.X (RET)
4.,750,7,3.8
3.7,380,6,2.7
2.2,580,5,2.3
3.8,510,6,3.2
3.8,620,3,3.5
2.8,440,2,2.4
3.0,540,8,2.6
3.4,650,2,3.0
3.3,480,7,2.7
3.0,550,4,2.8
LISTING(YES OR NO)?  YES
VARIABLE
 OBS.        1 X1         2 X2          3 X3          4 Y
   1      4.0000    750.0000       7.0000       3.8000
   2      3.7000    380.0000       6.0000       2.7000
   3      2.2000    580.0000       5.0000       2.3000
   4      3.8000    510.0000       6.0000       3.2000
   5      3.8000    620.0000       3.0000       3.5000
   6      2.8000    440.0000       2.0000       2.4000
   7      3.0000    540.0000       8.0000       2.6000
   8      3.4000    650.0000       2.0000       3.0000
   9      3.3000    480.0000       7.0000       2.7000
  10      3.0000    550.0000       4.0000       2.8000
DO YOU WISH TRANSFORMATIONS OF DATA?  NO
DO YOU WISH TO SAVE THIS DATA ON A DISK FILE?  NO
ENTER NUMBER OF INDEPENDENT VARIABLES IN THE FORM XX [RET]   3
ENTER COLUMN NUMBERS OF THE INDEPENDENT VARIABLES IN
THE FORM XX,XX,... ETC. [RET]  1,2,3
ENTER COLUMN NUMBER OF DEPENDENT VARIABLE IN THE FORM XX  4

 VARIABLE   MEAN     STANDARD   CORRELATION REGRESSION  STD. ERROR   COMPUTED
   NO.               DEVIATION    X VS Y    COEFFICIENT  OF REG.COEF.  T VALUE
   1  X̄1   3.300        .558 sX1    .849 rY1     .637 b1      .073        8.714
   2  X̄2 550.000     107.186 sX2    .667 rY2     .002 b2      .000        5.927
   3  X̄3   5.000       2.160 sX3    .075 rY3    -.007 b3      .018        -.401
 DEPENDENT
   4  Ȳ    2.900        .478 sY
INTERCEPT                                              -.386080 a
MULTIPLE CORRELATION COEFFICIENT                 RY.123 .979889
STANDARD ERROR OF ESTIMATE                              .116922 SY.123

WANT TABLE OF RESIDUALS(YES OR NO)?  YES
DO YOU WANT TO OUTPUT RESIDUALS TO A FILE(YES OR NO)?  NO

                            TABLE OF RESIDUALS

 OBSERVATION     Y OBSERVED     Y ESTIMATED      RESIDUAL   RES. PCNT. OF EST.
   1               3.800          3.775           -.025              -.657
   2               2.700          2.770            .070              2.519
   3               2.300          2.266           -.034             -1.494
   4               3.200          3.122           -.078             -2.492
   5           Y   3.500      Yc  3.389    Yc-Y   -.111             -3.286
   6               2.400          2.359           -.041             -1.719
   7               2.600          2.665            .065              2.424
   8               3.000          3.208            .208              6.482
   9               2.700          2.730            .030              1.091
  10               2.800          2.716           -.084             -3.080
WANT ANOTHER MULTIPLE LINEAR REGRESSION(YES OR NO)?  NO
```

FIGURE 10-2 Computer run for multiple regression, using sample student data.

made from the estimated regression equation

$$Y_C = -.386080 + .637X_1 + .002X_2 - .007X_3$$

Note that $b_3 = -.007$ is a small negative number and that the remaining regression coefficients have not changed much at all from the previous multiple regression before X_3 was incorporated into the analysis. This reflects the facts that X_3 correlates weakly with Y, $r_{Y3} = .075$, and (as may be established from additional computer runs) that X_3 exhibits tiny correlations with the other independent variables. Its worth as a predictor variable has not been supported by the sample data, and we may conclude that extracurricular activities in high school have little to do with a student's college performance.

As further evidence that X_3 is a poor predictor variable, the computer printout in Figure 10-2 shows that the resulting standard error of the estimate rounds to $S_{Y \cdot 123} = .117$, a higher figure than $S_{Y \cdot 12} = .111$ found in our earlier analysis. So the inclusion of high-school activities here only clouds our regression analysis, yielding a smaller amount of explainable variation in Y than before.

The printout contains more information than we have explained here. Also included is the *sample multiple correlation coefficient*, which will be explained later. Some intermediate data that would be necessary for hand computation are missing, and the printout gives fewer significant figures for the estimated regression coefficients than is usually desirable. In general, a canned program written by somebody else will give more in one way, and at the same time, less in another than is desired.

We will not give a detailed compilation of sources for canned multiple regression programs. These may be obtained from local computer centers or data processing advisors. Two good reasons not to provide such programs here are that they are rapidly changing and that they are totally dependent upon the particular kind of computer hardware available.

EXERCISES

10-6 For the family data given in the table below:

(a) Determine the coefficients for the estimated regression line $\bar{Y}_{X_1} = a + bX_1$ and calculate $s_{Y \cdot X_1}$.

Family	Total Spending Y	Income X_1	Size X_2	Additional Savings X_3
A	8,000	10,000	3	1,000
B	7,000	10,000	2	1,500
C	7,000	9,000	2	700
D	12,000	16,000	4	1,800
E	6,000	7,000	6	200
F	7,000	9,000	4	500
G	8,000	8,000	6	0
H	7,000	8,000	5	100
I	10,000	12,000	6	200
J	8,000	11,000	2	1,000

(b) Determine the coefficients for the estimated regression plane $Y_C = a + b_1 X_1 + b_2 X_2$ and calculate $S_{Y \cdot 12}$.

(c) Comparing the values $S_{Y \cdot 12}$ and $s_{Y \cdot X_1}$, do these suggest that including family size has been worthwhile in predicting total spending? Explain.

10-7 *Computer exercise:* Using the data in the table in Exercise 10-6:

(a) Determine the coefficients for the estimated regression hyperplane

$$Y_C = a + b_1 X_1 + b_2 X_2 + b_3 X_3$$

(b) Determine the value for $S_{Y \cdot 123}$.

(c) Compare the values of $S_{Y \cdot 123}$ and $S_{Y \cdot 12}$ (from Exercise 10-6). Does your comparison suggest that including the additional savings variable has proved worthwhile in predicting total spending? Explain your answers.

10-4 MULTIPLE CORRELATION

Sample Coefficient of Multiple Determination

The concept of correlation may be extended to multiple variables. As with the simple linear relationships discussed earlier in this chapter, we will begin by describing the *sample coefficient of multiple determination* as an index of association. When the multiple regression involves only two independent variables, X_1 and X_2, this is denoted symbolically by $R^2_{Y \cdot 12}$ and represents the ratio of variation in Y explained by the regression plane to the total variation:

$$R^2_{Y \cdot 12} = \frac{\text{Explained variation}}{\text{Total variation}} = 1 - \frac{\sum (Y - Y_C)^2}{\sum (Y - \bar{Y})^2} \qquad \text{(10-10)}$$

The *sample multiple correlation coefficient* is defined to be the square root of the coefficient of multiple determination, $R_{Y \cdot 12} = \sqrt{R^2_{Y \cdot 12}}$. The sign of $R_{Y \cdot 12}$ is always considered positive.

A shorter, equivalent version of expression (10-10) is used* to calculate the *sample coefficient of multiple determination*

$$R^2_{Y \cdot 12} = 1 - \frac{S^2_{Y \cdot 12}}{s^2_Y}\left(\frac{n-m}{n-1}\right) \qquad \text{(10-11)}$$

As before, m represents the number of variables used, and here $m = 3$.

* $R^2_{Y \cdot 12}$ is said to be unadjusted for degrees of freedom. The adjusted coefficient of multiple determination may be derived from

$$R^2_{aY \cdot 12} = 1 - \frac{S^2_{Y \cdot 12}}{s^2_Y}$$

For large n, $R^2_{aY \cdot 12}$ is nearly the same as $R^2_{Y \cdot 12}$.

To illustrate this computation, we again use the results found earlier for the multiple regression for college GPA Y when high-school GPA X_1 and verbal SAT score X_2 are the only independent variables. Then, we found

$$S_{Y \cdot 12} = .111$$

$$s_Y = .478$$

Thus

$$R^2_{Y \cdot 12} = 1 - \frac{(.111)^2}{(.478)^2}\left(\frac{10-3}{10-1}\right) = .958$$

The result $R^2_{Y \cdot 12} = .958$ indicates that the regression plane

$$Y_C = -.412 + .631X_1 + .00223X_2$$

explains 95.8 percent of the variation in Y.

We now have a basis for comparing the multiple regression to the earlier simple regression (performed in Chapter 9), where we found the simple coefficient of determination to be $r^2_{Y1} = .723$. Regression involving only Y and X_1 explains just 72.3 percent of the variation in college GPA. Including verbal SAT score in the analysis considerably improves the predictive power of the least squares regression.

Partial Correlation

To quantify the improvement achieved with a higher-dimensional regression analysis, we can calculate the proportional reduction in previously explained variation. This provides a further index of association, $r^2_{Y2 \cdot 1}$, called the *coefficient of partial determination*, that measures the correlation between Y and X_2, while the other independent variable X_1 is still considered but held constant. The subscript $Y2 \cdot 1$ indicates this. The coefficient is calculated from

$$r^2_{Y2 \cdot 1} = \frac{\text{Reduction in unexplained variation}}{\text{Previously unexplained variation}} = \frac{R^2_{Y \cdot 12} - r^2_{Y1}}{1 - r^2_{Y1}} \qquad (10\text{-}12)$$

Substituting the values found earlier into expression (10-12) gives us

$$r^2_{Y2 \cdot 1} = \frac{.958 - .723}{1 - .723} = .848$$

The difference, $R^2_{Y \cdot 12} - r^2_{Y1}$, represents the reduction in unexplained variation (which is also the increase in explained variation). The value $r^2_{Y2 \cdot 1} = .848$ tells us that 84.8 percent of the variation in college GPA Y that was left unexplained by a simple regression with high-school GPA X_1 alone can be explained by the regression plane obtained from including verbal SAT score X_2.

In multiple regression analysis, the coefficient of partial determination or its square root, the *partial correlation coefficient*, provides an index of the correlation between two variables *after* the effects of all the other variables have been considered. Thus, the value .848 expresses the net association between college GPA Y and verbal SAT score X_2 while high-school GPA X_1 is held constant but is still accounted for.

The value $r^2_{Y2\cdot 1} = .848$ may be compared to the *simple* coefficient of determination between Y and X_2, $r^2_{Y2} = .667$. The former provides a truer impact of including X_2 in the multiple regression analysis than would be indicated by a simple correlation analysis using only X_2. When evaluating the potential merits of a predictor variable, the coefficients of partial determination (or the partial correlation coefficients) are sometimes employed in determining which independent variables to include (a superior method to using the simple counterpart coefficient). We can illustrate how the coefficient of partial determination may be used to do this.

In discussing a third independent variable X_3 (the number of high-school activities) a multiple regression was run on the computer. Earlier in this illustration, we indicated X_3 is a poor predictor of college GPA. The coefficient of partial determination for Y and X_3, denoted by $r^2_{Y3\cdot 12}$, since X_1 and X_2 remain fixed, may be obtained by using the coefficient of multiple determination found for that regression in Figure 10-2:

$$R^2_{Y\cdot 123} = (.979889)^2 = .960$$

The partial coefficient of determination may be found from

$$r^2_{Y3\cdot 12} = \frac{\text{Reduction in unexplained variation}}{\text{Previously unexplained variation}} = \frac{R^2_{Y\cdot 123} - R^2_{Y\cdot 12}}{1 - R^2_{Y\cdot 12}} \tag{10-13}$$

Substituting the multiple coefficients of determination obtained from the previous two multiple regressions using the student data, we have

$$r^2_{Y3\cdot 12} = \frac{.960 - .958}{1 - .958} = .048$$

The value .048 means that by including X_3 in the analysis, there is only a 5-percent reduction in the variation in Y left unexplained by the earlier multiple regression employing only X_1 and X_2. Even including the relationship between X_3 and the other two independent variables, the value $r^2_{Y3\cdot 12} = .048$ indicates practically no correlation between Y and X_3. Because this value is so small, we may conclude that including the number of extracurricular high-school activities as one of the independent variables contributes very little to the multiple regression analysis.

EXERCISES

10-8 The following intermediate calculations have been made by an economist using sample data relating annual family spending Y to income X_1, size X_2, and annual savings X_3. A sample of size $n = 100$ was used, and all monetary figures are in thousands of dollars.

$$\bar{Y} = 10 \qquad \bar{X}_1 = 12 \qquad \bar{X}_2 = 5 \qquad \bar{X}_3 = 1$$

$$\sum Y^2 = 11{,}400 \qquad \sum X_1 Y = 13{,}000 \qquad \sum X_2 Y = 6{,}000 \qquad \sum X_3 Y = 500$$

Plane A: $Y_C = 1.7 + .4X_1 + .7X_2$ (using only X_1 and X_2)

Plane B: $Y_C = 4.3 + .3X_1 + .6X_2 - .9X_3$ (using X_1, X_2, and X_3)

(a) Calculate $S_{Y.12}$ and $R^2_{Y.12}$. What percentage of the variation in Y may be explained by the regression plane A?
(b) Suppose that $S_{Y.123} = .722$ and $R^2_{Y.123} = .964$. What percentage of the variation in Y may be explained by the regression hyperplane B?
(c) Has the standard error of the estimate for Y been reduced by the inclusion of X_3 in the regression analysis?
(d) Determine the proportions of unexplained variation in Y by using the regression equations for planes A and B. What proportional change in previously unexplained variation is achieved by adding X_3 to the analysis? Does this create an increase or a reduction in the unexplained variation?

10-9 Suppose that the economist in Exercise 10-8 has determined for *another sample* that the simple coefficient of determination between Y and X_1 is .50 and that it is .60 between Y and X_2. He has also found the coefficient of multiple determination to be .75, considering only X_1 and X_2.
(a) Calculate the coefficients of partial determination $r^2_{Y1.2}$ and $r^2_{Y2.1}$.
(b) State verbally the meaning of the values you obtained in (a).
(c) Compare the values in (a) to their respective simple coefficients of determination. Why may the corresponding values differ?

10-5 STEPWISE MULTIPLE REGRESSION

Modern data centers have computer programs in their libraries that can be used to handle a variety of statistical problems. One group of such programs is employed in *stepwise multiple regression*. In this procedure, additional predictor variables—one at a time in successive stages—raise the dimensions of the analysis by one. A great many independent variables can be handled on the computer using this procedure. One such program selects the most promising possible independent variable—the one that provides the greatest reduction in the unexplained variation in Y—at each stage. In doing this, the computer performs simple regression separately for each independent variable, printing the results for the best one. The next step of the program performs separate multiple regressions, each combining one of the remaining independent variables with those selected in the previous stages. Again, the regression that reduces unexplained variation the most is permanently included in all future stages. The process continues in successively higher dimensions either until every variable has been included in a multiple regression involving all variables or until no further reduction in the unexplained variation is possible. Such a program efficiently saves all the previous neccessary calculations for higher-dimensional analyses.

The output from such a program can be rich—providing at each step the various coefficients of multiple and partial determination. In addition, this type of program may provide simple correlations for all variable pairs.

Many assumptions concerning the linearity of the data are automatically made by these computer programs. As with any other application, in multivariate analysis, the computer does not eliminate a need for good judgment. It is best used in the intermediate stages of analysis. Computer printouts must be thoroughly evaluated to determine if there is a meaningful explanation as to why some variables are excluded and others are included. A variable may be rejected because it does not reduce the unexplained variation that is actuallv due to a strong curvilinear relation between the variable and Y.

11 Inferences Using Two Samples

The great tragedy of science—the slaying of a beautiful hypothesis by an ugly fact.

Thomas Huxley

A very important area of statistical inference involves using data from two samples, each representing a different population. A comparison of the two sample means may lead to the conclusion that one population tends to have larger values than the other. Two-sample inferences are important in a great many experiments where the impact of a change is being assessed. For example, a medical researcher may compare the effectiveness of a new treatment to that of an existing procedure. Thus, an experimental chemotherapy agent for leukemia may be compared to an earlier agent in terms of their respective remission rates. Two separate samples—one for each drug—would ordinarily be required, because this controls treatment conditions by allowing the researcher to limit any explanation of potential findings to differences in drugs rather than to other factors that could influence recovery.

Generalizations from two samples may take the form of an estimate or a test. Estimates may be made of the differences in the respective population means, proportions, or variances. Two-sample hypothesis testing procedures conclude either that one population is superior to the other or that the differences between the two are not significant enough to warrant this conclusion. Two-sample procedures follow the same dichotomies we identified in Chapter 8: one-sided versus two-sided tests; using the proportion versus using the mean; and small versus large samples. In addition to these, the existence of two populations may involve a further choice: whether to use independent or matched-pairs samples. As we saw in Chapter 3, matching the sample observations usually results in a considerable reduction in the number of observations that might otherwise be required.

11-1 CONFIDENCE INTERVAL FOR THE DIFFERENCE IN MEANS USING LARGE SAMPLES

The simplest two-sample inference involves estimating the difference between two population parameters. We begin by constructing a confidence interval estimate of the difference between the two population means.

For convenience, we will designate the populations by the letters A and B. Following this notation, ***the populations are represented as***

$$\mu_A = \text{Mean of population } A$$

$$\mu_B = \text{Mean of population } B$$

Which population corresponds to a particular letter is arbitrary. Usually, A designates the population presumed to have the larger mean. ***The difference in population means is***

$$D = \mu_A - \mu_B \tag{11-1}$$

This value may be estimated from the sample results. There are two ways of doing this, depending upon whether the observations are made independently or by matching pairs.

Independent Samples

The simplest case to consider is when the sample observations are selected independently. Sample means from the respective populations may be designated $\bar{X}_A$ and $\bar{X}_B$. ***The difference between sample means is***

$$d = \bar{X}_A - \bar{X}_B \tag{11-2}$$

This statistic is an unbiased estimator of the difference in population means D. The samples used in calculating expression (11-2) may differ in size, and the number of observations taken for the respective samples are represented by the symbols n_A and n_B.

As we have seen, the sampling distributions of $\bar{X}_A$ and $\bar{X}_B$ may each be approximated by the normal curve for sufficiently large samples. Because the samples are chosen independently, it may be established that *d will also be* *approximately normally distributed, with a mean of $D = \mu_A - \mu_B$ and a standard deviation* of*

$$\sigma_d = \sqrt{\frac{\sigma_A^2}{n_A} + \frac{\sigma_B^2}{n_B}} \tag{11-3}$$

where σ_A^2 and σ_B^2 are the respective population variances. Ordinarily, when μ_A and μ_B are of unknown value, σ_A^2 and σ_B^2 will also be unknown. The respective sample variances s_A^2 and s_B^2 may be used as estimators of these parameters. For *large samples* (in general, 30 or more observations for each), we may thereby obtain

$$s_d = \sqrt{\frac{s_A^2}{n_A} + \frac{s_B^2}{n_B}} \tag{11-4}$$

We are now ready to construct the confidence interval estimate for the difference in population means. We select a normal deviate z corresponding to the desired confidence level, so that we may estimate the difference in population means by the interval

$$D = d \pm z s_d \tag{11-5}$$

Since D and d are differences in means, a more convenient expression of ***the confidence interval for the difference in population means when using independent samples is***

$$\mu_A - \mu_B = \bar{X}_A - \bar{X}_B \pm z\sqrt{\frac{s_A^2}{n_A} + \frac{s_B^2}{n_B}} \tag{11-6}$$

Example 11-1 A statistics professor wishes to compare a new textbook to his old one. He teaches two sections of the same course. The professor continues to use the old book in section B, while he uses the new book in section A. Both sections of his course contain 30 students, and the respective classes are conducted in nearly identical fashions. Since the quality of the new book is being evaluated, we may refer to those students in section A as the experimental group and to those in section B as the control group. The students have been randomly assigned to Section A or B, and each may be viewed as a sample of all students taking statistics in the professor's classes.

The effectiveness of each textbook can be measured in terms of the combined examination scores achieved by the students on common tests given throughout the term. Population A represents the potential scores of all students at the university who might use the new textbook in a course taught by this particular

* Here, we use the fact that $\bar{X}_A$ and $\bar{X}_B$ are independent random variables, so that the variance of their sum or difference must equal the sum of the individual variances:

$$\sigma_d^2 = \sigma^2(\bar{X}_A - \bar{X}_B) = \sigma^2(\bar{X}_A) + \sigma^2(\bar{X}_B) = \frac{\sigma_A^2}{n_A} + \frac{\sigma_B^2}{n_B}$$

professor; population B represents the potential scores these same students would receive using the old textbook. Both sets of scores are target populations, because not all students will have the same statistics professor, use the same books, or take identical examinations.

TABLE 11-1 Sample Results for Students in Two Classes Using New and Old Textbooks.

Students Using New Book (A)			*Students Using Old Book (B)*		
Student Initials	*Combined Exam Score* X_A	X_A^2	*Student Initials*	*Combined Exam Score* X_B	X_B^2
C.A.	95	9,025	C.A.	85	7,225
L.A.	87	7,569	J.A.	74	5,476
A.B.	84	7,056	O.A.	59	3,481
I.B.	79	6,241	P.A.	71	5,041
P.B.	78	6,084	W.A.	48	2,304
B.C.	93	8,649	D.B.	78	6,084
M.C.	72	5,184	D.C.	94	8,836
D.D.	92	8,464	M.C.	75	5,625
E.E.	89	7,921	B.E.	92	8,464
S.F.	68	4,624	F.E.	85	7,225
T.G.	73	5,329	J.F.	76	5,776
V.G.	53	2,809	S.G.	69	4,761
G.I.	81	6,561	G.H.	82	6,724
F.L.	90	8,100	F.J.	76	5,776
M.L.	77	5,929	J.K.	85	7,225
N.L.	86	7,396	S.L.	70	4,900
G.M.	83	6,889	K.M.	69	4,761
H.M.	91	8,281	M.M.	71	5,041
T.M.	68	4,624	D.N.	80	6,400
N.N.	81	6,561	A.O.	91	8,281
Q.P.	75	5,625	L.P.	77	5,929
P.R.	76	5,776	A.S.	90	8,100
P.S.	64	4,096	B.S.	88	7,744
S.S.	63	3,969	T.S.	60	3,600
W.S.	61	3,721	W.S.	47	2,209
M.T.	83	6,889	L.T.	74	5,476
S.T.	72	5,184	N.T.	63	3,969
W.T.	69	4,761	T.T.	63	3,969
T.W.	59	3,481	L.V.	71	5,041
W.W.	53	2,809	P.W.	60	3,600
	2,295	179,607		2,223	169,043

The sample results are provided in Table 11-1. These data will be used to estimate the difference in mean population scores, $D = \mu_A - \mu_B$. For this study, $n_A = n_B = 30$. (It is not necessary for the two samples to be of the same size, even though they happen to be in our example.) The following statistics are obtained:

$$\bar{X}_A = 2{,}295/30 = 76.5 \qquad \bar{X}_B = 2{,}223/30 = 74.1$$

$$s_A^2 = \frac{179{,}607 - 30(76.5)^2}{29} = 139.29 \qquad s_B^2 = \frac{169{,}043 - 30(74.1)^2}{29} = 148.92$$

The professor wants to construct a 95-percent confidence interval estimate for $\mu_A - \mu_B$. The required normal deviate is $z = 1.96$, and from expression (11-6), we have

$$\mu_A - \mu_B = 76.5 - 74.1 \pm 1.96\sqrt{\frac{139.29}{30} + \frac{148.92}{30}}$$

$$= 2.4 \pm 6.08$$

or

$$-3.68 \leqslant \mu_A - \mu_B \leqslant 8.48$$

Thus, the mean difference in scores is estimated to lie between -3.68 and 8.48. The limit -3.68 indicates that the *disadvantage* in mean scores using book A might be 3.68 points (so that book B might be better), while 8.48 tells us that the *advantage* in mean scores using book A might be as great as 8.48 points.

The interpretation of the above interval is that were this experiment to be repeated several times, then this procedure would yield interval estimates containing the true $\mu_A - \mu_B$ difference about 95 percent of the time. Note that this estimate is not very precise and does not clearly indicate much difference between the effectiveness of the two books. Next, we will investigate another procedure which we might have used on this sample.

Matched-Pairs Sampling

A *matched-pairs sample* is obtained by making each population A observation in such a way that the selected elementary unit is matched with a "twin" from population B. In a medical experiment, the patient in the control group would have an experimental partner of the same sex, with a similar occupation, family environment, and medical history, and close in age, weight, height, build, and other physiological characteristics. In educational testing, partners would be matched by aptitudes and intelligence, family background, socioeconomic level, and achievement. In every case, matching would be accomplished in terms of those factors that might influence the characteristic being measured.

The individual differences for each pair can be used as the basis for estimating population differences. Matched-pairs sampling thus attempts to explain the differences in individual pairs in terms of differences due to the factor studied, while minimizing the variability in observations that may result from extraneous influences. The net effect is to reduce the impact of sampling error and thus to obtain greater sampling efficiency.

To illustrate how matched pairs may be obtained, we will expand the sampling study in Example 11-1. Suppose that the professor in our example decides to match students in the two classes. His matching criteria should be selected so that each student pair is closely alike in aptitude for statistics and in achievement in related areas. Since the students *should be matched before actual sample observations are made*—a safeguard that minimizes bias in selecting actual pairs—he cannot use the two classes' performances on statistics tests for this purpose.

The professor may use quantitative SAT scores as a matching criterion for statistical aptitude. These could easily be obtained from the registrar's office for the enrolled students. As a measure of achievement, he might use the students' grades in the prerequisite mathematics course—data also obtainable from

TABLE 11-2 Matching of Students in Terms of Mathematics Grades and Quantitative SAT Scores.

	Students Using New Book (A)			*Students Using Old Book (B)*		
Pair	*Student Initials*	*Mathematics Grade*	*Quantitative SAT*	*Student Initials*	*Mathematics Grade*	*Quantitative SAT*
1	A.B.	A	706	A.O.	A	713
2	B.C.	A	702	A.S.	A	698
3	C.A.	A	674	B.E.	A	685
4	D.D.	A	660	B.S.	A	659
5	E.E.	A	622	C.A.	A	610
6	F.L.	B	685	D.C.	A	507
7	G.M.	B	683	D.N.	B	671
8	G.I.	B	623	D.B.	B	654
9	H.M.	B	583	F.J.	B	612
10	I.B.	B	582	F.E.	B	596
11	L.A.	B	581	G.H.	B	575
12	N.L.	B	525	J.K.	B	533
13	M.T.	B	489	J.F.	B	477
14	M.L.	B	425	J.A.	B	454
15	M.C.	C	523	K.M.	B	426
16	N.N.	C	512	L.P.	C	544
17	O.P.	C	510	L.T.	C	523
18	P.R.	C	468	L.V.	C	481
19	P.B.	C	455	M.C.	C	454
20	P.S.	C	421	M.M.	C	409
21	S.F.	C	411	N.T.	C	408
22	S.S.	C	394	O.A.	C	402
23	S.T.	C	359	P.A.	C	383
24	T.M.	C	342	P.W.	C	336
25	T.G.	C	326	S.L.	C	326
26	T.W.	C	308	S.G.	C	315
27	W.S.	C	295	T.S.	D	435
28	W.T.	D	421	T.T.	D	386
29	V.G.	D	351	W.A.	D	321
30	W.W.	D	288	W.S.	D	317

student records. Suppose that the professor uses these indicators to match his students. Table 11-2 shows one matching assignment, using mathematics grades as the main ranking device and SAT scores to obtain a finer sorting of students with identical mathematics grades. In terms of these criteria, two "A" students having respective SAT scores of 706 and 713 constitute the first pair; similarly, two "C" students scoring 455 and 454 are the nineteenth pair.

The observations may be designated by the symbols

X_{A_i} = Sample value for population A partner in ith matched pair

X_{B_i} = Sample value for population B partner in ith matched pair

For each sample couple, we may find ***the matched-pair difference***

$$d_i = X_{A_i} - X_{B_i} \tag{11-7}$$

(d_i should not be confused with the unsubscripted d used earlier, which expresses the difference between the means of two independent samples rather than individual pairs.)

In the usual manner, we find ***the mean of the matched-pair differences***

$$\bar{d} = \frac{\sum d_i}{n} \tag{11-8}$$

where n is the *number of pairs* observed in the sampling study. Analogously, we determine ***the standard deviation of the matched-pairs differences***

$$s_{d\text{-paired}} = \sqrt{\frac{\sum d_i^2 - n\bar{d}^2}{n-1}} \tag{11-9}$$

Here $\bar{d}$ represents the mean of n separate matched-pair differences and has an expected value of $D = \mu_A - \mu_B$. This is the same fact noted for the d discussed previously for differences in means of entire independent samples. Both d and $\bar{d}$ may be used to draw inferences about the difference between the means of populations A and B. However, each matched-pair difference d_i can be treated like a single observation, so that $\bar{d}$ has the same properties described in earlier chapters for $\bar{X}$. Thus, the central limit theorem indicates that for a sufficiently large sample (usually 30 or more), $\bar{d}$ *is approximately normally distributed, with a mean of D and a standard deviation estimated from*

$$\frac{s_{d\text{-paired}}}{\sqrt{n}}$$

This makes it possible to select a normal deviate z corresponding to a desired confidence level, so that we may estimate $D = \mu_A - \mu_B$ by ***the confidence interval for the difference in population means when matched pairs are used***

$$\mu_A - \mu_B = \bar{d} \pm z\frac{s_{d\text{-paired}}}{\sqrt{n}} \tag{11-10}$$

Returning to the statistics professor's estimation problem of the difference in population mean scores using the new and old books, the values for the matched-pair differences appear in Table 11-3. For $n = 30$ pairs, the professor obtained the following results:

$$\bar{d} = 2.4 \qquad s_{d\text{-paired}} = 5.01$$

TABLE 11-3 Calculation of Matched-Pair Differences for Student Scores Using New and Old Books

	Student Pair		Combined Exam Scores			
i	Group A Student	Group B Student	Group A X_{A_i}	Group B X_{B_i}	Difference d_i	d_i^2
1	A.B.	A.O.	84	91	−7	49
2	B.C.	A.S.	93	90	3	9
3	C.A.	B.E.	95	92	3	9
4	D.D.	B.S.	92	88	4	16
5	E.E.	C.A.	89	85	4	16
6	F.L.	D.C.	90	94	−4	16
7	G.M.	D.N.	83	80	3	9
8	G.I.	D.B.	81	78	3	9
9	H.M.	F.J.	91	76	15	225
10	I.B.	F.E.	79	85	−6	36
11	L.A.	G.H.	87	82	5	25
12	N.K.	J.K.	86	85	1	1
13	M.T.	J.F.	83	76	7	49
14	M.L.	J.A.	77	74	3	9
15	M.C.	K.M.	72	69	3	9
16	N.N.	L.P.	81	77	4	16
17	Q.P.	L.T.	75	74	1	1
18	P.R.	L.V.	76	71	5	25
19	P.B.	M.C.	78	75	3	9
20	P.S.	M.M.	64	71	−7	49
21	S.F.	N.T.	68	63	5	25
22	S.S.	O.A.	63	59	4	16
23	S.T.	P.A.	72	71	1	1
24	T.M.	P.W.	68	60	8	64
25	T.G.	S.L.	73	70	3	9
26	T.W.	S.G.	59	69	−10	100
27	W.S.	T.S.	61	60	1	1
28	W.T.	T.T.	69	63	6	36
29	V.G.	W.A.	53	48	5	25
30	W.W.	W.S.	53	47	6	36
			2,295	2,223	72	900

$$\bar{d} = \frac{\sum d_i}{n} = \frac{72}{30} = 2.4$$

$$s_{d\text{-paired}} = \sqrt{\frac{\sum d_i^2 - n\bar{d}^2}{n-1}} = \sqrt{\frac{900 - 30\,(2.4)^2}{29}}$$

$$= \sqrt{25.0759} = 5.01$$

(Note that 2.4 is the same point etsimate for $\mu_A - \mu_B$ that we found earlier using the same data as independent samples.) A 95-percent confidence interval estimate for $\mu_A - \mu_B$ may be constructed, using expression (11-10) and the normal deviate value $z = 1.96$:

$$\mu_A - \mu_B = 2.4 \pm 1.96\left(\frac{5.01}{\sqrt{30}}\right)$$

$$= 2.4 \pm 1.79$$

or

$$.61 \leqslant \mu_A - \mu_B \leqslant 4.19$$

This confidence interval is more precise than the previous one using the student test data as independent samples. A comparison of these two procedures follows.

Matched Pairs Compared to Independent Samples

Which procedure—independent sampling or matched pairs—is best? As a first step in answering this question, we might compare the sample sizes necessary to provide identical levels of estimation reliability for the same tolerable error.

Generally, the comparative advantage of matched-pairs sampling depends upon how closely the matching scheme used correlates population A values with population B values. A good matching criterion should eliminate most extraneous sources of variation, so this correlation would ordinarily be quite high.

A sample-size comparson is easiest to make when there are just as many observations from A as from B, so that $n_A = n_B$. We may also presume that population variabilities are similar, so that σ_A and σ_B are approximately equal. Under these conditions, it is possible to mathematically prove that *matched-pairs sampling provides a reduction in sample size from the sample size required by independent sampling, and this reduction is proportional to the population correlation coefficient*. Thus, if A and B have a correlation coefficient of .9, then a 90-percent savings in sample size is possible by using matched pairs; that is, 10 times as many observations are needed for independent samples to yield the same precision and reliability obtained by using matched pairs. Should the correlation coefficient be .99, then independent sampling requires 100 times as many observations as matched-pairs sampling.

Returning to our original question, we might wonder why independent samples are even used at all. In part, the prevalence of independent sampling is both because pair matching is often difficult and because each observation takes a great deal of time. Sometimes it is even impossible to advantageously match elementary units. Successful matching generally requires a large amount of available data on every unit; not only must these data be studied just to establish relevant matching criteria—perhaps using multiple regression and correlation techniques—but the data must then be carefully applied when the sample units are paired. Once the matched-pairs machinery has been set in motion, further sampling costs are comparable, observation by observation, to those of independent sampling. For a study involving continued monitorship over a long period of time, matched-pairs samples have a distinct cost advantage. This makes them preferred in medical studies, where the fixed cost of establishing the sample pairs is small in comparison to total expenses.

EXERCISES

11-1 A psychometrist wishes to estimate the difference in mean IQ for students at two different schools in her district. Two independent random samples selected from each school produced the following data:

Woodwillow School	*Grant Elementary*
$n_A = 100$	$n_B = 150$
$\bar{X}_A = 111.4$	$\bar{X}_B = 108.3$
$s_A^2 = 334$	$s_B^2 = 397$

Construct a 95-percent confidence interval for the difference in population means.

11-2 A government meteorologist wishes to find the difference in mean annual tornado intensity (verified annual number of occurrences per 10,000 square miles) for two Midwestern states. Assuming that past tornado events over a ten-year period in the two states represent independent random samples of long-run weather patterns, he has plotted all reported tornadoes on a map and obtained the following intensity data:

State A	*State* B
$n_A = 96$	$n_B = 114$
$\bar{X}_A = .53$	$\bar{X}_B = .34$
$s_A^2 = 1.2$	$s_B^2 = 1.5$

Construct a 99-percent confidence interval for the difference in mean annual tornado intensity.

11-3 A Federal Aviation Agency statistician must determine how two airlines differ in their ability to meet flight schedules. From routes common to both airlines, he has randomly selected 100 actual arrival times for each. Each observation has been matched not only by route but by date and by time of day. From these data, the number of minutes late in every matched pair was determined for each airline (negative values were used for early arrivals). The sample mean and the standard deviation for the differences (the late time for airline *A* minus the late time for airline *B*) were determined as

$$\bar{d} = 2 \text{ minutes}$$

$$s_{d\text{-paired}} = 3 \text{ minutes}$$

Construct a 95-percent confidence interval estimate for the difference in mean amount of late time for the two airlines.

11-4 A city manager wants to know what gasoline savings would be made by converting the tires on all city cars. New tires were ordered for 200 cars; half of these cars used steel radial tires and the rest were equipped with the standard cord variety. Cars in the two groups were matched by make, model, age, use (police, building inspection, or whatever), and general condition. A month-long test revealed that the mean decrease in gasoline mileage when radial tires were used was 2 miles per gallon. The standard deviation in gasoline savings for the paired cars was also 2 miles per gallon. Construct a 95-percent confidence interval estimate for the savings advantage in mean gasoline mileage that radial tires would create if they were adopted for all city cars.

11-2 HYPOTHESIS TEST FOR COMPARING TWO MEANS USING LARGE SAMPLES

Rather than estimate the difference between two population means, it may be desirable to apply hypothesis-testing procedures to determine whether or not population A values are significantly larger or smaller than population B values. These methods would permit choices to be made. Thus, an experimental reading program may be adopted because the statistical test concludes that students using it attain a higher mean achievement score than those using the existing program. A new drug may be rejected by a hospital staff because it does not have a mean relief level which significantly exceeds the present one. A police department official may find sample results indicating that a new tire supplier should be awarded a contract because his tires wear longer than the current brand. In each of these cases, sample data are required for both the control and the experimental groups.

In two-sample tests, the null hypothesis may take any one of the following forms:

$$H_0 : \mu_A \geqslant \mu_B \text{ or } H_0 : D \geqslant 0 \text{ (lower-tailed test)}$$

$$H_0 : \mu_A \leqslant \mu_B \text{ or } H_0 : D \leqslant 0 \text{ (upper-tailed test)}$$

$$H_0 : \mu_A = \mu_B \text{ or } H_0 : D = 0 \text{ (two-sided test)}$$

The equivalent expressions on the right are found by subtracting μ_B on both sides of the original inequalities and by using the fact that $D = \mu_A - \mu_B$. As in estimating the difference between two means, two-sample hypothesis tests may be applied to independent samples as well as to matched pairs.

Independent Samples

When samples are selected independently, the appropriate test statistic is $d = \bar{X}_A - \bar{X}_B$. As we have seen, for sufficiently large samples d is approximately normally distributed, with a mean of D and a standard deviation of σ_d. We must choose a critical value for d, which we will denote as d^*, that provides a decision rule corresponding to the desired significance level. When population A values are presumed larger, so that $\mu_A \geqslant \mu_B$ and the null hypothesis is $H_0: D \geqslant 0$, extremely negative values for d will refute H_0. The type I error probability α is then represented by a lower-tailed area under the normal curve for d. Therefore, the decision rule takes the form:

$$\text{Accept } H_0 \text{ if } d \geqslant d^*$$

$$\text{Reject } H_0 \text{ if } d < d^*$$

In finding d^*, we use the fact that under H_0 the normal curve for d must be centered at $D = 0$. The critical normal deviate for the test is z_α, and d^* is therefore located z_α standard deviations away from this mean. Estimating σ_d by s_d, we have

$$d^* = z_\alpha s_d \tag{11-11}$$

Substituting the terms in expression (11-4) for s_d, we have ***the critical value for the difference in sample means***

$$d^* = z_\alpha \sqrt{\frac{s_A^2}{n_A} + \frac{s_B^2}{n_B}} \tag{11-12}$$

Should the values of the population variances be known in advance of sampling, then σ_A^2 and σ_B^2 can be used in place of s_A^2 and s_B^2 in expression (11-12).

Example 11-2 In evaluating the effectiveness of a new automobile engine in controlling exhaust emissions, the U.S. Environmental Protection Agency might conduct a sampling study like the one we will describe here. The experimental group (A) consisted of $n_A = 50$ prototype rotary engines of a type proposed to replace conventional engines. The control group (B) contained $n_B = 100$ piston engines equipped with catalytic converters. (Note that different sample sizes were used for the two groups.)

The EPA wishes to use the null hypothesis that the mean sulfur-dioxide emissions level in parts per million (ppm) from all rotary engines is at least as great as that from the current engines, $H_0: \mu_A \geqslant \mu_B$. Since rejection of this hypothesis will result in permitting a manufacturer to install the rotary engine in all of its cars, the EPA has established an $\alpha = .01$ significance level for the test; that is, there is a one-percent chance of making the incorrect decision and concluding that rotary engines are cleaner than piston engines when the opposite is true.

The sample results for sulfur-dioxide emissions were

Rotary Engine	*Piston Engine*
$\bar{X}_A = 25$ ppm $s_A^2 = 90$	$\bar{X}_B = 30$ ppm $s_B^2 = 100$

Substituting $z_{.01} = -2.33$ (a negative quantity because the test is lower-tailed) and the above results into expression (11-12), the critical value for the difference in sample means is computed as

$$d^* = -2.33\sqrt{\frac{90}{50}+\frac{100}{100}} = -3.90$$

The agency's decision rule for applying $d = \bar{X}_A - \bar{X}_B$ is therefore

Accept H_0 (refuse rotary engine) if $d \geqslant -3.90$

Reject H_0 (permit rotary engine) if $d < -3.90$

Since the actual difference in sample means is

$$d = \bar{X}_A - \bar{X}_B = 25 - 30 = -5$$

which is a value smaller than $d^* = -3.90$, the EPA must *reject* H_0 and allow the manufacturer to substitute the rotary engine.

The calculation of d^* may be bypassed, using the fact that $z = d/s_d$, so that we may calculate ***the normal deviate for the sample results***

$$z = \frac{\bar{X}_A - \bar{X}_B}{\sqrt{\frac{s_A^2}{n_A}+\frac{s_B^2}{n_B}}} \tag{11-13}$$

We may then compare this value to z_α. In the preceding example

$$z = \frac{25-30}{\sqrt{\frac{90}{50}+\frac{100}{100}}} = -2.99$$

which is smaller than $z_{.01} = -2.33$, and again we see that H_0 must be rejected.

When population A values are presumed smaller and $\mu_A \leqslant \mu_B$, so that $H_0 : D \leqslant 0$, the test is upper-tailed. The procedure is analogous to the one outlined above, except that z_α is positive and large positive values for d or z result in rejecting H_0. For a two-sided test, where it is assumed that $\mu_A = \mu_B$ and that the null hypothesis is $H_0 : D = 0$, the critical normal deviate is $z_{\alpha/2}$ (corresponding to an upper-tail area equal to one-half the desired significance level) and the critical values d_1^* and d_2^* may be found, so that our rule is

Accept H_0 if $d_1^* \leqslant d \leqslant d_2^*$

Reject H_0 either if $d < d_1^*$ *or* if $d > d_2^*$

The following computation provides ***the critical values for the difference in sample means***

$$d_1^* = -z_{\alpha/2}\sqrt{\frac{s_A^2}{n_A}+\frac{s_B^2}{n_B}}$$

$$d_2^* = z_{\alpha/2}\sqrt{\frac{s_A^2}{n_A}+\frac{s_B^2}{n_B}} \qquad (11\text{-}14)$$

Using the value for z calculated from expression (11-13) instead, H_0 would be accepted if z fell within $\pm z_{\alpha/2}$ and would be rejected outside these limits.

Often the two-sided test is conducted in conjunction with a confidence interval estimate of $D = \mu_A - \mu_B$. In Example 11-1, we found the 95-percent confidence interval for the difference in population mean statistics scores for students using a new book A and students using an old book B to be

$$-3.68 \leqslant \mu_A - \mu_B \leqslant 8.48$$

Since the above interval contains the value 0, the null hypothesis for those data $H_0: \mu_A = \mu_B$ (or $D = \mu_A - \mu_B = 0$) must be *accepted* at the $\alpha = 1 - .95 = .05$ significance level.

Matched-Pairs Sampling

When matched pairs are used, we find the pair differences $d_i = X_{A_i} - X_{B_i}$ and then calculate their mean $\bar{d}$ and standard deviation $s_{d\text{-paired}}$. Using $\bar{d}$ as the test statistic, for sufficiently large samples, we may presume that the normal curve applies. We may therefore represent critical values by $\bar{d}^*$ for a one-sided test or by $\bar{d}_1^*$ and $\bar{d}_2^*$ for a two-sided test. The hypothesis-testing decision rules may be summarized as follows:

	(1) *Using $\bar{d}$*	(2) *Using z*
(a) Upper-tailed tests	*Accept* H_0 if $\bar{d} \leqslant \bar{d}^*$	*Accept* H_0 if $z \leqslant z_\alpha$
$H_0: \mu_A \leqslant \mu_B$	*Reject* H_0 if $\bar{d} > \bar{d}^*$	*Reject* H_0 if $z > z_\alpha$
or		
$H_0: D \leqslant 0$		
(b) Lower-tailed tests	*Accept* H_0 if $\bar{d} \geqslant \bar{d}^*$	*Accept* H_0 if $z \geqslant z_\alpha$
$H_0: \mu_A \geqslant \mu_B$	*Reject* H_0 if $\bar{d} < \bar{d}^*$	*Reject* H_0 if $z < z_\alpha$
or		
$H_0: D \geqslant 0$		

(c) Two-sided tests	*Accept* H_0 if	*Accept* H_0 if
$H_0: \mu_A = \mu_B$	$\bar{d}_1^* \leqslant \bar{d} \leqslant \bar{d}_2^*$	$-z_{\alpha/2} \leqslant z \leqslant z_{\alpha/2}$
or	*Reject* H_0	*Reject* H_0
$H_0: D = 0$	either if $\bar{d} < \bar{d}_1^*$ or if $\bar{d} > \bar{d}_2^*$	either if $z < -z_{\alpha/2}$ or if $z > z_{\alpha/2}$

For one-sided tests:

$$\bar{d}^* = z_\alpha \frac{s_{d\text{-paired}}}{\sqrt{n}} \tag{11-15}$$

With two-sided tests:

$$\bar{d}_1^* = -z_{\alpha/2} \frac{s_{d\text{-paired}}}{\sqrt{n}}$$

$$\bar{d}_2^* = z_{\alpha/2} \frac{s_{d\text{-paired}}}{\sqrt{n}} \tag{11-16}$$

While z may be computed from

$$z = \frac{\bar{d}}{s_{d\text{-paired}}} \sqrt{n} \tag{11-17}$$

Example 11-3 Perhaps the most controversial application of statistics has been in cigarette-smoking studies. One researcher interested in the tendency among middle-aged American men toward angina pectoris conducted an elaborate matched-pairs sampling investigation to determine if the onset of this disease is accelerated by heavy smoking. Random samples of male smokers and nonsmokers were chosen from medical records. To speed her investigation, the researcher decided not to follow each patient until detection of the disease, but instead to study only those men already suffering from angina.

Each smoker was matched to a nonsmoker, first by age and then in accordance with life style, occupation, medical history, and general physical characteristics. The actual matching was determined from carefully prepared dossiers, where all references to the status of the circulatory system and the heart were deleted by an independent physician. (It is very important that any data directly relevant to the variable being observed not influence the matching, thereby eliminating an obvious source of bias.)

As her null hypothesis, the researcher assumed that the mean age at which angina was first detected was the same for smokers and nonsmokers:

$$H_0: \mu_A = \mu_B$$

A total of $n = 100$ pairs were obtained by matching, and the sample ages X_{A_i} for smokers and X_{B_i} for nonsmokers when angina was first detected were determined. The individual matched-pair differences d_i were then computed and the following statistics were obtained:

$$\bar{d} = -1.3 \text{ years}$$

$$s_{d\text{-paired}} = 2.4 \text{ years}$$

Using an $\alpha = .05$ significance level, the critical normal deviate for this two-sided test was $z_{\alpha/2} = z_{.025} = 1.96$. The critical values for $\bar{d}$ were computed to be

$$\bar{d}_1^* = -1.96\left(\frac{2.4}{\sqrt{100}}\right) = -.47$$

$$\bar{d}_2^* = 1.96\left(\frac{2.4}{\sqrt{100}}\right) = .47$$

The applicable decision rule was therefore

Accept H_0 (conclude smoking does not matter) if $-.47 \leqslant \bar{d} \leqslant .47$

Reject H_0 (conclude smoking affects angina) either if $\bar{d} < -.47$ *or* if $\bar{d} > .47$

Since $\bar{d} = -1.3$ years is smaller than $-.47$, the null hypothesis that smokers and nonsmokers contract angina at the same mean age had to be *rejected*, and the researcher had to conclude that smoking hastens the occurrence of angina.

Solved another way, we may calculate the normal deviate by using expression (11-17) to obtain

$$z = \frac{-1.3}{2.4}\sqrt{100} = -5.42$$

which is less than $-z_{.025} = -1.96$, and again we see that H_0 had to be rejected.

The value $\bar{d} = -1.3$ in the above example serves as a point estimate for the difference $D = \mu_A - \mu_B$, indicating that for the sample results obtained, angina occurs on the average 1.3 years sooner among smokers than among nonsmokers. A 95-percent confidence interval estimate of the difference in population mean ages could be obtained directly from the above results:

$$\mu_A - \mu_B = -1.3 \pm (-.47)$$

or

$$-1.77 \leqslant \mu_A - \mu_B \leqslant -.83 \text{ years}$$

The above confidence interval might have been used as the basis for testing the original hypothesis. We can see how this would work by continuing with an earlier example.

We may apply the earlier confidence-interval results for the matched pairs obtained by the college professor in a two-sided test where his null hypothesis was that a new statistics book yields mean scores no different than an old book does; that is, $H_0: \mu_A = \mu_B$. For the data in Section 11-1, we found the 95-percent confidence interval

$$.61 \leqslant \mu_A - \mu_B \leqslant 4.19$$

which does not contain the value 0. Thus, in this example, H_0 must be rejected at the $\alpha = .05$ significance level. Note that this result contradicts our earlier conclusion from the independent sample data, for in this particular application, matched-pairs sampling proved to be a more discriminating procedure. (Never use *both* methods on the *same* data; we did it here only to compare the techniques.)

The earlier arguments for and against matched-pairs sampling can be applied to hypothesis-testing situations as well. A matched-pairs test is much more efficient or powerful than a test from independent sampling. With matched pairs, a much smaller sample size can generally be used to provide similar protection against Type I and Type II errors. However, matched-pairs sampling is more costly and is applicable only in those situations where suitable matching criteria and supporting data are available.

EXERCISES

11-5 A physician must compare two treatments for venereal disease. Because new bacteria strains have evolved that are resistant to antibiotics, a formerly obsolete chemical cure has been revived. The doctor wishes to test the null hypothesis that patient recovery time using penicillin to treat gonorrhea is less than or equal to the duration of sulfa-drug treatment. If this hypothesis is rejected, he will use sulfa drugs on future patients; if the sample results are not significant, he will continue to administer penicillin. A random sample of $n_A = 100$ patients were selected from those treated with penicillin, and another group of $n_B = 50$ patients were given sulfa. The following results were obtained:

Penicillin	*Sulfa*
$\bar{X}_A = 10$ days	$\bar{X}_B = 9$ days
$s_A^2 = 10$	$s_B^2 = 3$ days

(a) Is this test lower- or upper-tailed?
(b) If an $\alpha = .01$ significance level is desired, find the doctor's critical value for the difference in sample means. Which treatment should he use in the future?
(c) Calculate the normal deviate for the sample results. What is the smallest significance level at which this value indicates the null hypothesis could have been rejected?

11-6 To combat aphid infestation, a farmer must decide whether he should spray his fruit trees with pesticide or inundate them with ladybugs. He sprays a sample of 250 trees and uses ladybugs on another sample of 250 trees. His null hypothesis is that spraying produces a yield as high as that obtained using the natural predators. His sample results for spraying provide a mean of 10 bushels of good fruit per tree, with a standard deviation of 3 bushels. With ladybugs, the mean is 10.5 bushels of good fruit per tree with a standard deviation of 1 bushel. Using as his test statistic the mean yield from spraying minus the mean yield from ladybugs, should the farmer accept or reject the null hypothesis at the .05 significance level?

11-7 A dietician wanted to determine whether or not noncaloric drinks help people who have weight problems. Although these drinks reduce caloric intake, they do not reduce overall appetite levels like sugared drinks do. The dietician matched 36 pairs of obese persons in terms of weight, sex, diet, life style, eating habits, and prior fattiness problems. The experimental group (A) was asked to use only diet soft drinks and to substitute artificial sweeteners. Both groups were to continue with their regular diets, but the control group (B) was asked to avoid noncaloric beverages of any kind. At the end of six months, the weight change was determined for each person. As her null hypothesis, the dietician assumes that the mean weight reduction will be the same for both groups. The mean of the matched-pairs differences was found to be a one-pound reduction in favor of non-caloric drinks. The standard deviation was 2 pounds. If an $\alpha = .01$ significance level is desired for this two-sided test:

(a) Find the critical values for the mean matched-pair difference.

(b) What conclusion should the dietician make?

(c) Calculate the normal deviate for the test results. What is the lowest significance level at which she may reject her null hypothesis.

11-8 In comparing the traits of first-born and last-born children, a researcher wishes to test the null hypothesis that mean personality index scores in the category of social inversion do not differ between these two groups. He has randomly chosen 100 families, testing the youngest and oldest siblings in each family, which constitute a matched pair. Subtracting the score for the last-born from the older sibling's score, the researcher obtained the following statistics:

$$\bar{d} = 2.5$$

$$s_{d\text{-paired}} = 10$$

For this two-sided test, what conclusion should the researcher draw if he uses an $\alpha = .05$ significance level?

11-9 Referring to the confidence interval found in Exercise 11-2 for the differences in mean annual tornado intensity in two states, should the meteorologist accept or reject the null hypothesis that the mean intensity is the same in both states if he uses an $\alpha = .01$ significance level?

11-10 Refer to the confidence interval found in Exercise 11-4 for the savings in mean gasoline mileage using radial tires. At the $\alpha = .05$ significance level, should the city manager accept or reject the null hypothesis that mean gasoline consumption is the same for radial and for standard tires?

11-3 HYPOTHESIS TESTS FOR COMPARING PROPORTIONS

Qualitative populations may be compared in terms of their respective proportions. We can extend the hypothesis-testing concepts of the preceding section to using sample proportions as the basis for comparison. Such testing may be useful to a politician who must determine whether or not the popularity of new legislation differs between two groups of constituents. Similarly, a television network might compare the proportions of potential audiences preferring its two comedies over a third, all shown on different nights, in deciding whether or not to reschedule programs.

As before, we employ the designations π_A and π_B to represent the two population proportions. *The difference between population proportions is*

$$D = \pi_A - \pi_B \tag{11-18}$$

The respective sample proportions, designated P_A and P_B, may be estimated by

the difference in sample proportions

$$d = P_A - P_B \tag{11-19}$$

Using d as the test statistic, a decision rule may be constructed to test one of the following null hypotheses:

$$H_0 : \pi_A \geqslant \pi_B \text{ or } H_0 : D \geqslant 0 \qquad \text{(lower-tailed test)}$$

$$H_0 : \pi_A \leqslant \pi_B \text{ or } H_0 : D \leqslant 0 \qquad \text{(upper-tailed test)}$$

$$H_0 : \pi_A = \pi_B \text{ or } H_0 : D = 0 \qquad \text{(two-sided test)}$$

Ordinarily, the sampling distributions of P_A and P_B can be approximated by the normal curve when sample sizes are large enough. Because the samples are selected independently, it follows that d can be assumed to be normally distributed, with a mean of $D = \pi_A - \pi_B$ and a standard deviation of*

$$\sigma_d = \sqrt{\frac{\pi_A(1-\pi_A)}{n_A} + \frac{\pi_B(1-\pi_B)}{n_B}} \tag{11-20}$$

Under any of the above null hypotheses, we may assume that $\pi_A = \pi_B$, so that $D = 0$ and the sample results may be treated as if they came from the same population. As an estimator of either π_A or π_B, the sample results may be pooled to compute ***the combined sample proportion***

$$P_C = \frac{n_A P_A + n_B P_B}{n_A + n_B} \tag{11-21}$$

The standard deviation for d may be estimated by

$$s_d = \sqrt{P_C(1-P_C)\left(\frac{1}{n_A} + \frac{1}{n_B}\right)} \tag{11-22}$$

Again, we may use the symbol d^* to represent the critical value for d. This must be chosen so that a decision rule is obtained which corresponds to the desired significance level. As in our earlier two-sample tests, the critical normal deviate z_α indicates how many standard deviations d^* lies above or below the mean $D = 0$. Thus, $d^* = z_\alpha s_d$, and we may express ***the critical value for the difference in sample proportions***

$$d^* = z_\alpha \sqrt{P_C(1-P_C)\left(\frac{1}{n_A} + \frac{1}{n_B}\right)} \tag{11-23}$$

* As with the difference in independent sample means, it follows that the variance of $d = P_A - P_B$ is the sum of the respective variances for the individual sample proportions.

The possible forms of the decision rule are analogous to those forms discussed in connection with the difference in independent sample means.

An alternative approach is to compute ***the normal deviate value for the sample results***

$$z = \frac{P_A - P_B}{\sqrt{P_C(1-P_C)\left(\frac{1}{n_A}+\frac{1}{n_B}\right)}} \tag{11-24}$$

which may be compared to z_α to determine whether H_0 must be accepted or rejected.

For a two-sided test analogously using $z_{\alpha/2}$ as the critical normal deviate, we may calculate ***the critical values for the difference in sample proportions***

$$\begin{aligned} d_1^* &= -z_{\alpha/2}\sqrt{P_C(1-P_C)\left(\frac{1}{n_A}+\frac{1}{n_B}\right)} \\ d_2^* &= z_{\alpha/2}\sqrt{P_C(1-P_C)\left(\frac{1}{n_A}+\frac{1}{n_B}\right)} \end{aligned} \tag{11-25}$$

Example 11-4 A television programming director has decided to replace his network's Monday night comedy show "Grundy's Mondays" with one of two candidate comedy series: "The Sky is Falling" or "Mr. McGregor's Garden." Although either of the new programs should have a wider appeal than the present one, the director wants to pick the new comedy that best suits the whims of the present Monday-night audience. To determine this, special pilots of "Sky" and "McGregor" were shown on nonsuccessive weeks in the "Grundy" time slot. Two independent random samples of regular "Grundy" viewers who also watched both comedy pilots were then selected, and the sample proportion favoring each of the new shows was determined. As his null hypothesis, the program director assumed that the respective population proportions of viewers preferring each of the program pilots over "Grundy's Mondays" are the same. Therefore, using A to designate "Sky" and B for "McGregor", $H_0{:}\pi_A = \pi_B$. In this problem, a two-sided test applies.

With $n_A = 150$ and $n_B = 100$, the respective sample proportions favoring the new programs were

$$P_A = .80 \text{ and } P_B = .70$$

These results provide a difference in proportions of

$$d = P_A - P_B = .80 - .70 = .10$$

The combined sample proportion favoring a change in programs is found by substituting the above into expression (11-21):

$$P_C = \frac{150(.80)+100(.70)}{150+100} = .76$$

Even though these results indicate that "Grundy" is not well liked by its current audience, the program director will use the limited sample results to choose a replacement only if one new show is significantly preferred over the other. Otherwise, further investigations will be required to pick a new show. The director therefore desires an $\alpha = .10$ significance level for this test, so that there is just a 10-percent chance of picking the new show that was not most preferred by "Grundy" addicts. Substituting $P_C = .76$ and $z_{.05} = 1.64$ for $z_{\alpha/2}$ in expression (11-25), the critical values for the difference in sample proportions are

$$d_1^* = -1.64\sqrt{.76(1-.76)\left(\frac{1}{150}+\frac{1}{100}\right)} = -.09$$

$$d_2^* = .09$$

and the decision rule for $d = P_A - P_B$ is

Accept H_0 (investigate further) if $-.09 \leqslant d \leqslant .09$

Reject H_0 either (use "McGregor") if $d < -.09$

or (use "Sky") if $d > .09$

Since $d = .10$ is greater than .09, H_0 must be *rejected* and the program director should replace "Grundy" with "The Sky is Falling".

A shorter procedure would be to find the normal deviate for the sample results by using expression (11-24):

$$z = \frac{.80 - .70}{\sqrt{.76(1-.76)\left(\frac{1}{150}+\frac{1}{100}\right)}} = 1.81$$

Since 1.81 exceeds $z_{.05} = 1.64$, we see again that H_0 must be rejected.

EXERCISES

11-11 Suppose that the television network programming director in Example 11-4 signed a contract for "Mr. McGregor's Garden" (B) but that he would be willing to use "The Sky is Falling" (A) instead if significantly more viewers favored it. For the same numerical results obtained in that example:

(a) What null hypothesis is being tested? Is the test two-sided, lower-tailed, or upper-tailed?

(b) At the $\alpha = .025$ significance level, determine the critical value(s) for the director's decision. Which program do the data indicate should be used?

(c) Using the normal deviate z for the sample results, what is the smallest significance level at which the null hypothesis can be rejected?

11-12 A political polling firm has been retained by a congressional candidate. In order to organize his strategy in the final weeks of the campaign, he needs to know whether he is stronger in the suburbs (A) or in the cities (B). Independent random samples of 100 voters in each category have been polled, and these data indicate the candidate is preferred by 48 percent in the suburbs and 53 percent in the cities. As his null hypothesis, he has assumed that he is equally strong in both areas. At the $\alpha = .05$ significance level, what conclusion should he make from the sample results?

11-13 A government statistician is evaluating a prototype aircraft radar system to see if it can detect a plane passing through the edges of the radar horizon better than the existing system can. Two independent samples of 100 test flights have been made using both systems, and the respective proportions detected have been computed. A 5-percent chance is desired as protection against incorrectly concluding that the new system (A) has a detection probability (a population proportion of flights detected at the horizon) higher than the old system (B) does. The test results yield $P_A = .80$ and $P_B = .70$.

(a) Formulate the null hypothesis. Is the test two-sided, lower-tailed, or upper-tailed?

(b) Based upon the sample results, can the statistician conclude that the new system is better?

11-4 INFERENCES FOR TWO MEANS USING SMALL SAMPLES

Thus far, we have considered inferences about two populations made from large samples. In such cases, the normal curve provides the sampling distributions for d and $\bar{d}$. But if small sample sizes must be used, the normal approximation proves inadequate in testing situations when the population variance is unknown. In these cases, the Student t distribution must be used instead.

Estimating the Difference in Means: Independent Samples

When independent samples are used, confidence intervals may be constructed for the difference in population means, $D = \mu_A - \mu_B$, in nearly the same manner we presented earlier. Again, we use $d = \bar{X}_A - \bar{X}_B$ to estimate D. In place of the normal deviate z, we use t_α (obtained from Appendix Table E). In addition, the standard error for d must be calculated somewhat differently.

When small samples are used (as a practical rule, whenever *both* n_A and n_B are less than 30), applicability of the Student t distribution requires that *both populations have the same variance*, so we must assume that $\sigma_A^2 = \sigma_B^2 = \sigma^2$. The standard error for d may therefore be expressed in a slightly different form than before:

$$\sigma_d = \sqrt{\frac{\sigma^2}{n_A} + \frac{\sigma^2}{n_B}} = \sigma\sqrt{\frac{1}{n_A} + \frac{1}{n_B}} \tag{11-26}$$

Under this requirement, s_A^2 and s_B^2 serve as unbiased estimators of σ^2. An even better estimate of σ^2 may be obtained by pooling the sample results. Thus, σ_d may be estimated by taking the square root of a weighted average of the sample variances. The following expression (the first term of which has its denominator reduced by 2 in order to yield an unbiased estimator) is used to estimate σ_d:

$$s_{d\text{-small}} = \sqrt{\frac{(n_A - 1)s_A^2 + (n_B - 1)s_B^2}{n_A + n_B - 2}}\sqrt{\frac{1}{n_A} + \frac{1}{n_B}} \tag{11-27}$$

Using independent samples, the confidence interval estimate for the difference in population means is

$$\mu_A - \mu_B = \bar{X}_A - \bar{X}_B \pm t_\alpha s_{d\text{-small}} \tag{11-28}$$

In finding t_α from Appendix Table E for a confidence level of C, we choose $\alpha = (1-C)/2$. The number of degrees of freedom is equal to $n_A + n_B - 2$. In effect, the combined sample size $n_A + n_B$ must be reduced by 2, because one degree of freedom is lost in calculating each of the two sample variances.

Example 11-5 An astronomer wishes to compare the resolution (magnification accuracy) of two large telescopes. One telescope, located in the Southern Hemisphere, is refractive and uses a thick transparent lens to focus light from distant galaxies. Of purported equal power is a reflective telescope in the Northern Hemisphere that focuses starlight by means of polished mirrors. Although different galaxies may be viewed from each hemisphere, the density of galaxies in space is considered the same in either case.

The astronomer has taken $n_A = 16$ pictures with the refractive telescope at randomly chosen coordinates in the South; a counterpart random sample of $n_B = 26$ frames has been exposed by the reflective telescope in the North. Camera and film type were the same at both sites, and all exposures were made for exactly one hour on clear nights. The number of detectable galaxies in each frame was counted, and the following sample means and variances for the respective counts were obtained:

$$\bar{X}_A = 29.5 \qquad \bar{X}_B = 21.4$$

$$s_A^2 = 96.3 \qquad s_B^2 = 112.4$$

The standard error for the differences in sample means, $d = \bar{X}_A - \bar{X}_B$, is

$$s_{d\text{-small}} = \sqrt{\frac{(16-1)96.3 + (26-1)112.4}{16+26-2}}\sqrt{\frac{1}{16}+\frac{1}{26}} = 3.277$$

Using these data, a 95-percent confidence interval was constructed for the difference in population mean galaxy counts. For $16+26-2 = 40$ degrees of freedom and using $\alpha = (1-.95)/2 = .025$, $t_{.025} = 2.021$. Thus

$$\mu_A - \mu_B = 29.5 - 21.4 \pm 2.021(3.277)$$

$$= 8.1 \pm 6.62$$

or

$$1.48 \leqslant \mu_A - \mu_B \leqslant 14.72$$

The advantage in resolution of the refractive telescope (A) over the reflective one (B) is therefore estimated to fall somewhere between 1.48 and 14.72 galaxies.

Estimating the Differences in Means: Matched Pairs

In matched-pairs sampling, we estimate $D = \mu_A - \mu_B$ by $\bar{d}$, the mean of the individual matched-pair differences. Using small samples (generally, when the number of pairs is $n < 30$), the only difference is that the normal deviate z is replaced by t_α. The standard error of $\bar{d}$ remains the same, and it is estimated from the standard deviation of individual paired differences $s_{d\text{-paired}}$. ***Using matched-pairs samples, the confidence interval estimate for the difference in population means is***

$$\mu_A - \mu_B = \bar{d} \pm t_\alpha \frac{s_{d\text{-paired}}}{\sqrt{n}} \tag{11-29}$$

We choose a value t_α that corresponds to the confidence level, using $n-1$ degrees of freedom

Example 11-6 A systems analyst for the Marlborough County Data Center has proposed adopting the dynamic-core allocation system used in neighboring Kent County. Since both counties have identical computers and otherwise identical software (program systems), a special test is to be administered to estimate the savings in mean "throughput" time for processing regular job batches under the proposed system. Designating Marlborough's present system A and the new system B, a random sample of $n = 25$ batches has been run twice, once in each county. Thus, the two processing times for one batch constitute the observations of a matched pair. For the ith batch, d_i represents the throughput time on the Marlborough computer run minus the analogous figure from the Kent operation. The following results were obtained:

$$\bar{d} = 25 \text{ minutes}$$

$$s_{d\text{-paired}} = 20 \text{ minutes}$$

A 95-percent confidence interval estimate of the savings in mean batch throughput time from using dynamic-core allocation can then be constructed. For $25 - 1 = 24$ degrees of freedom with $\alpha = (1 - .95)/2 = .025$, $t_{.025} = 2.064$, so that the estimated mean time advantage for using dynamic-core allocation is

$$\mu_A - \mu_B = 25 \pm 2.064 \frac{20}{\sqrt{25}}$$

$$= 25 \pm 8.3$$

or

$$16.7 \leqslant \mu_A - \mu_B \leqslant 33.3 \text{ minutes per batch}$$

The above interval indicates that Marlborough's present computer software can be improved by using dynamic-core allocation, which will reduce mean batch processing time between 16.7 and 33.3 minutes.

Hypothesis Tests for Comparing Two Means: Independent Samples

When small independent samples are used in testing hypotheses regarding μ_A and μ_B, the value

$$t = \frac{d - D}{s_{d\text{-small}}} \tag{11-30}$$

is the test statistic. Under the various null hypotheses typically encountered, the value for $D = \mu_A - \mu_B$ which always applies is zero. Thus, using the expression for $s_{d\text{-small}}$ given earlier, we may calculate ***the t statistic for independent samples***

$$t = \frac{\bar{X}_A - \bar{X}_B}{\sqrt{\frac{(n_A - 1)s_A^2 + (n_B - 1)s_B^2}{n_A + n_B - 2}}\sqrt{\frac{1}{n_A} + \frac{1}{n_B}}} \tag{11-31}$$

The critical value corresponding to the desired significance level α may be found from Appendix Table E for $n_A + n_B - 2$ degrees of freedom. For lower-tailed tests, where the null hypothesis $H_0 : \mu_A \geqslant \mu_B$, t_α is negative and H_0 must be rejected when $t < t_\alpha$. With $H_0 : \mu_A \leqslant \mu_B$, the test is upper-tailed and t_α will be positive, so that H_0 must be rejected when $t > t_\alpha$. With $H_0 : \mu_A = \mu_B$, the critical value is $t_{\alpha/2}$. Here, H_0 must be accepted when t lies within $\pm t_{\alpha/2}$; otherwise it must be rejected.

Example 11-7 Drinking coffee affects many people's sleep. One investigator compared the effects of coffee on habitual drinkers with the effects on occasional drinkers. Independent random samples of $n_A = 10$ light and $n_B = 15$ heavy coffee drinkers were obtained. One night's sleep was monitored for each subject, and before retiring, every person was given a cup of coffee. As her null hypothesis, the researcher assumes that the light coffee drinkers will go to sleep at least as fast as the heavy coffee drinkers. In terms of mean time to onset of sleep, $H_0 : \mu_A \leqslant \mu_B$. The following test data were obtained:

Light Coffee Drinkers	*Heavy Coffee Drinkers*
$\bar{X}_A = 45$ minutes	$\bar{X}_B = 36$ minutes
$s_A^2 = 706$	$s_B^2 = 654$

This is an upper-tailed test. The number of degrees of freedom is $10 + 15 - 2 = 23$. For an $\alpha = .05$ significance level, $t_{.05} = 1.714$. Substituting the above results into expression (11-31), the following value is obtained for the test statistic:

$$t = \frac{45 - 36}{\sqrt{\frac{9(706) + 14(654)}{23}}\sqrt{\frac{1}{10} + \frac{1}{15}}} = .85$$

Since .85 is smaller than $t_{.05} = 1.714$, H_0 must be *accepted.* The results are not statistically significant, and the sample data do not refute the null hypothesis that light drinkers will go to sleep at least as quickly as heavy drinkers after one cup of coffee.

Hypothesis Tests for Comparing Two Means: Matched Pairs

Matched-pairs tests with small samples involve the test statistic

$$t = \frac{\bar{d} - D}{s_{d\text{-paired}}}\sqrt{n} \tag{11-32}$$

where $\bar{d}$ and $s_{d\text{-paired}}$ are the mean and the standard deviation of the sample matched-pair differences. Under the typical null hypothesis, we may assume that $D = \mu_A - \mu_B$ is zero and we may therefore simplify expression (11-32) in calculating ***the t statistic for matched pairs***

$$t = \frac{\bar{d}}{s_{d\text{-paired}}} \sqrt{n} \tag{11-33}$$

The critical value t_α for the desired significance level α is found from Appendix Table E by using $n-1$ degrees of freedom. For a two-sided test, we use $t_{\alpha/2}$ as the critical value.

To illustrate this procedure, suppose that the systems analyst in Example 11-6 desired to test the null hypothesis that the Kent County dynamic-core allocation system was no better than Marlborough's present one. In this case, the analyst would use $\mu_A \leqslant \mu_B$ for H_0, indicating that the new system's mean processing time is at least as great as the present system's. Here, a one-sided, upper-tailed test would apply.

Suppose that an $\alpha = .05$ significance level is desired. For 24 degrees of freedom, the critical value for the test statistic is $t_{.05} = 1.711$. Comparing this to the test statistic calculated from the earlier data

$$t = \frac{25}{20}\sqrt{25} = 6.25$$

we see that this is larger than the critical value and H_0 must therefore be *rejected.* The analyst would conclude that his present data processing system is not as fast as Kent County's system.

EXERCISES

11-14 Estimate the difference in mean gasoline mileage in two car models by a 95-percent confidence interval if the following results apply for independent samples.

Car A	*Car B*
$n_A = 10$	$n_B = 10$
$\bar{X}_A = 14$ mpg	$\bar{X}_B = 12.5$ mpg
$s_A^2 = 2$	$s_B^2 = 2.5$

11-15 Suppose that the drivers of the two cars in Exercise 11-14 were matched according to their driving skills. The mean and the standard deviation for the differences in gasoline mileage found by subtracting car *B*'s mpg from car *A*'s are

$$\bar{d} = 1.5 \text{ mpg}$$

$$s_{d\text{-paired}} = 1.2 \text{ mpg}$$

Construct a 95-percent confidence interval for the difference in mean gasoline mileage.

11-16 A city manager wanted to estimate the advantage in the mean lifetime of a new type of tire compared to an existing tire. A taxicab company cooperated in a test where 10 cars were equipped with two of each type of tire. A record was kept of each car's total mileage until both tires of the same type had to be replaced. For each car, the replacement mileage for the present tire (B) was subtracted from the corresponding figure for the new tire (A). The mean difference in mileage was 2,000 miles, with a standard deviation of 2,000 miles.

(a) Construct a 95-percent confidence interval estimate for the difference in mean tire lifetimes.

(b) Should the null hypothesis that the mean lifetimes are identical be accepted or rejected at the 5-percent significance level?

11-17 Suppose that a test similar to the one in Exercise 11-16 was performed on another group of cars. Here, 15 cars were fully equipped with the new tire (A) and 10 cars were equipped with the old tire (B). No attempt was made to match the cars. The sets of four tires were replaced when driving each car became hazardous. The mileage results showed

$$\bar{X}_A = 35{,}000 \text{ miles} \qquad \bar{X}_B = 33{,}000 \text{ miles}$$

$$s_A^2 = 6{,}500{,}000 \qquad s_B^2 = 7{,}000{,}000$$

(a) Construct a 95-percent confidence interval estimate for the difference in mean tire lifetimes.

(b) Should the null hypothesis that the mean tire lifetimes are identical be accepted or rejected at the 5-percent significance level?

11-18 One aspect of horticulture is to find and nurture genetic strains of plants with favorable characteristics. In trees grown for timber, it is desirable to select fast-growing seeds. Two parent strains of a certain species are being compared in test plots. The following data have been obtained for the third-year growth of seedlings:

Strain A	*Strain B*
$n_A = 14$	$n_B = 28$
$\bar{X}_A = 2.4$ feet	$\bar{X}_B = 2.6$ feet
$s_A^2 = .25$	$s_B^2 = .35$

Should the null hypothesis that seedling growth for strain A is at least as great as seedling growth for strain B be accepted or rejected at the 5-percent significance level?

11-19 A pediatrician has decided to compare firstborn infants to infants who have older siblings when born. Two random samples of 15 firstborn and 15 later-born babies have been independently selected. All are full-term, breast-fed infants. As his null hypothesis, the doctor assumes that firstborns have a mean first-month weight gain at least as great as the later-born babies. The following data have been obtained:

Later-born Infants	*Firstborn Infants*
$\bar{X}_A = 2.3$ pounds	$\bar{X}_B = 2.1$ pounds
$s_A^2 = .22$	$s_B^2 = .19$

Can the doctor reject his null hypothesis if he uses a 5-percent significance level?

11-20 Suppose that the astronomer in Example 11-5 wishes to compare the near-range resolution of the Southern Hemisphere refractive telescope (A)

with the resolution of the Northern Hemisphere reflective telescope (B). Since Jupiter can be seen from both sites simultaneously, shots of this planet from both telescopes were made at the same time on 15 different nights throughout the year. The photographs were then analyzed to determine the number of Jovian moons that could be detected in each. The mean difference in moon counts was found to be .3 moon, with a standard deviation of .7 moon. Must the null hypothesis that the reflective telescope provides at least as high a moon count be accepted or rejected at the 5-percent significance level?

11-21 Suppose that a psychologist wishes to compare firstborn to last-born siblings in terms of an index of hypomania; large scores on this scale express a high degree of nonconformist behavior. As her null hypothesis, the psychologist assumes that the mean hypomania score will be the same for both groups. In each of 25 sample families, she tests both the firstborn and the last-born siblings, whose scores constitute a matched pair. Subtracting the last-born's score from the firstborn's, she obtains a mean difference of .5 point, with a standard deviation of 2.5 points. At the 5-percent significance level, what should the psychologist conclude?

11-5 FURTHER REMARKS

In this chapter, we have considered two-sample inferences regarding population means and proportions. In testing means, either independent or matched-pairs samples may be employed, and each procedure has its advantages as well as its drawbacks. Generally, matched-pairs sampling makes more efficient use of sample information, but it is more elaborate and costly.

Again, we have encountered the basic dichotomy between large and small samples. With large samples, we make use of the fact that when the population standard deviations are known, the central limit theorem indicates that the respective estimators and test statistics are approximately normally distributed. Ordinarily, however, σ_A^2 and σ_B^2 are unknown and may be estimated by s_A^2 and s_B^2. In all such cases, the Student t distribution applies instead of the normal distribution. But for large samples, the normal curve is nearly identical to the Student t distribution, so that we use it anyway.

A requirement of the Student t distribution is that the parent population be normally distributed. As we saw in Chapter 8, this usually presents no problem unless the population is highly skewed. In independent, two-sample tests which have small ns, there is the additional requirement that both population variances be equal. If reason to believe that σ_A^2 is much different than σ_B^2 exists, the t statistic should not be used on small, independent samples.

Further statistical procedures can be employed in making inferences regarding σ_A^2 and σ_B^2. Up to this point, we have largely skirted the issue of making inferences regarding variances. Procedures for doing this will be presented in Chapters 12 and 13. The material was deferred until the necessary sampling distributions were described in their more traditional applications. These later chapters will primarily focus upon testing for independence and comparing three or more population parameters.

REVIEW EXERCISES

11-22 A statistician who is also a baseball fan is interested in how major league batting averages differ between home and away games. He has selected 25 batters at random and has kept detailed records of their performances. As his null hypothesis, the statistician assumes that the mean batting average of hitters away from home is at least as large as the mean average they attain playing on their home field. After compiling his data and separately computing the batting averages, the statistician found the mean difference in a player's batting average to be .028 in favor of the home games. The standard deviation in the differences was computed to be .056.

(a) At the $\alpha = .05$ significance level, what conclusion should the statistician make?

(b) Construct a 95-percent confidence interval for the advantage in mean batting average attained by playing at home.

11-23 Consider the sample data shown below for the weights and ages of athletic and nonathletic men, all approximately 6 feet tall and all of medium build.

Athletic Men (A)		*Nonathletic Men (B)*	
Weight	*Age*	*Weight*	*Age*
151	22	152	21
148	24	153	26
156	31	149	33
155	37	162	35
157	41	165	39
161	42	168	44
158	49	157	47
168	51	178	49
149	55	161	57
174	61	186	59

(a) Calculate the sample means and variances for the weights of the two groups.

(b) Treating the weights as independent random samples, should the null hypothesis that the population mean weight for athletic men will be at least as large as that for nonathletic men be accepted or rejected at the $\alpha = .10$ significance level?

11-24 Referring to Exercise 11-23, assume that the data represent observation pairs matched by age.

(a) Compute the weight differences for each pair, subtracting the non-athlete's weight from the athlete's. Then find the mean and the standard deviation of these matched-pair differences.

(b) Treating the weights as matched-pairs samples, should the null hypothesis that the population mean weight for athletic men is at least as large as that for nonathletic men be accepted or rejected at the $\alpha = .10$ significance level?

11-25 A company tested two cereals to determine the taste preferences of potential buyers. Two different panels of persons were selected; the individuals on one were asked to taste brand A, while the others were asked to taste brand B. Each person was then asked if he would buy the product he tasted. The following results were obtained:

	Cereal A	*Cereal B*
Number who would buy:	75	80
Number who would not buy or who were undecided:	50	60
	125	140

Consider the null hypothesis that there is no difference in the proportion of potential buyers who would desire to buy either product. At the 5-percent significance level, should H_0 be accepted or rejected?

11-26 An agronomist compared the corn-crop yields per acre on which a nitrate-based fertilizer was used with corn yields from plots on which a sulfur-based fertilizer was used. A sample of 100 acres using nitrates (A) yielded an average of 56.2 bushels per acre, with a variance of 156.25. The sample of 150 acres fertilized with sulfurs (B) yielded an average of 52.6 bushels per acre, with a variance of 190.44. For each of the following null hypotheses: (1) express H_0, using the appropriate symbols; (2) construct the decision rule for the difference in sample means for an $\alpha = .05$ significance level; and (3) find the value of the test statistic and indicate whether H_0 must be accepted or rejected.

(a) There is no difference in resulting crop yield between the two fertilizers.
(b) The nitrates provide at least the yield of the sulfurs.
(c) The sulfurs provide at least the yield of the nitrates.

11-27 A government agency wishes to determine which of two types of gasoline will provide the most miles per gallon. Two independent random samples of 15 cars were selected. Each car was driven 2,000 miles by employees on normal business, but a different type of gasoline was used in each sample. The results were

Type A	*Type B*
$\bar{X}_A = 15.0$ mpg	$\bar{X}_B = 14.8$ mpg
$s_A^2 = 1.44$	$s_B^2 = .81$

(a) Construct a 95-percent confidence interval estimate for the difference in mean gasoline mileages.
(b) Should the null hypothesis that there is no difference between the gasolines be accepted or rejected at the $\alpha = .05$ significance level?

11-28 Suppose that the plots used to test the nitrate-and sulfur-based fertilizers in Exercise 11-26 were selected so that the two types of fertilizers were placed on neighboring one-acre plots and that the yield of each sulfur-fertilized acre (B) was subtracted from the yield of its neighboring nitrate-fertilized acre (A). From a total of 100 pairs, the mean matched-pairs difference was 1.3 bushels, with a standard deviation of 3.2 bushels. Should the null hypothesis that there is no difference in yields due to choice of fertilizer be accepted or rejected at the $\alpha = .05$ significance level?

11-29 Suppose that the government agency in Exercise 11-27 matched the cars using gasoline A with those using gasoline B, that the cars in each pair were nearly identical in many respects (make, age, engine size, accessories, and so on), and that both groups of cars were driven over the same route by the same professional driver. The results showed that the mean difference in gasoline mileage between the cars using gasoline A and the cars using gasoline B was .2 miles per gallon, with a standard deviation of .1.

(a) Construct a 95-percent confidence interval estimate for the difference in mean gasoline mileages.

(b) Should the null hypothesis that there is no difference in the gasolines be accepted or rejected at the $\alpha = .05$ significance level?

12 Chi-Square Applications

Some people hate the very name of statistics, but I find them full of beauty and interest. Whenever they are not brutalized, but delicately handled by the higher methods, and are warily interpreted, their power in dealing with complicated phenomena is extraordinary. They are the only tools by which an opening can be cut through the formidable thicket of difficulties that bars the path of those who pursue the Science of man.

Sir Francis Galton (1889)

Thus far, we have described procedures using the normal and the Student t distributions, but many important statistical applications go beyond these distributions. In this chapter, we will introduce an additional distribution—the *chi-square distribution*—which will considerably expand our basic repertoire, for it applies to many further areas of inference.

The first new area we will investigate involves using sample data *in testing whether or not two variables are independent*. As we have seen in applications of probability, statistical independence is an important concept, because it allows us to considerably simplify the multiplication law. Even more important, independence between sample observations permits us to streamline procedures considerably. Perhaps it is most important to know if variables are independent when alternative methods and treatments are evaluated.

When population variables are qualitative characteristics (such as marital status, political affiliation, sex, state of health, type of treatment, or kind or response), then the presence or absence of independence between variables can be used to draw important conclusions. For example, if a doctor knows that preventive measures (vaccinated, unvaccinated) and resistance (diseased, not diseased) are dependent, then he may conclude that a vaccine is effective. A politician who learns that highest education level (elementary, high-school, college) and voter preference (candidates A, B, C) are dependent might find it necessary to reorient his campaign. If a public safety director finds that severity of automobile accidents (property damage, injury, fatalities) is dependent upon where they occur (city streets, rural roads, highways), he may decide to revise traffic law-enforcement procedures.

A second area of statistical inference we will investigate is *comparing several population proportions*. This application is closely related to testing for independence. A procedure comparing two populations was described in Chapter 11, but when three or more populations are involved, the testing procedure makes use of the chi-square distribution.

Until now, we have not applied statistical procedures in making inferences regarding the population variance in terms of *confidence intervals for* σ^2, nor have we described *hypothesis tests regarding* σ^2. These applications provide the third area of statistical inference to be described in this chapter. Although important, inferences regarding σ^2 have a lower priority than those for μ and π, where we use $\bar{X}$ and P as estimators and test statistics. Here, we will use s^2 in the same fashion. We have waited until now to do this because of the role the chi-square distribution plays in the sampling distribution for s^2.

12-1 INDEPENDENCE AND THE CHI-SQUARE DISTRIBUTION

The first new application of statistical inference we will investigate is testing to determine if two qualitative population variables are independent. In doing this, we will describe a special test statistic whose sampling distribution is the chi-square.

Independence Between Qualitative Variables

To describe what we mean by independence between two qualitative variables or population characteristics, we will briefly consider a basic concept of probability. Recall that two events A and B are statistically independent if the occurrence of either event does not affect the probability of the other event. More formally, this states that A and B are independent if

$$P[A \mid B] = P[A]$$

or, in other words, if the conditional and unconditional probabilities are equal. Thus, we see that drawing an ace from a fully shuffled deck of 52 ordinary playing cards is statistically independent of drawing a club because

$$P[\text{ace} \mid \text{club}] = 1/13 = 4/52 = P[\text{ace}]$$

Viewed another way, the proportion of aces in the club suit (1/13) is the same as the proportion of aces in the entire deck (4/52).

We can extend this concept of statistical independence to populations and samples. Consider a population whose elementary units may be classified in terms of two qualitative variables, A and B. Variable A might represent a person's sex, so that it would have two possible attributes—male and female; likewise, variable B might be political party, so that its attributes would be Democrat, Republican, and so on. Let us assume that men occur in the same proportion throughout the entire population as they do among Democrats. Under that assumption, we could then conclude that for the attributes of a randomly selected person

$$P[\text{male} \mid \text{Democrat}] = P[\text{male}]$$

so that the *events* male and Democrat are statistically independent. If this is the case, then the multiplication law of probability for independent events tells us that

$$P[\text{male } \textit{and} \text{ Democrat}] = P[\text{male}] \times P[\text{Democrat}]$$

If a similar fact holds for all attribute combinations, the population variables will exhibit a very important property:

DEFINITION Two qualitative population variables A and B are ***independent*** if the proportion of the population having any particular attribute of A is the same in the total population as it is in the part of the population having a particular attribute of B, no matter which attributes are considered. This implies that the frequency of units having any particular attribute pair may be found by multiplying the respective frequencies for the individual attributes and dividing by the population size.

Example 12-1 A symphony orchestra has 100 members. Of these, 70 are men and 30 are women. Each member is categorized by the type of instrument he or she plays: woodwind, brass, string, or percussion. The table below shows the frequencies of the members in terms of two characteristics: sex and instrument. Here, the two characteristics are independent. The ratio of male woodwind players to the total number of woodwind players is 21/30 = .7. The same value is obtained for male brass, string, or percussion player ratios, respectively: 14/20, 21/30, and 14/20 all equal .7, the proportion of men in the orchestra. Likewise, the ratio of female woodwind players to the total woodwinds is .3, which is the same as the ratio for any other instrument played by females and is also the proportion of females in the orchestra.

	Instrument Category				
Sex	(1) *Woodwind*	(2) *Brass*	(3) *String*	(4) *Percussion*	Total
(1) *Male*	21	14	21	14	70
(2) *Female*	9	6	9	6	30
Total	30	20	30	20	100

Since sex and instrument are independent, the frequency of male string players is equal to the product of the frequency for the respective attributes divided by the total number of players in the orchestra:

$$\frac{70 \times 30}{100} = 21$$

Likewise, the frequency of female string players is found by multiplying the number of females by the number of string players and dividing by 100:

$$\frac{30 \times 30}{100} = 9$$

Testing for Independence

In situations where the population frequencies are unknown, the presence or absence of independence between two variables must also be unknown. A sample may be used to test for independence, but to do this, we must extend the principles of hypothesis testing used previously. It will be helpful if we proceed by developing the following illustration.

TABLE 12-1 Contingency Table for Actual Sample Results from a Marijuana Study.

Dominant Antisocial Behavior	*Level of Marijuana Use*			Total
	(1) *Light*	(2) *Moderate*	(3) *Heavy*	
(1) *Insomnia*	10	5	7	22
(2) *Aggressiveness*	11	7	18	36
(3) *Transient psychosis*	6	11	7	24
(4) *None apparent*	10	6	2	18
Total	37	29	34	100

In the late 1960s and early 1970s, a great deal of research was conducted on the effects of marijuana, and tests for independence may be applied to some of these findings. One researcher attempted to determine if one variable, the level of marijuana use, affects a second variable, the incidence of antisocial behavior. Table 12-1 shows the results he obtained from a random sample of 100 marijuana users. Each subject was categorized as a light, moderate, or heavy smoker. The dominant antisocial behavior exhibited by each person was identified in one of four categories: insomnia, aggressiveness, transient psychosis, or none apparent.

The Contingency Table

In Table 12-1, the respective attributes for the marijuana usage variable are represented by columns and each row corresponds to the dominant type of antisocial behavior. For each column and row, there is a cell representing a possible category classification for a subject. The value within each cell is the tally for those subjects classified as having that cell's corresponding attributes. These cell numbers are referred to as *actual frequencies*. Such an arrangement of data is called a *contingency table*, because it accounts for all combinations of the factors being investigated—or, in other words, for all contingencies.

Using the researcher's data, we wish to find out whether or not the level of marijuana usage affects antisocial behavior. The null hypothesis is that the undesirable personal traits are unaffected by how heavily the drug is used. Stated another way, *the null hypothesis is that the two variables are independent.* The first step in accepting or rejecting H_0 is to determine the kind of results that might be expected if the variables were truly independent.

Expected Frequencies

The sample results which would be obtained on the average if the null hypothesis of independence were true serve as a basis for comparison. We refer to these hypothetical data as the *expected sample results*. To see how these deviate from the actual results, another contingency table must be constructed. The differences between the actual and the expected results can then be summarized by comparing the two contingency tables.

TABLE 12-2 Contingency Table for Expected Sample Results Under the Null Hypothesis of Independence for Marijuana Variables

Dominant Antisocial Behavior	*Level of Marijuana Use*			Total
	(1) *Light*	(2) *Moderate*	(3) *Heavy*	
(1) *Insomnia*	**8.14**	**6.38**	**7.48**	22
(2) *Aggressiveness*	**13.32**	**10.44**	**12.24**	36
(3) *Transient psychosis*	**8.88**	**6.96**	**8.16**	24
(4) *None apparent*	**6.66**	**5.22**	**6.12**	18
Total	37	29	34	100

Table 12-2 shows the contingency table for the expected results of this study. The total numbers of observations for each type of behavior and level of marijuana use are the same. Because they are calculated in a manner consistent with the null hypothesis of independence, the cell entries are referred to as *expected frequencies.* These appear in color in this contingency table. Since these quantities are *expected* rather than actual values, it is not necessary for the cell entries to be whole numbers. (Generally, two places of accuracy after the decimal point provide adequate accuracy.)

To facilitate our comparison, we will represent the actual and the expected frequencies symbolically. We use the letter f for frequency, with the subscripts a for actual and e for expected. Thus, for row i and column j, $f_a(i, j)$ denotes the actual frequency and $f_e(i, j)$ represents the expected frequency.

The actual frequency f_a values for the marijuana study results were provided initially in Table 12-1. The expected frequency $f_e(1, 1) = 8.14$ for the cell in row (1) and column (1) of Table 12-2 was found by multiplying the row (1) and the column (1) totals and dividing by the sample size

$$8.14 = \frac{22 \times 37}{100}$$

In general, the following procedure is used to calculate ***the expected frequency of the cell in the ith row and jth column***

$$f_e(i, j) = \frac{r_i c_j}{n} \tag{12-1}$$

where r_i is the marginal total of row i, c_j is the marginal total of column j, and n is the sample size. All of the expected frequencies in Table 12-2 were calculated in the above manner. For example, the expected number of subjects who exhibit a dominant antisocial behavior of transient psychosis and who use marijuana moderately is 6.96 and appears in the cell of row (3) and column (2). Here, $r_3 = 24$ and $c_2 = 29$, so that

$$f_e(3, 2) = \frac{24 \times 29}{100} = 6.96$$

In either the actual or the expected contingency tables, r_i and c_j are identical and both have the restriction that

$$\sum r_i = \sum c_j = n$$

The Chi-Square Statistic

The actual and the expected frequencies must be compared so that the null hypothesis of independence can be accepted or rejected. A decision rule must be established which provides a desirable balance between the probabilities of making the Type I error of rejecting independence when it indeed exists and of making the Type II error of accepting independence when it is not so. In practice, when the sample size is fixed in advance—as in our marijuana illustration—we are free to control only the Type I error probability α.

A test statistic must be used that does two things: it must measure the amount of deviation between the actual and the expected results, and it must have a sampling distribution that enables us to determine the Type I error probability α.

Such a statistic may be computed from the cell entries in the actual and expected contingency tables. It is based upon the individual differences between actual and expected frequencies in each cell. The following expression is used to calculate the test statistic which we call the ***chi-square statistic***

$$\chi^2 = \sum \frac{(f_a - f_e)^2}{f_e} \tag{12-2}$$

where the symbol χ is the Greek lower case *chi* (pronounced kie) and χ^2 is read as "chi square." The summation is taken over all cells in the contingency table. Table 12-3 shows the χ^2 calculations for the marijuana study results. Here, we find that $\chi^2 = 13.012$.

The possible values of χ^2 range upwards from zero. If the deviation between f_a and f_e is large for a particular cell, then so is the squared deviation $(f_a - f_e)^2$. In calculating χ^2, the squared deviations are divided by f_e before they are summed, ensuring that any differences are not exaggerated simply because a large number of observations are obtained. Large $(f_a - f_e)^2/f_e$ ratios occur when actual and expected results differ considerably, making the size of χ^2 large. Therefore, the more the sample results deviate from what would be expected if independence were true, the larger the value of χ^2 will be, and vice versa.

TABLE 12-3 Chi-Square Calculations for Marijuana Study Results

(Row, Column)	*Actual Frequency* f_a	*Expected Frequency* f_e	$f_a - f_e$	$(f_a - f_e)^2$	$\frac{(f_a - f_e)^2}{f_e}$
(1, 1)	10	8.14	1.86	3.4596	.425
(1, 2)	5	6.38	−1.38	1.9044	.298
(1, 3)	7	7.48	− .48	.2304	.031
(2, 1)	11	13.32	−2.32	5.3824	.404
(2, 2)	7	10.44	−3.44	11.8336	1.133
(2, 3)	18	12.24	5.76	33.1776	2.711
(3, 1)	6	8.88	−2.88	8.2944	.934
(3, 2)	11	6.96	4.04	16.3216	2.345
(3, 3)	7	8.16	−1.16	1.3456	.165
(4, 1)	10	6.66	3.34	11.1556	1.675
(4, 2)	6	5.22	.78	.6084	.117
(4, 3)	2	6.12	−4.12	16.9744	2.774
	100	100.00	0.00		$\chi^2 = 13.012$

Before we can draw any conclusions regarding the influence of marijuana use on behavior, we must determine whether or not the value for χ^2 obtained from the sample is inconsistent with H_0 that these variables are independent. As with earlier hypothesis-testing procedures, the sampling distribution for χ^2 will be used to obtain a critical value for this test statistic.

The Chi-Square Distribution

The chi-square distribution is a theoretical probability distribution that, under the proper conditions, may be used as the sampling distribution of χ^2. It is described by a single parameter—the number of degrees of freedom—which means much the same as it did in our previous discussion of the Student t distribution. (The procedure for determining the number of degrees of freedom will be given a little later.)

Figure 12-1 shows curves for the chi-square distributions when the degrees of freedom are 2, 4, 10, and 20. Note that these curves are positively skewed but that as the degrees of freedom increase, the degree of skew declines. As the number of degrees of freedom becomes large, the chi-square approaches the normal distribution.

Appendix Table F provides upper-tailed areas for the chi-square distribution. Denoting the value of the χ^2 statistic for which the upper-tailed area is α by χ^2_α, we have

$$\alpha = P[\chi^2_\alpha \leqslant \chi^2]$$

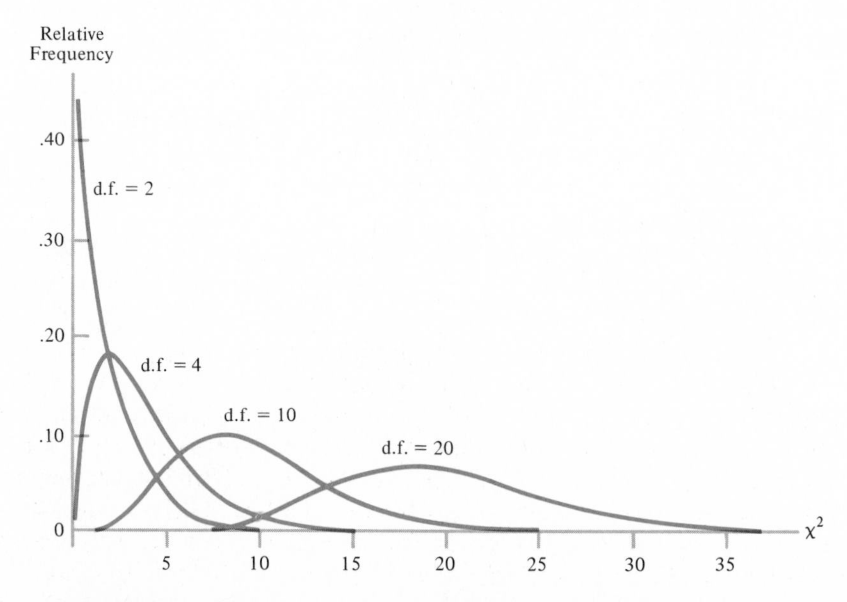

FIGURE 12-1 Various curves for the chi-square distribution.

As with the Student t distribution, there is a separate distribution for each number of degrees of freedom, and a row corresponding to each of these numbers appears in the table of upper-tail areas. The tail areas are given at the head of each column, and the entries in the body of the table are the corresponding values of χ^2_α. The curve for the chi-square distribution with 6 degrees of freedom is shown in Figure 12-2. To find the upper 5-percent value of χ^2, denoted by $\chi^2_{.05}$, we read the entry in the 6 degrees of freedom row and in the column for area .05, obtaining $\chi^2_{.05} = 12.592$. Thus, we have

$$.05 = P[12.592 \leq \chi^2]$$

We can also use Appendix Table F to find areas above points lying in the lower tail. For example, for an upper-tail area of .90, we obtain the value $\chi^2_{.90} = 2.204$ at 6 degrees of freedom (see Figure 12-2).

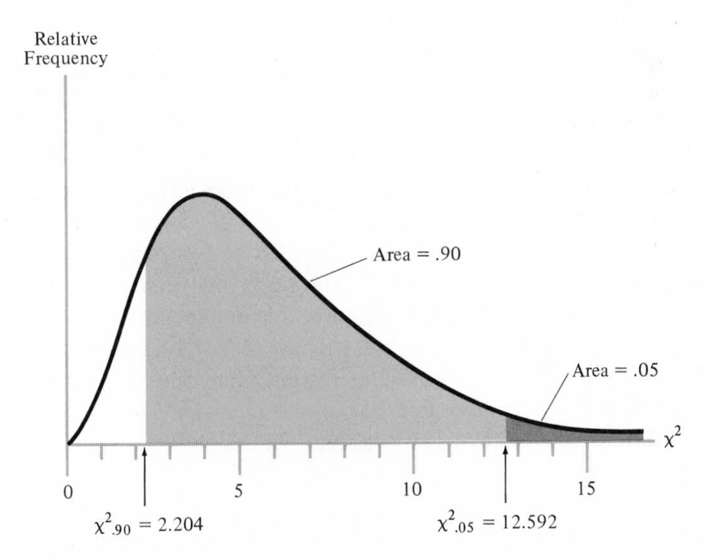

FIGURE 12-2 Chi-square curve for 6 degrees of freedom.

Degrees of Freedom for the Contingency Table

Recall that row and column totals were used to estimate the expected frequencies in Table 12-2 for the marijuana study. The cell entries in that contingency table were found by multiplying the respective marginal totals for the row and column and then dividing by the sample size. Because the four row totals must sum to n, the values of any three rows automatically fix the value of the fourth. Knowing r_1, r_2, r_3, and n, the value for r_4 follows directly. We can say that three of the row totals are "free," while the fourth is "fixed." Likewise, only two of three column totals are free. Although there are $4 \times 3 = 12$ cells, we are free to specify only $3 \times 2 = 6$ of these cells, where the free rows and columns intersect. Our number of degrees of freedom is therefore 6. In general, we establish the following

RULE The number of degrees of freedom for a contingency table is determined by

$$(\text{Number of rows} - 1) \times (\text{Number of columns} - 1)$$

The Decision Rule

Suppose that our researcher wants to protect himself at the $\alpha = .05$ significance level against making the Type I error of concluding that the levels of marijuana usage and dominant antisocial behavior are dependent when these variables are in fact independent. From the table of chi-square areas, we have already found that $\chi^2_{.05} = 12.592$ using 6 degrees of freedom. The appropriate decision rule would then be

Accept H_0 (conclude that the variables are independent) if $\chi^2 \leqslant 12.592$

Reject H_0 (conclude that the variables are dependent) if $\chi^2 > 12.592$

The interpretation of $\chi^2_{.05} = 12.592$ is that due to chance sampling error, this number will be exceeded on the average by only 5 percent of all χ^2 values calculated from repeated samples, each taken from a population where the null hypothesis is true. Thus, if the two variables are independent, about 5 percent of the sample results will disagree so much that independence will be rejected. Each of these samples will have a value, such as 13.5 or 14.2, that exceeds 12.592.

The above decision rule illustrates a convenient feature of this procedure: *all chi-square tests that compare frequencies are upper-tailed.* Stated another way, large χ^2 values tend to refute H_0, while small values confirm the null hypothesis. Since our researcher's computed value of $\chi^2 = 13.012$ is larger than $\chi^2_{.05}$, *the null hypothesis that the level of marijuana usage is independent of dominant type of antisocial behavior must be rejected* at the .05 significance level. The researcher must assume that the type of antisocial behavior resulting from marijuana is influenced by how heavily the drug is used.

Had a smaller significance level such as $\alpha = .01$ been desired, then the researcher would have obtained $\chi^2_{.01} = 15.033$ from Appendix Table F for 6 degrees of freedom. Since this value is larger than 13.012, the null hypothesis of independent variables would have been accepted. In such a case, there may be considerable chance of committing the Type II error of accepting H_0 when the variables are not independent. (In our illustration, this would mean concluding that level of marijuana use does not influence antisocial behavior when it does.) There are many ways in which such errors may occur, and finding β probabilities is a task too difficult to consider here. Only by making the sample size large can the chance of a Type II error be kept within reasonable boundaries.

Limitation on Using the Chi-Square Distribution

The true sampling distribution of χ^2 calculated from expression (12-2) is only approximated by the chi-square distribution. The nature of this approximation is somewhat analogous to substituting the normal distribution in place of the binomial distribution in certain instances. Ordinarily, the approximation is adequate if the sample size is sufficiently large. In practice, the sample will be large enough when the expected frequencies for each class interval are of size 5 or more. Should some of these be smaller than 5, this requirement may be met by combining two rows or columns before calculating χ^2. A corresponding reduction in the degrees of freedom would then have to be made to account for the lower number of cells.

EXERCISES

12-1 Find the value of the chi-square statistic which holds for each upper-tail area α and each number of degrees of freedom (d.f.) stated below:

(a)	(b)	(c)	(d)	(e)
$\alpha = .05$	$\alpha = .01$	$\alpha = .99$	$\alpha = .10$	$\alpha = .80$
d.f. = 10	d.f. = 20	d.f. = 18	d.f. = 6	d.f. = 29

12-2 A hospital dietician is testing to determine if a person's sex is independent of his preference for fruits. For each of the following situations where the sample results have been summarized, find the degrees of freedom and the critical value. Then indicate whether the null hypothesis of independence should be accepted or rejected.

(a) Bananas, apples, and pears are considered. $\chi^2 = 7.85$ and $\alpha = .05$.

(b) Pineapples, guavas, papayas, and passionfruit are considered. $\chi^2 = 15.23$ and $\alpha = .01$.

(c) Boysenberries, blueberries, huckleberries, blackberries, and strawberries are considered. $\chi^2 = 7.801$ and $\alpha = .10$.

(d) Plums, apricots, peaches, cherries, persimmons, and nectarines are considered. $\chi^2 = 10.99$ and $\alpha = .05$.

12-3 Various random samples have been selected and the chi-square statistics have been computed for each of the following independence-testing situations. For the stated significance level, determine the applicable number of degrees of freedom and the corresponding critical value in each situation. Then indicate whether or not the null hypothesis of independence should be accepted for each of the actual chi-square statistics obtained.

(a) Sex (male, female) versus marital status (single, married), with $\alpha = .05$ and $\chi^2 = 3.62$.

(b) College major (liberal arts, science, social science, professional) versus type of employment (manufacturing, service, government), with $\alpha = .01$ and $\chi^2 = 23.885$.

(c) Number of siblings (0, 1, 2, 3 or more) versus desired family size (0, 1, 2, 3 or more), with $\alpha = .10$ and $\chi^2 = 14.753$.

(d) Political affiliation (Republican, Democrat, other) versus sexual attitude (repressive, permissive), with $\alpha = .01$ and $\chi^2 = 8.063$.

12-4 A contingency table of actual sample frequencies is given below:

Preference	Marital Status (1) *Single*	(2) *Married*	Total
(1) *Candidate* A	20	10	30
(2) *Candidate* B	20	50	70
Total	40	60	100

(a) Complete the contingency table for the expected frequencies under the null hypothesis of independence between marital status and candidate preference.
(b) Calculate the chi-square statistic.
(c) Find the number of degrees of freedom and the critical value for the test at the $\alpha = .01$ significance level.
(d) Should the null hypothesis that marital status and candidate preference are independent be accepted or rejected?

12-5 A sociologist wishes to determine whether there are any significant differences in regular series program preferences between male and female television viewers. A random sample of persons has been interviewed. Each person has been asked to indicate which one of five program types he or she prefers. The results are provided below:

Preferred Program Type	Viewer's Sex (1) *Male*	(2) *Female*	Total
(1) *Western*	32	18	50
(2) *Situation comedy*	17	13	30
(3) *Drama*	27	33	60
(4) *Comedy*	13	7	20
(5) *Variety*	24	16	40
Total	113	87	200

The null hypothesis is that a viewer's sex and preference are independent.
(a) Determine the contingency table of the expected results.
(b) Calculate the χ^2 statistic.
(c) How many degrees of freedom are associated with this test statistic?
(d) Assuming that the sociologist wishes to protect himself against the Type I error of incorrectly concluding that sex is dependent upon preference at a significance level of $\alpha = .01$. Determine the critical value for the test statistic. Should the hypothesis of independence be accepted?
(e) Would your conclusion in (d) change if an $\alpha = .05$ significance level were chosen instead? Explain.

12-6 Repeat Exercise 12-5, but suppose that the following data were obtained instead:

Preferred Program Type	Viewer's Sex (1) *Male*	(2) *Female*	Total
(1) *Western*	37	13	50
(2) *Situation comedy*	21	23	44
(3) *Drama*	26	32	58
(4) *Comedy*	19	15	34
(5) *Variety*	22	18	40
Total	125	101	226

12-2 TESTING FOR EQUALITY OF SEVERAL PROPORTIONS

An important statistical question involves a comparison of several population parameters. In Chapter 11, we described the inferences for two populations and the tests comparing *two* population means or proportions. There, we considered the problem of deciding which of two television programs was preferred by using the proportions of two independent samples taken from different populations. But what if more than two programs were to be compared? It would still be possible to make pair comparisons as we did in Chapter 11, but sometimes it is better to test the population proportions simultaneously to see if they really differ. With more than two populations, we must take a different approach. For comparing several proportions, the procedures used earlier in testing for independence may be applied.

Illustration of Procedure

We will illustrate the procedure for testing several proportions with a situation commonly encountered in the advertising field.

An advertising agency wishes to determine whether there are any differences in terms of recall among three kinds of magazine advertisements. One ad is humorous, the second is quite technical, while the third provides a pictorial comparison with competing brands. A national magazine with three regional editions is chosen for the test, and a different quarter-page advertisement is placed in each of its editions. A random sample of persons is chosen from a list of subscribers in each region, and one month after the ads have appeared, each person in the sample is visited. All are shown five ads of similar format, four of which are fakes, and are asked if they remember any of the five ads. Those selecting the correct ad are included in the tally of rememberers, while those unable to select the correct ad are classified as nonrememberers. Following the identification test, a brief quiz is administered to determine whether or not the magazine was read (thus avoiding prestige bias by persons falsely claiming to have read the ad), and nonreaders are eliminated from the sample.

Null Hypothesis

The null hypothesis is that there are no differences in the mnemonic properties of the three kinds of advertisements. This may be expressed in terms of the proportions of the readers of the three magazine editions who remembered the advertisements. Thus, letting π_1 represent the proportion remembering the first ad, π_2 the second, and π_3 the third, we may express the null hypothesis

$$H_0 : \pi_1 = \pi_2 = \pi_3$$

The alternative hypothesis suitable for this test is that at least two proportions differ.

Test Statistic

The results obtained are provided in Table 12-4. Recall that in testing for independence, the expected cell frequencies are determined by assuming equal ratios or proportions. In testing for independence, we are therefore testing for

TABLE 12-4 Results of Mnemonic Advertising Samples

	Type of Advertisement		
	(1) *Humorous*	(2) *Technical*	(3) *Comparative*
Number of rememberers	25	10	7
Number of nonrememberers	73	93	108
Number of readers	98	103	115
Proportion of rememberers	$P_1 = .255$	$P_2 = .097$	$P_3 = .061$

equality of proportions. The same procedures may also be used here. From the sample results, we obtain the contingency table shown below (Table 12-5). Applying expression (12-1), the contingency table for the results to be expected under the null hypothesis can then be determined (Table 12-6).

The χ^2 test statistic may be calculated in the same manner as in testing for independence. The calculations are provided in Table 12-7. The number of degrees of freedom is

$$(\text{No. of rows} - 1) \times (\text{No. of columns} - 1) = (2-1) \times (3-1) = 2$$

TABLE 12-5 Contingency Table for Actual Observed Frequencies

	Type of Advertisement			
	(1) *Humorous*	(2) *Technical*	(3) *Comparative*	Total
(1) *Rememberers*	25	10	7	42
(2) *Nonrememberers*	73	93	108	274
Total	98	103	115	316

TABLE 12-6 Contingency Table for Expected Frequencies

	Type of Advertisement			
	(1) *Humorous*	(2) *Technical*	(3) *Comparative*	Total
(1) *Rememberers*	13.025	13.690	15.285	42
(2) *Nonrememberers*	84.975	89.310	99.715	274
Total	98	103	115	316

TABLE 12-7 χ^2 Calculations for Mnemonic Advertisement Study

(Row, Column)	*Actual Frequency* f_e	*Expected Frequency* f_e	$f_a - f_e$	$(f_a - f_e)^2$	$\frac{(f_a - f_e)^2}{f_e}$
(1, 1)	25	13.025	11.975	143.4006	11.010
(1, 2)	10	13.690	−3.690	13.6161	.995
(1, 3)	7	15.285	−8.285	68.6412	4.491
(2, 1)	73	84.975	−11.975	143.4006	1.688
(2, 2)	93	89.310	3.690	13.6161	.152
(2, 3)	108	99.715	8.285	68.6412	.688
	316	316.000	0.000		$\chi^2 = 19.024$

Testing the Hypothesis

The chosen level of significance is $\alpha = .01$. Appendix Table F shows us that this corresponds to a critical value of $\chi^2_{.01} = 9.210$. This is smaller than the calculated value of the test statistic, $\chi^2 = 19.024$, so that *the null hypothesis of equal proportions must be rejected* at the .01 significance level. The agency must conclude that the three ads are not equally easy to remember. (The results are also significant at the .001 level, since $\chi^2_{.001} = 13.815$.)

EXERCISES

12-7 In planning campaign strategy regarding how to treat a particular issue, a congressional candidate wishes to determine if there are any differences in the proportion of voters who favor the issue among his rural, suburban, and urban constituents. He has collected sample data of opinions and has obtained the following results:

	Rural	*Suburban*	*Urban*
In favor	65	63	52
Not in favor	35	37	48

(a) Construct a contingency table for the expected frequencies.
(b) Calculate the chi-square statistic.
(c) Find the number of degrees of freedom and the critical value associated with this test at the $\alpha = .01$ significance level.
(d) Must the null hypothesis that the proportion in favor is equal among constituent types be accepted or rejected?

12-8 A pastry chef wishes to determine whether the proportion of unsatisfactory bear claws is affected by oven temperature. He has baked batches of them at 350°, 400°, and 425°, and has obtained the following results:

	Temperature		
	350°	400°	425°
Number satisfactory	132	128	111
Number unsatisfactory	14	17	35

If the chef wishes to test the null hypothesis of identical proportions at an $\alpha = .05$ significance level, what conclusion should he reach?

12-9 The quality control manager of an electronics assembly plant wants to know if the day of the week influences the number of erroneous cable harness assemblies. He suspects that on Mondays and Fridays the error rate is significantly higher. If he finds that the day of the week does make a difference, he will conduct a more detailed study to determine which days are worse and will recommend that cables be assembled only on special days. He has collected a random sample of assemblies produced on different days of the week, and obtained the following results:

	Day of the Week				
	Monday	*Tuesday*	*Wednesday*	*Thursday*	*Friday*
Number of erroneous assemblies	32	12	15	18	27
Number of correct assemblies	95	87	91	79	73

If the manager wishes to test the null hypothesis of identical proportions of erroneous assemblies at an $\alpha = .05$ significance level, what conclusion should he reach?

12-3 INFERENCES REGARDING THE POPULATION VARIANCE

An important area of statistics is concerned with making inferences regarding the population variance σ^2. As we have seen, knowledge of population variability is an important element of statistical analysis. For instance, an educator may choose a particular teaching method because it provides low dispersion in student achievement levels, thus making the classroom teacher's job less demanding than a high-variability procedure would. Likewise, a postal service policy might favor a single waiting line that feeds into several service windows rather than separate lines, because it has been estimated that the single line will minimize variability in patron waiting times—even though the mean time spent in line is the same in either case. For a government car pool, tires with a low variability in wear may be preferred to more durable tires with greater variability in useful lifetime, simply because it is cheaper to replace an entire set of tires periodically for each car.

Like the population mean, *σ^2 is ordinarily unknown and its value must be estimated using sample data.* Until now, we have been using the sample variance

$$s^2 = \frac{\sum X^2 - n\bar{X}^2}{n-1}$$

in making point estimates of σ^2. These estimates were made as an adjunct to inferences regarding μ, where confidence intervals and hypothesis-testing decision rules have been constructed using s^2 in place of σ^2.

Although less common than inferences for μ, inferences for σ^2 may be similarly made. To this end, s^2 may be used as an unbiased and consistent estimator of the population variance, playing the analogous role with σ^2 that $\bar{X}$ does with μ. Thus, s^2 serves as the basis either for constructing confidence interval estimates or for testing hypotheses regarding σ^2.

Probabilities for the s^2 Variable

In the *planning stage* of a sampling study, s^2 must be treated as a random variable, since its value is yet to be determined and is subject to chance variation. In discussing how to apply probability analysis to s^2, it will be helpful to briefly review our earlier treatment of $\bar{X}$.

Recall that when σ^2 is unknown, the Student t distribution is used to make inferences regarding the population mean. This is because the random variable $t = (\bar{X}-\mu)\sqrt{n}/s$ is so distributed that probability statements regarding $\bar{X}$ must be made in terms of the Student t distribution. For instance

$$\alpha = P\left[t_\alpha \leqslant \frac{\bar{X}-\mu}{s}\sqrt{n}\right]$$

The chi-square distribution assigns a role to s^2 that is similar to the role the Student t assigns to $\bar{X}$. *Before the sample data are collected, s^2 must be viewed as a random variable* (just like $\bar{X}$ is). Multiplying s^2 by $n-1$ and dividing by σ^2 converts it into ***the chi-square random variable***

$$\chi^2 = \frac{(n-1)s^2}{\sigma^2} \qquad (12\text{-}4)$$

which has the chi-square distribution.* Thus, it is possible to find probabilities for this variable:

$$\alpha = P\left[\chi_\alpha^2 \leqslant \frac{(n-1)s^2}{\sigma^2}\right] \qquad (12\text{-}5)$$

when the number of degrees of freedom is $n-1$.

To illustrate this, suppose that a sample of $n = 25$ heights has been randomly selected from a population of men where the population standard deviation is $\sigma = 3$ inches. We are interested in finding the value for s^2 that only 5 percent of such samples is expected to exceed. From Appendix Table F, we have $\chi_{.05}^2 = 36.415$ for $25-1 = 24$ degrees of freedom. Thus, using the fact that $\sigma^2 = 3^2 = 9$

$$.05 = P\left[36.415 \leqslant \frac{24s^2}{9}\right]$$

Multiplying both sides of this inequality by 9 and dividing by 24, we have

$$.05 = P\left[\frac{9(36.415)}{24} \leqslant s^2\right] = P[13.66 \leqslant s^2]$$

We see that there is a probability of .05 that the sample variance will exceed 13.66. Stated differently, there is only a 5-percent chance that the variability in the sample will be so extreme that the sample standard deviation s will exceed $\sqrt{13.66} = 3.7$ inches.

To find the probability that s^2 falls within a range of numbers, two limits are required for the chi-square variable. These may be chosen so that χ^2 is just as likely to fall below its *lower limit*, denoted by χ_L^2, as it is to fall above its *upper limit* χ_U^2. For example, if these limits are to be chosen for a .90 probability:

$$.90 = P\left[\chi_L^2 \leqslant \frac{(n-1)s^2}{\sigma^2} \leqslant \chi_U^2\right]$$

In this case, the χ^2 values are found from Appendix Table F for $n-1 = 25-1 = 24$ degrees of freedom. The upper-tail areas are .95 for the lower limit and .05 for

* Like the Student t distribution, applicability of the chi-square distribution requires that the *population* be normally distributed.

the upper limit, and the area under that portion of the chi-square curve between the limits is .90:

$$\chi_L^2 = \chi_{.95}^2 = 13.848$$

$$\chi_U^2 = \chi_{.05}^2 = 36.415$$

Two separate tabled values are required to find the limits χ_L^2 and χ_U^2, because the chi-square distribution is *positively skewed.* This is in contrast to our earlier applications involving the normal and the Student t curves, both of which are symmetrical.

Again, using $\sigma = 3$ inches, we have

$$.90 = P\left[13.848 \leqslant \frac{24s^2}{9} \leqslant 36.415\right]$$

so that

$$.90 = P\left[\frac{9(13.848)}{24} \leqslant s^2 \leqslant \frac{9(36.415)}{24}\right] = P[5.19 \leqslant s^2 \leqslant 13.66]$$

Thus, there is a 90-percent chance that a sample of $n = 25$ observations from a population where $\sigma = 3$ inches will provide a sample variance lying between 5.19 and 13.66 (in other words, that s will lie between $\sqrt{5.19} = 2.3$ and $\sqrt{13.66} =$ 3.7 inches).

Confidence Interval Estimate of σ^2

The foregoing probability analysis presumes that the value of the population variance is known. This is not usually the case in a sampling situation. *After the sample data are collected,* s^2 can be calculated and can serve as the basis for making inferences about the unknown σ^2. Treating σ^2 as the unknown and s^2 as the known quantities, we may transform the probability interval for s^2

$$\chi_L^2 \leqslant \frac{(n-1)s^2}{\sigma^2} \leqslant \chi_U^2 \qquad (12\text{-}6)$$

into the form for making an interval estimate of σ^2. Separate algebraic manipulations of the two inequality portions of the above expression provide the following procedure for constructing a ***confidence interval estimate for σ^2***

$$\frac{(n-1)s^2}{\chi_U^2} \leqslant \sigma^2 \leqslant \frac{(n-1)s^2}{\chi_L^2} \qquad (12\text{-}7)$$

Here, χ_L^2 and χ_U^2 must be chosen from Appendix Table F, using $n-1$ degrees of freedom, to correspond to the confidence level desired.

Example 12-2 A sociologist estimated the variability of IQ, as measured on the Stanford-Binet scale, for the inmates at a certain state prison. For a sample of $n = 30$ prisoners, he obtained a standard deviation of $s = 10.2$ points, considerably below the corresponding figure of 16 points for all test takers.

The sample variance is $s^2 = (10.2)^2 = 104.04$. Using this result, we may construct a 90-percent confidence interval estimate for the variance in IQ of the prison population. For $30 - 1 = 29$ degrees of freedom, Table F provides the following lower and upper limits for the chi-square variable:

$$\chi_L^2 = \chi_{.95}^2 = 17.708$$

$$\chi_U^2 = \chi_{.05}^2 = 42.557$$

Substituting these and the sample results into expression (12-7), we obtain the following interval estimate for σ^2:

$$\frac{(30-1)104.04}{42.557} \leqslant \sigma^2 \leqslant \frac{(30-1)104.04}{17.708}$$

or

$$70.8969 \leqslant \sigma^2 \leqslant 170.3840$$

To estimate the population standard deviation of these results, we may take the square root of the above terms and obtain the following 90-percent confidence interval for σ:

$$\sqrt{70.8969} \leqslant \sigma \leqslant \sqrt{170.3840}$$

or

$$8.42 \leqslant \sigma \leqslant 13.05$$

Since this interval lies well below 16, the sociologist might safely conclude that the prisoners have a lower variability in IQ than the population does as a whole. Later in this section, we will see how such comparisons may be made using hypothesis-testing concepts.

In explaining why inmates have less variability in IQ, the sociologist concludes: You can't be very dull and still be a serious criminal; besides, retarded persons who commit crimes are incarçerated in other kinds of institutions. On the other hand, the smartest people have, by and large, avoided a life of crime. In effect, the tails of the normal curve for IQ are underrepresented in prisons, which explains the lower variability in measured intelligence.

Testing Hypotheses Regarding σ^2

Hypothesis tests regarding σ^2 may be conducted in a manner similar to the tests described in earlier chapters. The null hypothesis regarding σ^2 may be either one- or two-sided. Decision rules may be constructed in terms of the chi-square variable, which serves as the test statistic. Using the hypothesized value of the population variance, denoted by σ_0^2, the appropriate *test statistic is*

$$\chi^2 = \frac{(n-1)s^2}{\sigma_0^2} \tag{12-8}$$

The significance level α establishes the critical values for this statistic.

One-Sided Tests

For one-sided tests, the decision rule is based either upon χ^2_α, which must correspond to the upper-tail of the chi-square curve, or upon $\chi^2_{1-\alpha}$ for the lower-tail of that curve. This is because for a lower-tailed test, the critical value for χ^2 must be that point *below* which the area is α. The cases may be summarized as follows:

Lower-tailed test	*Accept* H_0 if $\chi^2 \geqslant \chi^2_{1-\alpha}$
$H_0 : \sigma^2 \geqslant \sigma_0^2$	*Reject* H_0 if $\chi^2 < \chi^2_{1-\alpha}$
Upper-tailed test	*Accept* H_0 if $\chi^2 \leqslant \chi^2_\alpha$
$H_0 : \sigma^2 \leqslant \sigma_0^2$	*Reject* H_0 if $\chi^2 > \chi^2_\alpha$

Example 12-3 That predictability and variability are related can be illustrated by the attitudes of people who wait in lines. A lengthy waiting time is more acceptable if the variability is smaller, even though the average wait may be the same. When the variability is smaller, the inconvenience of waiting becomes more predictable. This accounts for the fact that many businesses and government offices dealing with the public have instituted a "single-line" policy as a replacement for the earlier, chaotic procedure of having independent lines form at various service areas (like windows at a post office). Although the mean waiting time is not greatly affected by the single-line policy, the variability in waiting time is.

A particular postmaster has determined that the current procedure of separate lines yields a standard deviation in waiting times on late December mornings of 10 minutes per customer. He has decided to implement the single-line policy on a trial basis to see if a reduction in waiting-time variability occurs.

A sample of 30 customers was monitored, and their waiting times were determined. The sample standard deviation was $s = 5$ minutes per customer. As his null hypothesis, the postmaster assumes that the variance in waiting times will be at least as great under the experimental procedure

$$H_0 : \sigma^2 \geqslant \sigma_0^2 \qquad (10^2 = 100)$$

This test is lower-tailed. At an $\alpha = .01$ significance level, we find for $30 - 1 = 29$ degrees of freedom that the critical value for the chi-square statistic is

$$\chi^2_{1-\alpha} = \chi^2_{.99} = 14.256$$

The postmaster's decision rule is

Accept H_0 (and retain the old procedure) if $\chi^2 \geqslant 14.256$

Reject H_0 (and adopt the new procedure) if $\chi^2 < 14.256$

From expression (12-8), the test statistic has the value

$$\chi^2 = \frac{(30-1)(5)^2}{(10)^2} = 7.25$$

Since 7.25 is smaller than $\chi^2_{.99} = 14.256$, H_0 must be *rejected* and the postmaster should adopt the new single-line system.

Two-Sided Tests

When the null hypothesis takes the form $H_0 : \sigma^2 = \sigma_0^2$, a two-sided test applies. Large or small values for the sample variance tend to refute H_0. Thus, either a large or a small calculation for χ^2 will result in rejecting H_0. As with the two-sided tests we discussed in earlier chapters, a convenient procedure for making a decision is to construct a confidence interval corresponding to the significance level. If σ_0^2 falls outside this interval, H_0 must be rejected; if the interval contains σ_0^2, H_0 must be accepted.

To illustrate this procedure, we again consider the variance in prisoner IQs in Example 12-2. There, we obtained the 90-percent confidence interval

$$70.8969 \leqslant \sigma^2 \leqslant 170.3840$$

Suppose that the sociologist in our example wished to test the null hypothesis that the variability in prisoner IQs is the same as the IQ variability in the general population as a whole, where the Stanford-Binet test has a standard deviation of 16. To do this, he would use

$$H_0 : \sigma^2 = \sigma_0^2 \quad (16^2 = 256)$$

Since $\sigma_0^2 = 256$ lies above the upper limit of the 90-percent confidence interval, H_0 must be *rejected* at the $\alpha = .10$ significance level. The sociologist must conclude that prisoner IQs have a lower variability than IQs in the general population. In fact, H_0 could be rejected at the lower significance level $\alpha = .02$. (As an exercise, you may verify this by constructing a 98-percent confidence level.)

Normal Approximation for Large Sample Sizes

The chi-square distributions in Appendix Table F are not given above 30 degrees of freedom. ***For values of n larger than 30, we may approximate the chi-square distribution by the normal curve.*** It may be shown that χ^2 has a mean equal to the degrees of freedom and a variance equal to twice that amount. Thus, we may compute the applicable normal deviate from:

$$z = \frac{\chi^2 - (n-1)}{\sqrt{2(n-1)}} \tag{12-9}$$

Expression (12-9) is useful for testing hypotheses regarding σ^2. In order to construct a confidence interval, however, it is more convenient to express the critical values of χ^2 in terms of the tabled value of z by the equivalent expressions

$$\chi_L^2 = \chi_{1-\alpha}^2 = n - 1 - z\sqrt{2(n-1)}$$

and (12-10)

$$\chi_U^2 = \chi_\alpha^2 = n - 1 + z\sqrt{2(n-1)}$$

Example 12-4 A medical researcher ordered a sampling study to see how effectively a particular tranquilizer induces sleep. The drug was administered to a sample of $n = 100$ patients. The sample standard deviation in times required until onset of sleep was computed as $s = 7.3$ minutes. Based upon this result, the researcher may construct a 95-percent confidence interval for the variance in time to sleep. Using $z = 1.96$ (so that the area under the normal curve between $\pm z$ is .95 and the area in each of the two tails is $\alpha = .025$) the following limits for the chi-square variable can be found from expression (12-10):

$$\chi_L^2 = \chi_{.975}^2 = 99 - 1.96\sqrt{2(99)}$$
$$= 99 - 27.58 = 71.42$$
$$\chi_U^2 = \chi_{.025}^2 = 99 + 27.58 = 126.58$$

Substituting these values into expression (12-7), with $s = 7.3$ and $n = 100$:

$$\frac{99(7.3)^2}{126.58} \leqslant \sigma^2 \leqslant \frac{99(7.3)^2}{71.42}$$

or

$$41.68 \leqslant \sigma^2 \leqslant 73.87$$

Taking the square root of the above reveals a 95-percent confidence that the population standard deviation will fall between 6.46 and 8.59 minutes.

Suppose that our researcher wanted to test the null hypothesis that the variability in the new sleeping drug is the same as the variability in the old one, where previous records show that the standard deviation is 8 minutes. Thus, he would have

$$H_0 : \sigma^2 = \sigma_0^2 \quad (8^2 = 64)$$

Since this value falls inside the 95-percent confidence interval for σ^2, the researcher must *accept* H_0 at the $\alpha = .05$ significance level and conclude that the two drugs are identical in variability of time necessary to induce sleep.

EXERCISES

12-10 For each of the following situations, construct a 90-percent confidence interval estimate for the population variance:

(a) $s^2 = 20.3$, $n = 25$ (b) $s^2 = 101.6$, $n = 12$ (c) $s^2 = .53$, $n = 15$ (d) $s^2 = 7.78$, $n = 20$

12-11 For each of the following hypothesis-testing situations: (1) indicate whether the test is lower-tailed or upper-tailed; (2) find the critical value for the chi-square statistic for the indicated significance level; and (3) calculate χ^2, and state whether H_0 must be accepted or rejected.

(a) H_0:$\sigma^2 \geqslant 16$; $n = 25$; $\alpha = .05$; $s^2 = 19$.
(b) H_0:$\sigma^2 \leqslant 100$; $n = 10$; $\alpha = .01$; $s^2 = 105$.
(c) H_0:$\sigma^2 \geqslant .64$; $n = 19$; $\alpha = .10$; $s^2 = .59$.
(d) H_0:$\sigma^2 \leqslant 6.1$; $n = 17$; $\alpha = .05$; $s^2 = 10.1$.

12-12 A biological researcher wished to publish her findings regarding the effects of a special diet on the variability in litter size of laboratory rats. To do this, she randomly selected a sample of $n = 25$ infant females who were raised on the new diet up to and through pregnancy. The mean and the variance for the litter sizes of this group were $\bar{X} = 6.3$ rats, with $s^2 = 3.38$.

(a) Construct a 90-percent confidence interval for the population variance for the litter size of rats under the special diet.
(b) Use your answer to (a) to find what two values the *standard deviation* in litter size falls between at the 90-percent confidence level.
(c) Suppose that the variance in litter size for rats fed an ordinary diet is 5.1. Should the researcher conclude that the new diet yields the same variance? Use $\alpha = .10$ as the significance level.

12-13 A sociologist has administered the Stanford-Binet test to 30 randomly chosen persons in a particular city in order to draw conclusions regarding the IQ scores of professionals. The sample standard deviation obtained was $s = 12.7$.

(a) Assuming that the population standard deviation for the IQs of nonprofessional persons is 14, should the null hypothesis that professionals have at least as great a variability in IQ be accepted or rejected at the $\alpha = .05$ significance level?

(b) Suppose that the population standard deviation for attorneys is 10.3. Should the null hypothesis that professional people as a whole have a variability in IQ that is no larger be accepted or rejected at the $\alpha = .05$ significance level?

(c) Construct a 90-percent confidence interval estimate for the variance in professional IQ scores. Does your answer indicate that the variability in IQ for this group is different from the variability for the general population, where the standard deviation is 16? (Use $\alpha = .10$.)

12-14 Suppose that the researcher in Example 12-4 used another drug on $n = 200$ patients, where the standard deviation in time to sleep was $s = 8.2$ minutes.

(a) Construct a 95-percent confidence interval for the population variance for the time to sleep.

(b) Suppose the researcher wished to test the null hypothesis that this drug will have the same variability as the one presently used, where the standard deviation is 8 minutes. What conclusion should he reach at the $\alpha = .05$ significance level?

12-15 The researcher in Exercise 12-12 performed a similar experiment to assess the effect of a hormone on the variability in litter size of laboratory rats. Using a sample size of $n = 100$ females, she found a mean litter size of $\bar{X} = 5.8$ rats and a sample variance of $s^2 = 6.3$. Untreated rats have a population variance in litter size of 5.1. At the $\alpha = .05$ significance level, should the researcher accept or reject the null hypothesis that the variability in litter size for treated rats will be at most as large as that for untreated ones?

REVIEW EXERCISES

12-16 In developing nations, birth control is not yet widespread. A government population planning official in an Asian country must determine if traditional contraceptive procedures will work with the rural poor. Using a sample of 250 women, each practicing one of four methods of contraception for two years, he has obtained the following results:

	Contraceptive Method				
	(1) *IUD*	(2) *Pill*	(3) *Mechanical*	(4) *None*	Total
(1) *Pregnant*	20	10	30	30	90
(2) *Not pregnant*	40	50	50	20	160
Total	60	60	80	50	250

Should the population planner conclude that traditional contraceptive procedures will work? (Use $\alpha = .01$.)

12-17 A psychologist has administered a personality profile test to a random sample of $n = 25$ welfare mothers. For the index of depression, a mean of $\bar{X} = 65$ and a standard deviation of $s = 12$ were obtained from the sample data.

(a) Construct a 90-percent confidence interval for the population mean in depression scores.

(b) Construct a 90-percent confidence interval for the population variance in depression scores.

(c) If the population mean depression index is 55 for mothers in general, should the psychologist conclude that this index differs for welfare mothers? (Use $\alpha = .10$.)

(d) If the population standard deviation in scores for all mothers is 10 points, should the psychologist conclude that variability in the index differs for welfare mothers? (Use $\alpha = .10$.)

12-18 A consumer agency has tested a random sample of $n = 100$ sets of radial tires. The sample standard deviation in tire-set lifetimes was determined to be $s = 2{,}000$ miles.

(a) Construct a 99-percent confidence interval estimate for the population variance in lifetimes of radial tires.

(b) Should the agency conclude that radial tires have a different lifetime variability than nonradial tires, which have a population standard deviation of 2,500 miles? (Use $\alpha = .01$.)

12-19 A psychology student wishes to test the null hypothesis that people's sexual attitudes are independent of the kinds of cars they own. A random sample of 100 persons was selected, and each was interviewed and placed into a "dominant" sexual-attitude category. The following results were obtained:

Dominant Sexual Attitude	*Previously Purchased Car*			
	(1) *Foreign*	(2) *Small*	(3) *Large*	Total
(1) *Repressed*	3	6	11	20
(2) *Ho-hum*	13	25	12	50
(3) *Swinger*	14	9	7	30
Total	30	40	30	100

At the $\alpha = .05$ significance level, what conclusion should the student make?

12-20 A presidential candidate is formulating a strategy to win the election. To do this, he must determine whether his appeal is generally broad, or whether it varies from region to region. A sampling study provided him with the following regional data:

	East	*Central*	*South*	*West*
Number preferring	21	29	23	16
Number not preferring	39	41	27	24

Should the candidate conclude that he is equally preferred in all regions? (Use $\alpha = .01$.)

12-21 Suppose that the candidate in Exercise 12-20 wishes to test the null hypothesis that he is equally strong in the East and in the West. Using an $\alpha = .05$ significance level, what should he conclude?

13 Analysis of Variance and Related Topics

Nasrudin was walking on the main street of a town, throwing out bread crumbs. His neighbors asked, "What are you doing, Nasrudin?"

"Keeping the tigers away."

"There have not been tigers in these parts for hundreds of years."

"Exactly. Effective, isn't it?"

This chapter describes procedures for analyzing several quantitative populations. The central focus is a statistical application called *analysis of variance*, which is actually a method for comparing the means of more than two populations.

This kind of analysis is very useful in many areas of research. In medicine, it can help to determine whether a patient's recovery is influenced by different treatments. It can be used by psychologists to assess the influence of environment on human behavior. In agriculture, analysis of variance may be helpful in determining if crop yields differ according to the kinds of fertilizers or pesticides used.

The basic questions considered in this chapter have been posed earlier in Chapter 11, where we compared two population means using two samples—control and experimental groups. When more than two samples are involved, however, a radically different approach is required.

Since analysis of variance deals with means, it may appear to be misnamed, but as we will see, *this procedure achieves its goal by comparing sample variances.* In doing this, a new probability distribution—the *F distribution*, must be used.

Like the chi-square distribution introduced in Chapter 12, the *F* distribution has applications beyond its central role in the analysis of variance. Later in this chapter, we will describe one of these additional uses—the *comparison of two population variances.*

13-1 ANALYSIS OF VARIANCE AND THE *F* DISTRIBUTION

Analysis of variance uses sample data to compare several *treatments* in order to determine if they achieve different results. Here, we use the word "treatment" in a broad sense that includes not only medical therapy but other factors a researcher might investigate. Thus, an agronomist may consider several concentrations of fertilizer as "treatments" for crops. Similarly, each reading program in elementary education may represent a different "treatment" of the developing child. Sample data can be obtained by applying the respective treatments to different samples. These sample data may then be analyzed to determine if the treatments differ.

Testing for Equality of Means

As an illustration, consider the data in Table 13-1 obtained for three

TABLE 13-1 Sample Fertilizer Treatment Yields (square yards) for Three Fertilizing Schemes

		Fertilizer Treatment		
Sample Observation Number		*(1) Quarter Dosage Once Weekly*	*(2) Half Dosage Every Two Weeks*	*(3) Full Dosage Every Four Weeks*
1		77	83	80
2		79	91	82
3		87	94	86
4		85	88	85
5		78	85	80
	Totals	406	441	413
	Means	$\bar{X}_1 = 81.2$	$\bar{X}_2 = 88.2$	$\bar{X}_3 = 82.6$

$$\bar{\bar{X}} = \frac{406+441+413}{5(3)} = 84$$

different fertilizing schemes being evaluated by a lawn-sod farmer. Here, the total dosages are the same, but the quantities applied and the fertilizing frequencies have been varied. The sample observations represent the number of square yards of marketable sod obtained from random plots of 100 square yards, all seeded with the same variety of grass mix. In each case, a random sample of five plots was subjected to one of the three treatments.

The analysis of variance procedure used to determine whether fertilizing affects sod yield involves two variables, and we may note some similarities between this situation and regression analysis. The method of fertilizing is a qualitative variable, which is sometimes referred to as a *factor*, and each treatment is a *factor level*. The factor is analogous to the independent variable in regression analysis. The sod yield is the *response variable*, which must be quantitative, and since the response achieved may depend upon the particular treatment used, this variable plays a role similar to the dependent variable used in regression analysis. Although analysis of variance is primarily concerned with *qualitative factors*, it can be used with quantitative factors as well; in such cases, the factors are fixed at a few key levels and constitute *quantitative categories* rather than continuous variates.

The sod farmer must compare the alternative fertilizing schemes to see if they provide different mean yields of marketable sod. In effect, he wishes to test the null hypothesis that the population mean yield per 100 square yards of seeded surface is the same under each treatment. Using subscripts 1, 2, and 3 for the respective treatments, we may express the **null hypothesis**

$$H_0 : \mu_1 = \mu_2 = \mu_3$$

The corresponding alternative hypothesis is that the means are not equal; that is, that at least one pair of the μs differ.

The procedures described in this chapter actually test a somewhat stronger null hypothesis—that the treatment populations are identical, or in other words, that they have the same frequency distribution form. In particular, this assumption means that each treatment population has the same common value for its variance.

The concepts underlying this procedure are illustrated in Figure 13-1. When the sample data are combined, they appear to be observations from a single, highly disperse population, as shown in (a). But when each treatment is viewed separately, these same sod yields appear to belong to three separate populations with smaller variances, as indicated in (b). Under the null hypothesis, however, the treatment populations have identical means and the same variance, so that an identical frequency curve like the one in (c) applies for each method of fertilizing.

As with earlier hypothesis-testing procedures, we must convert the sample data into a test statistic and see whether or not the value achieved refutes H_0. Before we do this, it is necessary to establish some notation and concepts.

Summarizing the Data

Each sample plot yield in Table 13-1 may be represented by a symbol X_{ij}, where i refers to the row or observation number and j refers to the column. In each column, the values are the sample observations made from the corresponding treatment populations. For example, $X_{32} = 94$ square yards—the sod yield for the third test plot in the second sample, where treatment (2), half the fertilizer

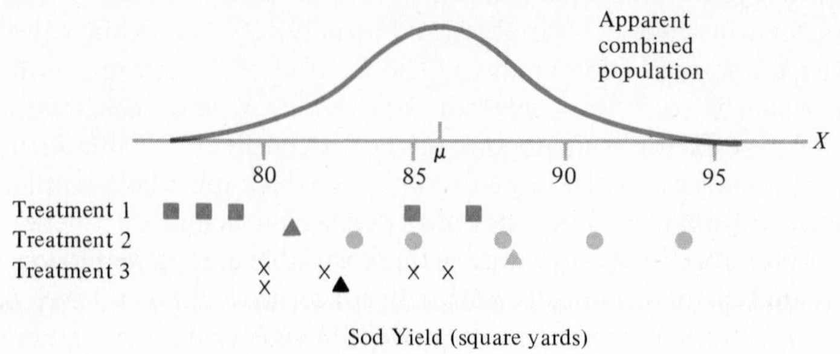

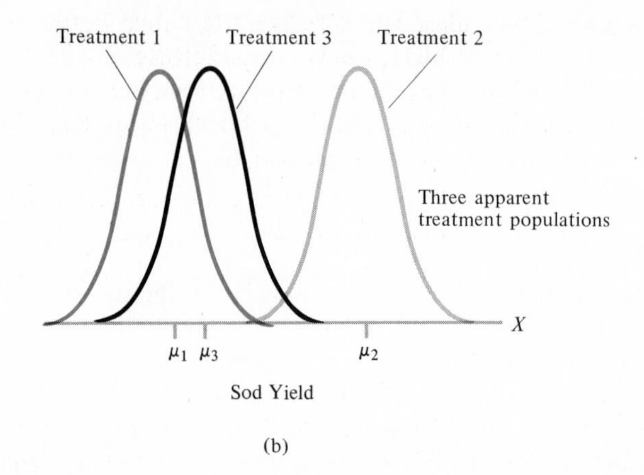

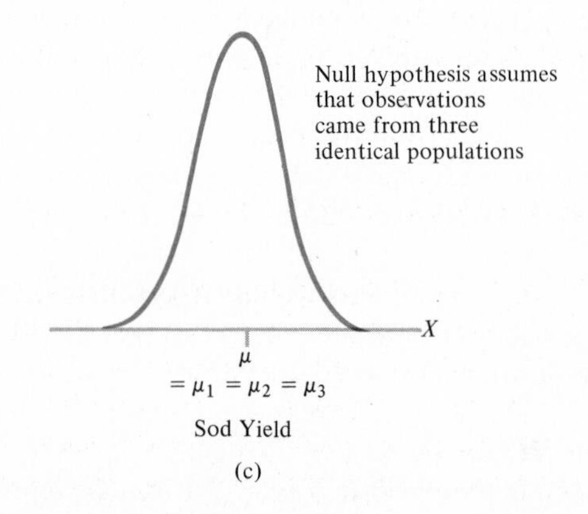

FIGURE 13-1 Concepts underlying analysis of variance.

dosage every two weeks, was used. In the usual manner, we can calculate the ***sample mean for the jth treatment column***

$$\bar{X}_j = \frac{\sum_i X_{ij}}{r} \tag{13-1}$$

where all the observations in the jth column are summed and divided by the number of observations made under that treatment, r = the number of rows. In our example, $j = 1, 2,$ or 3, depending upon which treatment is being considered. For the first treatment, the sample mean is found by summing the values in column (1) and dividing by $r = 5$:

$$\bar{X}_1 = \frac{77+79+87+85+78}{5} = 81.2 \text{ square yards}$$

The other sample means can be similarly calculated as $\bar{X}_2 = 88.2$ and $\bar{X}_3 = 82.6$ square yards.

To facilitate testing the null hypothesis, we require that the data from the three samples be pooled in calculating the ***grand mean***

$$\bar{\bar{X}} = \frac{\sum_j \sum_i X_{ij}}{rc} \tag{13-2}$$

Here, we denote the grand mean by $\bar{\bar{X}}$ ("X double bar"). The double summation indicates that first we obtain the column totals and then we sum these figures for all treatments. The resultant total is divided by the combined sample size. We denote the number of treatments or columns by the letter c; in this illustration, there are $c = 3$ treatments. The combined sample size is therefore $rc = 5(3) = 15$. The grand mean for this illustration was computed in Table 13-1 to be $\bar{\bar{X}} = 84$ square yards.

Using Variability to Identify Differences

As with any testing problem, a test statistic is desired that will (1) highlight the differences between the observed sample results and what would be expected under the null hypothesis, and (2) have a convenient sampling distribution to measure the effect of chance sampling error. As we have seen, the amount of sampling error may be easily estimated from the variability of the sample results. We may also use the variability of the results to express differences. For instance, if several values are unalike, then we have seen that their dispersion— expressed by a range, variance, or standard deviation—will be greater than it will if the values are nearly the same. When many values are involved, their collective differences can be summarized by one of these measures of variability.

Since we are dealing with several populations, it is convenient to use the sample data to measure three sources of variability: (1) *treatments* variation, which measures how the sample results differ under the various treatments; (2) *error*, which collectively summarizes how the observations vary within their respective samples; and (3) *total* variation of the sample observations without regard to the populations to which they belong.

Treatments Sum of Squares

To summarize the variability between sample results, we use the following for the ***treatments sum of squares***

$$SST = r\sum(\bar{X}_j - \bar{\bar{X}})^2 \tag{13-3}$$

In calculating SST, we sum squared deviations of the sample treatment means from the grand mean and then multiply by the number of observations r made under each treatment. Using the earlier results from our sod-fertilizing experiment

$$\begin{aligned} SST &= 5[(81.2-84)^2 + (88.2-84)^2 + (82.6-84)^2] \\ &= 5(7.84+17.64+1.96) = 137.2 \end{aligned}$$

We multiply by $r = 5$, the number of observations per treatment (or the number of rows), so that with $c = 3$ treatments or columns, all of the $rc = 15$ observations are represented.

The treatments sum of squares expresses the variation between columns, which is often referred to as *explained variation*. This is because SST is obtained from differences in the sample means. Thus, SST summarizes those differences in sample results that might be due to inherent differences in the treatment populations rather than to chance alone.

Error Sum of Squares

We summarize the variability within samples by a number found by summing the squared deviations of the individual observations about their respective sample means. This variability is referred to as the ***error sum of squares***

$$SSE = \sum\sum(X_{ij} - \bar{X}_j)^2 \tag{13-4}$$

Table 13-2 shows the calculation of the error sum of squares for the results of the sod-fertilizing experiment. There, we obtain $SSE = 190.80$. In performing this calculation, it is convenient to maintain the same column arrangement we used initially: first, totaling the squared deviations for each treatment and then summing these to obtain SSE. As with SST, SSE accounts for each individual sample observation made.

The error sum of squares expresses variation within the columns. This is sometimes called *unexplained variation*, because the error sum of squares measures

TABLE 13-2 Calculation of the Error Sum of Squares for the Sod-Fertilizing Experiment

i	$(X_{i1}-\bar{X}_1)^2$	$(X_{i2}-\bar{X}_2)^2$	$(X_{i3}-\bar{X}_3)^2$
1	$(77-81.2)^2 = 17.64$	$(83-88.2)^2 = 27.04$	$(80-82.6)^2 = 6.76$
2	$(79-81.2)^2 = 4.84$	$(91-88.2)^2 = 7.84$	$(82-82.6)^2 = .36$
3	$(87-81.2)^2 = 33.64$	$(94-88.2)^2 = 33.64$	$(86-82.6)^2 = 11.56$
4	$(85-81.2)^2 = 14.44$	$(88-88.2)^2 = .04$	$(85-82.6)^2 = 5.76$
5	$(78-81.2)^2 = 10.24$	$(85-88.2)^2 = 10.24$	$(80-82.6)^2 = 6.76$
	80.80	78.80	31.20

$$SSE = \sum\sum (X_{ij}-\bar{X}_j)^2 = 80.80+78.80+31.20 = 190.80$$

differences between sample values that are due to chance (or residual) variation, for which no identifiable cause can be found. This is in contrast to *SST*, which explains the variation between samples in terms of differences in treatment populations.

Total Sum of Squares

If we initially ignore the groupings of the sample observations, then we can determine the sum of squares for a single combined sample. The result obtained is called the ***total sum of squares***

$$\text{Total } SS = \sum\sum (X_{ij}-\bar{\bar{X}})^2 \tag{13-5}$$

We may calculate the total sum of squares for our 15 observations:

$$\text{Total } SS = (77-84)^2 + (79-84)^2 + \cdots + (80-84)^2 = 328$$

Note that if we add the treatments and error sums of squares, we obtain the same result:

$$SST + SSE = 137.20 + 190.80 = 328$$

In general, it can be mathematically shown that

$$\text{Total } SS = SST + SSE \tag{13-6}$$

for any set of sample results.

Computational Difficulties

When sample sizes are large, sum of squares calculations can be quite burdensome. In such cases, computations may be considerably simplified by using shortcut expressions for *SST* and *SSE*. Since the more straightforward procedures presented here work just as well for small samples, short-cut expressions have not been included in this book. Instead, *it is recommended that large problems be run on a computer.* Canned programs that perform analysis of variance for a variety of situations are widely available from most large computer centers and time-sharing facilities.

The Test Statistic

Expression (13-6) shows that the two components of total variation are explained (SST) and the unexplained (SSE) variations. Our task is to determine whether the explained variation is significant enough to warrant rejecting the null hypothesis that the treatment populations have identical means.

Comparing Variations

As our test statistic, we must use a summary measure to express how much the sample results deviate from what is expected when the null hypothesis is true. This may be achieved by comparing the explained and the unexplained variations. Regardless of whether or not H_0 is true, we should expect some error or unexplained variation within each sample, as this is natural in any random sampling experiment. But according to H_0—that population means are identical for each method of fertilization—the amount of explained treatments variation between samples should be small; that is, the respective sample means should be about the same. If, in fact, the population means are equal, then the explained and the unexplained variation components should be of comparable size.

To compare variation explained by treatments to the error or unexplained variation, we will use their ratio.

Mean Squares

To find this ratio, we cannot immediately divide the sums of squares. Recall that SSE is the sum of $r \times c$ squared differences (for the 15 observations in our example), while only c squares (representing the 3 samples in our example) are used to calculate SST. Each sum of squares must be converted into an average before SSE and SST are comparable. We then have sample variances, which we will call *mean squares* to avoid confusion with population variances. Mean squares may be viewed as estimators of the population variances, which, under the assumption that the samples are taken from identical populations, are equal to a common value of σ^2. These estimators are unbiased when the proper divisors are chosen. We express the ***treatments mean square***

$$MST = \frac{SST}{c-1} \tag{13-7}$$

In a similar manner, we define the ***error mean square***

$$MSE = \frac{SSE}{(r-1)c} \tag{13-8}$$

Returning to our sod-fertilizing illustration, we have

$$MST = \frac{137.2}{3-1} = 68.6 \qquad \text{and} \qquad MSE = \frac{190.8}{(5-1)\,3} = 15.9$$

Note that the treatments mean square is more than four times as large as the error mean square. Under the null hypothesis of identical population means, they should be nearly the same. Such a large difference seems unlikely, but it may be "explained" by differences between populations. We have yet to determine just how unlikely this large discrepancy would be if the populations were in fact identical.

The ANOVA Table

It is helpful to summarize the computations for analysis of variance in the format of Table 13-3. For simplicity, this is referred to as an ANOVA table (a contraction which combines the beginning letters from "Analysis Of Variance"). This table conveniently arranges the intermediate calculations. As we will see later in this chapter, the ANOVA table is important in organizing the computations for more complicated experiments, because it provides the degrees of freedom, the sum of squares, and the mean square for each source of variation. The degrees of freedom are the divisors we use to calculate the mean squares. The test statistic, which will be discussed next, is the value in the column labeled F.

TABLE 13-3 ANOVA Table for Sod-Fertilizing Experiment

Variation	*Degrees of Freedom*	*Sum of Squares*	*Mean Square*	*F*
Explained by Treatments (between columns)	$c-1=2$	$SST = 137.2$	$MST = 137.2/2 = 68.6$	$MST/MSE = 68.6/15.9 = 4.31$
Error or Unexplained (within columns)	$(r-1)c = 12$	$SSE = 190.8$	$MSE = 190.8/12 = 15.9$	
Total	$rc-1 = 14$	$SS = 328$		

The F Statistic

We now have two sample variances, MST and MSE. In keeping with our earlier discussions, we may refer to the treatments mean square as the variance explained by treatments and to the error mean square as the unexplained variance. In calculating the ratio of these variances, we obtain our *test statistic*

$$F = \frac{\text{Variance explained by treatments}}{\text{Variance left unexplained}} = \frac{MST}{MSE} \tag{13-9}$$

For our example, we calculate

$$F = \frac{68.6}{15.9} = 4.31$$

Under the null hypothesis, we would expect values for F to be close to 1, because MST and MSE are both unbiased estimators of the common population variance σ^2. Since they have the same expected value, the sample should yield like values for MST and MSE, and the ratio of these values should be near 1. In order to formulate a decision rule, we must establish the sampling distribution of this test statistic. From this, we can find a critical value that will tell us whether the calculated value of F is large enough to cause rejection of the null hypothesis. The probability distribution we use to do this is called the F distribution. Before describing this distribution, we must clarify what is meant by degrees of freedom.

Degrees of Freedom

We may view the divisor $c-1$ in the calculation for MST as the number of degrees of freedom associated with using MST to estimate σ^2. Here, finding the sum of squares involves calculating $\overline{\overline{X}}$, which may be expressed in terms of the $\overline{X}_j$s. For a fixed $\overline{\overline{X}}$, all but one of the c $\overline{X}_j$s are free to vary.

Analogously, the $(r-1)c$ divisor in the calculation for MSE is the number of degrees of freedom associated with using MSE to estimate σ^2. This is due to the fact that in finding the sum of squares, each term involves an $\overline{X}_j$ calculated from the X_{ij} values. For a given value of $\overline{X}_j$, only $r-1$ of the X_{ij}s are free to assume any value. For c treatments, the number of free variables is therefore only $(r-1)c$.

Thus, a pair of degrees of freedom is associated with the F statistic. This pair sums to the total number of observations minus one. In our sod-fertilizing example, $r = 5$ and $c = 3$, so that the following degrees of freedom apply:

$$\text{For the numerator: } 3 - 1 = 2$$

$$\text{For the denominator: } (5-1)3 = 12$$

and $2+12 = 14$, a value 1 less than the total number of observations.

The *F* Distribution

Under the proper conditions, we may employ the F distribution to obtain probabilities for possible values of F. Like those for the chi-square and t distributions, the F distribution is characterized by degrees of freedom. Because the F statistic is defined as a ratio, the F distribution has two kinds of degrees of freedom. We associate one of these with the numerator and one with the denominator.

Figure 13-2 shows curves for the F distribution when the degrees of freedom for the numerator and denominator, respectively, are: 6 and 6; 20 and 6; and 30 and 30. Note that the F distribution curve is positively skewed, with possible values ranging from zero to infinity. There is a different distribution and curve for each pair of degrees of freedom.

Since the F distribution is continuous, probabilities for the values of F are provided by the areas under the curves. The critical values for upper-tail areas under the F distribution are provided in Appendix Table G. The table is constructed in the same manner as the tables for the t and chi-square distributions. Due to space limitations, only two upper-tail areas are considered.

In Appendix Table G, the rows correspond to the number of degrees of freedom in the denominator and the columns correspond to the number of degrees of freedom in the numerator. The entries in the body of the table are critical values, designated as F_α, where $\alpha = .01$ or $.05$. To find $F_{.01}$ and $F_{.05}$ when the numerator and the denominator degrees of freedom are 6 and 6, we read the entries from column 6 and row 6. The boldface type is the value for $F_{.01}$

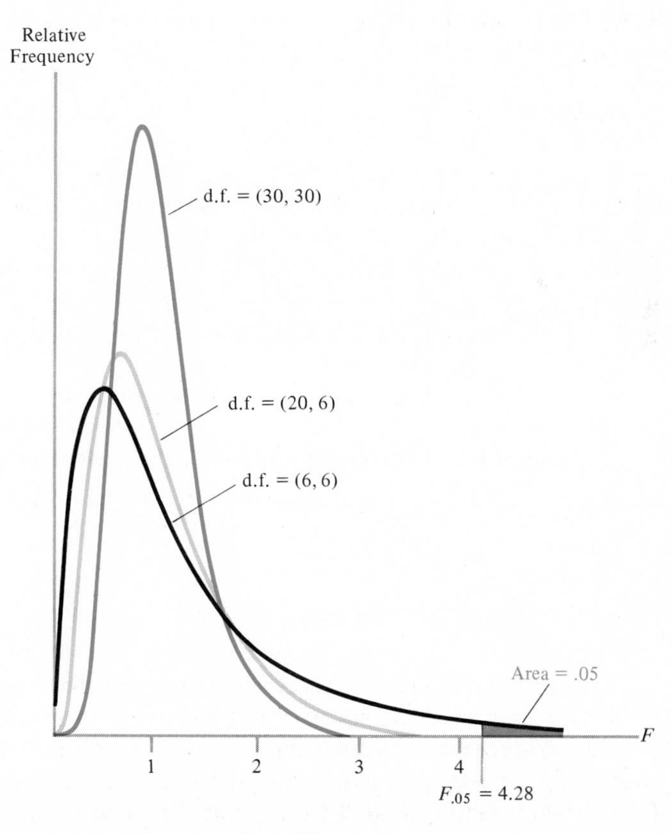

FIGURE 13-2 Various F distribution curves.

and the lightface type is the value for $F_{.05}$, so the values are $F_{.01} = 8.47$ and $F_{.05} = 4.28$. Thus

$$P[8.47 \leqslant F] = .01$$

$$P[4.28 \leqslant F] = .05$$

The second probability is represented by the colored area under the curve for this distribution, beginning at $F_{.05} = 4.28$. Because the tails of the F curve are so long and narrow, the graph in Figure 13-2 does not show the area above $F_{.01}$. These values signify that in repeated experiments, values of F exceeding 4.28 will be obtained on an average of 5 percent of the time and values exceeding 8.47 will be obtained about 1 percent of the time.

Testing the Hypothesis

Large values for the F statistic tend to refute the null hypothesis of equal population means, For a significance level established at $\alpha = .01$ or $\alpha = .05$, Appendix Table G provides the critical value F_α for the following decision rule:

$$\textit{Accept } H_0 \text{ if } F \leqslant F_\alpha$$

$$\textit{Reject } H_0 \text{ if } F > F_\alpha$$

Suppose that an $\alpha = .05$ Type I error probability is desired for correctly concluding that the population means for the various sod-fertilizing treatments are not identical. Using 2 as the degrees of freedom for the numerator and 12 for the denominator, Table G provides $F_{.05} = 3.88$. Since the calculated value $F = 4.31$ exceeds this critical value, the null hypothesis must be *rejected* and we must conclude that the methods of fertilizing differ.

(If the decision maker required greater protection against the Type I error and used the smaller significance level of $\alpha = .01$, the opposite conclusion would be reached, since $F_{.01} = 6.93$, which is larger than the calculated value for F.)

Additional Comments

What should the sod farmer do after rejecting the null hypothesis of identical population means? Obviously, he should pick the fertilizer treatment that will maximize mean sod yield. But his results are only sample data, which are subject to sampling error. He could arrive at a decision by comparing individual treatments. (A procedure for doing this will be described in Section 13-2.)

The analysis of variance procedure we have described thus far has considered just one type of treatment—sod fertilization. This one factor explained enough of the variation in sod yield to justify our conclusion that the fertilizer treatments have different effects. But there may be other explanations for differences in sample yields: for example, varying soil conditions or watering schedules. In Section 13-3 procedures for analyzing two factors will be discussed, and later we will see how as many as three factors may be incorporated into the analysis. As more factors are included, the testing procedures become more discriminating and efficient.

The theoretical conditions under which the F distribution applies to our problem are (1) the populations for each sample must be *normally distributed* with identical mean and variance (standard deviation), and (2) all sample observations are *independent*. The requirement of normality is shared with the Student t distribution. As we noted in our discussion of the t distribution, as long as the populations are not highly skewed, departures from normality are not considered serious. In Chapter 14, an alternative procedure which does not require the assumption of normality—the Kruskal-Wallis test—will be discussed.

Because we have already established the sample sizes, we are free only to set the value of the Type I error probability α for falsely rejecting the null hypothesis of identical fertilizer treatment means. As we saw in Chapter 8, no matter what the sample size, a critical value may be obtained that always guards against erroneous rejection of the null hypothesis. But when H_0 is accepted after comparing F to the critical value, there may be a very large chance of committing the Type II error and accepting a false hypothesis. To protect against erroneous acceptance, much larger samples should generally be used than those in our sod-fertilizing experiment.

EXERCISES

13-1 For each pair of degrees of freedom and each significance level provided below, indicate the critical value F_α for the F statistic above which the stated tail area holds.

	(a)	(b)	(c)	(d)
α	.01	.05	.05	.01
Numerator d.f.	10	12	5	10
Denominator d.f.	10	8	26	5

13-2 A chemical engineer is investigating different pressure settings to determine whether pressure affects particular synthetics in terms of quantity produced. Once for every chemical, three sample batches, each of size 5, are run at low, medium, and high pressures. For each synthetic, H_0 is that the mean output is identical under all pressure settings. In the following cases, find the appropriate critical value and then indicate whether H_0 should be accepted or rejected.

(a) For synthetic X, $F = 5.32$ and $\alpha = .01$.
(b) For synthetic Y, $F = 4.97$ and $\alpha = .05$.
(c) For synthetic Z, $F = 6.43$ and $\alpha = .05$.

13-3 A statistics instructor teaches three small experimental sections of the same course. In each of these, he uses a different book. The instructor wishes to test the null hypothesis that the mean scores achieved on a standard examination by all students using a particular testbook at his university will be identical. He has administered this test to his students, and the 9 scores obtained from each class constitute a random sample from the respective population of university-wide scores. In applying the F test, find the respective degrees of freedom. Then, in each of the situations below, find whether the null hypothesis should be accepted or rejected at the stated significance level.

(a) $F = 4.86$ and $\alpha = .01$.
(b) $F = 4.59$ and $\alpha = .05$.
(c) $F = 5.91$ and $\alpha = .01$.
(d) $F = 3.19$ and $\alpha = .05$.

13-4 Quicker Oats is contemplating changing the shape of its box from the quaint cylinder presently in use. In an experiment, different random samples were selected from five stores of similar size in the same region, and for several days, one of three candidate boxes was substituted for the cylinder. Total boxes sold have been determined as follows:

	Box Shape		
Sample Store	(1) *Pyramid*	(2) *Rectangle*	(3) *Cube*
1	110	57	92
2	85	65	81
3	69	73	66
4	97	49	71
5	78	77	70

(a) Calculate the individual sample means and the grand mean.
(b) Determine the treatments and error sums of squares. Use these values to find the total sum of squares.
(c) Construct the appropriate ANOVA table.
(d) At the $\alpha = .05$ significance level, should the null hypothesis that the mean sales are identical regardless of box shape be accepted or rejected?

13-5 In assessing the impact of the level of impurities in a particular ingredient upon the solubility of his company's aspirin tablets, a statistician wishes to test the null hypothesis that the mean dissolving time is the same regardless of the impurity level. In test batches, the following dissolving times measured in seconds were obtained:

	Level of Impurities		
Observation	(1) *1 percent*	(2) *5 percent*	(3) *10 percent*
1	2.01	1.95	2.30
2	1.82	2.21	2.29
3	1.74	2.14	2.17
4	1.90	1.93	2.06
5	2.03	2.07	2.58

(a) Construct the applicable ANOVA table.
(b) At the $\alpha = .05$ significance level, what should the statistician conclude?

13-6 A detergent manufacturer advertises that its product will remove all stains except oil-based paint in any kind of water. A consumer information service, reporting on detergent quality, is testing this claim. Batches of washings were run in five randomly chosen homes having a particular type of water—hard, soft, or moderate. Each batch contained an assortment of rags and cloth scraps stained with food products, grease, and dirt over a 100 square inch area. After washing, the number of square inches still stained was determined, and the following results were obtained:

Observation	Level of Water Hardness (1) Hard	(2) Moderate	(3) Soft
1	5	4	4
2	3	7	0
3	2	8	1
4	10	3	3
5	6	2	2

(a) Using an $\alpha = .01$ level of significance, will the consumer service conclude that the type of water affects the effectiveness of the detergent?
(b) Are there any factors besides water type that might "explain" some of the sources of variation?

13-2 ESTIMATING TREATMENT MEANS AND DIFFERENCES

Once the null hypothesis of equal treatment population means has been rejected, what do we do next? Continuing with the sod farmer's experiment, we have concluded that the sod yields differ according to the fertilizing treatment applied. But what should the farmer do? He might simply select that procedure with the greatest mean yield. But in doing so, he must acknowledge that chance sampling error still clouds the results. In this section, we will extend the analysis of variance results by describing further procedures that can help us translate them into action.

There are two basic aspects to this extended analysis. First, we will individually estimate the means of the treatment populations. Then, we will compare treatments by estimating the differences between pairs of population means. In both cases, the confidence interval estimation techniques encountered earlier in this book may be applied.

Confidence Intervals for Treatment Population Means

Recall that for a single sample, a confidence interval may be constructed for the population mean by

$$\mu = \overline{X} \pm t_\alpha \frac{s}{\sqrt{n}}$$

where α is selected so that the chosen confidence level equals $1-2\alpha$ (and t_α is the critical value for which the upper-tail area under the Student t curve is α), s is the sample standard deviation, and n is the sample size. When several samples have been taken from different populations, only slight modifications are necessary to obtain a similar expression for each of the respective population means $\mu_1, \mu_2, \ldots$.

Remember that the treatment populations were presumed to have identical variances σ^2. Thus, a better estimator for σ^2 can be achieved by pooling the sample results, using the unexplained variance MSE in place of a single sample variance s^2. In estimating μ_j, the population mean for the jth treatment, we may use sample treatment mean $\overline{X}_j$ as the estimator. In place of n, we use the number of observations r made under that treatment. Finally, the number of degrees of freedom used in finding t_α is $(r-1)c$, the same figure we used in computing MSE. Thus, we obtain the following expression for computing the ***confidence interval estimate of*** μ_j

$$\mu_j = \overline{X}_j \pm t_\alpha \sqrt{\frac{MSE}{r}} \tag{13-10}$$

Returning to the fertilizing experiment results in the previous section, we may construct a 95-percent confidence interval estimate for the mean sod yield for treatment (1). For $(r-1)c = (5-1)3 = 12$ degrees of freedom, Appendix Table E provides $t_{.025} = 2.179$. With $\overline{X}_1 = 81.2$ and $MSE = 15.9$, we obtain the following estimated mean yield of marketable sod per 100 square yards:

$$\begin{aligned}\mu_1 &= \overline{X}_1 \pm t_{.025} \sqrt{\frac{MSE}{5}} \\ &= 81.2 \pm 2.179 \sqrt{\frac{15.9}{5}} \\ &= 81.2 \pm 3.9\end{aligned}$$

or

$$77.3 \leqslant \mu_1 \leqslant 85.1 \text{ square yards}$$

Similarly, the confidence intervals for the other treatments are

$$84.3 \leqslant \mu_2 \leqslant 92.1 \text{ square yards}$$

$$78.7 \leqslant \mu_3 \leqslant 86.5 \text{ square yards}$$

Comparing Treatment Means Using Differences

In Chapter 11, we investigated various procedures for using two sample means to estimate the difference in population means. In the case of independent samples, we used a confidence interval of the form

$$\mu_A - \mu_B = \overline{X}_A - \overline{X}_B \pm t_\alpha s_{d\text{-small}}$$

where $s_{d\text{-small}}$ represents an estimate of the standard error of the difference $d = \bar{X}_A - \bar{X}_B$ that is found by pooling the sample standard deviations. Although here we use number subscripts, the procedure may be extended to any treatment pair. In place of $s_{d\text{-small}}$, we may use $\sqrt{2MSE/r}$, again reflecting the fact that the common population variance is estimated by MSE. The presence of the 2 indicates that the variability is additive for two samples. Thus, we have the following ***confidence interval estimate for the difference $\mu_2 - \mu_1$***

$$\mu_2 - \mu_1 = \bar{X}_2 - \bar{X}_1 \pm t_\alpha \sqrt{\frac{2MSE}{r}} \qquad (13\text{-}11)$$

where t_α is chosen to correspond to the confidence level and the degrees of freedom are $(r-1)c$ as before. Applying this expression to our fertilizing experiment results, the 95-percent confidence interval for the difference in mean sod yields between treatments (2) and (1) is

$$\mu_2 - \mu_1 = 88.2 - 81.2 \pm 2.179 \sqrt{\frac{2(15.9)}{5}}$$

$$= 7 \pm 5.5$$

or

$$1.5 \leqslant \mu_2 - \mu_1 \leqslant 12.5 \text{ square yards}$$

Thus, we estimate that the advantage in mean sod yield of using treatment (2) over treatment (1) is somewhere between 1.5 and 12.5 square yards.

We may use the above confidence interval to test the null hypothesis that the treatment means are equal. The fact that the interval does not overlap zero indicates that $H_0 : \mu_1 = \mu_2$ can be *rejected*, and we can only conclude that μ_1 and μ_2 differ at the $1 - .95 = .05$ significance level. Thus, we can recommend that the sod farmer choose treatment (2) over treatment (1), with only a .05 probability that this action could be incorrect if the treatments actually yield identical results.

Other 95-percent confidence intervals may be similarly constructed for the remaining pairs:

$$\mu_3 - \mu_1 = 1.4 \pm 5.5 \quad \text{or} \quad -4.1 \leqslant \mu_3 - \mu_1 \leqslant 6.9 \text{ square yards}$$

$$\mu_3 - \mu_2 = -5.6 \pm 5.5 \quad \text{or} \quad -11.1 \leqslant \mu_3 - \mu_2 \leqslant -.1 \text{ square yard}$$

The first confidence interval contains zero, indicating that $H_0 : \mu_3 = \mu_1$ must be accepted at the 5-percent significance level. However, the null hypothesis that $\mu_3 = \mu_2$ must be rejected.

Multiple Comparisons

The preceding methods are simple extensions of earlier concepts to the analysis of variance results. But their interpretation may be somewhat misleading. The above confidence and significance levels apply only to the *single* estimate or test and *not to the entire series* of estimates or tests. It would be incorrect to tie the preceding three confidence intervals together in a single statement, such as: "Treatment (2) yields greatest mean sod yield, while treatments (1) and (3) are similar to each other but are both inferior to (2)." This is because *as a set*, these three confidence intervals correspond to less than a 95-percent confidence level. Stated another way, there is greater than a .05 chance that there will be at least one erroneous rejection of the individual null hypotheses that each pair of treatments has equal means. The separate inferences are interdependent.

This problem can be alleviated by constructing somewhat wider confidence intervals, a procedure referred to as *multiple comparisons*. Because of their relative complexity, a discussion of these methods is beyond the scope of this book.*

EXERCISES

13-7 The following ANOVA table and mean calculations apply to the results of a study seeking to determine if a person's attitude toward his present job influences the number of years he has remained in that position. The treatments were three job attitudes: (1) dislike, (2) apathetic, and (3) enjoy. A sample of 4 persons was used for each treatment.

Variation	*Degrees of Freedom*	*Sum of Squares*	*Mean Square*	*F*
Treatments	2	104	52	8.7
Error	9	54	6	
Total	11	158		

$\bar{X}_1=2$ years $\quad \bar{X}_2=4$ years $\quad \bar{X}_3=9$ years $\quad \bar{\bar{X}}=5$ years

(a) Construct 95-percent confidence intervals for the treatment population means.

(b) Construct 95-percent confidence intervals for the differences $\mu_2-\mu_1$, $\mu_3-\mu_2$, and $\mu_3-\mu_1$.

(c) Referring to your answers to (b), do the population means for attitudes (3) and (1) appear significantly different at the 5-percent level?

13-8 Referring to the data in Exercise 13-4 (page 474) and to your answers to that exercise:

(a) Construct a 95-percent confidence interval estimate for the mean sales of the pyramid-shaped box.

* A complete description of three multiple comparison procedures, due to J. W. Tukey, Henry Scheffé, and Bonferroni, respectively, are described in Neter, John, and Wasserman, William, *Applied Linear Statistical Models* (Homewood, Ill.: Richard D. Irwin, 1974), Chapter 14.

(b) Construct a 95-percent confidence interval for the difference found by subtracting the mean sales of the rectangular box from the mean sales of the pyramid box. Do the two styles have different mean sales at the $\alpha = .05$ significance level?

13-9 Referring to Exercise 13-5 (page 474) and to your answers to that exercise, determine whether the one-percent and 10-percent impurity levels provide significantly different mean dissolving times. (Use $\alpha = .05$.)

13-3 TWO-FACTOR ANALYSIS OF VARIANCE

In analyzing the sod-yield data in Table 13-1, we saw that the test results were significant enough (but barely so) to warrant rejecting H_0 of identical treatment means at the 5-percent level. For $\alpha = .01$, however, H_0 would have to be accepted, because we found too much unexplained variation in sod yields to justify rejecting H_0 at this lower level. Is it possible to lower the level of unexplained variation by explaining a portion of it in terms of another factor? Perhaps some of the differences in yield, could be explained in terms of varying soil qualities, the slopes of plots, or different watering methods, for example. All of these could be influencing factors if the plots came from homogeneous parcels of land.

Suppose that every sample observation consisted of three neighboring plots in each of five separate parcels owned by the farmer and that every treatment was randomly assigned to one plot in each parcel. We could then treat the parcel as a second factor in the analysis. Although the sod farmer is interested in finding out how yields are affected by fertilizer treatments—and not by the parcels themselves (which are permanent features of the farm)—consideration of parcels as a second factor may still help to explain some of the differences achieved in sod yields. In this section, we will describe a procedure for doing this which is called *two-factor analysis of variance.*

There are two basic forms of two-factor analysis of variance. We will illustrate the first form by expanding our analysis of the sod-fertilizing experiment, where inferences will be made regarding the original factor only. Later in this section, we will investigate situations where inferences are made regarding both factors.

The Randomized Block Design

Each parcel of land is referred to as a *block*. Table 13-4 arranges the original sod-yield data by block and treatment. This experiment has a *randomized block design*, since treatments have been randomly assigned to units within each block. Although our illustration represents a block as a contiguous, presumably rectangular area (like a "city block"), the term actually refers to a second factor in the analysis that is used primarily to reduce the unexplained variation by having homogeneous sample units in each block. Here, the blocking factor is of no inherent interest, and its primary use is to reduce the unexplained variation in

TABLE 13-4 Two-Factor Layout for Sod-Fertilizing Experiment Using Parcels as Blocks and with Sod Yield (square yards) as the Response Variable

	Fertilizer Treatment			
Blocks	(1) *Quarter Dosage Once Weekly*	(2) *Half Dosage Every Two Weeks*	(3) *Full Dosage Every Four Weeks*	*Mean Yield (square yards)*
(1) *Parcel A*	77	83	80	$\bar{X}_{1\cdot} = 80$
(2) *Parcel B*	79	91	82	$\bar{X}_{2\cdot} = 84$
(3) *Parcel C*	87	94	86	$\bar{X}_{3\cdot} = 89$
(4) *Parcel D*	85	88	85	$\bar{X}_{4\cdot} = 86$
(5) *Parcel E*	78	85	80	$\bar{X}_{5\cdot} = 81$
Mean Yield (square yards	$\bar{X}_{\cdot 1} = 81.2$	$\bar{X}_{\cdot 2} = 88.2$	$\bar{X}_{\cdot 3} = 82.6$	$\bar{\bar{X}} = 84$

the response variable. This same design can also be used to evaluate several reading programs in terms of achievement; but if the students in a variety of schools are to be tested, the experimental results will be more discriminating if the schools are blocked according to size and region (either of which can influence achievement), with an equal representation of each block in the various reading programs.

The sod farmer is still faced with his earlier question: Are mean sod yields affected by fertilizer treatments? His null hypothesis is the same as before:

$$H_0 : \mu_1 = \mu_2 = \mu_3$$

Extending the procedure outlined in Section 13-1, which we now refer to as *one-factor analysis of variance*, we have another source of variation which may be explained by differences between blocks. In addition to treatments and error variation, *blocks variation* is a third component of total variation. We may measure each of these three component variations by computing a sum of squares and a mean square. But before we do this, we must expand upon our earlier notation.

Sample Mean Calculations

As before, each observation is denoted by X_{ij}, but here i refers to the ith block and j to the treatment. In addition to the column means used for treatments, we now find row means for each block. To help distinguish between these, the row means are designated by symbols of the form $\bar{X}_{i\cdot}$ and the column means are represented as $\bar{X}_{\cdot j}$s. The subscripted dots signify that more than one factor is being considered. The following calculation provides the ***sample mean for the*** i**th *block***

$$\bar{X}_{i\cdot} = \frac{\sum_j X_{ij}}{c} \tag{13-12}$$

where all the observations in the ith row are summed and divided by the number of entries in that row, which is the number of columns c. In our example, $i = 1, 2, 3, 4$, or 5, depending upon which block we are considering. For block (4), representing plots in parcel D, the sample mean is found by summing the values in row (4) and then dividing by $c = 3$:

$$\bar{X}_{4\cdot} = \frac{85+88+85}{3} = \frac{258}{3} = 86$$

The sample means for the remaining rows are provided in the margins of Table 13-4.

Likewise, we may calculate the ***sample mean for the*** j**th *treatment***

$$\bar{X}_{\cdot j} = \frac{\sum_i X_{ij}}{r} \tag{13-13}$$

These values are also provided in Table 13-4.

As before, the mean of all observations in the combined sample—which may be computed by averaging either the row or the column means—serves as the ***grand mean***

$$\bar{\bar{X}} = \frac{\sum \bar{X}_{i\cdot}}{r} = \frac{\sum \bar{X}_{\cdot j}}{c} \tag{13-14}$$

The Two-Factor ANOVA Table

The two-factor ANOVA table for our sod-fertilizing experiment is provided in Table 13-5, which is similar to Table 13-3 but contains an additional row for the blocks variation. For blocks, the number of degrees of freedom is $r-1$, the number of rows minus one. Also, note that the number of degrees of freedom for the error row is $(r-1)(c-1)$, accounting for the fact that one degree of freedom each is lost from the rows and columns.

Here, the treatments sum of squares SST expresses variation between the column means; it is found as before, by summing the squared deviations of the column means from the grand mean and then multiplying by the number of rows. Analogously, the variation between the row means is expressed by summing their squared deviations from the grand mean and then multiplying by the number of columns to obtain the ***blocks sum of squares***

$$SSB = c \sum (\bar{X}_{i\cdot} - \bar{\bar{X}})^2 \tag{13-15}$$

TABLE 13-5 Two-Factor ANOVA Table for Sod-Fertilizing Experiment Using Randomized Block Design

Variation	*Degrees of Freedom*	*Sum of Squares*	*Mean Square*	*F*
Explained by Treatments (between columns)	$c-1 = 2$	$SST = 137.2$	$MST = 137.2/2 = 68.6$	$MST/MSE = 68.3/3.6 = 19.06$
Explained by Blocks (between rows)	$r-1 = 4$	$SSB = 162$	$MSB = 162/4 = 40.5$	*
Error or Unexplained (residual)	$(r-1)(c-1) = 8$	$SSE = 28.8$	$MSE = 28.8/8 = 3.6$	
Total	$rc-1 = 14$	$SS = 328$		

*In the randomized block design, F is not ordinarily calculated for blocks.

In our example

$$\begin{aligned} SSB &= 3[(80-84)^2 + (84-84)^2 + (89-84)^2 + (86-84)^2 + (81-84)^2] \\ &= 3(16+0+25+4+9) = 162 \end{aligned}$$

A New Source of Explained Variation

Together, SST and SSB account for the explained variation in sod yields. SST explains this variation in terms of differences in fertilizer treatments; SSB, in terms of differences between parcels of land. Left unexplained by either fertilizing or parcels is the residual variation in sod yield, which is expressed by the error sum of squares SSE. It may be shown mathematically that the total sum of squares is the sum of the three components:

$$\text{Total } SS = SST + SSB + SSE \qquad (13\text{-}16)$$

By calculating the total sum of squares and then subtracting SSB and SST, we may determine the ***error sum of squares***

$$SSE = \text{Total } SS - SST - SSB \qquad (13\text{-}17)$$

Earlier, the total sum of squares for sod yields was found to be 328. Thus, for that experiment

$$SSE = 328 - 137.2 - 162 = 28.8$$

Note that SSE is smaller than it was in our earlier one-factor analysis of variance, because much of the formerly unexplained variation in sod yields can now be attributed to differences in parcels or blocks.

The mean squares for each component of variation are computed by dividing every respective sum of squares by the applicable number of degrees of freedom. We may calculate the ***blocks mean square***

$$MSB = \frac{SSB}{r-1} \tag{13-18}$$

For our example

$$MSB = \frac{162}{5-1} = 40.5$$

The value of the treatments mean square remains at $MST = 68.6$, unchanged from our earlier one-factor analysis of variance. Dividing SSE by a smaller number of degrees of freedom than before, we calculate the ***error mean square for two-factor analysis of variance***

$$MSE = \frac{SSE}{(r-1)(c-1)} \tag{13-19}$$

For the sod-fertilizing experiment

$$MSE = \frac{28.8}{(5-1)(3-1)} = 3.6$$

Recall that each mean square is really a sample variance and serves as an estimator of the common variance of the treatment populations.

Testing the Null Hypothesis

We are now ready to test the null hypothesis. First we calculate the appropriate ***F statistic***

$$F = \frac{\text{Variance explained by treatment factor}}{\text{Variance left unexplained}} \tag{13-20}$$

This is accomplished by dividing the respective mean squares. The F statistic for the fertilizer treatments is

$$F = \frac{MST}{MSE} = \frac{68.6}{3.6} = 19.06$$

To see if these results are significant, we must find the critical value for 2 numerator and 8 denominator degrees of freedom in Appendix Table G. We see that the calculated value for F is greater than both $F_{.05} = 4.46$ and $F_{.01} = 8.65$, indicating that we must *reject* the null hypothesis that population fertilizer treatment means are identical at the one-percent significance level.

Although it is possible to calculate a value of F for blocks using MSB, this variable is of secondary interest and is used primarily to allow finer discrimination between treatments. Later in this section, we will investigate two-factor experiments where inferences may be made regarding both factors.

Increased Efficiency of Two-Factor Analysis

Our example illustrates how two-factor analysis of variance can be more efficient than one-factor analysis. We obtained a much larger value of F for treatments than before, allowing us to reject H_0 (equal fertilizing population means) at a smaller significance level than we could in the one-factor analysis. This is because including parcels as a second factor considerably reduced the previously unexplained variation in sod yield.

Making Inferences About Both Factors

Another kind of two-factor analysis of variance is performed when inferences using both factors are made. To illustrate this, consider the data in Table 13-6, where factor A (acidity level) and factor B (chlorine concentration) are used to explain the monthly drop in alkalinity (in parts per million or ppm) for a random sample of swimming pools. Pool acidity and chlorine are set at various levels to satisfy sanitation requirements, but for each combination of these factor levels, the drop in alkalinity must be periodically offset by adding soda ash.

Here, we wish to make inferences regarding the underlying populations for the alkalinity-drop response variable at various levels for both factors. In this example, the sample units (swimming pools) have been randomly assigned with equal probability to each factor combination. Such a two-factor study has a *completely randomized design*, as opposed to the randomized block design described earlier, where the unit assignments to blocks are fixed by their nature because of location, characteristics, or some other fixed factor. In a completely randomized design involving two factors, there is a separate treatment for each combination of levels for the two factors, represented by the cells in Table 13-6.

The particular design used here allows us to *test the two null hypotheses* described below.

1. The response (alkalinity-drop) means for factor A populations (acidity levels) are identical.
2. The response (alkalinity-drop) means for factor B populations (chlorine concentrations) are identical.

A separate test may be performed on each hypothesis using the same data. Again, this shows that a two-factor analysis can be more efficient than separate one-factor analyses. (As we indicate in Section 13-4, the two-factor analysis sometimes answers questions that cannot even be considered in one-factor analyses.)

TABLE 13-6 Two-Factor Layout and Sample Results for Relating Monthly Drop in Swimming Pool Alkalinity (ppm) to Acidity Levels and Chlorine Concentrations

	Factor A (Acidity Level)				
Factor B (*Chlorine Concentration*)	(1) *pH* 7.2	(2) *pH* 7.4	(3) *pH* 7.6	(4) *pH* 7.8	*Sample Mean*
(1) *Low*	23	18	9	7	$\bar{X}_{1\cdot} = 14.25$
(2) *Medium*	10	12	8	4	$\bar{X}_{2\cdot} = 8.50$
(3) *High*	9	9	7	4	$\bar{X}_{3\cdot} = 7.25$
Sample Mean	$\bar{X}_{\cdot 1} = 14$	$\bar{X}_{\cdot 2} = 13$	$\bar{X}_{\cdot 3} = 8$	$\bar{X}_{\cdot 4} = 5$	$\bar{\bar{X}} = 10$

$$\text{Total } SS = \sum \sum (X_{ij} - \bar{\bar{X}})^2:$$

$(23-10)^2 = 169$	$(18-10)^2 = 64$	$(9-10)^2 = 1$	$(7-10)^2 = 9$
$(10-10)^2 = 0$	$(12-10)^2 = 4$	$(8-10)^2 = 4$	$(4-10)^2 = 36$
$(9-10)^2 = 1$	$(9-10)^2 = 1$	$(7-10)^2 = 9$	$(4-10)^2 = 36$
170	69	14	81

$$\text{Total } SS = 170 + 69 + 14 + 81 = 334$$

$$\begin{aligned} SSA &= r \sum (\bar{X}_{\cdot j} - \bar{\bar{X}})^2 \text{ (between-columns variation)} \\ &= 3[(14-10)^2 + (13-10)^2 + (8-10)^2 + (5-10)^2] \\ &= 3(16+9+4+25) = 162 \end{aligned}$$

$$\begin{aligned} SSB &= c \sum (\bar{X}_{i\cdot} - \bar{\bar{X}})^2 \text{ (between-rows variation)} \\ &= 4[(14.25-10)^2 + (8.50-10)^2 + (7.25-10)^2] \\ &= 4(18.0625 + 2.25 + 7.5625) = 111.5 \end{aligned}$$

$$SSE = \text{Total } SS - SSA - SSB = 334 - 162 - 111.5 = 60.5$$

Since there is a separate treatment for each cell, it is necessary to modify our previous notation somewhat. For this purpose, we use *SSA* and *SSB* to denote the factor *A* and the factor *B* sums of squares. These are calculated in Table 13-6 in the same manner as their counterparts in the randomized block design, where *SSA* is based upon the deviations of *column means* and *SSB* is based upon the deviations of *row means*. The total and the error sums of squares are also calculated in Table 13-6 in the same way as before.

The ANOVA table for this experiment is provided in Table 13-7. The respective mean squares, *MSA*, *MSB*, and *MSE*, are found by dividing the corresponding sum of squares by the applicable number of degrees of freedom. We find the values of the *F* statistic

$$\textit{For factor A} \text{ (acidity level): } F = \frac{MSA}{MSE} = 5.36$$

and

$$\textit{For factor B} \text{ (chlorine concentration): } F = \frac{MSB}{MSE} = 5.53$$

TABLE 13-7 Two-Factor ANOVA Table for Completely Randomized Design for Experiment for Effects of Acidity and Chlorine Levels on Drop in Swimming Pool Alkalinity

Variation	*Degrees of Freedom*	*Sum of Squares*	*Mean Square*	*F*
Explained by Factor A—Acidity Level (between columns)	$c-1 = 3$	$SSA = 162$	$MSA = 162/3 = 54$	$MSA/MSE = 54/10.08 = 5.36$
Explained by Factor B—Chlorine Concentration (between rows)	$r-1 = 2$	$SSB = 111.5$	$MSB = 111.5/2 = 55.75$	$MSB/MSE = 55.75/10.08 = 5.53$
Error or Unexplained (residual)	$(r-1)(c-1) = 6$	$SSE = 60.5$	$MSE = 60.5/6 = 10.08$	
Total	$rc-1 = 11$	$SS = 334$		

In the case of factor A, the degrees of freedom are 3 for the numerator and 6 for the denominator. Appendix Table G provides $F_{.05} = 4.76$. Since this value is smaller than the calculated F, we must *reject* the null hypothesis that the acidity level population means are identical at the $\alpha = .05$ significance level. Similarly, for 2 and 6 degrees of freedom for factor B, we have $F_{.05} = 5.14$, and the second null hypothesis of identical population means for chlorine concentration must also be *rejected* at the $\alpha = .05$ significance level.

An Important Assumption

In the above example, we make the important assumption that the factor effects are *additive*, so that the population treatment mean is raised or lowered from some "background" level by a different constant amount for each level of every treatment. But this may not be true, for there may be an *interaction* between the two factors that cannot be explained by either one alone. For example, the mean alkalinity drop in swimming pools may be more severe if the acidity level is high when the chlorine concentration is decreased than would be indicated by combining the separate effects. By not accounting for such interactions, the testing procedure outlined above may be less discriminating.

EXERCISES

13-10 A randomized block design is being used to test the null hypothesis that mean responses are identical under five treatments, with four levels used for the blocking factor. The following data are obtained:

$$SST = 84$$
$$SSB = 132$$
$$\text{Total } SS = 288$$

(a) Construct the ANOVA table.

(b) Should the null hypothesis of identical population means be accepted or rejected at the $\alpha = .01$ significance level?

13-11 Suppose that the sample data in Exercise 13-4 (page 474) represent three observations from stores in each of five regions. Treating "region" as the blocking variable and referring to your answers to that exercise:

(a) Determine the sample means for each row. Then compute the blocks sum of squares.

(b) Construct the ANOVA table for this experiment.

(c) Comparing the value of *SSE* to that obtained in Exercise 13-4, what can you conclude regarding the effect upon unexplained variation of adding the blocking variable?

(d) At the $\alpha = .01$ significance level, should the null hypothesis of identical mean sales for each box shape be accepted or rejected?

13-12 A sociologist wishes to assess the effects of factor *A* (education) with five levels and factor *B* (intelligence) with four levels upon a person's annual earnings. The following data have been obtained in each category for 20 randomly chosen persons:

$$SSA = 800{,}000$$
$$SSB = 900{,}000$$
$$\text{Total } SS = 2{,}000{,}000$$

(a) Construct a one-factor ANOVA table, using education as the only treatment. At the $\alpha = .05$ significance level, can you conclude that the treatment means differ?

(b) Construct a one-factor ANOVA table, using intelligence as the only treatment. At the $\alpha = .05$ significance level, can you conclude that the treatment means differ?

(c) Construct a two-factor ANOVA table, using both education and intelligence. What can you conclude regarding the respective null hypotheses for identical mean incomes for education levels and for intelligence levels? (Use $\alpha = .01$.)

(d) Do you notice any discrepancy between the one-factor and the two-factor results? Explain.

13-13 A computer programming instructor conducted an experiment to determine the effects upon student achievement of the computer languages he taught and of the types of computers he used. Over a period of four terms, he taught 16 classes, and each class was given a standard achievement test. The mean scores achieved by the respective classes are provided below:

	Language			
Types of Computer	(1) *Basic with No FORTRAN*	(2) *Basic with Some FORTRAN*	(3) *FORTRAN with No Basic*	(4) *FORTRAN with Some Basic*
(1) *Batch/No Time Share*	64	74	68	69
(2) *Batch/Some Time Share*	86	83	85	84
(3) *Time Share/ No Batch*	88	90	84	87
(4) *Time Share/ Some Batch*	84	92	69	89

(a) Compute the row, column, and grand means.

(b) Construct the two-factor ANOVA table.

(c) At the $\alpha = .01$ significance level, can the instructor conclude that *languages* make a difference in achievement?
(d) At the $\alpha = .01$ significance level, can the instructor conclude that *computer types* make a difference in achievement?

13-4 LATIN SQUARES AND OTHER ANOVA DESIGNS

In the preceding sections of this chapter, we have considered basic analysis of variance procedures involving one and two factors. But analysis of variance is a very broad topic, and we have barely scratched the surface. In this section, we will touch upon the breadth and depth of this topic, and we will conclude by discussing a very useful procedure for analyzing three factors.

Important ANOVA Considerations

Analysis of variance situations may be classified in a number of ways. Some of the important ones are discussed below.

Factorial Design

Thus far, we have encountered examples where all combinations of the factors have been covered. In our sod-fertilizing experiment, we investigated three levels for the fertilizing factor and five levels for the parcel blocking factor. Each level for both factors was represented in the sample data. Such experiments are called *complete factorial designs.* But sometimes only a fraction of the combinations can be studied, either because of expense or because some are infeasible (as, for example, would be the case in our example if some parcels had soil that could not absorb heavy applications of fertilizer). Such experiments are said to have a *fractional factorial design.* At times, it may be useful to omit certain factor combinations but to still incorporate all factorial levels in the experiment. This is called an *incomplete factorial design.* A useful experiment in this category is one conducted with a *latin square design*, the procedure we will describe later in this section for analyzing three factors.

Factor Levels Considered

In our illustrations thus far, we have established in advance the levels for the factors we have investigated. For instance, the three fertilizing policies were established before the data were collected. That study involved a *fixed-effects experiment.* This is in contrast to a *random-effects experiment* where the factor levels are not established in advance and are subject to chance, as would be the case when the levels constitute a sample from a larger population. For example, suppose that a medical society is studying doctors' incomes, using "specialty" and "region where educated" as factors. Data might be compiled from one medical school in each region, so that the school selected is just a sample and the levels of the education factor are themselves samples. Because of the added complexity, a detailed discussion of random-effects experiments is beyond the scope of this book.

Interactions

As we saw in Section 13-2, analysis of variance can be extended to answer other important questions. There, we saw how pairs of treatments may be compared. Often of considerable importance are the interactions between one or more factor levels that exist when together they have an added influence that they do not have individually. Techniques for analyzing interactions are so complex that we have limited the experiments in this book to applications where interactions can be ignored.

Sample Size and Replication

Statisticians use the word *replication* to indicate that an experiment is repeated. The number of repetitions is the number of replications. When a single sample is used, this number is the sample size. Larger sample sizes have the advantage of reducing sampling error. In hypothesis-testing situations, the chance of the Type II error of accepting a false null hypothesis becomes smaller as the level of n is raised. When two or more samples are used, we compare population parameters or estimate their differences. In either case, the usual advantages to large sample sizes are accrued, but of course, large sample sizes are more expensive.

In our analysis of variance illustrations, replication occurred only in the one-factor application where five plots were used for each fertilizer treatment. Our two-factor experiments involved a single observation for each factor combination, and the designs for these experiments did not involve replication. More complicated designs (not described in this book) incorporate replication when multiple observations are made for each cell. Doing so can be useful. To investigate interaction effects, such a replication scheme is required.

Randomization

We classified our two-factor analysis of the sod-fertilizing treatments as a randomized block design. There, the sample units within each parcel were assigned to treatments randomly. The advantage of doing this is that slight differences between plots within parcels were allowed no systematic influence—as might be the case, for instance, if the plot receiving the best irrigation was always heavily fertilized. Often an important element in sampling studies, such an assignment of sample units to treatments is referred to as *randomization*. For example, in evaluating the effectiveness of training programs, it is important to account for the differences among instructors. This can be accomplished by randomly assigning a particular training program to each instructor. Such a randomization is less discriminating than treating "instructors" as a separate blocking variable, but the latter would force each instructor to train once under each treatment program, which would be both time consuming and burdensome.

The Latin Square Design

Until this point, we have discussed only one- and two-factor designs. We will now consider a *three-factor analysis* involving one treatment factor of interest and two blocking variables.

A psychologist conducting an experiment to evaluate the effects on achievement-test performance of three different training methods may choose a design that incorporates other factors that might explain performance variability, such as a trainee's aptitude and age. Thus, the psychologist might block the sample subjects in terms of three levels of aptitude and three levels of age. A total of 9 blocks would be necessary—one for each aptitude and age-level combination. In a complete design, three treatments must be considered for each case, so that a minimum 27 subjects would be required. In the case of four treatments and two blocking variables, each also having four levels, a minimum of $4^3 = 64$ subjects would be required for a complete design. For a moderate number of levels, a very large number of sample units may be required, and such experiments can be very expensive.

One way of reducing the number of sample units is to use an incomplete design where not all combinations of treatments and blocks are represented. An efficient procedure for doing this is the *latin square design*. Here, the number of treatments is limited to just the number of levels used for the blocks. In the case of the above training-method evaluation, this can be done with just 9 instead of 27 subjects. Table 13-8 shows the data obtained in this experiment for a sample of 9 persons. The letters in the cells pertain to the particular program applied. These same letters form the ***latin square***

$$\begin{matrix} A & B & C \\ B & C & A \\ C & A & B \end{matrix}$$

This arrangement of latin letters is designed so that each letter appears exactly once in each column and in each row. Other latin squares for 4 and 5 letters are

$$\begin{matrix} A & B & C & D \\ B & A & D & C \\ C & D & A & B \\ D & C & B & A \end{matrix} \qquad \begin{matrix} A & B & C & D & E \\ B & A & E & C & D \\ C & D & A & E & B \\ D & E & B & A & C \\ E & C & D & B & A \end{matrix}$$

Since a letter represents a particular treatment and since a different level of the blocking variables corresponds to each column and row, *the latin square design forces each treatment to be applied exactly once under each level of both blocking variables.*

TABLE 13-8 Layout and Sample Data for Latin Square Design Using Achievement Test Scores To Evaluate Training Programs

Row Blocking Factor (Aptitude)	*Column Blocking Factor (Age)* (1) *Young*	(2) *Middle*	(3) *Old*	*Sample Mean for Row*
(1) *Low*	*A* 82	*B* 87	*C* 80	$\bar{X}_{1\cdot} = 83$
(2) *Medium*	*B* 92	*C* 82	*A* 81	$\bar{X}_{2\cdot} = 85$
(3) *High*	*C* 90	*A* 83	*B* 88	$\bar{X}_{3\cdot} = 87$
Sample Mean for Column	$\bar{X}_{\cdot 1} = 88$	$\bar{X}_{\cdot 2} = 84$	$\bar{X}_{\cdot 3} = 83$	$\bar{\bar{X}} = 85$
Treatment: *Training Program* *Sample Mean*	*A* $\bar{X}_A = 82$	*B* $\bar{X}_B = 89$	*C* $\bar{X}_C = 84$	

The sample means in Table 13-8 have been calculated in the usual manner for the rows and columns. The treatment sample means are found by summing the cell responses having the corresponding letter and then dividing by the number of cells having that letter. The respective means are denoted as $\bar{X}_A$, $\bar{X}_B$, and $\bar{X}_C$, and their values are given in Table 13-8. For training program A, the following sample mean calculation applies:

$$\bar{X}_A = \frac{82+81+83}{3} = 82$$

In the usual way, we calculate the ***sums of squares for the row and column blocking factors***

$$\begin{aligned} SSROW &= r\sum(\bar{X}_{i\cdot}-\bar{\bar{X}})^2 \\ &= 3[(83-85)^2+(85-85)^2+(87-85)^2] \\ &= 3(4+0+4) = 24 \end{aligned} \tag{13-21}$$

$$\begin{aligned} SSCOL &= c\sum(\bar{X}_{\cdot j}-\bar{\bar{X}})^2 \\ &= 3[(88-85)^2+(84-85)^2+(83-85)^2] \\ &= 3(9+1+4) = 42 \end{aligned} \tag{13-22}$$

In latin square designs, the number of rows and columns are the same, so that $c = r(=3$, here). The number of treatments must also be equal to the number of rows. We have the ***treatments sum of squares***

$$\begin{aligned} SST &= r[(\bar{X}_A - \bar{\bar{X}})^2 + (\bar{X}_B - \bar{\bar{X}})^2 + (\bar{X}_C - \bar{\bar{X}})^2] \\ &= 3[(82-85)^2 + (89-85)^2 + (84-85)^2] \\ &= 3(9+16+1) = 78 \end{aligned} \tag{13-23}$$

The total sum of squares is found as before, by squaring the deviations of the column responses about the grand mean and then summing these totals:

(1)	(2)	(3)
$(82-85)^2 = 9$	$(87-85)^2 = 4$	$(80-85)^2 = 25$
$(92-85)^2 = 49$	$(82-85)^2 = 9$	$(81-85)^2 = 16$
$(90-85)^2 = 25$	$(83-85)^2 = 4$	$(88-85)^2 = 9$
83	17	50

$$\text{Total } SS = 83 + 17 + 50 = 150$$

The following difference provides the ***error sum of squares***

$$\begin{aligned} SSE &= \text{Total } SS - SST - SSROW - SSCOL \\ &= 150 - 78 - 24 - 42 = 6 \end{aligned} \tag{13-24}$$

The ANOVA table is provided in Table 13-9. The degrees of freedom are $r-1=2$ for all factors and $(r-1)(r-2)=2$ for the error. The mean squares are calculated in the usual fashion.

TABLE 13-9 Latin Square Design ANOVA Table

Variation	*Degrees of Freedom*	*Sum of Squares*	*Mean Square*	*F*
Explained by Treatments (between programs)	$r-1=2$	$SST = 78$	$MST = 78/2 = 39$	$MST/MSE = 39/3.0 = 13.0$
Explained by Column Blocks (between ages)	$r-1=2$	$SSCOL = 42$	$MSCOL = 42/2 = 21.0$	*
Explained by Row Blocks (between aptitudes)	$r-1=2$	$SSROW = 24$	$MSROW = 24/2 = 12$	*
Error or Unexplained (residual)	$(r-1)(r-2)=2$	$SSE = 6$	$MSE = 6/2 = 3.0$	
Total	$r^2-1=8$	$SS = 150$		

*For blocking variables F is not ordinarily calculated.

The test statistic is

$$F = \frac{MST}{MSE} = 13.0$$

From Appendix Table G, for 2 degrees of freedom in the numerator and 2 in the denominator, we see that $F_{.01} = 99.00$ and $F_{.05} = 19.00$. Since both critical values are larger than the calculated value for F, the null hypothesis of identical treatment population means must be *accepted*, and we must conclude that the training programs do not differ.

The primary advantage of the latin square design is that it reduces the number of sample units needed in testing with three factors. But our illustration also points out an important disadvantage to this design: the latin square design has a small number of degrees of freedom, making it very hard to reject the null hypothesis of identical population means unless SST is very large in relation to SSE. This type of test becomes more discriminating, however, if several observations are made for each cell or if the number of treatments is increased. Another drawback to the latin square design is that it is limited to situations where exactly the same number of levels are used for both the blocking variables and the treatments.

EXERCISES

13-14 Indicate whether or not each of the following arrangements forms a latin square:

(a)

A B
B A

(b)

A B C D
B C D A
C D A B

(c)

A B C
B C A
C A C

(d)

E A D C B
A C B D E
D E C B A
B D A E C
C B E A D

(e)

D C B A
C D A B
B A C D
A B D C

13-15 The following sums of squares were obtained, using a 4-by-4 latin square design:

$SST = 97$
$SSCOL = 53$
$SSROW = 48$
Total $SS = 236$

Construct the ANOVA table. What can you conclude regarding the treatment means? (Use $\alpha = .05$.)

13-16 A latin square design can be used to make inferences regarding three separate factors. Suppose that the following sums of squares were obtained for a five-by-five design:

$SST = 157$ (factor X)
$SSCOL = 77$ (factor Y)
$SSROW = 143$ (factor Z)
Total $SS = 452$

(a) Construct the ANOVA table, calculating the appropriate F statistics for each factor.
(b) For which factors are the population means significantly different? (In each case, use $\alpha = .01$.)

13-17 The sample data below represent scores achieved on a classical music appreciation test. The "treatment" of interest is the age of the subject:

A = preteen
B = teen-ager to 25
C = over 25

Two blocking variables, "family background" and "intelligence", were used:

	Intelligence		
Family Background	(1) *Low*	(2) *Medium*	(3) *High*
(1) *Blue Collar*	*A* 78	*B* 88	*C* 68
(2) *White Collar*	*B* 90	*C* 82	*A* 74
(3) *Professional*	*C* 87	*A* 79	*B* 92

(a) Construct the ANOVA table for the latin square design.
(b) At the $\alpha = .05$ significance level, can you conclude that mean music appreciation scores differ for the various age levels?

13-5 COMPARING TWO VARIANCES USING F DISTRIBUTION

In Chapter 11, we indicated the need for comparing the variances of two populations. These comparisons are an important auxiliary feature when testing two population means using small samples, since the t test in these cases requires that $\sigma_A^2 = \sigma_B^2$. Such comparisons are also useful in making statistical decisions where the population variance is a pivotal factor (as we saw in Chapter 12, when we compared the populations of aptitude or intelligence test scores and to choose a method to service waiting lines).

The F distribution may be employed here to compute the ratio of the sample variances s_A^2 and s_B^2. For samples from normally distributed populations, it can be shown that the F distribution provides probabilities of the following form:

$$\alpha = P\left[F_\alpha \leqslant \frac{s_A^2}{s_B^2}\right] \tag{13-25}$$

where the appropriate F curve has degrees of freedom $n_A - 1$ for the numerator and $n_B - 1$ for the denominator.

One-Sided Tests

The above discussion suggests that we may test a one-sided null hypothesis of the form

$$H_0 : \sigma_A^2 \leqslant \sigma_B^2$$

by calculating the test statistic

$$F = \frac{s_A^2}{s_B^2} \qquad (13\text{-}26)$$

and then applying the following decision rule for an upper-tailed test for the desired significance level:

$$\textit{Reject } H_0 \text{ if } F > F_\alpha$$

$$\textit{Accept } H_0 \text{ if } F \leqslant F_\alpha$$

We can only construct upper-tailed tests by applying the F distribution. If the reverse situation were to be tested, then we would only have to relabel the populations, changing designation A to B, and vice versa.

Example 13-1 A sociologist wishes to test the null hypothesis that the variability in IQ for capital criminals (A) is no larger than that for persons who have committed lesser felonies (B) at the $\alpha = .05$ significance level. From a nearby prison, he has obtained a random sample of IQ scores for $n_A = 10$ and $n_B = 25$ prisoners and has found the sample standard deviations to be $s_A = 17.4$ and $s_B = 10.7$. The computed value for the test statistic is therefore

$$F = \frac{(17.4)^2}{(10.7)^2} = 2.64$$

For $10 - 1 = 9$ numerator and $25 - 1 = 24$ denominator degrees of freedom, Appendix Table G provides $F_{.05} = 2.30$. Since this value is smaller than 2.64, the sociologist must *reject* his null hypothesis and conclude that the variance in IQ for capital criminals is higher than it is for lesser felons.

Two-Sided Tests

Null hypotheses of the form

$$H_0 : \sigma_A^2 = \sigma_B^2$$

involve a two-sided alternative and must be rejected if σ_A^2 is much larger or smaller than σ_B^2. Because we lack critical values for lower-tail areas of the F curve, a roundabout procedure is required which generally involves *two curves*. One curve provides the sampling distribution s_A^2/s_B^2. To illustrate, suppose that $n_A = 10$ and $n_B = 8$, so that we have $n_A - 1 = 10 - 1 = 9$ degrees of freedom for the numerator and $n_B - 1 = 8 - 1 = 7$ degrees of freedom for the denominator. For an upper-tailed area of .01, Appendix Table G provides the critical value $F_{.01} = 6.71$. It will be convenient to represent this value by a special symbol $F_{A/B} = 6.71$. The other curve provides s_B^2/s_A^2, with $n_B - 1 = 7$ numerator and $n_A - 1 = 9$ denominator degrees of freedom. Again using .01 for the upper-tailed area, Table G provides $F_{.01} = 5.62$ for this curve. Since the second curve applies when the degrees of freedom are reversed, we distinguish its critical value by the symbol $F_{B/A} = 5.62$.

The following probability statements apply for the first and second curves, respectively:

$$.01 = P\left[6.71 < \frac{s_A^2}{s_B^2}\right] = P\left[F_{A/B} < \frac{s_A^2}{s_B^2}\right]$$

$$.01 = P\left[5.62 < \frac{s_B^2}{s_A^2}\right] = P\left[F_{B/A} < \frac{s_B^2}{s_A^2}\right]$$

Using our original statistic s_A^2/s_B^2, the second equation may be equivalently written

$$.01 = P\left[\frac{s_A^2}{s_B^2} < \frac{1}{5.62}\right] = P\left[\frac{s_A^2}{s_B^2} < \frac{1}{F_{B/A}}\right]$$

The above facts suggest that the appropriate procedure is to calculate $F = s_A^2/s_B^2$. Then, for the desired significance level α, there are two critical values—each corresponding to upper-tail F-curve areas equal to $\alpha/2$—one for each curve. The null hypothesis of equal variances thus has a probability α of incorrectly being rejected, either because $F > F_{A/B}$ or because $F < 1/F_{B/A}$. This suggests the following decision rule:

$$\textit{Accept } H_0 \text{ if } \frac{1}{F_{B/A}} \leq F \leq F_{A/B}$$

$$\textit{Reject } H_0 \text{ if otherwise}$$

Because Table G only represents two upper-tailed areas, .01 and .05, then it may be used only for testing with two significance levels: 2(.01) = .02 and 2(.05) = .10.

Example 13-2 In studying the effects of coffee drinking on sleep, a researcher obtained the following data for the time taken to get to sleep for two independent samples of monitored subjects:

Light Coffee Drinkers	*Heavy Coffee Drinkers*
$\bar{X}_A = 45$ minutes	$\bar{X}_B = 36$ minutes
$s_A = 25$ minutes	$s_B = 20$ minutes
$n_A = 12$	$n_B = 10$

She wishes to test the null hypothesis that the population variances are equal at the $\alpha = .10$ significance level. The critical values desired are

$$F_{A/B} = F_{.05} = 3.10$$

using $12-1=11$ numerator and $10-1=9$ denominator degrees of freedom, and

$$F_{B/A} = F_{.05} = 2.90$$

with the degrees of freedom reversed. The null hypothesis will be accepted if F lies between $1/F_{B/A} = 1/2.90 = .34$ and $F_{A/B} = 3.10$. Using the given data, we compute

$$F = \frac{(25)^2}{(20)^2} = 1.56$$

Since this result lies between .34 and 3.10, H_0 must be *accepted* and the researcher must conclude that the population variance in time to sleep is the same for light and heavy coffee drinkers.

EXERCISES

13-18 For each of the following situations, determine whether the null hypothesis that $\sigma_A^2 \leqslant \sigma_B^2$ should be accepted or rejected. (Use $\alpha = .05$.)

(a)	(b)	(c)	(d)
$s_A=10$	$s_A=5$	$s_A=4$	$s_A=123$
$s_B=5$	$s_B=3$	$s_B=3$	$s_B=98$
$n_A=10$	$n_A=12$	$n_A=8$	$n_A=7$
$n_B=8$	$n_B=20$	$n_B=14$	$n_B=10$

13-19 The null hypothesis of identical population variances is to be tested at the 10-percent significance level. There are $n_A = 12$ and $n_B = 8$ sample observations.

(a) Formulate the applicable decision rule.

(b) For each of the following situations, determine whether H_0 must be accepted or rejected:

	(1)	(2)	(3)	(4)
s_A	10	7	4.5	12
s_B	5.5	6	9	15

13-20 A researcher wishes to test the null hypothesis that the variability in the litter sizes of rats is the same under diets A and B. She has obtained the following data:

$s_A = 4.1$	$s_B = 6.2$
$n_A = 13$	$n_B = 8$

Can she conclude that the population variances are different? (Use $\alpha = .10$.)

REVIEW EXERCISES

13-21 A researcher conducting a study involving human behavior offered a contest to the students in four psychology classes. Points were given for the number of nonsense syllables memorized during a class period.

Rewards varied between classes. Using a random sample of five students from each class, the following numbers of correct syllables were obtained:

	Number of Points in Reward			
Student	(1) $\frac{1}{2}$	(2) 1	(3) $1\frac{1}{2}$	(4) 2
1	18	32	46	52
2	34	48	58	73
3	27	25	37	46
4	31	26	48	63
5	20	39	61	56

(a) Determine the sample means, calculate the sums of squares, and construct the ANOVA table for this experiment.
(b) Should the researcher conclude that motivation to perform meaningless tasks is affected by the level of reward? (Use $\alpha = .01$.)
(c) Construct 95-percent confidence interval estimates for $\mu_4 - \mu_1$. Do rewards of $\frac{1}{2}$ and 2 points provide significantly different motivation at the 5-percent level?

13-22 Consider the following results for a two-factor experiment where six levels were used for factor A and five levels were used for factor B:

SSA	78
SSB	65
Total SS	183

(a) Construct the ANOVA table for this experiment.
(b) At the $\alpha = .01$ significance level, what can you conclude regarding the null hypotheses of equal population means for the levels of the respective factors?

13-23 A job-performance evaluator is concerned with how workers' performance ratings are affected by their skill levels. Three levels are considered:

Unskilled (A)
Semiskilled (B)
Skilled (C)

Using a sample of three workers from each level, with productivity and attitude serving as blocking variables, the following data were obtained:

	Attitude		
Productivity	(1) Poor	(2) Fair	(3) Good
(1) Low	A 50	B 79	C 105
(2) Medium	B 74	C 106	A 63
(3) High	C 98	A 61	B 93

(a) Calculate the sample means and the sums of squares.
(b) Construct the ANOVA table for the latin square design.
(c) At the $\alpha = .01$ significance level, what can you conclude regarding the effect of skill level on the mean performance ratings?

13-24 A school psychometrist arranged for a special study to evaluate the effects of testing environment on the performance of high-school students taking the SAT. Group A students were allowed to take the test in their homerooms in the presence of their regular teachers. Group B students also took the test in their homerooms, but they were proctored by strangers. Group C students were examined in distant cities with their regular teachers serving as proctors. Group D students took the test in a distant city, under the supervision of strangers. The students in the experiment were carefully blocked in terms of motivation and academic performance. The following data were obtained:

	Motivation			
Academic Performance	(1) *Poor*	(2) *Fair*	(3) *Good*	(4) *Excellent*
(1) *Poor*	*A* 342	*B* 313	*C* 325	*D* 324
(2) *Fair*	*B* 368	*A* 406	*D* 349	*C* 401
(3) *Good*	*C* 377	*D* 348	*A* 512	*B* 493
(4) *Excellent*	*D* 455	*C* 482	*B* 575	*A* 634

(a) Calculate the sample means and the sums of squares.
(b) Construct the ANOVA table.
(c) Do the data suggest that test surroundings affect SAT performance? (Use $\alpha = .01$.)

13-25 Referring to Exercise 13-24 and to your answers to that exercise:
(a) Construct a one-factor ANOVA table, using "test surroundings" as the treatment and ignoring blocking factors. Can you reject the null hypothesis of identical population means at the $\alpha = .05$ significance level?
(b) Using only the "academic performance" blocking variable, construct a two-factor ANOVA table for a randomized block design, again with "test surroundings" as the treatment. Can you reject the null hypothesis of identical population means? (Use $\alpha = .05$.)
(c) What does a comparison of your one-, two-, and three-factor ANOVA tables suggest regarding the discrimination of testing when "academic performance" and "motivation" are used as the blocking variables?

14 Nonparametric Statistics

INDUCTION, analogy, hypotheses founded upon facts and rectified continually by new observations, a happy tact given by nature and strengthened by numerous comparisons of its indications with experience, such are the principle means for arriving at truth.

Marquis de Laplace (1820)

Most of the hypothesis-testing procedures discussed earlier in this book involve inferences regarding population parameters such as the mean. These tests are therefore referred to as *parametric*, and their test statistics are sometimes called *parametric statistics.* The sampling distributions for the statistics used in these tests usually depend upon assumptions regarding the *populations* from which the samples are obtained. Particularly stringent are the assumptions of the Student t statistic, used for tests of means when the sample sizes are small, and of the F statistic, used in analysis of variance. Both the t and the F sampling distributions assume that the sampled populations are normal.

Without complete enumeration, there is no way to tell *for certain* that the populations are actually normal. Even if the population distribution is known, the normal distribution is a theoretical "nicety" that serves only as a mathematically convenient representation, for we have seen that no real population is truly normal. When a sample must be used, further complications arise due to chance error. Slight deviations from normality may be tolerated in using the t and the F tests, but in certain cases, a population not meeting the assumption of normality can invalidate the results of the statistical test.

14-1 THE NEED FOR NONPARAMETRIC STATISTICS

In this chapter, we will present statistical tests that make no assumptions regarding the shape of population distributions. For this reason, they are often referred to as being "distribution free." Many of these same statistics require no assumptions regarding population parameters. Thus, they have come to be more commonly referred to as *nonparametric statistics*. A nonparametric test has been defined as one that makes no hypothesis about the value of a population parameter.* We have already used a nonparametric statistic, the chi-square, in testing for independence. There, we made no assumptions regarding the distribution of the populations nor did our hypotheses involve suppositions about parameters.†

Advantages of Nonparametric Statistics

Nonparametric statistics have many nice properties; among these is their ease of calculation. They often have sampling distributions that can be easily explained by the simplest laws of probability. But their major advantage is that they are encumbered by few restrictive assumptions, which is particularly important when, for various reasons, samples cannot be large. For instance, nonparametric statistics are vital to behavioral scientists, because large samples are the exception in long-term experiments involving people.

In addition to explicit assumptions regarding the shape of a population, certain parametric tests make assumptions about the population parameters *not being tested.* For example, equal variances are assumed in using the Student t statistic to test for differences in population means with independent samples (see Section 11-4). Implicitly, another assumption often made by parametric tests involving means is that the population values are continuous. A great many populations, although they involve variates, have values that are inherently not continuous. A large class of such values cannot be combined arithmetically.

Many practical research problems require subjective ratings on a numerical scale. For example, if a subject rates the taste of brand A beets as 5, brand C beets as 4, and brand B as 3, he is really expressing his order of preference: A is better than C, which is better than B. The numbers have no real meaning in themselves (ratings of 999, 2, and -30, respectively, would express the same thing), so that the mean of such numbers is not very meaningful.

Numbers that primarily express preference, ranking, hierarchy, and so on, are convenient symbols that convey order, and their numerical values are said to

* Some tests not strictly adhering to this requirement are nevertheless classified as nonparametric.

† The χ^2 is a hybrid in that it is sometimes employed as a parametric statistic. For example, χ^2 may be used as a test statistic for hypotheses regarding a population's variance.

belong to an *ordinal scale*. Distance between values has no inherent meaning. Thus, standard arithmetic operations are not consistent when applied to ordinal numbers. Any statistical test requiring the calculations of means or variances would be largely invalid when applied to samples from populations whose variates are ordinal. *Usually, for such populations only nonparametric tests are valid.*

Another deficiency of parametric tests is their preoccupation with the mean and the variance, much of which stems from the nice mathematical properties of these measures. Except for symmetrical distributions, the median is a more meaningful measure of central tendency than the mean. Various interfractile ranges depict dispersion more vividly than the standard deviation. For example, to the unsophisticated (that is, to almost everybody), knowing that the incomes of 90 percent of all doctors in a particular region lie between \$20,000 and \$150,000 is more meaningful than knowing that $\sigma = \$25,000$. Some nonparametric tests are well suited for testing hypotheses regarding medians, while the popular parametric tests are limited to the less meaningful mean. Furthermore, medians are equally valid measures whether the variates are continuous or ordinal.

Disadvantages of Nonparametric Statistics

There are, of course, disadvantages to using nonparametric statistics. Their primary drawback is that they tend to ignore much sample information that is gleaned by their parametric counterparts. This makes them generally less efficient. But it has been argued that although they are less efficient, a researcher using nonparametric statistics has more confidence in his methodology than he does if he must adhere to the unsubstantiable assumptions inherent in parametric tests like the Student *t*. One handicap of nonparametric tests is that there are so many from which to choose that the researcher must pay more heed to the extra dimension of efficiency in designing a statistical study. Another disadvantage is the lack of universal dissemination in probability tables, which are also often woefully inadequate in their coverage of the breadth of circumstances for which some statistics might otherwise be employed.

Topics Covered

In this chapter, we will introduce some of the more popular and relatively efficient nonparametric tests. Most of these will be nonparametric counterparts to earlier procedures discussed in this book. We will begin by describing a series of tests that compare two populations. In analyzing independent samples, the Wilcoxon rank-sum test and the Mann-Whitney *U* test will be introduced. Then the sign test and the Wilcoxon signed-rank test will be described for matched-pairs sampling, followed by a discussion of the number-of-runs test for randomness. The Spearman rank correlation coefficient will also be discussed. The chapter will conclude with an alternative to the F test—the Kruskal-Wallis procedure for one-factor analysis of variance.

14-2 COMPARING TWO POPULATIONS USING INDEPENDENT SAMPLES: THE WILCOXON RANK-SUM TEST

Two methods (independent samples and matched pairs) were proposed in Chapter 11, for making a decision that involves the choice of the population with the greater mean, but these tests are adequate only when the populations are normally distributed, or closely so. Here we introduce another procedure to use with independent samples, the *Wilcoxon rank-sum test*, which is free of the possibly invalid assumptions of normality. This test is named after the statistician Frank Wilcoxon, who first proposed it in 1945.

Description of Test

The Wilcoxon test compares two samples—a control and an experimental group—taken from two populations. All null hypotheses share a common assumption than the samples were selected from identical populations, which is more stringent than assuming that they have identical means. The test is therefore based upon the principle that the two samples may be treated as though they came from a common population. Under the various null hypotheses encountered, the data for the two samples may be combined. The observed values in the pooled sample are then ranked from smallest to largest; the smallest value is assigned a rank of 1, the next smallest value is ranked 2, and so forth. After the rankings are obtained, the samples are separated and the sums of the ranks are calculated for each. The rank sums obtained are used as test statistics.

As an illustration, we will compare type A to type B fertilizer for effectiveness in increasing the yield of corn. In this experiment, ten corn-belt farmers set aside one-acre plots for testing. Five acres chosen randomly are to use type A and five are to use type B. It is assumed that the two fertilizers are equally effective, so that the hypotheses are

H_0 : Types A and B fertilizers are equally effective.

H_1 : Types A and B fertilizers differ in effectiveness.

Based upon the yields in the sample plots, H_0 will either be accepted or rejected. In effect, the *null hypothesis states that the population of yields for plots fertilized with type A is identical to the corresponding population using type B*. The alternative hypothesis is that the two populations differ somehow. Since the alternative might be true either if A is better than B or if B is better than A, this particular experiment involves a *two-sided test*.

TABLE 14-1 Corn Yields Obtained in Test of Two Fertilizers

	Yield (bushels)				
Type A	42.3	38.7	42.8	35.6	47.2
Type B	61.4	45.3	46.4	53.1	50.1

TABLE 14-2 Ranks of Fertilizer Test Yields

Type	*A*	*A*	*A*	*A*	*B*	*B*	*A*	*B*	*B*	*B*
Yield	35.6	38.7	42.3	42.8	45.3	46.4	47.2	50.1	53.1	61.4
Rank	1	2	3	4	5	6	7	8	9	10

The yields obtained from the sample plots are provided in Table 14-1, and the pooled sample results are ranked in Table 14-2. If the fertilizers are equally effective, then we would expect type A to be ranked low and high about as many times as type B. A convenient comparison can be made in terms of the sums of the ranks obtained for the respective samples. *For this purpose, the sum of the ranks for sample A is the test statistic.* Using the ranks in Table 14-2 and letting W represent this rank sum, we have

$$W = 1 + 2 + 3 + 4 + 7 = 17$$

We can calculate the rank sum for sample B similarly, but its value is more easily obtained by subtracting W from the total rank sum, which is 55 here.* Thus, the sample B rank sum is $55 - 17 = 38$. A comparison of the two rank sums shows that type A fertilizer has a value less than half as large as type B, indicating that type A will result in predominately lower corn yields. Although this seems to contradict the null hypothesis that the two types of fertilizer are equally effective, we still must determine if these results are significant enough to warrant rejecting H_0. To do this, we need only investigate the properties of W obtained for sample A. (Because the sample B rank sum is automatically determined by the value of W, we only need to consider the rank sum for sample A.)

The Sampling Distribution of the Rank Sum W

Under the hypothesis of identical effectiveness, any five of the ten ranks for the pooled samples can belong to population A; that is, the first rank is no more likely to belong to sample A than to B. This holds for any rank. In other words, if the fertilizers are equally effective, then the smallest yield is just as likely to have been obtained from a plot where type A has been applied as from a plot where type B has been used, and likewise for the highest yield or for any of

* The sum of N integers, $1, 2, \ldots, N$, is obtained from $N(N+1)/2$. When $N = 10$, the sum is $10(10+1)/2 = 55$.

the yields in between. Thus, assuming a true null hypothesis, we may view the sample results as only one possible outcome from a random experiment having as many equally likely outcomes as there are ways to select five ranks for sample A out of a total of ten ranks.

The sampling distribution for W could be found by listing all of the rank combinations. Treating these elementary events as equally likely, we could then determine exact probabilities for rank sums such as 16, 17, 18, 19, and so on. For most test situations, however, this would be tedious.

Instead, we can approximate the distribution of W by the normal curve, which works well for most sample sizes encountered. For this purpose, the following expression* is used to calculate ***the normal deviate for the sample results***

$$z = \frac{W - \dfrac{n_A(n_A + n_B + 1)}{2}}{\sqrt{\dfrac{n_A n_B(n_A + n_B + 1)}{12}}} \tag{14-1}$$

where n_A and n_B represent the number of sample observations made from populations A and B, respectively.

Substituting $n_A = n_B = 5$ and $W = 17$ into expression (14-1), the normal deviate for the fertilizer test results is

$$z = \frac{17 - \dfrac{5(5+5+1)}{2}}{\sqrt{\dfrac{5(5)(5+5+1)}{12}}} = \frac{17 - 27.5}{\sqrt{22.9167}} = -2.19$$

Small values of W, and hence negative zs, reflect that smaller ranks have been assigned to sample A, which would be consistent with a situation where the values of population A are smaller than the values of B. For large Ws and positive zs, the reverse holds. The value $z = -2.19$ strongly indicates that fertilizer A generally produces lower yields than B does. Before we can determine the appropriate action to take, we must find out whether or not -2.19 significantly refutes the null hypothesis that both fertilizers are equally effective.

The usual decision rules apply for the rank-sum test. For one-sided tests, the critical normal deviate z_α which corresponds to the desired significance level is obtained from Appendix Table D. For two-sided tests, $z_{\alpha/2}$ is used.

Suppose that a significance level of $\alpha = .05$ is desired in the fertilizer experiment. We find $z_{.025} = 1.96$, so the decision rule is

* Here, we use the fact that $z = [W - E(W)]/\sigma(W)$, where the mean of W is half of the total rank sum $E(W) = n_A(n_A + n_B + 1)/2$ and it can be shown that the variance of W is $\sigma^2(W) = n_A n_B(n_A + n_B + 1)/12$.

Accept H_0 (conclude fertilizers equally effective) if $-1.96 \leqslant z \leqslant 1.96$

Reject H_0 either (conclude A most effective) if $z > 1.96$
or (conclude B most effective) if $z < -1.96$

Since the computed value of the normal deviate $z = -2.19$ is smaller than -1.96, H_0 must be *rejected.* The sample results indicate that fertilizer B is more effective at the $\alpha = .05$ significance level.

Application to One-Sided Tests

To show how this procedure may be applied in a one-sided test, consider the following example.

Example 14-1 A school psychologist has collected the sample data in Table 14-3 from an experiment comparing the IQ scores of visually handicapped fourth-grade children (A) to those fourth graders with normal vision (B). Since IQ scores so heavily reflect verbal skills, it is commonly believed that the visually handicapped achieve lower IQ scores than other children. However, the psychologist feels that the special attention given such children tends to result in a *higher* rather than a lower measurable intelligence. As her null hypothesis, she assumes that the population of IQ scores for the handicapped is not larger:

H_0: IQs of population A are less than or equal to those of population B.

Since a preponderance of high ranks for sample A should refute H_0, this test is *upper-tailed.* The combined sample data were ranked as shown in Table 14-3. (The special handling of tying scores is explained below.) The rank sum for the sample A IQs is

$$W = 5.5 + 9 + 11 + 13 + 14 + 15 + 16 + 18 + 20 + 22 = 143.5$$

Applying expression (14-1) with sample sizes $n_A = 10$ and $n_B = 12$, the normal deviate is

$$z = \frac{143.5 - \dfrac{10(10+12+1)}{2}}{\sqrt{\dfrac{10(12)(10+12+1)}{12}}} = \frac{28.5}{\sqrt{230}} = 1.88$$

The psychologist desires an $\alpha = .05$ significance level. Thus, $z_{.05} = 1.64$. Since the computed value for z is larger than 1.64, she must *reject* H_0 that population A IQ scores are no larger than those for population B and she must conclude that visually handicapped children have higher IQs than normal children.

The Problem of Ties

Several ties were encountered in ranking the data in Table 14-3. In sample A, two children shared the IQ score 110 and in sample B two had IQs of 103. *As long as ties occur within the same sample group, successive ranks may be assigned arbitrarily to the sample observations.* Thus, we ranked the two 103 scores as 7 and 8 and the two 110 scores as 13 and 14.

A difficulty arises however when ties occur between sample groups, because the choice of ranks affects W. *When ties occur between sample groups, each observation is assigned the average of the ranks for that value.* The ranks of 5 and 6 would have applied to the two IQs of 102 had they been in the same sample. However, since one child in *each* sample achieved the same score, the scores receive equal ranks of $(5+6)/2 = 5.5$.

TABLE 14-3 Sample Data for Experiment Comparing IQs of Visually Handicapped Children to IQs of Children with Normal Vision

IQs of Handicapped Children X_A	*IQs of Normal Children* X_B	X_A^2	X_B^2
104	94	10,816	8,836
110	103	12,100	10,609
106	114	11,236	12,996
113	126	12,769	15,876
115	95	13,225	9,025
111	102	12,321	10,404
102	100	10,404	10,000
128	98	16,384	9,604
110	103	12,100	10,609
117	116	13,689	13,456
1,116	105	125,044	11,025
	107		11,449
	1,263		133,889

Sample	B	B	B	B	A	B	B	B	A	B	A	B	A
IQ	94	95	98	100	102	102	103	103	104	105	106	107	110
Rank	1	2	3	4	5.5	5.5	7	8	9	10	11	12	13

A	A	A	B	A	B	A	B	A
110	111	113	114	115	116	117	126	128
14	15	16	17	18	19	20	21	22

$n_A = 10$ $\qquad$ $n_B = 12$

$$\bar{X}_A = \frac{1{,}116}{10} = 111.60 \qquad \bar{X}_B = \frac{1{,}263}{12} = 105.25$$

$$s_A^2 = \frac{125{,}044 - 10\,(111.60)^2}{9} = 55.38 \qquad s_B^2 = \frac{133{,}889 - 12\,(105.25)^2}{11} = 87.11$$

Comparison of Wilcoxon and Student t Tests

The data from Example 14-1 may be used to compare the Wilcoxon rank-sum test to the Student t test for independent samples. Substituting the intermediate calculations from Table 14-3 into expression (11-31), we obtain

$$t = \frac{111.60 - 105.25}{\sqrt{\dfrac{9(55.38)+11(87.11)}{10+12-2}}\sqrt{\dfrac{1}{10}+\dfrac{1}{12}}}$$

$$= \frac{6.35}{\sqrt{72.8315}\sqrt{.1833}} = 1.738$$

Using $\alpha = .05$ with $n_A + n_B - 2 = 20$ degrees of freedom, Appendix Table E provides $t_{.05} = 1.725$. Since the computed value for t is larger than 1.725, the t test also indicates that H_0 must be *rejected.*

Similar results were obtained using both tests. (Once again, it should be emphasized that a particular test would normally be chosen in advance of sampling and that only one test would ordinarily be used on the same data. Both tests were applied here only to compare the two techniques.) Generally, the t test is more efficient or powerful than the Wilcoxon, so using the same data, the t test should provide a lower probability (α) of incorrectly rejecting H_0. However, the Student t test requires that the *populations* be normally distributed, which is often not the case (although it would be for most IQ populations).* Also, the Wilcoxon test can be used for experiments where the observations are measured on an ordinal scale, whereas the t test cannot. *The Wilcoxon rank-sum test can therefore be more widely used.*

EXERCISES

14-1 A county hospital must decide whether to replace its brand A hearing-aide batteries with brand B. It will do so if testing shows that brand B is more effective. The null hypothesis is that brand A is at least as effective as B. Ten new brand A batteries were installed for a random sample of deaf welfare patients, and 15 new brand B batteries were installed for another randomly chosen group. The number of days until each battery required replacement were determined:

Brand A 323, 178, 246, 195, 402, 603 496, 328, 213, 187
Brand B 421, 327, 609, 433, 519, 504, 183, 455, 365, 615, 504, 312, 513, 497, 723

(a) Is an upper-tailed or a lower-tailed test appropriate here?
(b) Rank the sample results and calculate W.
(c) Suppose that the hospital staff wishes to protect itself against incorrectly deciding to change battery brands at a .05 level of significance. Find the critical value for the test statistic. Should the battery brands be changed?

14-2 Suppose that the hospital in Exercise 14-1 installed 100 brand A batteries

* Both procedures require equal population variances. The techniques outlined in Chapter 13 can be applied in testing for this condition. For the data in Example 14-1, no significant difference in variances could be found.

and 150 brand B batteries, and that $W = 11{,}500$. Using the normal approximation, should the hospital reject the null hypothesis that brand A is at least as effective as B at the .05 significance level?

14-3 Suppose that the physician in Exercise 11-5 (page 419) uses the Wilcoxon rank-sum test on a sample of eight patients treated with penicillin and on a group of ten patients to whom sulfa was administered and that he obtains the following results for the number of days required to cure a virulent strain of venereal disease:

(A) Penicillin	15, 9, 12, 22, 14, 9, 10, 15
(B) Sulfa	7, 8, 10, 6, 7, 7, 4, 13, 11, 5

(a) Rank the sample data and calculate W.
(b) The doctor's null hypothesis is that the treatments are equally effective. Formulate his decision rule, assuming that in the future he will use the treatment he finds most effective but that otherwise he will test further. (Use $\alpha = .05$.)
(c) What action should the doctor take?

14-4 A farm cooperative wishes to select the fastest railroad for shipments to Central City, served by both the CW & B and Sun Belt lines. A random sample of less-than-carload shipments via each line were selected for detailed monitoring, and the following results were obtained for hours required to complete shipments:

(A) CW & B	10, 18, 12, 15, 27
(B) Sun Belt	14, 19, 22, 23, 25, 19, 24, 31, 26

Shipments are currently made on the Sun Belt line. The null hypothesis is that the Sun Belt is at least as fast as CW & B.
(a) Is an upper- or a lower-tailed test appropriate here?
(b) Rank the sample results and calculate W.
(c) Suppose that protection against erroneously choosing the wrong railroad must be at the $\alpha = .025$ significance level. Should the null hypothesis be accepted or rejected?

14-5 Two security analysts for a mutual fund want to compare their trading strategies. Each takes a random selection of ten stocks and trades them for a year. Analyst A's procedure is a bit revolutionary, so the null hypothesis is that analyst B's method will provide percentage rates of return at least as great as A's. The following results (in percentage rates of return) have been obtained:

A	10	−5	15	23	113	57	−51	203	33	44
B	11	27	9	9	18	−4	−8	53	112	6

(a) Is an upper- or a lower-tailed test appropriate here?
(b) Find the critical value for the $\alpha = .10$ significance level. Formulate the decision rule.
(c) Calculate W.
(d) Should H_0 be accepted or rejected at the $\alpha = .10$ significance level?

14-3 THE MANN-WHITNEY *U* TEST

The *Mann-Whitney U test* is often used in comparing two populations with independent samples. This procedure is actually equivalent to the Wilcoxon rank-sum test, and both procedures lead to identical conclusions.

The Mann-Whitney U test proceeds like the Wilcoxon, by ranking the combined sample data and then calculating the sum W of the sample A ranks. The following test statistic is used:

$$U = n_A n_B + \frac{n_A(n_A+1)}{2} - W \tag{14-2}$$

The normal curve also serves as the approximate sampling distribution for U. The following calculation provides the normal deviate for the sample results:

$$z = \frac{U - \frac{n_A n_B}{2}}{\sqrt{\frac{n_A n_B(n_A+n_B+1)}{12}}} \tag{14-3}$$

If expression (14-2) for U is substituted into expression (14-3) and the terms are collected, the same expression for z that is used in the Wilcoxon rank-sum test is obtained but with reversed signs. Since the Wilcoxon procedure is equivalent to the U test and involves fewer steps, its use is recommended.

14-4 COMPARING TWO POPULATIONS USING MATCHED PAIRS: THE SIGN TEST

In this section, we will discuss a nonparametric test for the differences between two populations that involves samples of *matched pairs*. This is the *sign test*, so named because it considers only the direction of difference in each sample pair, which may be expressed by either a plus or a minus sign. Like the Wilcoxon rank-sum test for independent samples, the sign test may be applied to a wider variety of situations than the parametric t test we described in Chapter 11. The sign test evaluates null hypotheses where population A values are (1) at least as great as B's; (2) at most as large as B's; or (3) equal to B's. The test makes no assumptions whatsoever about the shape or the parameters for the population frequency distributions. (The Wilcoxon test involved null hypotheses where the targeted experimental and the control populations were considered identical.)

Description of the Test

The sign test will be described in terms of the following example.

The Blue-Beard Razor Blade Company wishes to compare its prototype "rapier" blade with a competitor's "scimitar" brand to determine whether the rapier is of superior quality. A sampling study is to be conducted, and the rapier brand will be marketed only if testing shows that rapier is superior to scimitar; otherwise, more effort will be devoted to improving rapier.

A random sample of a representative cross section of 20 men is chosen for the test. Each man will shave one side of his face with the rapier blade and the other side with the scimitar for five consecutive days. At the end of the test, each man will rate the two blades in each of five categories: closeness of shave, shaving comfort, durability of sharpness, ease of effort, and residual facial discomfort during the day. The highest rating is 10 points, and the lowest is a point value of zero. The points in each category are added together, so that a blade receiving top ratings in all five categories will be given a score of 50 and a blade rating lowest in all categories will be given a score of zero. Fractional scoring is allowed.

Blade assignment is randomly made by tossing a coin to determine which side of each man's face will be shaved with the rapier blade: The identities of the blades are not known to the men participating in the experiment. The same blade will be used on the same side of a subject's face throughout the experiment.

The hypotheses are expressed as:

H_0 : Rapier is not superior to scimitar.

H_1 : Rapier is superior to scimitar.

Suppose that the data provided in Table 14-4 were obtained from the experiment. We must determine a test statistic that will enable us to accept or reject the null hypothesis.

The Test Statistic

The null hypothesis allows us to consider the special case where the blades may be of equal quality. If the blades are identical in quality, the scores obtained for rapier and scimitar would be due only to the side of face shaved with rapier. The first subject in Table 14-4 has assigned a higher score to rapier. This may be due to the blade, but it may also be due to the fact that he is right-handed and that he shaved the left side of his face with scimitar, which would be more cumbersome. If this is the case, then the coin flip, not the blade quality, resulted in the difference in ratings. Under the presumption of equal quality, the opposite conclusion would be made if the coin toss resulted in a tail instead of a head. Then instead, the score for rapier would be 37.0 and the score for scimitar would be 48.0.

For each matched pair, the difference in sample values is determined. In our illustration, the differences in Table 14-4 have been obtained by subtracting the scimitar score from the rapier score.

We can now determine whether the data support or refute the null hypothesis. For this purpose, we use only the signs of the difference for each pair. The rating scheme used in the experiment resulted in three ties. In these cases, it is

TABLE 14-4 Experimental Results of Razor Blade Test

Test Subject Number	Score Received: Rapier	Scimitar	Difference	Sign
1	48.0	37.0	+11.0	+
2	33.0	41.0	−8.0	−
3	37.5	23.4	+14.1	+
4	48.0	17.0	+31.0	+
5	42.5	31.5	+11.0	+
6	40.0	40.0	0.0	tie
7	42.0	31.0	+11.0	+
8	36.0	36.0	0.0	tie
9	11.3	5.7	+5.6	+
10	22.0	11.5	+10.5	+
11	36.0	21.0	+15.0	+
12	27.3	6.1	+21.2	+
13	14.2	26.5	−12.3	−
14	32.1	21.3	+10.8	+
15	52.0	44.5	+7.5	+
16	38.0	28.0	+10.0	+
17	17.3	22.6	−5.3	−
18	20.0	20.0	0.0	tie
19	21.0	11.0	+10.0	+
20	46.1	22.3	+23.8	+

assumed that there are truly some differences too subtle to assess, but since they are impossible to identify, the tied outcomes are eliminated from the analysis. This leaves us with $20-3=17$ pairs with sign differences.

Note the preponderance of positive signs in Table 14-4. Since a larger number of positive signs is evidence that rapier is favored over scimitar, the decision may be based upon the number of positive signs. Therefore, an appropriate ***test statistic for the sign test is***

$$R = \text{number of positive signs}$$

In our illustration, $R=14$. Although this value is larger because a majority of the sample rated rapier blades higher than scimitar blades, we must still determine whether R is large enough to warrant rejecting H_0. Our first step will be to find the sampling distribution for R.

Sampling Distribution of Test Statistic R

Under the presumption of identical blade quality, each of the signs in Table 14-4 could have been reversed; that is, the probability of a positive sign difference for each matched pair is .5, just as if the results had been determined by the toss of a coin. Thus, we may determine the probability of how untypical our results are under the null hypothesis by using the *binomial distribution*. The probability of obtaining a positive sign difference for any particular pair would be $\pi=.5$, and we have $n=17$ pairs. The probabilities for the possible number of positive signs R may be found by using Appendix Table C, which gives cumulative probabilities for the binomial distribution. Using $\pi = .5$ and $n = 17$, for example, we find that

$$P[R > 11] = 1 - P[R \leqslant 11] = 1 - .9283 = .0717$$

The Decision Rule

As in all hypothesis-testing situations, a decision rule must be found here that coincides with the significance level desired. This decision rule will determine a *critical value* R^* for the number of positive signs. Since large values of R favor rapier and tend to refute H_0 that rapier is not superior to scimitar, H_0 must be rejected for large values of R. The probability of incorrectly rejecting H_0 for large values of R is represented by upper-tail probabilities of the binomial distribution, making the test in this illustration *upper-tailed.*

To find R^*, we assume that the Blue-Beard Razor Blade Company has established the significance level at $\alpha = .05$. In this case, *we must select* R^* *so that the probability that the number of positive signs exceeds* R^* *is no larger than* α. We have already found that 11 is too small to be the critical value, since $P[R > 11] = .0717$, which is greater than .05. The next larger value is 12, for which Appendix Table C provides

$$P[R > 12] = 1 - P[R \leqslant 12] = 1 - .9755 = .0245$$

Since 12 is the smallest value yielding a probability less than .05, we use $R^* = 12$ as the critical value.

The appropriate decision rule is

Accept H_0 (do not market rapier) if $R \leqslant 12$

Reject H_0 (market rapier) if $R > 12$

Because the actual number of positive signs $R = 14$ exceeds $R^* = 12$, H_0 that rapier is not superior to scimitar must be *rejected* at the $\alpha = .05$ significance level. The Blue-Beard Razor Blade Company should therefore market its new rapier blade. There is less than a .05 chance that this action will prove incorrect.

A decision rule for a lower-tailed test is reversed. In such a case, the smallest value for R^* is chosen, so that $P[R < R^*]$ is no larger than α. It is also possible to apply the procedure to two-sided hypothesis tests. This would involve two critical values for R.

The Normal Approximation

When the sample size is large, the probabilities for extreme values of R may be determined by using the normal approximation to the binomial distribution. In Chapter 5, we established that the mean and the standard deviation for the number of successes R (here, a success is a positive sign) were

$$E(R) = n\pi = n(.5) = n/2$$

$$\sigma(R) = \sqrt{n\pi(1-\pi)} = \sqrt{n(.5)(1-.5)} = \sqrt{n/4}$$

so that substituting these parameters into $z = [R - E(R)]/\sigma(R)$ and simplifying the numerator and the denominator gives us the following expression for the *normal deviate for the sample results*

$$z = \frac{2R-n}{\sqrt{n}} \tag{14-4}$$

The usual decision rules apply, using the critical normal deviates z_α for one-sided tests and $z_{\alpha/2}$ for two-sided tests. The following example illustrates how the normal approximation may be applied.

Example 14-2 An educational researcher conducted an experiment to determine if elementary-school language programs might be improved by a new approach to vocabulary and spelling. Rather than the usual memorization methods, the researcher believes that the same amount of time and effort devoted to additional reading would be more effective. Fourth-grade students from various classes in several schools were divided into two groups. Children in the control group A took the standard language program in their home rooms. Students in the experimental group B went to special classes during language periods, where they read, submitted book reports, and discussed assigned outside reading.

As her null hypothesis, the researcher assumes that the new procedure does not improve language skills, as measured by scores on an achievement test. Thus:

H_0 : A scores are at least as great as B scores (new procedure is no improvement).

H_1 : A scores are lower than B scores (new procedure is an improvement).

Each child in group B was matched to a group A counterpart in his or her own classroom who had similar skills and aptitudes. Toward the end of the school year, both groups were tested. After subtracting the B child's score from the A child's score, the number of positive sign differences were then determined. Since a small value for R will refute the null hypothesis, this test is *lower-tailed.* Using a significance level of $\alpha = .05$, the critical normal deviate is $z_{.05} = -1.64$ and the decision rule is

Accept H_0 (do not recommend new procedure) if $z \geq -1.64$

Reject H_0 (recommend new procedure) if $z < -1.64$

The data in Table 14-5 were obtained. Here, we see that there were two ties; eliminating these leaves $n = 23$ pairs to consider. The number of positive signs was $R = 8$. Thus, expression (14-4) provides

$$z = \frac{2(8)-23}{\sqrt{23}} = -1.46$$

Since this value is greater than $z_{.05} = -1.64$, the researcher must *accept* H_0 that the new procedure is no improvement and must not recommend its adoption.

TABLE 14-5 Sample Data for Experiment Comparing Two Language-Skill Teaching Procedures

Pair i	Group A Score X_{A_i}	Group B Score X_B	Difference $d_i = X_{A_i} - X_{B_i}$	Sign	d_i^2
1	65	67	−2	−	4
2	72	66	+6	+	36
3	74	77	−3	−	9
4	81	90	−9	−	81
5	76	72	+4	+	16
6	95	95	0	tie	0
7	63	68	−5	−	25
8	85	95	−10	−	100
9	90	98	−8	−	64
10	64	60	+4	+	16
11	78	72	+6	+	36
12	86	94	−8	−	64
13	89	96	−7	−	49
14	75	73	+2	+	4
15	93	96	−3	−	9
16	78	90	−12	−	144
17	92	92	0	tie	0
18	67	63	+4	+	16
19	88	95	−7	−	49
20	79	88	−9	−	81
21	60	75	−15	−	225
22	90	95	−5	−	25
23	82	78	+4	+	16
24	73	85	−12	−	144
25	86	82	+4	+	16
			−81		1,229

$$\bar{d} = \frac{-81}{25} = -3.24$$

$$s_{d\text{-paired}} = \sqrt{\frac{1{,}229 - 25\,(-3.24)^2}{25-1}} = 6.35$$

Comparison of Sign and Student *t* Tests

The sign test may be compared to the analogous parametric procedure described in Chapter 11 for matched-pairs testing. There, the Student t test was used with small samples. Substituting the mean of the matched-pairs differences and their estimated standard error (both calculated in Table 14-5) and using $n = 25$ as the sample size, expression (11-33) provides

$$t = \frac{-3.24}{6.35}\sqrt{25} = -2.551$$

Using $\alpha = .05$, from Appendix Table E for $25 - 1 = 24$ degrees of freedom, the critical value is $t_{.05} = -1.711$. Since the computed value for t is smaller than -1.711, the Student t test indicates that H_0 should be *rejected.* This conclusion is opposite to the decision we reached using the sign test for the same data.

As with the Wilcoxon test this reflects the fact that *when applicable* the t test is more powerful and therefore more efficient. Its greater discrimination is due to the fact that the relative sizes of the matched-pairs differences—as well as their signs—are used in calculating t. However, the t test requires that the *populations* of all potential matched-pairs differences be normally distributed (or nearly so). *When this requirement is not met, the sign test is the valid test to use.*

EXERCISES

14-6 In each of the following situations, the sign test will be used to determine the appropriate action.

(1) Indicate whether the test is lower- or upper-tailed.

(2) Use Appendix Table C to find the critical value R^* for the number of positive sign differences.

(3) Formulate the appropriate decision rule.

(4) Indicate whether H_0 should be accepted or rejected at the stated significance level for the result provided.

(a)	(b)	(c)	(d)
$H_0: As \geq Bs$	$H_0: As \leq Bs$	$H_0: As \geq Bs$	$H_0: As \leq Bs$
$n = 18$	$n = 100$	$n = 50$	$n = 19$
$\alpha = .05$	$\alpha = .01$	$\alpha = .01$	$\alpha = .05$
$R = 4$	$R = 58$	$R = 13$	$R = 13$

14-7 A police department must determine whether its cars should burn leaded (A) or unleaded (B) gasoline. Fifteen pairs of cars are selected: one car from each pair uses leaded gasoline; the other, unleaded. The cars in each pair are nearly identical in essential respects. These cars were driven in the same manner over the same routes. The null hypothesis is that unleaded gasoline provides at least as good mileage as leaded gasoline. The following sample results were obtained:

	Miles per Gallon			Miles per Gallon	
Pair	Leaded (A)	Unleaded (B)	Pair	Leaded (A)	Unleaded (B)
1	15.3	14.7	9	15.0	13.8
2	18.1	17.9	10	16.2	16.1
3	14.9	15.0	11	15.9	14.2
4	17.3	17.3	12	21.3	20.6
5	13.7	12.3	13	18.4	17.7
6	11.8	9.2	14	19.3	17.4
7	20.3	19.7	15	9.8	7.6
8	15.2	14.7			

(a) Determine the sign difference (leaded minus unleaded mileage) for each pair.

(b) Find the value for the test statistic R, the number of positive signs. Is an upper- or a lower-tailed test required?

(c) Under the assumption that both gasolines provide equally good gasoline mileage, determine the probability that a result would be obtained that is as untypical as the actual one. (Use Appendix Table C.)

(d) Assuming that leaded gas will be used if it provides better mileage, would it be used if the desired significance level is $\alpha = .001$?

14-8 For the gasoline mileages in Exercise 14-7, suppose that the normal approximation is used to test the same null hypothesis at the $\alpha = .05$ significance level when the number of automobile pairs is $n = 100$. Assume that $R = 60$. Which type of gasoline would be used?

14-9 Suppose that the researcher in Example 11-3 (page 417) found the signs of differences in mean age for contracting angina pectoris by subtracting the onset age of the nonsmoker (B) from that of the smoker (A), obtaining $R = 30$ positive signs. As her null hypothesis, the researcher assumes that smoking makes no difference in the speed of the disease. Using an $\alpha = .01$ significance level, what should she conclude regarding the affect of smoking on the age when angina occurs?

14-10 A doctor wishes to determine if a new appetite suppressor supplement reduces weight more effectively than the one he now uses. If so, the new supplement will replace the current one. A random sample of 20 pairs of persons chosen for testing are paired so that all factors reasoned to influence weight buildup are nearly the same for each pair: women are paired with women, smokers are paired with smokers, physically inactive types are paired, and so forth. One pair member is administered the new supplement, and the other is given the old supplement. The null hypothesis is that the new supplement is at most as effective as the old one. A supplement is said to be more effective if it brings about a greater percentage weight reduction in three months of use. The results are:

	Reduction Percentage			*Reduction Percentage*	
Pair	*New Supplement*	*Old Supplement*	*Pair*	*New Supplement*	*Old Supplement*
1	10	−2	11	15	8
2	5	3	12	13	12
3	7	1	13	14	10
4	8	10	14	13	5
5	4	2	15	7	8
6	15	11	16	11	3
7	12	13	17	−2	−3
8	18	5	18	0	−2
9	3	−2	19	16	9
10	8	12	20	9	8

(a) Determine the sign difference (percentage for new minus that for old) for each pair.

(b) Find the test statistic R, the number of positive signs. Is an upper- or a lower-tailed test required?

(c) Calculate the normal deviate for the sample results.

(d) Making the normal approximation, express the decision rule at the $\alpha = .05$ significance level. What course of action is indicated by the sample results?

14-11 A chemical supplier wishes to determine whether a new preservative will provide a longer shelf life for the bread of its bakery customers. Ten bakeries have used the new preservative in some of their dough and have provided the supplier with two fresh loaves of standard bread, one baked with their regular preservative and the other baked with the new preservative. The following shelf lives have been obtained:

20

(b) $R = 16$ upper-tailed

(c) $z = \dfrac{2R - n}{\sqrt{n}} = \dfrac{12}{\sqrt{20}} = 2.68$

(d) $z_{.05} = 1.96$ Reject H_0 if $z > 1.96$

$2.68 > 1.96$, so reject H_0

6

7

8 .5

(b) $R = 13$

upper-tailed

(c) $P(R \geq 13) = 1 - P(R \leq 12) = 1 - .9991 = .0009$

(d) $R^* = 12$ $R^* - 1 = 12$ yes

Rej. if $R \geq 13$

~~14~~ 10

-51	-8	-5	-1	6	9	9	10	11	15	18	23	27	33	44	53	57	112	113	203
1	2	3	4	5	6	7	8	9	10	11	12	13	14	15	16	17	18	19	20

$W = 1+3+8+10+12+14+15+17+19+20 = 119$

(d) $$z = \frac{119 - \frac{10(10+10+1)}{2}}{\sqrt{\frac{10(10)(10+10+1)}{12}}} = \frac{14}{13.23} = 1.06$$

accept H_0 at the $\alpha = .10$ significance level

(b) H_0: [illegible]

Accept H_0 if $-z_{.025} \le z \le z_{.025}$.

$-1.96 \le z \le 1.96$

Reject H_0 if $z > z_{.025}$ or if $z < -z_{.025}$

$z > 1.96$ $\quad$ $z < -1.96$

$$z = \frac{W - \frac{n_A(n_A+n_B+1)}{2}}{\sqrt{\frac{n_A n_B(n_A+n_B+1)}{12}}} = \frac{106.5 - \frac{8(8+10+1)}{2}}{\sqrt{\frac{(8)(10)(8+10+1)}{12}}}$$

$$= \frac{30.5}{11.25} = 2.71$$

(c) $2.71 > 1.96$, so reject H_0. Large z means higher ranks have been assigned to A. We're looking for

14 - 3, 5, 7, 10

14-3

(i)

Type	B	B	B	B	B	B	B	A	A	A	B	B	A	B	A	A	A	A
	4	5	6	7	7	7	8	9	9	10	10	11	12	13	14	15	15	22
Rank	1	2	3	4	5	6	7	8	9	10.5	10.5	12	13	14	15	16	17	18

$W = 8 + 9 + 10.5 + 13 + 15 + 16 + 17 + 18 = 106.5$

Treatment A = Treatment B

small numbers (no. of days for cure to take effect), so doctor should choose Treatment B, Sulfa.

14-5

Should you use z (not t) even with small samples?

(a) H_0: $A \le B$

H_1: $A > B$

upper tailed →

(b) $z_{.10} = 1.64$

Reject H_0 if $z > 1.64$

(c)

| A | B | A | B | A | B | A | B | A | B | A | B | A | A | B | A | B | A |

Kathy - Cookies

14-7 H_0: $A \le B$

(a)

Pair	Difference		Pair	Difference
1	.6		9	1.2
2	.2		10	.1
3	-.1		11	1.7
4	0	tie	12	-.3
5	1.4		13	.7
[illegible]	2.6		14	1.9
[illegible]	.6		15	[illegible]

14-10

H_0: $A \leq B$

(a)

Pair	Diff	Pair	Diff
1	12	11	7
2	2	12	1
3	6	13	4
4	−2	14	8
5	2	15	−1
6	4	16	8
7	−1	17	1
8	13	18	2
9	5	19	7

	Shelf Life (days)	
Bakery	*Old Preservative*	*New Preservative*
1	5.7	6.3
2	4.2	3.9
3	6.5	6.7
4	3.4	3.6
5	6.1	6.3
6	5.3	5.7
7	4.9	5.2
8	3.7	3.7
9	2.8	3.7
10	4.3	4.5

(a) Use the sign test with binomial probabilities to determine whether to accept or reject the null hypothesis that the new preservative yields shelf lives that are at most as long as the old preservative at an $\alpha = .05$ level of significance.

(b) Repeat this test using the t statistic.

14-5 THE WILCOXON SIGNED-RANK TEST

Thus far, we have described three nonparametric procedures. The first of these is based upon rank sums and applies to independent samples. The second is equivalent to the first. The third is applicable when the sample data are matched into pairs. The first test is concerned only with the ranking of sample data and not with the relative magnitudes of the differences between sample groups. The third considers the direction but not the size of differences in sample pairs.

A test that considers both the direction and the magnitude of differences in matched sample pairs would exhibit both features of the above procedures. Such a test was proposed by Frank Wilcoxon. In this section, we will describe his *Wilcoxon signed-rank test*, a procedure based upon both rank sums and the signs of paired differences.

Description of the Test

The Wilcoxon signed-rank test is based upon matched-pairs differences. Ignoring signs, the *absolute values* of the differences are ranked from low to high. Once these rankings have been assigned, those ranks corresponding to the original positive matched-pairs differences are summed. This procedure can be summarized in the following steps, which provide ***the test statistic V***:

1. Calculate the differences $d_i = X_{A_i} - X_{B_i}$ for all sample pairs.
2. Ignoring signs, rank the absolute values of the d_is. Do not rank zero differences.
3. Calculate V = sum of ranks for positive d_is.

The statistic V may be used to test the same hypotheses as the sign test. In every application, H_0 includes the special case where *A and B values are generated by identical populations.* If this is true, then each matched-pairs difference has a 50-50 chance of having a positive or a negative sign. Furthermore, positive or negative differences of the same absolute size should be equally likely. This indicates that under H_0, about half of the ranks ought to correspond to positive differences and that the sum of these ranks V should be close to half of the total rank-sum value. To determine whether the value of V significantly refutes H_0, we must examine its sampling distribution.

The Sampling Distribution of the Test Statistic V

As with the earlier Wilcoxon test, a detailed probability accounting will provide the exact sampling distribution of V. However, for most sample sizes encountered, we may use the normal curve instead. The following expression* is used to calculate the ***normal deviate for the sample results***

$$z = \frac{V - \dfrac{n(n+1)}{4}}{\sqrt{\dfrac{n(n+1)(2n+1)}{24}}} \qquad (14\text{-}5)$$

The value for n is the number of nonzero differences found in the sample. As in the sign test, we ignore ties with sample pairs, since they provide matched-pairs differences of zero.

The computed z may be compared to the appropriate critical normal deviate to determine the course of action to take. As an illustration, we will reconsider Example 14-2, where an educational researcher conducted an experiment to compare two methods of language-skill instruction. Table 14-6 shows the data she obtained. Recall that A observations represent the scores achieved by fourth-grade students under a traditional memory-drill method designed to expand vocabulary and improve spelling. Scores for the experimental group B were achieved by fourth-graders taking a special outside reading program in lieu of the regular procedure. The hypothesis to be tested is $H_0: A$ scores are at least as great as B scores (new procedure is no improvement over old procedure).

Ignoring the signs, the ranks for the nonzero matched-pairs differences in Table 14-6 were assigned to the absolute amounts, starting with 1 for the lowest. Assigning absolute values to pairs of opposite signs is handled exactly as we discussed in the Wilcoxon rank-sum test. Thus, the pairs with differences -2 and $+2$, which tie for the smallest absolute value, both receive the average of the two lowest ranks $(1+2)/2 = 1.5$. As before, ties between positive or negative differences only are ignored.

* Here, we use the fact that $z = [V - E(V)]/\sigma(V)$, where the mean of V is half of the total rank sum $n(n+1)/2$ and it can be shown that the variance of V is $\sigma^2(V) = n(n+1)(2n+1)/24$.

TABLE 14-6 Sample Data for Experiment Comparing Two Language-Skill Teaching Procedures with Assignment of Signed Ranks

Pair i	Group A Score X_{A_i}	Group B Score X_{B_i}	Difference $d_i = X_{A_i} - X_{B_i}$	Sign	Rank of Absolute Values of Matched-Pairs Differences
1	65	67	−2	−	1.5
2	72	66	+6	+	12
3	74	77	−3	−	3
4	81	90	−9	−	18
5	76	72	+4	+	5
6	95	95	0	tie	—
7	63	68	−5	−	10
8	85	95	−10	−	20
9	90	98	−8	−	16
10	64	60	+4	+	6
11	78	72	+6	+	13
12	86	94	−8	−	17
13	89	96	−7	−	14
14	75	73	+2	+	1.5
15	93	96	−3	−	4
16	78	90	−12	−	21
17	92	92	0	tie	—
18	67	63	+4	+	7
19	88	95	−7	−	15
20	79	88	−9	−	19
21	60	75	−15	−	23
22	90	95	−5	−	11
23	82	78	+4	+	8
24	73	85	−12	−	22
25	86	82	+4	+	9

The sum of the ranks for the eight positive differences is

$$V = 12 + 5 + 6 + 13 + 1.5 + 7 + 8 + 9 = 61.5$$

The corresponding normal deviate is found from expression (14-5), using $n = 23$ nonzero pair differences:

$$z = \frac{61.5 - \dfrac{23(24)}{4}}{\sqrt{\dfrac{23(24)(46+1)}{24}}} = \frac{-76.5}{\sqrt{1{,}081}} = -2.33$$

Since small values of V and resultant negative zs indicate that the B values are preponderately larger than the A values—the opposite of what the null hypothesis implies—this test is *lower-tailed.* At the $\alpha = .05$ significance level, $z_{.05} = -1.64$.

Thus, the Wilcoxon signed-rank test indicates that the null hypothesis should be *rejected* and that the researcher should conclude that the experimental language program is an improvement over the traditional procedure.

Comparison of the Wilcoxon Signed-Rank and the Sign Tests

The conclusion just reached is contradictory to the one we reached previously. Using these data, the sign test accepted the same null hypothesis at the identical significance level. With the sign test, we obtained $z = -1.46$, which corresponds to a lower-tail normal curve area of $.5 - .4279 = .0721$, the lowest significance level at which H_0 could have been rejected using that test. The Wilcoxon signed-rank test provides $z = -2.33$, corresponding to a lower-tail normal curve area of $.5 - .4901 = .0099$.Thus, the probability of incorrectly rejecting H_0 is much lower with the Wilcoxon than with the sign test. This reflects the fact that the Wilcoxon signed-rank test is more powerful; that is, for the same sample size, it is more discriminating or more efficient than the sign test.

The greater efficiency of the Wilcoxon procedure results from the greater amount of information it derives from the sample data. The test considers the sizes of the differences as well as their signs. Why, then, is the less efficient sign test used at all? One reason for its popularity is that the sign test is simpler to use. Another is that when there are very many ties between differences of opposite signs, the Wilcoxon signed-rank test needs adjustment to be applicable. *A further advantage of the sign test is that its assumptions are less restrictive than the Wilcoxon's.* It does not assume that A and B values have identical population frequency distributions (and, hence, equal variances), as does the Wilcoxon signed-rank test.

EXERCISES

14-12 In each of the following situations, the Wilcoxon signed-rank test will be used to determine the appropriate action:

(a)	(b)	(c)	(d)
$H_0: As \leqslant Bs$	$H_0: As \geqslant Bs$	$H_0: As \leqslant Bs$	$H_0: As \geqslant Bs$
$n = 18$	$n = 100$	$n = 50$	$n = 19$
$\alpha = .05$	$\alpha = .01$	$\alpha = .01$	$\alpha = .05$
$V = 40$	$V = 2{,}065$	$V = 890$	$V = 44$

(1) Indicate whether the test is lower- or upper-tailed.
(2) Find the appropriate decision rule.
(3) Indicate whether H_0 should be accepted or rejected at the stated significance level for the result provided.

14-13 Consider the data in Table 11-3 (page 410) for the examination scores achieved by two groups of students, one using a new statistics book (A) and the other using an old statistics book (B). There, the professor assumed the null hypothesis that the scores would be identical for the two groups.
(a) Is this test one- or two-sided?

(b) Formulate the professor's decision rule at the $\alpha = .05$ significance level.

(c) Determine the value of the test statistic V. What conclusion does this value indicate the professor should make?

14-14 Using the data in Exercise 14-11 (page 518), compute the test statistic V. Should the null hypothesis that the new preservative yields shelf lives at most as long as the old preservative yields be accepted or rejected at the $\alpha = .05$ significance level?

14-15 Use the sample data provided in Exercise 14-7 (page 517) to apply the Wilcoxon signed-rank test. Assuming that leaded gasoline will be used only if it provides significantly better mileage than unleaded gasoline, determine which gasoline the data indicate will be used. (Use $\alpha = .0010$.)

14-6 TEST FOR RANDOMNESS: THE NUMBER-OF-RUNS TEST

The Need to Test Randomness

We have seen that in order to use probabilities to qualify inferences about populations based on sample results, the observations must be randomly obtained. The question of whether a sample is random is especially critical in those instances when sample evidence from rare occurrences is collected over time. For example, production delays due to equipment malfunction, the causes or costs of aircraft accidents, or the IQs of identical twins reared apart are data for which evidence can be collected only historically. It is not possible to collect sample evidence through random sampling in a short time. Thus, the samples available are not random in the usual sense, but they may be analyzed as if they were, provided the data appear to be random.

Sample data collected over time are not random if there is some sort of serial dependency, so that the order in which particular attributes or variates of similar size occur is affected by what happened previously. To apply statistical methodology in such circumstances, it must be verified that the order in which the observations are obtained is similar to the order expected from a truly random sample.

The Number-of-Runs Test

A very useful procedure is to separate the sample results into two mutually exclusive categories: defective, nondefective; above 80, 80 or below; or fatal, nonfatal. Then the data can be represented chronologically as a string of category designations. For example, suppose that a university department wishes to base its judgments regarding future admissions on the test scores recorded for 20 graduate students admitted over the last two years. Letting a represent a score below the median and b a score above, the results can be represented by a string of as and bs, as in

$$\text{Sequence 1: } \underbrace{a}\ \underbrace{b\ b}\ \underbrace{a\ a\ a}\ \underbrace{b\ b}\ \underbrace{a}\ \underbrace{b\ b\ b\ b}\ \underbrace{a\ a\ a}\ \underbrace{b\ b}\ \underbrace{a\ a}$$

The braces indicate *runs* of a particular category: in this sequence, 5 of a and 4 of b. *A run is a succession of one or more observations in the same category.* The hypothesis-testing procedure uses the number of runs as a basis for determining randomness.

The underlying rationale of using runs to test for randomness is that too few runs or too many runs are unlikely if the sample is truly random over time. For instance, consider the following two-run result in

$$\text{Sequence 2: } \underbrace{a\ a\ a\ a\ a\ a\ a\ a\ a\ a}\ \underbrace{b\ b\ b\ b\ b\ b\ b\ b\ b\ b}$$

The first ten students admitted all scored poorly, while the later ten all received scores above the median. Such a result is indeed peculiar and is quite rarely achieved from a random process. It might be explained by a change in screening procedure: for instance, a tougher admissions policy after the tenth student. On the other hand, a run of 20 results is also highly unlikely, as in

$$\text{Sequence 3: } b\ a\ b\ a\ b\ a\ b\ a\ b\ a\ b\ a\ b\ a\ b\ a\ b\ a\ b\ a$$

Such a history could be obtained by a policy attempting to maintain a balance in abilities or by altruistic motives on the part of the department, such as having a fixed ratio of disadvantaged students.

The number-of-runs test is based upon the following null hypothesis:

H_0: The sampling is random. Therefore, each sequence position has the same prior chance of obtaining an a as any other.

The alternative is that the sample is not random.

The test statistic is the number of runs for category a, denoted R_a. (We could just as well use the number of runs for category b, or the total number of runs; all provide nearly the same results.) For the screening test results of sequence 1, the number of a runs is $R_a = 5$; likewise, we have $R_a = 1$ for sequence 2 and $R_a = 10$ for sequence 3. To facilitate formulating an appropriate decision rule, we will first describe the sampling distribution for R_a.

The Sampling Distribution for the Number of Runs R_a

Basic concepts of probability may be applied to show that R_a has the *hypergeometric distribution.* Like the binomial, probability values for this distribution are tedious to compute and are usually tabled. It is more convenient to approximate the probabilities by using the normal curve. The following expression* may be used to calculate the ***normal deviate for the sample results***

$$z = \frac{R_a - \dfrac{n_a(n_b+1)}{n_a+n_b}}{\sqrt{\dfrac{n_a(n_b+1)(n_a-1)}{(n_a+n_b)^2}\left(\dfrac{n_b}{n_a+n_b-1}\right)}} \tag{14-6}$$

where n_a and n_b denote the number of *a*s and *b*s in the sample. For the three test-score sequences, $n_a = n_b = 10$ and the following normal deviates apply:

$$\text{Sequence 1:} \quad z = \frac{5 - \dfrac{10(10+1)}{10+10}}{\sqrt{\dfrac{10(10+1)(10-1)}{(10+10)^2}\left(\dfrac{10}{10+10-1}\right)}}$$

$$= \frac{5-5.5}{\sqrt{1.3026}} = \frac{5-5.5}{1.14} = -.44$$

$$\text{Sequence 2:} \quad z = \frac{1-5.5}{1.14} = -3.95$$

$$\text{Sequence 3:} \quad z = \frac{10-5.5}{1.14} = 3.95$$

The Decision Rule

Continuing with our screening-test illustration, suppose that we desired an $\alpha = .05$ significance level for the probability of incorrectly concluding that the test scores were randomly generated. *The testing procedure is two-sided,* since H_0 is refuted by either too few or too many *a* runs. The critical normal deviate is $z_{.025} = 1.96$, and the appropriate decision rule is

Accept H_0 (conclude test scores are random) if $-1.96 \leqslant z \leqslant 1.96$

Reject H_0 (conclude test scores are not random) either if $z < -1.96$

or if $z > 1.96$

Applying the above rule to sequence 1, the normal deviate $z = -.44$ lies

* Here, we use the relation $z = [R_a - E(R_a)]/\sigma(R_a)$, where the mean of R_a is

$$E(R_a) = n_a(n_b+1)/(n_a+n_b)$$

and

$$\sigma^2(R_a) = \frac{n_a(n_b+1)(n_a-1)}{(n_a+n_b)^2}\left(\frac{n_b}{n_a+n_b-1}\right)$$

within ± 1.96, so H_0 must be *accepted.* At the 5-percent significance level, we may conclude that the test scores in sequence 1 were randomly generated. Since the normal deviates for sequences 2 and 3 lie outside the acceptance region, we must *reject* H_0 for either sequence, concluding that the test scores in each case were not randomly generated.

Testing the 1970 Draft Lottery for Randomness

Example 14-3 In December 1969, the U.S. Selective Service initiated a new procedure for determining the priorities for inducting young men into compulsory military service. Starting with the 19-year-olds born in 1950, a revised policy was implemented which drafted these younger men before the older ones. A 19-year-old man who was not drafted in 1970 would have a lower priority of being drafted in 1971 than the new crop of men born in 1951. As the manpower requirements for draftees in 1970 were much less than the number of available men, a method was devised to determine the induction priorities for 1970 by lottery.

Capsules were to be drawn from a barrel. Those men whose birthdays were drawn first would almost certainly be drafted in 1970. Those born on dates chosen near the end of the lottery would be almost certain of never having to enter military service.

Newspapers devoted wide coverage to the lottery. Some drew an imaginary line at 183 (half the number of days in a leap year), concluding that those whose numbers were 183 or below were "vulnerable," while those having numbers greater than 183 were "safe."

Table 14-7 gives the calendar dates and the corresponding draft-priority numbers obtained from the lottery in December 1969. Notice that birthdays early in the year tend to be preponderantly "safe," while those late in the year comprise a majority of "vulnerables." This leads us to question whether the lottery was truly random. To help identify the runs in priority numbers, safe values are marked with a square on the left and vulnerable values are marked with a square on the right.

We may use the number-of-runs test to answer this question, because the draft lottery fits all the necessary assumptions. We will make one slight interpretation. The capsules contained *dates*, so that the sequence in which a date was drawn determined the draft priority for that date. It will be simpler if we envision a completely analogous experiment in which capsules containing numbers from 1 through 366 are drawn one at a time, the first capsule corresponding to January 1, the second to January 2, and so forth throughout the year. The number inside each capsule represents the draft priority for that date. All possible sequences of priority numbers are equally likely, so that our underlying assumption regarding the sampling distribution of R_a is met.

Suppose we test the null hypothesis that the lottery is random at a significance level of $\alpha = .01$. Our decision rule may be found by determining the critical value $z_{\alpha/2} = 2.57$ from Appendix Table D. We may express our decision rule as

Accept H_0 (conclude lottery was random) if $-2.57 \leq z \leq 2.57$

Reject H_0 (conclude lottery was not random) either if $z < -2.57$ *or* if $z > 2.57$

From Table 14-7 we observe that there are 86 runs of safe (low-priority) *a*s and 86 runs of vulnerable (high-priority) *b*s. Thus, $R_a = 86$. From expression

TABLE 14-7 Determination of Number of Runs of Low Priority Numbers from 1970 Draft Lottery

January Birthday	Draft-priority number	*February* Birthday	Draft-priority number	*March* Birthday	Draft-priority number	*April* Birthday	Draft-priority number	*May* Birthday	Draft-priority number	*June* Birthday	Draft-priority number
1	·305	1	86·	1	108·	1	32·	1	·330	1	·249
2	159·	2	144·	2	29·	2	·271	2	·298	2	·228
3	·251	3	·297	3	·267	3	83·	3	40·	3	·301
4	·215	4	·210	4	·275	4	81·	4	·276	4	20·
5	101·	5	·214	5	·293	5	·269	5	·364	5	28·
6	·224	6	·347	6	139·	6	·253	6	155·	6	110·
7	·306	7	91·	7	122·	7	147·	7	35·	7	85·
8	·199	8	181·	8	·213	8	·312	8	·321	8	·366
9	·194	9	·338	9	·317	9	·219	9	·197	9	·335
10	·325	10	·216	10	·323	10	·218	10	65·	10	·206
11	·329	11	150·	11	136·	11	14·	11	37·	11	134·
12	·221	12	68·	12	·300	12	·346	12	133·	12	·272
13	·318	13	152·	13	·259	13	124·	13	·295	13	69·
14	·238	14	4·	14	·354	14	·231	14	178·	14	·356
15	17·	15	89·	15	169·	15	·273	15	130·	15	180·
16	121·	16	·212	16	166·	16	148·	16	55·	16	·274
17	·235	17	·189	17	33·	17	·260	17	112·	17	73·
18	140·	18	·292	18	·332	18	90·	18	·278	18	·341
19	58·	19	25·	19	·200	19	·336	19	75·	19	104·
20	·280	20	·302	20	·239	20	·345	20	183·	20	·360
21	·186	21	·363	21	·334	21	62·	21	·250	21	60·
22	·337	22	·290	22	·265	22	·316	22	·326	22	·247
23	118·	23	57·	23	·256	23	·252	23	·319	23	109·
24	59·	24	·236	24	·258	24	2·	24	31·	24	·358
25	52·	25	179·	25	·343	25	·351	25	·361	25	137·
26	92·	26	·365	26	170·	26	·340	26	·357	26	22·
27	·355	27	·205	27	·268	27	74·	27	·296	27	64·
28	77·	28	·299	28	·223	28	·262	28	·308	28	·222
29	·349	29	·285	29	·362	29	·191	29	·226	29	·353
30	164·			30	·217	30	·208	30	103·	30	·209
31	·211			31	30·			31	·313		

July Birthday	Draft-priority number	*August* Birthday	Draft-priority number	*September* Birthday	Draft-priority number	*October* Birthday	Draft-priority number	*November* Birthday	Draft-priority number	*December* Birthday	Draft-priority number
1	93·	1	111·	1	·225	1	·359	1	19·	1	129·
2	·350	2	45·	2	161·	2	125·	2	34·	2	·328
3	115·	3	·261	3	49·	3	·244	3	·348	3	157·
4	·279	4	145·	4	·322	4	·202	4	·266	4	165·
5	·188	5	54·	5	82·	5	24·	5	·310	5	56·
6	·327	6	114·	6	6·	6	87·	6	76·	6	10·
7	50·	7	168·	7	8·	7	·234	7	51·	7	12·
8	13·	8	48·	8	·184	8	·283	8	97·	8	105·
9	·277	9	106·	9	·263	9	·342	9	80·	9	43·
10	·284	10	21·	10	71·	10	·220	10	·282	10	41·
11	·248	11	·324	11	158·	11	·237	11	46·	11	39·
12	15·	12	142·	12	·242	12	72·	12	66·	12	·314
13	42·	13	·307	13	175·	13	138·	13	126·	13	163·
14	·331	14	·198	14	1·	14	·294	14	127·	14	26·
15	·322	15	102·	15	113·	15	171·	15	131·	15	·320
16	120·	16	44·	16	·207	16	·254	16	107·	16	96·
17	98·	17	154·	17	·255	17	·288	17	143·	17	·304
18	·190	18	141·	18	·246	18	5·	18	146·	18	128·
19	·227	19	·311	19	177·	19	·241	19	·203	19	·240
20	·187	20	·344	20	63·	20	·192	20	·185	20	135·
21	27·	21	·291	21	·204	21	·243	21	156·	21	70·
22	153·	22	·339	22	160·	22	117·	22	9·	22	53·
23	172·	23	116·	23	119·	23	·201	23	182·	23	162·
24	23·	24	36·	24	·195	24	·196	24	·230	24	95·
25	67·	25	·286	25	149·	25	176·	25	132·	25	84·
26	·303	26	·245	26	18·	26	7·	26	·309	26	173·
27	·289	27	·352	27	·233	27	·264	27	47·	27	78·
28	88·	28	167·	28	·257	28	94·	28	·281	28	123·
29	·270	29	61·	29	151·	29	·229	29	99·	29	16·
30	·287	30	·333	30	·315	30	38·	30	174·	30	3·
31	·193	31	11·			31	79·			31	100·

SOURCE: *U.S. News & World Report,* December 15, 1969, p. 34. Reprinted from *U.S. News & World Report.* Copyright 1969 U.S. News & World Report, Inc.

(14-6), we calculate the corresponding normal deviate, using $n_a = n_b = 183$:

$$z = \frac{86 - \dfrac{183(183+1)}{(183+183)}}{\sqrt{\dfrac{183(183+1)(183-1)}{(183+183)^2}\left(\dfrac{183}{183+183-1}\right)}} = -1.253$$

Since z lies between the critical values -2.57 and 2.57, this value falls into the acceptance region and we can conclude that *the draft lottery numbers were randomly selected.*

Referring to the table of areas under the normal curve, $z = -1.253$ corresponds to a lower-tail area of about .105, so that the null hypothesis can only be rejected at a level of significance of twice this amount, or approximately $\alpha = .21$.

Meaningfulness of Results

The draft lottery example was included in this book partly because of the controversy the lottery stirred in 1970. As we indicated in Chapter 3, the U.S. Selective Service was vehemently criticized because the capsules were selected in such a haphazard way. The criticism resulted from the disparity of the numbers in Table 14-7. Investigations of the procedure used showed that the capsules were placed into the barrel by month, starting with January and ending with December. The mixing of the capsules was superficial, so that more December dates than January dates, for example, were on the top. It has been convincingly argued that the lottery was unfair to young men having birthdays late in the year. Several law suits were even initiated to invalidate the lottery for this reason.

But our number-of-runs test does not provide nearly sufficient evidence to refute the hypothesis of random selection. Our lowest possible significance level leading to rejection is very high—greater than .20. In many scientific applications, for example, a significance level as low as .01 or .001 is usually required before results are worthy of inclusion in the body of the theory. Most persons employing statistics would regard a significance level of about .20 as highly "insignificant" in establishing that the null hypothesis is untrue and would therefore either reserve judgment or accept the null hypothesis.

This example points out a major inadequacy of using hypothesis testing as a basis for drawing conclusions about how well data fit a particular assumption. This inadequacy is fundamental to the entire statistical approach of explaining outcomes in terms of probabilities. Our number-of-runs model is not refined enough to incorporate all information (for example, the manner in which the capsules were supposedly mixed and then selected). It focuses on results only; all other relevant information is ignored. The same statistician who would conclude that the *results* are not significant enough to warrant rejection would, upon hearing a *description of the procedure* for capsule selection, be almost certain to dismiss the same null hypothesis with the full strength of his convictions, *regardless of what results were obtained.*

One important question remains. Why was such a typical run statistic value calculated for the 1970 draft lottery, when it was so convincingly argued that it was far from random? We note that although circumstances were far from ideal, there was *some mixing* of the capsules. Capsules were often withdrawn by plunging the fingers deeply into the barrel, so that even the top capsules were not guaranteed an early withdrawal and the bottom capsules were not immune from an early grasp. Thus, these operating factors contributed some aspects of randomness. As the numbers in Table 14-7 indicate, the number of vulnerable dates for February, May, June, July, September, and October did not deviate by very much from half the number of days in these months.

Undoubtedly embarrassed and goaded by public pressure, the Selective Service revised its procedures for the 1971 draft lottery. Two sets of capsules were used—one containing dates and the other containing priority numbers. Mixing machines were employed, and the drawings consisted of simultaneously selecting a capsule from each barrel and matching the respective date and priority number. The randomness of this procedure was beyond reproach.

Other Tests for Randomness

We have only considered a single test for randomness, based upon the *number* of runs. A similar test not described here is based upon the *lengths* of runs. Number-of-runs tests consider only the *sequence* of numbers. In some situations, it is desirable to test for other features, such as the *frequency* of various values or their *serial correlation* (which measures the tendency of certain numbers to be followed by other numbers).

To illustrate the need to consider all these features, examine the problem of testing how suitable it would be to use computer-generated random numbers (which only appear random and are properly called pseudorandom numbers) to select sample units. A number-of-runs test would accept as random the following sequence of digits:

1 9 9 9 1 1 1 9 9 1 1 1 9 9 9 1 1 1 9 9

Clearly, such a sequence would be inadequate in selecting a random sample. The number-of-runs test does not detect the frequency with which each digit (in this case, zero through 9) occurs. For a list of true random numbers, each digit is expected to occur 10 percent of the time, so that the uniform distribution represents the underlying population of values. Testing procedures (the goodness-of-fit test) exist to determine whether or not the above sequence came from a population having that distribution. Such a test would therefore result in eliminating the particular computer-generation scheme as unsuitable.

A test for serial correlation could be applied to the following sequence

2 7 6 9 5 3 2 1 0 7 6 9 5 4 3 8 7 6 4 0

to detect unrandomlike patterns—that each occurrence of 7 is always followed by a 6, for example. Such quirks are ignored by both number-of-runs and goodness-of-fit tests, either of which would lead to accepting the above sequence as being consistent with its respective null hypothesis.

A detailed discussion of the various tests of randomness and how they may be combined as a series is beyond the scope of this book.*

EXERCISES

14-16 The following four sequences of men and women represent the order in which persons of the respective sexes were admitted to various graduate programs. In each case below, there are 10 men and 10 women.

(1) *M W M W W W M M W M M W M W M W W M M W*
(2) *M W M M W M W W M W M M W M W W M W M W*
(3) *M M M M W W W W W M M M M M M W W W W W*
(4) *M W W M W M M W M W M W M W M W M W W M*

(a) Using an $\alpha = .05$ significance level, formulate the appropriate decision rule for testing the null hypothesis that the sexes of the persons admitted were randomly determined for each successive admission. Let R_a represent the number of runs of women.
(b) Applying your decision rule from (a), indicate whether an H_0 of randomness must be accepted or rejected for each sequence.

14-17 A fair coin tossed 16 times produced a total of 8 heads and 8 tails. There is some doubt as to whether these successive tosses were random events. Using the number of head runs as R_a, apply the normal approximation to determine whether H_0 that the tosses were random should be accepted or rejected for each of the following sequences. (Use $\alpha = .05$.)

(a) *H T T T H T T H T T H H T H H H*
(b) *H H H T T T T T H H H H H T T T T*
(c) *H T H T H T T H H T H T H T H T*
(d) *H H T T H H T T T H H H T H T T*

14-18 The results of the 1971 draft lottery are provided in Table 14-8.
(a) Determine the number of runs of vulnerable dates (those having priority 183 or less).
(b) Remembering that there are only 365 dates, calculate the normal deviate corresponding to your result.
(c) What is the lowest significance level at which the null hypothesis of random selection can be rejected? (Recall that the runs test is two-sided.)

14-19 Toss a coin 30 times, recording whether a head or a tail occurs for each toss. Then test your results for randomness, using the number-of-runs test with the normal approximation. At a significance level of .10, can you reject the null hypothesis that you are a fair coin tosser?

14-20 The successive terms in the expansions of certain constants, such as $\pi = 3.1416\ldots$, have been proposed as substitutes for random numbers. In testing for one property of the numbers obtained in this fashion, a

* For a comprehensive discussion for testing the randomness of number lists, see J.W. Schmidt and R. E. Taylor, *Simulation and Analysis of Industrial Systems* (Homewood, Ill.: Richard D. Irwin, 1970), Chapter 6.

TABLE 14-8 1971 Draft Lottery Results

January		*February*		*March*		*April*		*May*		*June*	
Birth-day	*Draft-priority number*	*Birth-day*	*Draft-priority number*	*Birth-day*	*Draft-priority number*	*Birth-day*	*Draft-priority number*	*Birth-day*	*Draft-priority number*	*Birth-day*	*Draft-priotriy number*
1	133	1	335	1	14	1	224	1	179	1	65
2	195	2	354	2	77	2	216	2	96	2	304
3	336	3	186	3	207	3	297	3	171	3	135
4	99	4	94	4	117	4	37	4	240	4	42
5	33	5	97	5	299	5	124	5	301	5	233
6	285	6	16	6	296	6	312	6	268	6	153
7	159	7	25	7	141	7	142	7	29	7	169
8	116	8	127	8	79	8	267	8	105	8	7
9	53	9	187	9	278	9	223	9	357	9	352
10	101	10	46	10	150	10	165	10	146	10	76
11	144	11	227	11	317	11	178	11	293	11	355
12	152	12	262	12	24	12	89	12	210	12	51
13	330	13	13	13	241	13	143	13	353	13	342
14	71	14	260	14	12	14	202	14	40	14	363
15	75	15	201	15	157	15	182	15	344	15	276
16	136	16	334	16	258	16	31	16	175	16	229
17	54	17	345	17	220	17	264	17	212	17	289
18	185	18	337	18	319	18	138	18	180	18	214
19	188	19	331	19	189	19	62	19	155	19	163
20	211	20	20	20	170	20	118	20	242	20	43
21	129	21	213	21	246	21	8	21	225	21	113
22	132	22	271	22	269	22	256	22	199	22	307
23	48	23	351	23	281	23	292	23	222	23	44
24	177	24	226	24	203	24	244	24	22	24	236
25	57	25	325	25	298	25	328	25	26	25	327
26	140	26	86	26	121	26	137	26	148	26	308
27	173	27	66	27	254	27	235	27	122	27	55
28	346	28	234	28	95	28	82	28	9	28	215
29	277			29	147	29	111	29	61	29	154
30	112			30	56	30	358	30	209	30	217
31	60			31	38			31	350		

July		*August*		*September*		*October*		*November*		*December*	
Birth-day	*Draft-priority number*	*Birth-day*	*Draft-priority number*	*Birth-day*	*Draft-priority number*	*Birth-day*	*Draft-priority number*	*Birth-day*	*Draft-priority number*	*Birth-day*	*Draft-priority number*
1	104	1	326	1	283	1	306	1	243	1	347
2	322	2	102	2	161	2	191	2	205	2	321
3	30	3	279	3	183	3	134	3	294	3	110
4	59	4	300	4	231	4	266	4	39	4	305
5	287	5	64	5	295	5	166	5	286	5	27
6	164	6	251	6	21	6	78	6	245	6	198
7	365	7	263	7	265	7	131	7	72	7	162
8	106	8	49	8	108	8	45	8	119	8	323
9	1	9	125	9	313	9	302	9	176	9	114
10	158	10	359	10	130	10	160	10	63	10	204
11	174	11	230	11	288	11	84	11	123	11	73
12	257	12	320	12	314	12	70	12	255	12	19
13	349	13	58	13	238	13	92	13	272	13	151
14	156	14	103	14	247	14	115	14	11	14	348
15	273	15	270	15	291	15	310	15	362	15	87
16	284	16	329	16	139	16	34	16	197	16	41
17	341	17	343	17	200	17	290	17	6	17	315
18	90	18	109	18	333	18	340	18	280	18	208
19	316	19	83	19	228	19	74	19	252	19	249
20	120	20	69	20	261	20	196	20	98	20	218
21	356	21	50	21	68	21	5	21	35	21	181
22	282	22	250	22	88	22	36	22	253	22	194
23	172	23	10	23	206	23	339	23	193	23	219
24	360	24	274	24	237	24	149	24	81	24	2
25	3	25	364	25	107	25	17	25	23	25	361
26	47	26	91	26	93	26	184	26	52	26	80
27	85	27	232	27	338	27	318	27	168	27	239
28	190	28	248	28	309	28	28	28	324	28	128
29	4	29	32	29	303	29	259	29	100	29	145
30	15	30	167	30	18	30	332	30	67	30	192
31	221	31	275			31	311			31	126

SOURCE: *U.S. News & World Report,* July 13, 1970, p. 27. Reprinted from *U.S. News & World Report.*

statistician's resultant data were 45 successive four-digit numbers 5,000 or above (a) and 55 numbers below (b) 5,000. Testing the null hypothesis of randomness (or, more correctly, the appearance of randomness) at the $\alpha = .05$ significance level, he found $R_a = 26$. What must the statistician conclude regarding the "randomness" of successive terms in the π expansion?

14-21 Before random numbers tables were constructed, a group of British statisticians used the London telephone directory to select random samples. You are to determine whether your local directory can be used for this purpose.

(a) Select a page from the telephone directory and determine the number of runs of even last digits of the telephone numbers. Using the normal approximation, can you conclude that the sequence of odd and even last digits is not random at a significance level of .01?

(b) Regardless of your results in (a), do you think that telephone numbers would be suitable random numbers? Discuss.

14-7 THE RANK CORRELATION COEFFICIENT

In Chapter 9, the sample correlation coefficient r was introduced as an index measuring the degree of association between two variables X and Y. Nonparametric statistics can be employed to provide alternative measures of correlation. One such statistic is the ***Spearman rank correlation coefficient***

$$r_s = 1 - \frac{6 \sum (X-Y)^2}{n(n^2-1)} \tag{14-7}$$

where X and Y are the *ranks* of the two variables measured.

The rank correlation coefficient was derived directly from the conventional correlation coefficient [as calculated in expression (9-28), page 372], except that instead of the observation values themselves ranks are used for X and Y. The fact that the sums of the ranks and their squares and the rank means are automatically known for each sample size n (for example, when $n = 5$, $\sum X = 1+2+3+4+5 = 15$ and $\bar{X} = 15/5 = 3$, always) permits us to use the equivalent and simpler calculation in expression (14-7).

To illustrate the rank correlation coefficient, consider the data in Table 14-9. Here, observations have been made of the average number of weekly hours a sample of $n = 10$ university students spent studying and their grade-point averages for the term.

Beginning with 1 for the lowest value, the observations for each variable in Table 14-9 were ranked. *Tying observations were given the average of the successive ranks that would have been assigned had the values been different.* Thus, the two figures of 17 study hours were both equally ranked as $(2+3)/2 = 2.5$, and each of the two GPAs of 3.6 received a rank of $(7+8)/2 = 7.5$.

TABLE 14-9 Rank Correlation Calculations for Study Hours and GPA

Variables		Ranks			
Study Hours	GPA	for Study Hours X	for GPA Y	Rank Difference X−Y	(X−Y)²
24	3.6	6	7.5	−1.5	2.25
17	2.0	2.5	1	1.5	2.25
20	2.7	4	4	0.0	0.00
41	3.6	8	7.5	.5	.25
52	3.7	10	9	1.0	1.00
23	3.1	5	5	0.0	0.00
46	3.8	9	10	−1.0	1.00
17	2.5	2.5	3	− .5	.25
15	2.1	1	2	−1.0	1.00
29	3.3	7	6	1.0	1.00
				0.0	9.00

The differences in ranks were determined and their squares were calculated. The sum of the squared rank differences is $\sum(X-Y)^2 = 9.00$. The rank correlation coefficient between average weekly study hours and GPA is

$$r_s = 1 - \frac{6(9.00)}{10(10^2-1)} = 1 - \frac{54}{990} = .946$$

This indicates a high correlation between study time and grades.

Usefulness of Rank Correlation

The rank correlation coefficient is useful in many situations where the conventional correlation coefficient is not suitable. Recall from Chapter 9 that the conventional r is based upon the assumption that the underlying relationship is linear. Consider the study time and GPA data in Table 14-9 when they are plotted in Figure 14-1. The graph shows a pronounced curvilinear relationship between these variables, summarized by the colored curve. Here, GPA increases with study time, but each additional hour of study time produces a progressively smaller improvement in grades. A least squares regression would fit the black line to the sample data, and the conventional correlation coefficient (through the coefficient of determination r^2) would express the proportion of variation in GPA explained by this line. Obviously, the regression line poorly fits the data here, as confirmed by the much lower conventional correlation value of $r = .58$ obtained.

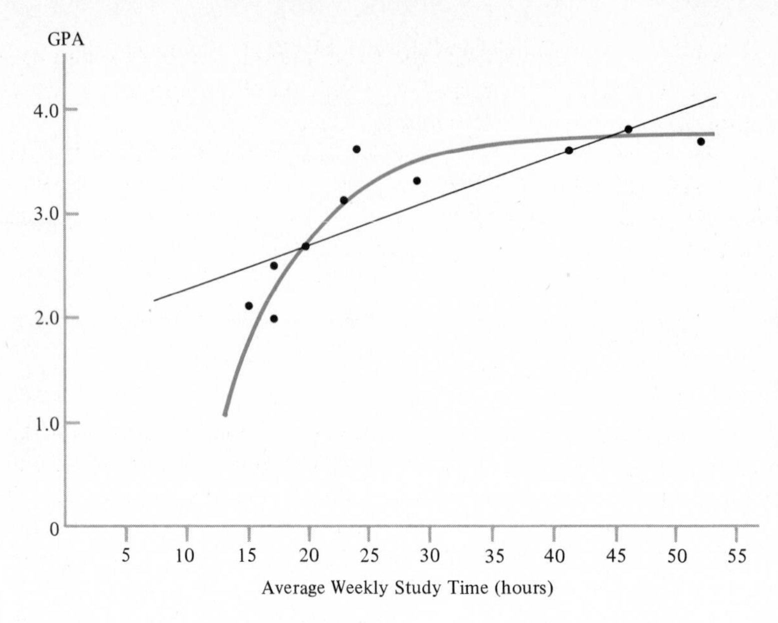

FIGURE 14-1 Illustration of inadequacy of conventional correlation coefficient when data are curvilinear.

EXERCISES

14-22 The following daily data were obtained for a sample of $n = 10$ smokers who drink coffee:

Packs Smoked	*Cups Drunk*
.5	3
1.0	4
.5	5
1.5	6
2.0	4
2.5	6
1.0	3
.5	2
1.5	7
2.0	5

Calculate the rank correlation coefficient.

14-23 Referring to Exercise 9-19 on page 376, determine the rank correlation coefficient for prior and current reading level scores.

14-24 Referring to Exercise 9-23 on page 377, calculate the rank correlation coefficient for mother's age and mean litter size for mice.

14-8 ONE-FACTOR ANALYSIS OF VARIANCE: THE KRUSKAL-WALLIS TEST

Alternative to the *F* Test

In Chapter 13, we presented a testing procedure for determining whether differences exist among several means, using the F test to perform analysis of variance. A very serious drawback to the F test is that it requires the populations to be normal. As noted in our preceding discussion of nonparametric tests, this assumption may invalidate the testing procedure.

An alternative to the F test is the Kruskal-Wallis one-factor analysis of variance, named after W. H. Kruskal and W. A. Wallis, who introduced it in 1952. It is really an extension of the Wilcoxon rank-sum test to the analysis of several samples.

Description of the Test

Underlying Principle

The primary difference from the F test is that the Kruskal-Wallis test is based upon a test statistic computed from ranks determined for pooled sample observations. Its null hypothesis is that the rank assigned to a particular observation has an equal chance of being any number between 1 and n, regardless of the sample group to which it belongs. We will illustrate this procedure with the following detailed example.

A college computer instructor wishes to determine if there is really any difference in programming aptitudes between students in different majors. The students in his college can be classified into four major groups: science, liberal arts, business, and engineering. He selected a random sample of students in each category and administered a computer programming aptitude test to them. Their scores and score ranks appear in Table 14-10. (Although no ties were found for these data, ranks are only averaged when tying observations occur between two or more sample groups.)

TABLE 14-10 Test Scores and Ranks for Computer Programming by Major

(1) *Science*		(2) *Liberal Arts*		(3) *Business*		(4) *Engineering*	
Score	*Rank*	*Score*	*Rank*	*Score*	*Rank*	*Score*	*Rank*
85	12	95	19	67	3	90	15
73	7	54	1	74	8	65	2
96	20	72	6	84	11	92	17
91	16	81	10	68	4	94	18
88	14	69	5	87	13		
				77	9		
$T_1 = 69$		$T_2 = 41$		$T_3 = 48$		$T_4 = 52$	
$T_1^2 = 4{,}761$		$T_2^2 = 1{,}681$		$T_3^2 = 2{,}304$		$T_4^2 = 2{,}704$	
$n_1 = 5$		$n_2 = 5$		$n_3 = 6$		$n_4 = 4$	
$\frac{T_1^2}{n_1} = 952.2$		$\frac{T_2^2}{n_2} = 336.2$		$\frac{T_3^2}{n_3} = 384.0$		$\frac{T_4^2}{n_4} = 676.0$	

Score:	54	65	67	68	69	72	73	74	77	81	84	85	87	88	90	91	92	94	95	96
Rank:	1	2	3	4	5	6	7	8	9	10	11	12	13	14	15	16	17	18	19	20

At a significance level of $\alpha = .05$, the instructor wishes to test the null hypothesis

H_0 : There are no programming aptitude differences between major groups.

The alternative hypothesis is that college major makes a difference in aptitude.

As with the Wilcoxon tests, the null hypothesis implies that the four sample groups are obtained from the same population. Under the null hypothesis, therefore, each score in Table 14-10 has the same prior probability of receiving any rank of 1 through 20.

Test Statistic

For our test statistic, we choose to compare variabilities of the ranks in each column. As we saw with the F test, the sum of squares may be used for this purpose. For convenience in obtaining a meaningful statistic, the sum of squares is not directly calculated. Instead, the sum of ranks for each category T_j is computed. A sum of the squares of these sums, each term weighted by the reciprocal of the sample size n_j of the jth group, is obtained. The following calculation provides the ***Kruskal-Wallis test statistic***

$$K = \frac{12}{n(n+1)}\left(\sum \frac{T_j^2}{n_j}\right) - 3(n+1) \tag{14-8}$$

where $n = \sum n_j$.

The sampling distribution of K is approximately chi-square with $m-1$ degrees of freedom, where m is the number of categories. Thus, in order to find the critical value, we use Appendix Table F, which provides critical values for specified tail areas.

In our example, $m = 4$, so there are $m-1 = 4-1 = 3$ degrees of freedom. Letting $\alpha = .05$, our critical value found from the chi-square table is $\chi^2_{.05} = 7.815$. Our decision rule is therefore

Accept H_0 (conclude major makes no difference) if $K \leqslant 7.815$

Reject H_0 (conclude major makes a difference) if $K > 7.815$

Substituting the values from Table 14-10 into expression (14-8), we calculate K, using $n = 20$:

$$\begin{aligned} K &= \frac{12}{20(20+1)}(952.2+336.2+384.0+676.0) - 3(20+1) \\ &= 4.097 \end{aligned}$$

Since K is smaller than $\chi^2_{.05}$, *we must accept the null hypothesis that there are no differences.*

EXERCISES

14-25 A chemical engineer wishes to know whether the mean time required to complete a chemical reaction is affected by the proportion of impurities present. Samples of reactions have been conducted with several different proportions of impurities. The reaction times obtained are provided below:

Proportion of Impurities			
.001	.01	.05	.10
Reaction Times (minutes)			
103	104	153	207
111	113	127	183
107	117	143	173
105	120	119	
	113	138	
		143	

(a) Rank the sample results.
(b) Calculate the test statistic K.
(c) How many degrees of freedom are associated with this test statistic?
(d) Using a significance level of $\alpha = .05$, can the null hypothesis that the proportion of impurities does not affect the reaction times be rejected?

14-26 Referring to the sod-yield data in Table 13-1 on page 462, apply the Kruskal-Wallis test to the null hypothesis that mean yields are identical under all fertilizer treatments. (Use $\alpha = .05$.) What conclusion should be made?

14-27 An insurance company wishes to determine if there is any difference due to the type of profession in the mean amount of whole life insurance held by its professional policyholders. A random sample of policyholders is selected for a thorough study. The results obtained are:

Insurance Coverage		
Physicians	*Lawyers*	*Dentists*
$200,000	$ 50,000	$ 80,000
150,000	100,000	45,000
40,000	95,000	155,000
35,000	10,000	325,000
110,000	300,000	
	75,000	

Determine whether the null hypothesis of no difference must be accepted or rejected at a .05 significance level.

REVIEW EXERCISES

14-28 In testing the null hypothesis that drug A yields relief from sinus headaches for at least as long as drug B, a statistician selects two independent samples of 30 sinus sufferers each and asks a physician to administer drug A to one group and drug B to the other. The number of pills taken over a fixed time period is used to compute each sufferer's mean relief duration score. In ranking the scores of the two groups, he found that $W = 826$. Should H_0 be accepted or rejected at the $\alpha = .05$ significance level?

14-29 Suppose that the statistician grouped the patients in Exercise 14-28 into 30 pairs, matched on the basis of medical history, age, sex, and life style. Also, assume that exactly the same individuals received the respective drugs. (*Note*: We would never use both the Wilcoxon rank-sum test and the sign test for matched pairs on the same data. This is an illustration only of how results may vary between the two tests.) The difference in relief duration for each pair was found by subtracting the mean time of the drug B patient from the corresponding figure of his partner. A total of $R = 5$ positive differences resulted, and there were 5 ties. Should H_0 be accepted or rejected at the $\alpha = .05$ significance level?

14-30 Repeat Exercise 14-29, applying the Wilcoxon signed-rank test when the sample data provide $V = 30$.

14-31 Referring to the data in Exercise 13-5 on page 474, apply the Kruskal-Wallis test to determine whether the dissolving time of aspirin tablets differs with the level of impurities. (Use $\alpha = .01$.)

14-32 Referring to the data in Exercise 9-5 on page 349, calculate the rank correlation coefficient.

14-33 A statistics professor asked his students to use their judgment in constructing their own random number lists from scratch. The purpose of the experiment was to demonstrate how faulty this procedure can be. One student obtained the following list. Proceeding down one column at a time, determine the number of runs of small values (0 through 4) and then determine whether the null hypothesis of randomness should be accepted or rejected. (Use $\alpha = .05$.)

4	6	9	9	0
8	1	2	3	6
3	9	8	8	2
7	2	3	2	3
9	4	7	7	4
0	8	1	5	8
2	7	7	1	5
1	6	3	7	3
8	0	4	0	8
5	3	6	5	0

Selected References

The Role of Statistics

Careers in Statistics. The American Statistical Association (no date available).
Huff, Darrell. *How to Lie with Statistics.* New York: W.W. Norton, 1954.
Moroney, M.J. *Facts from Figures.* Baltimore: Penguin Books, 1965.
Reichmann, W.J. *Use and Abuse of Statistics.* New York: Oxford University Press, 1962.
Wallis, W.A., and H.V. Roberts. *The Nature of Statistics.* New York: The Free Press, 1965.

General Introductory Textbooks

Dixon, Wilfrid J., and Frank J. Massey, Jr. *Introduction to Statistical Analysis*, 3rd ed. New York: McGraw-Hill, 1969.
Freund, John E. *Modern Elementary Statistics*, 3rd ed. Englewood Cliffs, N.J.: Prentice-Hall, 1967.
Hodges, J.L., Jr., and E.L. Lehmann. *Basic Concepts of Probability and Statistics.* San Francisco: Holden-Day, 1964.
Hoel, P.G. *Elementary Statistics*, 2nd ed. New York: John Wiley & Sons, 1966.
Mendenhall, William. *Introduction to Probability and Statistics*, 4th ed. Belmont, Calif.: Duxbury Press, 1975.
Snedecor, George W., and William G. Cochran. *Statistical Methods*, 6th ed. Ames, Iowa: Iowa State University Press, 1967.
Wallis, W.A., and H.V. Roberts. *Statistics: A New Approach.* New York: Macmillan, 1956.
Wonnacott, Thomas H., and Ronald J. Wonnacott. *Introductory Statistics.* New York: John Wiley & Sons, 1969.
Yamane, Taro. *Statistics—An Introductory Analysis*, 2nd ed. New York: Harper & Row, 1967.

Probability

Feller, William. *An Introduction to Probability Theory and Its Applications*, Vol. 1. New York: John Wiley & Sons, 1957.
Hodges, J.L., Jr., and E.L. Lehmann. *Elements of Finite Probability.* San Francisco: Holden-Day, 1965.
Laplace, Pierre Simon, Marquis de. *A Philosophical Essay on Probabilities.* New York: Dover Publications, 1951.
Lindgren, B.W., and G.W. McElrath. *Introduction to Probability and Statistics*, 3rd ed. New York: Macmillan, 1969.
Mosteller, F., R. Rourke, and G. Thomas, Jr. *Probability and Statistics.* Reading, Mass.: Addison-Wesley, 1961.
Parzen, Emmanuel. *Modern Probability Theory and Its Applications.* New York: John Wiley & Sons, 1960.

Regression and Correlation Analysis

Ezekiel, Mordecai, and Karl A. Fox. *Methods of Correlation and Regression Analysis*, 3rd ed. New York: John Wiley & Sons, 1959.
Johnston, J. *Econometric Methods*. New York: McGraw-Hill, 1963.
Neter, John, and William Wasserman. *Applied Linear Statistical Models*. Homewood, Ill.: Richard D. Irwin, 1974.
Williams, E.J. *Regression Analysis*. New York: John Wiley & Sons, 1959.
Wonnacott, Ronald J., and Thomas H. Wonnacott. *Econometrics*. New York: John Wiley & Sons, 1970.

Analysis of Variance and Design of Experiments

Cochran, William G., and Gertrude M. Cox. *Experimental Designs*, 2nd ed. New York: John Wiley & Sons, 1957.
Cox, D.R. *Planning of Experiments*. New York: John Wiley & Sons, 1958.
Guenther, W.C. *Analysis of Variance*. Englewood Cliffs, N.J.: Prentice-Hall, 1964.
Mendenhall, William. *An Introduction to Linear Models and the Design and Analysis of Experiments*. Belmont, Calif.: Wadsworth, 1967.
Neter, John, and William Wasserman. *Applied Linear Statistical Models*. Homewood, Ill.: Richard D. Irwin, 1974.
Scheffé, Henry. *The Analysis of Variance*. New York: John Wiley & Sons, 1959.

Nonparametric Statistics

Bradley, James V. *Distribution-Free Statistical Tests*. Englewood Cliffs, N.J.: Prentice-Hall, 1968.
Conover, W.J. *Practical Nonparametric Statistics*. New York: John Wiley & Sons, 1971.
Gibbons, J. *Nonparametric Statistical Inference*. New York: McGraw-Hill, 1971.
Hájek, Jaroslav. *Nonparametric Statistics*. San Francisco: Holden-Day, 1969.
Kraft, Charles H., and Constance van Eeden. *A Nonparametric Introduction to Statistics*. New York: Macmillan, 1968.
Noether, Gottfried E. *Introduction to Statistics: A Fresh Approach*. Boston: Houghton Mifflin, 1971.
Siegel, Sydney. *Nonparametric Statistics for the Behavioral Sciences*. New York: McGraw-Hill, 1956.

Statistical Tables

Beyer, William H. (ed.). *Handbook of Tables for Probability and Statistics*. Cleveland, Ohio: The Chemical Rubber Co., 1966.
Burington, Richard S., and Donald C. May. *Handbook of Probability and Statistics with Tables*. New York: McGraw-Hill, 1953.
Fisher, R.A., and F. Yates. *Statistical Tables for Biological, Agricultural and Medical Research*. London: Longman Group, 1963.
National Bureau of Standards. *Tables of the Binomial Distribution*. Washington, D.C.: U.S. Government Printing Office, 1950.
Owen, D.B. *Handbook of Statistical Tables*. Reading, Mass.: Addison-Wesley, 1962.
Pearson, E.S., and H.O. Hartley. *Biometrika Tables for Statisticians*, 3rd ed. Cambridge, England: Cambridge University Press, 1966.
The Rand Corporation. *A Million Random Digits with 100,000 Normal Deviates*. New York: The Free Press, 1955.

Appendix Tables

TABLE A Squares, Square Roots, and Reciprocals

N	$\sqrt{N}$	N^2	$\sqrt{10N}$	$1000/N$	N	$\sqrt{N}$	N^2	$\sqrt{10N}$	$1000/N$
					50	7.07107	2500	22.36068	20.00000
1	1.00000	1	3.16228	1000.00000	51	7.14143	2601	22.58318	19.60784
2	1.41421	4	4.47214	500.00000	52	7.21110	2704	22.80351	19.23077
3	1.73205	9	5.47723	333.33333	53	7.28011	2809	23.02173	18.86792
4	2.00000	16	6.32456	250.00000	54	7.34847	2916	23.23790	18.51852
5	2.23607	25	7.07107	200.00000	55	7.41620	3025	23.45208	18.18182
6	2.44949	36	7.74597	166.66667	56	7.48331	3136	23.66432	17.85714
7	2.64575	49	8.36660	142.85714	57	7.54983	3249	23.87467	17.54386
8	2.82843	64	8.94427	125.00000	58	7.61577	3364	24.08319	17.24138
9	3.00000	81	9.48683	111.11111	59	7.68115	3481	24.28992	16.94915
10	3.16228	100	10.00000	100.00000	60	7.74597	3600	24.49490	16.66667
11	3.31662	121	10.48809	90.90909	61	7.81025	3721	24.69818	16.39344
12	3.46410	144	10.95445	83.33333	62	7.87401	3844	24.89980	16.12903
13	3.60555	169	11.40175	76.92308	63	7.93725	3969	25.09980	15.87302
14	3.74166	196	11.83216	71.42857	64	8.00000	4096	25.29822	15.62500
15	3.87298	225	12.24745	66.66667	65	8.06226	4225	25.49510	15.38462
16	4.00000	256	12.64911	62.50000	66	8.12404	4356	25.69047	15.15152
17	4.12311	289	13.03840	58.82353	67	8.18535	4489	25.88436	14.92537
18	4.24264	324	13.41641	55.55556	68	8.24621	4624	26.07681	14.70588
19	4.35890	361	13.78405	52.63158	69	8.30662	4761	26.26785	14.49275
20	4.47214	400	14.14214	50.00000	70	8.36660	4900	26.45751	14.28571
21	4.58258	441	14.49138	47.61905	71	8.42615	5041	26.64583	14.08451
22	4.69042	484	14.83240	45.45455	72	8.48528	5184	26.83282	13.88889
23	4.79583	529	15.16575	43.47826	73	8.54400	5329	27.01851	13.69863
24	4.89898	576	15.49193	41.66667	74	8.60233	5476	27.20294	13.51351
25	5.00000	625	15.81139	40.00000	75	8.66025	5625	27.38613	13.33333
26	5.09902	676	16.12452	38.46154	76	8.71780	5776	27.56810	13.15789
27	5.19615	729	16.43168	37.03704	77	8.77496	5929	27.74887	12.98701
28	5.29150	784	16.73320	35.71429	78	8.83176	6084	27.92848	12.82051
29	5.38516	841	17.02939	34.48276	79	8.88819	6241	28.10694	12.65823
30	5.47723	900	17.32051	33.33333	80	8.94427	6400	28.28427	12.50000
31	5.56776	961	17.60682	32.25806	81	9.00000	6561	28.46050	12.34568
32	5.65685	1024	17.88854	31.25000	82	9.05539	6724	28.63564	12.19512
33	5.74456	1089	18.16590	30.30303	83	9.11043	6889	28.80972	12.04819
34	5.83095	1156	18.43909	29.41176	84	9.16515	7056	28.98275	11.90476
35	5.91608	1225	18.70829	28.57143	85	9.21954	7225	29.15476	11.76471
36	6.00000	1296	18.97367	27.77778	86	9.27362	7396	29.32576	11.62791
37	6.08276	1369	19.23538	27.02703	87	9.32738	7569	29.49576	11.49425
38	6.16441	1444	19.49359	26.31579	88	9.38083	7744	29.66479	11.36364
39	6.24500	1521	19.74842	25.64103	89	9.43398	7921	29.83287	11.23596
40	6.32456	1600	20.00000	25.00000	90	9.48683	8100	30.00000	11.11111
41	6.40312	1681	20.24846	24.39024	91	9.53939	8281	30.16621	10.98901
42	6.48074	1764	20.49390	23.80952	92	9.59166	8464	30.33150	10.86957
43	6.55744	1849	20.73644	23.25581	93	9.64365	8649	30.49590	10.75269
44	6.63325	1936	20.97618	22.72727	94	9.69536	8836	30.65942	10.63830
45	6.70820	2025	21.21320	22.22222	95	9.74679	9025	30.82207	10.52632
46	6.78233	2116	21.44761	21.73913	96	9.79796	9216	30.98387	10.41667
47	6.85565	2209	21.67948	21.27660	97	9.84886	9409	31.14482	10.30928
48	6.92820	2304	21.90890	20.83333	98	9.89949	9604	31.30495	10.20408
49	7.00000	2401	22.13594	20.40816	99	9.94987	9801	31.46427	10.10101
50	7.07107	2500	22.36068	20.00000	100	10.00000	10000	31.62278	10.00000

TABLE A *(continued)*

N	$\sqrt{N}$	N^2	$\sqrt{10N}$	$1000/N$	N	$\sqrt{N}$	N^2	$\sqrt{10N}$	$1000/N$
100	10.00000	10000	31.62278	10.00000	150	12.24745	22500	38.72983	6.66667
101	10.04988	10201	31.78050	9.90099	151	12.28821	22801	38.85872	6.62252
102	10.09950	10404	31.93744	9.80392	152	12.32883	23104	38.98718	6.57895
103	10.14889	10609	32.09361	9.70874	153	12.36932	23409	39.11521	6.53595
104	10.19804	10816	32.24903	9.61538	154	12.40967	23716	39.24283	6.49351
105	10.24695	11025	32.40370	9.52381	155	12.44990	24025	39.37004	6.45161
106	10.29563	11236	32.55764	9.43396	156	12.49000	24336	39.49684	6.41026
107	10.34408	11449	32.71085	9.34579	157	12.52996	24649	39.62323	6.36943
108	10.39230	11664	32.86335	9.25926	158	12.56981	24964	39.74921	6.32911
109	10.44031	11881	33.01515	9.17431	159	12.60952	25281	39.87480	6.28931
110	10.48809	12100	33.16625	9.09091	160	12.64911	25600	40.00000	6.25000
111	10.53565	12321	33.31666	9.00901	161	12.68858	25921	40.12481	6.21118
112	10.58301	12544	33.46640	8.92857	162	12.72792	26244	40.24922	6.17284
113	10.63015	12769	33.61547	8.84956	163	12.76715	26569	40.37326	6.13497
114	10.67708	12996	33.76389	8.77193	164	12.80625	26896	40.49691	6.09756
115	10.72381	13225	33.91165	8.69565	165	12.84523	27225	40.62019	6.06061
116	10.77033	13456	34.05877	8.62069	166	12.88410	27556	40.74310	6.02410
117	10.81665	13689	34.20526	8.54701	167	12.92285	27889	40.86563	5.98802
118	10.86278	13924	34.35113	8.47458	168	12.96148	28224	40.98780	5.95238
119	10.90871	14161	34.49638	8.40336	169	13.00000	28561	41.10961	5.91716
120	10.95445	14400	34.64102	8.33333	170	13.03840	28900	41.23106	5.88235
121	11.00000	14641	34.78505	8.26446	171	13.07670	29241	41.35215	5.84795
122	11.04536	14884	34.92850	8.19672	172	13.11488	29584	41.47288	5.81395
123	11.09054	15129	35.07136	8.13008	173	13.15295	29929	41.59327	5.78035
124	11.13553	15376	35.21363	8.06452	174	13.19091	30276	41.71331	5.74713
125	11.18034	15625	35.35534	8.00000	175	13.22876	30625	41.83300	5.71429
126	11.22497	15876	35.49648	7.93651	176	13.26650	30976	41.95235	5.68182
127	11.26943	16129	35.63706	7.87402	177	13.30413	31329	42.07137	5.64972
128	11.31371	16384	35.77709	7.81250	178	13.34166	31684	42.19005	5.61798
129	11.35782	16641	35.91657	7.75194	179	13.37909	32041	42.30839	5.58659
130	11.40175	16900	36.05551	7.69231	180	13.41641	32400	42.42641	5.55556
131	11.44552	17161	36.19392	7.63359	181	13.45362	32761	42.54409	5.52486
132	11.48913	17424	36.33180	7.57576	182	13.49074	33124	42.66146	5.49451
133	11.53256	17689	36.46917	7.51880	183	13.52775	33489	42.77850	5.46448
134	11.57584	17956	36.60601	7.46269	184	13.56466	33856	42.89522	5.43478
135	11.61895	18225	36.74235	7.40741	185	13.60147	34225	43.01163	5.40541
136	11.66190	18496	36.87818	7.35294	186	13.63818	34596	43.12772	5.37634
137	11.70470	18769	37.01351	7.29927	187	13.67479	34969	43.24350	5.34759
138	11.74734	19044	37.14835	7.24638	188	13.71131	35344	43.35897	5.31915
139	11.78983	19321	37.28270	7.19424	189	13.74773	35721	43.47413	5.29101
140	11.83216	19600	37.41657	7.14286	190	13.78405	36100	43.58899	5.26316
141	11.87434	19881	37.54997	7.09220	191	13.82027	36481	43.70355	5.23560
142	11.91638	20164	37.68289	7.04225	192	13.85641	36864	43.81780	5.20833
143	11.95826	20449	37.81534	6.99301	193	13.89244	37249	43.93177	5.18135
144	12.00000	20736	37.94733	6.94444	194	13.92839	37636	44.04543	5.15464
145	12.04159	21025	38.07887	6.89655	195	13.96424	38025	44.15880	5.12821
146	12.08305	21316	38.20995	6.84932	196	14.00000	38416	44.27189	5.10204
147	12.12436	21609	38.34058	6.80272	197	14.03567	38809	44.38468	5.07614
148	12.16553	21904	38.47077	6.75676	198	14.07125	39204	44.49719	5.05051
149	12.20656	22201	38.60052	6.71141	199	14.10674	39601	44.60942	5.02513
150	12.24745	22500	38.72983	6.66667	200	14.14214	40000	44.72136	5.00000

TABLE A (*continued*)

N	$\sqrt{N}$	N^2	$\sqrt{10N}$	$1000/N$
200	14.14214	40000	44.72136	5.00000
201	14.17745	40401	44.83302	4.97512
202	14.21267	40804	44.94441	4.95050
203	14.24781	41209	45.05552	4.92611
204	14.28286	41616	45.16636	4.90196
205	14.31782	42025	45.27693	4.87805
206	14.35270	42436	45.38722	4.85437
207	14.38749	42849	45.49725	4.83092
208	14.42221	43264	45.60702	4.80769
209	14.45683	43681	45.71652	4.78469
210	14.49138	44100	45.82576	4.76190
211	14.52584	44521	45.93474	4.73934
212	14.56022	44944	46.04346	4.71698
213	14.59452	45369	46.15192	4.69484
214	14.62874	45796	46.26013	4.67290
215	14.66288	46225	46.36809	4.65116
216	14.69694	46656	46.47580	4.62963
217	14.73092	47089	46.58326	4.60829
218	14.76482	47524	46.69047	4.58716
219	14.79865	47961	46.79744	4.56621
220	14.83240	48400	46.90416	4.54545
221	14.86607	48841	47.01064	4.52489
222	14.89966	49284	47.11688	4.50450
223	14.93318	49729	47.22288	4.48430
224	14.96663	50176	47.32864	4.46429
225	15.00000	50625	47.43416	4.44444
226	15.03330	51076	47.53946	4.42478
227	15.06652	51529	47.64452	4.40529
228	15.09967	51984	47.74935	4.38596
229	15.13275	52441	47.85394	4.36681
230	15.16575	52900	47.95832	4.34783
231	15.19868	53361	48.06246	4.32900
232	15.23155	53824	48.16638	4.31034
233	15.26434	54289	48.27007	4.29185
234	15.29706	54756	48.37355	4.27350
235	15.32971	55225	48.47680	4.25532
236	15.36229	55696	48.57983	4.23729
237	15.39480	56169	48.68265	4.21941
238	15.42725	56644	48.78524	4.20168
239	15.45962	57121	48.88763	4.18410
240	15.49193	57600	48.98979	4.16667
241	15.52417	58081	49.09175	4.14938
242	15.55635	58564	49.19350	4.13223
243	15.58846	59049	49.29503	4.11523
244	15.62050	59536	49.39636	4.09836
245	15.65248	60025	49.49747	4.08163
246	15.68439	60516	49.59839	4.06504
247	15.71623	61009	49.69909	4.04858
248	15.74802	61504	49.79960	4.03226
249	15.77973	62001	49.89990	4.01606
250	15.81139	62500	50.00000	4.00000
250	15.81139	62500	50.00000	4.00000
251	15.84298	63001	50.09990	3.98406
252	15.87451	63504	50.19960	3.96825
253	15.90597	64009	50.29911	3.95257
254	15.93738	64516	50.39841	3.93701
255	15.96872	65025	50.49752	3.92157
256	16.00000	65536	50.59644	3.90625
257	16.03122	66049	50.69517	3.89105
258	16.06238	66564	50.79370	3.87597
259	16.09348	67081	50.89204	3.86100
260	16.12452	67600	50.99020	3.84615
261	16.15549	68121	51.08816	3.83142
262	16.18641	68644	51.18594	3.81679
263	16.21727	69169	51.28353	3.80228
264	16.24808	69696	51.38093	3.78788
265	16.27882	70225	51.47815	3.77358
266	16.30951	70756	51.57519	3.75940
267	16.34013	71289	51.67204	3.74532
268	16.37071	71824	51.76872	3.73134
269	16.40122	72361	51.86521	3.71747
270	16.43168	72900	51.96152	3.70370
271	16.46208	73441	52.05766	3.69004
272	16.49242	73984	52.15362	3.67647
273	16.52271	74529	52.24940	3.66300
274	16.55295	75076	52.34501	3.64964
275	16.58312	75625	52.44044	3.63636
276	16.61325	76176	52.53570	3.62319
277	16.64332	76729	52.63079	3.61011
278	16.67333	77284	52.72571	3.59712
279	16.70329	77841	52.82045	3.58423
280	16.73320	78400	52.91503	3.57143
281	16.76305	78961	53.00943	3.55872
282	16.79286	79524	53.10367	3.54610
283	16.82260	80089	53.19774	3.53357
284	16.85230	80656	53.29165	3.52113
285	16.88194	81225	53.38539	3.50877
286	16.91153	81796	53.47897	3.49650
287	16.94107	82369	53.57238	3.48432
288	16.97056	82944	53.66563	3.47222
289	17.00000	83521	53.75872	3.46021
290	17.02939	84100	53.85165	3.44828
291	17.05872	84681	53.94442	3.43643
292	17.08801	85264	54.03702	3.42466
293	17.11724	85849	54.12947	3.41297
294	17.14643	86436	54.22177	3.40136
295	17.17556	87025	54.31390	3.38983
296	17.20465	87616	54.40588	3.37838
297	17.23369	88209	54.49771	3.36700
298	17.26268	88804	54.58938	3.35570
299	17.29162	89401	54.68089	3.34448
300	17.32051	90000	54.77226	3.33333

TABLE A *(continued)*

N	$\sqrt{N}$	N^2	$\sqrt{10N}$	$1000/N$
300	17.32051	90000	54.77226	3.33333
301	17.34935	90601	54.86347	3.32226
302	17.37815	91204	54.95453	3.31126
303	17.40690	91809	55.04544	3.30033
304	17.43560	92416	55.13620	3.28947
305	17.46425	93025	55.22681	3.27869
306	17.49286	93636	55.31727	3.26797
307	17.52142	94249	55.40758	3.25733
308	17.54993	94864	55.49775	3.24675
309	17.57840	95481	55.58777	3.23625
310	17.60682	96100	55.67764	3.22581
311	17.63519	96721	55.76737	3.21543
312	17.66352	97344	55.85696	3.20513
313	17.69181	97969	55.94640	3.19489
314	17.72005	98596	56.03570	3.18471
315	17.74824	99225	56.12486	3.17460
316	17.77639	99856	56.21388	3.16456
317	17.80449	100489	56.30275	3.15457
318	17.83255	101124	56.39149	3.14465
319	17.86057	101761	56.48008	3.13480
320	17.88854	102400	56.56854	3.12500
321	17.91647	103041	56.65686	3.11526
322	17.94436	103684	56.74504	3.10559
323	17.97220	104329	56.83309	3.09598
324	18.00000	104976	56.92100	3.08642
325	18.02776	105625	57.00877	3.07692
326	18.05547	106276	57.09641	3.06748
327	18.08314	106929	57.18391	3.05810
328	18.11077	107584	57.27128	3.04878
329	18.13836	108241	57.35852	3.03951
330	18.16590	108900	57.44563	3.03030
331	18.19341	109561	57.53260	3.02115
332	18.22087	110224	57.61944	3.01205
333	18.24829	110889	57.70615	3.00300
334	18.27567	111556	57.79273	2.99401
335	18.30301	112225	57.87918	2.98507
336	18.33030	112896	57.96551	2.97619
337	18.35756	113569	58.05170	2.96736
338	18.38478	114244	58.13777	2.95858
339	18.41195	114921	58.22371	2.94985
340	18.43909	115600	58.30952	2.94118
341	18.46619	116281	58.39521	2.93255
342	18.49324	116964	58.48077	2.92398
343	18.52026	117649	58.56620	2.91545
344	18.54724	118336	58.65151	2.90698
345	18.57418	119025	58.73670	2.89855
346	18.60108	119716	58.82176	2.89017
347	18.62794	120409	58.90671	2.88184
348	18.65476	121104	58.99152	2.87356
349	18.68154	121801	59.07622	2.86533
350	18.70829	122500	59.16080	2.85714
350	18.70829	122500	59.16080	2.85714
351	18.73499	123201	59.24525	2.84900
352	18.76166	123904	59.32959	2.84091
353	18.78829	124609	59.41380	2.83286
354	18.81489	125316	59.49790	2.82486
355	18.84144	126025	59.58188	2.81690
356	18.86796	126736	59.66574	2.80899
357	18.89444	127449	59.74948	2.80112
358	18.92089	128164	59.83310	2.79330
359	18.94730	128881	59.91661	2.78552
360	18.97367	129600	60.00000	2.77778
361	19.00000	130321	60.08328	2.77008
362	19.02630	131044	60.16644	2.76243
363	19.05256	131769	60.24948	2.75482
364	19.07878	132496	60.33241	2.74725
365	19.10497	133225	60.41523	2.73973
366	19.13113	133956	60.49793	2.73224
367	19.15724	134689	60.58052	2.72480
368	19.18333	135424	60.66300	2.71739
369	19.20937	136161	60.74537	2.71003
370	19.23538	136900	60.82763	2.70270
371	19.26136	137641	60.90977	2.69542
372	19.28730	138384	60.99180	2.68817
373	19.31321	139129	61.07373	2.68097
374	19.33908	139876	61.15554	2.67380
375	19.36492	140625	61.23724	2.66667
376	19.39072	141376	61.31884	2.65957
377	19.41649	142129	61.40033	2.65252
378	19.44222	142884	61.48170	2.64550
379	19.46792	143641	61.56298	2.63852
380	19.49359	144400	61.64414	2.63158
381	19.51922	145161	61.72520	2.62467
382	19.54482	145924	61.80615	2.61780
383	19.57039	146689	61.88699	2.61097
384	19.59592	147456	61.96773	2.60417
385	19.62142	148225	62.04837	2.59740
386	19.64688	148996	62.12890	2.59067
387	19.67232	149769	62.20932	2.58398
388	19.69772	150544	62.28965	2.57732
389	19.72308	151321	62.36986	2.57069
390	19.74842	152100	62.44998	2.56410
391	19.77372	152881	62.52999	2.55754
392	19.79899	153664	62.60990	2.55102
393	19.82423	154449	62.68971	2.54453
394	19.84943	155236	62.76942	2.53807
395	19.87461	156025	62.84903	2.53165
396	19.89975	156816	62.92853	2.52525
397	19.92486	157609	63.00794	2.51889
398	19.94994	158404	63.08724	2.51256
399	19.97498	159201	63.16645	2.50627
400	20.00000	160000	63.24555	2.50000

TABLE A (*continued*)

N	$\sqrt{N}$	N^2	$\sqrt{10N}$	$1000/N$	N	$\sqrt{N}$	N^2	$\sqrt{10N}$	$1000/N$
400	20.00000	160000	63.24555	2.50000	450	21.21320	202500	67.08204	2.22222
401	20.02498	160801	63.32456	2.49377	451	21.23676	203401	67.15653	2.21729
402	20.04994	161604	63.40347	2.48756	452	21.26029	204304	67.23095	2.21239
403	20.07486	162409	63.48228	2.48139	453	21.28380	205209	67.30527	2.20751
404	20.09975	163216	63.56099	2.47525	454	21.30728	206116	67.37952	2.20264
405	20.12461	164025	63.63961	2.46914	455	21.33073	207025	67.45369	2.19780
406	20.14944	164836	63.71813	2.46305	456	21.35416	207936	67.52777	2.19298
407	20.17424	165649	63.79655	2.45700	457	21.37756	208849	67.60178	2.18818
408	20.19901	166464	63.87488	2.45098	458	21.40093	209764	67.67570	2.18341
409	20.22375	167281	63.95311	2.44499	459	21.42429	210681	67.74954	2.17865
410	20.24846	168100	64.03124	2.43902	460	21.44761	211600	67.82330	2.17391
411	20.27313	168921	64.10928	2.43309	461	21.47091	212521	67.89698	2.16920
412	20.29778	169744	64.18723	2.42718	462	21.49419	213444	67.97058	2.16450
413	20.32240	170569	64.26508	2.42131	463	21.51743	214369	68.04410	2.15983
414	20.34699	171396	64.34283	2.41546	464	21.54066	215296	68.11755	2.15517
415	20.37155	172225	64.42049	2.40964	465	21.56386	216225	68.19091	2.15054
416	20.39608	173056	64.49806	2.40385	466	21.58703	217156	68.26419	2.14592
417	20.42058	173889	64.57554	2.39808	467	21.61018	218089	68.33740	2.14133
418	20.44505	174724	64.65292	2.39234	468	21.63331	219024	68.41053	2.13675
419	20.46949	175561	64.73021	2.38663	469	21.65641	219961	68.48357	2.13220
420	20.49390	176400	64.80741	2.38095	470	21.67948	220900	68.55655	2.12766
421	20.51828	177241	64.88451	2.37530	471	21.70253	221841	68.62944	2.12314
422	20.54264	178084	64.96153	2.36967	472	21.72556	222784	68.70226	2.11864
423	20.56696	178929	65.03845	2.36407	473	21.74856	223729	68.77500	2.11416
424	20.59126	179776	65.11528	2.35849	474	21.77154	224676	68.84766	2.10970
425	20.61553	180625	65.19202	2.35294	475	21.79449	225625	68.92024	2.10526
426	20.63977	181476	65.26868	2.34742	476	21.81742	226576	68.99275	2.10084
427	20.66398	182329	65.34524	2.34192	477	21.84033	227529	69.06519	2.09644
428	20.68816	183184	65.42171	2.33645	478	21.86321	228484	69.13754	2.09205
429	20.71232	184041	65.49809	2.33100	479	21.88607	229441	69.20983	2.08768
430	20.73644	184900	65.57439	2.32558	480	21.90890	230400	69.28203	2.08333
431	20.76054	185761	65.65059	2.32019	481	21.93171	231361	69.35416	2.07900
432	20.78461	186624	65.72671	2.31481	482	21.95450	232324	69.42622	2.07469
433	20.80865	187489	65.80274	2.30947	483	21.97726	233289	69.49820	2.07039
434	20.83267	188356	65.87868	2.30415	484	22.00000	234256	69.57011	2.06612
435	20.85665	189225	65.95453	2.29885	485	22.02272	235225	69.64194	2.06186
436	20.88061	190096	66.03030	2.29358	486	22.04541	236196	69.71370	2.05761
437	20.90454	190969	66.10598	2.28833	487	22.06808	237169	69.78539	2.05339
438	20.92845	191844	66.18157	2.28311	488	22.09072	238144	69.85700	2.04918
439	20.95233	192721	66.25708	2.27790	489	22.11334	239121	69.92853	2.04499
440	20.97618	193600	66.33250	2.27273	490	22.13594	240100	70.00000	2.04082
441	21.00000	194481	66.40783	2.26757	491	22.15852	241081	70.07139	2.03666
442	21.02380	195364	66.48308	2.26244	492	22.18107	242064	70.14271	2.03252
443	21.04757	196249	66.55825	2.25734	493	22.20360	243049	70.21396	2.02840
444	21.07131	197136	66.63332	2.25225	494	22.22611	244036	70.28513	2.02429
445	21.09502	198025	66.70832	2.24719	495	22.24860	245025	70.35624	2.02020
446	21.11871	198916	66.78323	2.24215	496	22.27106	246016	70.42727	2.01613
447	21.14237	199809	66.85806	2.23714	497	22.29350	247009	70.49823	2.01207
448	21.16601	200704	66.93280	2.23214	498	22.31591	248004	70.56912	2.00803
449	21.18962	201601	67.00746	2.22717	499	22.33831	249001	70.63993	2.00401
450	21.21320	202500	67.08204	2.22222	500	22.36068	250000	70.71068	2.00000

TABLE A (*continued*)

N	$\sqrt{N}$	N^2	$\sqrt{10N}$	$1000/N$	N	$\sqrt{N}$	N^2	$\sqrt{10N}$	$1000/N$
500	22.36068	250000	70.71068	2.00000	550	23.45208	302500	74.16198	1.81818
501	22.38303	251001	70.78135	1.99601	551	23.47339	303601	74.22937	1.81488
502	22.40536	252004	70.85196	1.99203	552	23.49468	304704	74.29670	1.81159
503	22.42766	253009	70.92249	1.98807	553	23.51595	305809	74.36397	1.80832
504	22.44994	254016	70.99296	1.98413	554	23.53720	306916	74.43118	1.80505
505	22.47221	255025	71.06335	1.98020	555	23.55844	308025	74.49832	1.80180
506	22.49444	256036	71.13368	1.97628	556	23.57965	309136	74.56541	1.79856
507	22.51666	257049	71.20393	1.97239	557	23.60085	310249	74.63243	1.79533
508	22.53886	258064	71.27412	1.96850	558	23.62202	311364	74.69940	1.79211
509	22.56103	259081	71.34424	1.96464	559	23.64318	312481	74.76630	1.78891
510	22.58318	260100	71.41428	1.96078	560	23.66432	313600	74.83315	1.78571
511	22.60531	261121	71.48426	1.95695	561	23.68544	314721	74.89993	1.78253
512	22.62742	262144	71.55418	1.95313	562	23.70654	315844	74.96666	1.77936
513	22.64950	263169	71.62402	1.94932	563	23.72762	316969	75.03333	1.77620
514	22.67157	264196	71.69379	1.94553	564	23.74868	318096	75.09993	1.77305
515	22.69361	265225	71.76350	1.94175	565	23.76973	319225	75.16648	1.76991
516	22.71563	266256	71.83314	1.93798	566	23.79075	320356	75.23297	1.76678
517	22.73763	267289	71.90271	1.93424	567	23.81176	321489	75.29940	1.76367
518	22.75961	268324	71.97222	1.93050	568	23.83275	322624	75.36577	1.76056
519	22.78157	269361	72.04165	1.92678	569	23.85372	323761	75.43209	1.75747
520	22.80351	270400	72.11103	1.92308	570	23.87467	324900	75.49834	1.75439
521	22.82542	271441	72.18033	1.91939	571	23.89561	326041	75.56454	1.75131
522	22.84732	272484	72.24957	1.91571	572	23.91652	327184	75.63068	1.74825
523	22.86919	273529	72.31874	1.91205	573	23.93742	328329	75.69676	1.74520
524	22.89105	274576	72.38784	1.90840	574	23.95830	329476	75.76279	1.74216
525	22.91288	275625	72.45688	1.90476	575	23.97916	330625	75.82875	1.73913
526	22.93469	276676	72.52586	1.90114	576	24.00000	331776	75.89466	1.73611
527	22.95648	277729	72.59477	1.89753	577	24.02082	332929	75.96052	1.73310
528	22.97825	278784	72.66361	1.89394	578	24.04163	334084	76.02631	1.73010
529	23.00000	279841	72.73239	1.89036	579	24.06242	335241	76.09205	1.72712
530	23.02173	280900	72.80110	1.88679	580	24.08319	336400	76.15773	1.72414
531	23.04344	281961	72.86975	1.88324	581	24.10394	337561	76.22336	1.72117
532	23.06513	283024	72.93833	1.87970	582	24.12468	338724	76.28892	1.71821
533	23.08679	284089	73.00685	1.87617	583	24.14539	339889	76.35444	1.71527
534	23.10844	285156	73.07530	1.87266	584	24.16609	341056	76.41989	1.71233
535	23.13007	286225	73.14369	1.86916	585	24.18677	342225	76.48529	1.70940
536	23.15167	287296	73.21202	1.86567	586	24.20744	343396	76.55064	1.70648
537	23.17326	288369	73.28028	1.86220	587	24.22808	344569	76.61593	1.70358
538	23.19483	289444	73.34848	1.85874	588	24.24871	345744	76.68116	1.70068
539	23.21637	290521	73.41662	1.85529	589	24.26932	346921	76.74634	1.69779
540	23.23790	291600	73.48469	1.85185	590	24.28992	348100	76.81146	1.69492
541	23.25941	292681	73.55270	1.84843	591	24.31049	349281	76.87652	1.69205
542	23.28089	293764	73.62065	1.84502	592	24.33105	350464	76.94154	1.68919
543	23.30236	294849	73.68853	1.84162	593	24.35159	351649	77.00649	1.68634
544	23.32381	295936	73.75636	1.83824	594	24.37212	352836	77.07140	1.68350
545	23.34524	297025	73.82412	1.83486	595	24.39262	354025	77.13624	1.68067
546	23.36664	298116	73.89181	1.83150	596	24.41311	355216	77.20104	1.67785
547	23.38803	299209	73.95945	1.82815	597	24.43358	356409	77.26578	1.67504
548	23.40940	300304	74.02702	1.82482	598	24.45404	357604	77.33046	1.67224
549	23.43075	301401	74.09453	1.82149	599	24.47448	358801	77.39509	1.66945
550	23.45208	302500	74.16198	1.81818	600	24.49490	360000	77.45967	1.66667

TABLE A (*continued*)

N	$\sqrt{N}$	N^2	$\sqrt{10N}$	$1000/N$	N	$\sqrt{N}$	N^2	$\sqrt{10N}$	$1000/N$
600	24.49490	360000	77.45967	1.66667	650	25.49510	422500	80.62258	1.53846
601	24.51530	361201	77.52419	1.66389	651	25.51470	423801	80.68457	1.53610
602	24.53569	362404	77.58866	1.66113	652	25.53429	425104	80.74652	1.53374
603	24.55606	363609	77.65307	1.65837	653	25.55386	426409	80.80842	1.53139
604	24.57641	364816	77.71744	1.65563	654	25.57342	427716	80.87027	1.52905
605	24.59675	366025	77.78175	1.65289	655	25.59297	429025	80.93207	1.52672
606	24.61707	367236	77.84600	1.65017	656	25.61250	430336	80.99383	1.52439
607	24.63737	368449	77.91020	1.64745	657	25.63201	431649	81.05554	1.52207
608	24.65766	369664	77.97435	1.64474	658	25.65151	432964	81.11720	1.51976
609	24.67793	370881	78.03845	1.64204	659	25.67100	434281	81.17881	1.51745
610	24.69818	372100	78.10250	1.63934	660	25.69047	435600	81.24038	1.51515
611	24.71841	373321	78.16649	1.63666	661	25.70992	436921	81.30191	1.51286
612	24.73863	374544	78.23043	1.63399	662	25.72936	438244	81.36338	1.51057
613	24.75884	375769	78.29432	1.63132	663	25.74879	439569	81.42481	1.50830
614	24.77902	376996	78.35815	1.62866	664	25.76820	440896	81.48620	1.50602
615	24.79919	378225	78.42194	1.62602	665	25.78759	442225	81.54753	1.50376
616	24.81935	379456	78.48567	1.62338	666	25.80698	443556	81.60882	1.50150
617	24.83948	380689	78.54935	1.62075	667	25.82634	444889	81.67007	1.49925
618	24.85961	381924	78.61298	1.61812	668	25.84570	446224	81.73127	1.49701
619	24.87971	383161	78.67655	1.61551	669	25.86503	447561	81.79242	1.49477
620	24.89980	384400	78.74008	1.61290	670	25.88436	448900	81.85353	1.49254
621	24.91987	385641	78.80355	1.61031	671	25.90367	450241	81.91459	1.49031
622	24.93993	386884	78.86698	1.60772	672	25.92296	451584	81.97561	1.48810
623	24.95997	388129	78.93035	1.60514	673	25.94224	452929	82.03658	1.48588
624	24.97999	389376	78.99367	1.60256	674	25.96151	454276	82.09750	1.48368
625	25.00000	390625	79.05694	1.60000	675	25.98076	455625	82.15838	1.48148
626	25.01999	391876	79.12016	1.59744	676	26.00000	456976	82.21922	1.47929
627	25.03997	393129	79.18333	1.59490	677	26.01922	458329	82.28001	1.47710
628	25.05993	394384	79.24645	1.59236	678	26.03843	459684	82.34076	1.47493
629	25.07987	395641	79.30952	1.58983	679	26.05763	461041	82.40146	1.47275
630	25.09980	396900	79.37254	1.58730	680	26.07681	462400	82.46211	1.47059
631	25.11971	398161	79.43551	1.58479	681	26.09598	463761	82.52272	1.46843
632	25.13961	399424	79.49843	1.58228	682	26.11513	465124	82.58329	1.46628
633	25.15949	400689	79.56130	1.57978	683	26.13427	466489	82.64381	1.46413
634	25.17936	401956	79.62412	1.57729	684	26.15339	467856	82.70429	1.46199
635	25.19921	403225	79.68689	1.57480	685	26.17250	469225	82.76473	1.45985
636	25.21904	404496	79.74961	1.57233	686	26.19160	470596	82.82512	1.45773
637	25.23886	405769	79.81228	1.56986	687	26.21068	471969	82.88546	1.45560
638	25.25866	407044	79.87490	1.56740	688	26.22975	473344	82.94577	1.45349
639	25.27845	408321	79.93748	1.56495	689	26.24881	474721	83.00602	1.45138
640	25.29822	409600	80.00000	1.56250	690	26.26785	476100	83.06624	1.44928
641	25.31798	410881	80.06248	1.56006	691	26.28688	477481	83.12641	1.44718
642	25.33772	412164	80.12490	1.55763	692	26.30589	478864	83.18654	1.44509
643	25.35744	413449	80.18728	1.55521	693	26.32489	480249	83.24662	1.44300
644	25.37716	414736	80.24961	1.55280	694	26.34388	481636	83.30666	1.44092
645	25.39685	416025	80.31189	1.55039	695	26.36285	483025	83.36666	1.43885
646	25.41653	417316	80.37413	1.54799	696	26.38181	484416	83.42661	1.43678
647	25.43619	418609	80.43631	1.54560	697	26.40076	485809	83.48653	1.43472
648	25.45584	419904	80.49845	1.54321	698	26.41969	487204	83.54639	1.43266
649	25.47548	421201	80.56054	1.54083	699	26.43861	488601	83.60622	1.43062
650	25.49510	422500	80.62258	1.53846	700	26.45751	490000	83.66600	1.42857

TABLE A *(continued)*

N	$\sqrt{N}$	N^2	$\sqrt{10N}$	$1000/N$
700	26.45751	490000	83.66600	1.42857
701	26.47640	491401	83.72574	1.42653
702	26.49528	492804	83.78544	1.42450
703	26.51415	494209	83.84510	1.42248
704	26.53300	495616	83.90471	1.42045
705	26.55184	497025	83.96428	1.41844
706	26.57066	498436	84.02381	1.41643
707	26.58947	499849	84.08329	1.41443
708	26.60827	501264	84.14274	1.41243
709	26.62705	502681	84.20214	1.41044
710	26.64583	504100	84.26150	1.40845
711	26.66458	505521	84.32082	1.40647
712	26.68333	506944	84.38009	1.40449
713	26.70206	508369	84.43933	1.40252
714	26.72078	509796	84.49852	1.40056
715	26.73948	511225	84.55767	1.39860
716	26.75818	512656	84.61678	1.39665
717	26.77686	514089	84.67585	1.39470
718	26.79552	515524	84.73488	1.39276
719	26.81418	516961	84.79387	1.39082
720	26.83282	518400	84.85281	1.38889
721	26.85144	519841	84.91172	1.38696
722	26.87006	521284	84.97058	1.38504
723	26.88866	522729	85.02941	1.38313
724	26.90725	524176	85.08819	1.38122
725	26.92582	525625	85.14693	1.37931
726	26.94439	527076	85.20563	1.37741
727	26.96294	528529	85.26429	1.37552
728	26.98148	529984	85.32292	1.37363
729	27.00000	531441	85.38150	1.37174
730	27.01851	532900	85.44004	1.36986
731	27.03701	534361	85.49854	1.36799
732	27.05550	535824	85.55700	1.36612
733	27.07397	537289	85.61542	1.36426
734	27.09243	538756	85.67380	1.36240
735	27.11088	540225	85.73214	1.36054
736	27.12932	541696	85.79044	1.35870
737	27.14774	543169	85.84870	1.35685
738	27.16616	544644	85.90693	1.35501
739	27.18455	546121	85.96511	1.35318
740	27.20294	547600	86.02325	1.35135
741	27.22132	549081	86.08136	1.34953
742	27.23968	550564	86.13942	1.34771
743	27.25803	552049	86.19745	1.34590
744	27.27636	553536	86.25543	1.34409
745	27.29469	555025	86.31338	1.34228
746	27.31300	556516	86.37129	1.34048
747	27.33130	558009	86.42916	1.33869
748	27.34959	559504	86.48699	1.33690
749	27.36786	561001	86.54479	1.33511
750	27.38613	562500	86.60254	1.33333
750	27.38613	562500	86.60254	1.33333
751	27.40438	564001	86.66026	1.33156
752	27.42262	565504	86.71793	1.32979
753	27.44085	567009	86.77557	1.32802
754	27.45906	568516	86.83317	1.32626
755	27.47726	570025	86.89074	1.32450
756	27.49545	571536	86.94826	1.32275
757	27.51363	573049	87.00575	1.32100
758	27.53180	574564	87.06320	1.31926
759	27.54995	576081	87.12061	1.31752
760	27.56810	577600	87.17798	1.31579
761	27.58623	579121	87.23531	1.31406
762	27.60435	580644	87.29261	1.31234
763	27.62245	582169	87.34987	1.31062
764	27.64055	583696	87.40709	1.30890
765	27.65863	585225	87.46428	1.30719
766	27.67671	586756	87.52143	1.30548
767	27.69476	588289	87.57854	1.30378
768	27.71281	589824	87.63561	1.30208
769	27.73085	591361	87.69265	1.30039
770	27.74887	592900	87.74964	1.29870
771	27.76689	594441	87.80661	1.29702
772	27.78489	595984	87.86353	1.29534
773	27.80288	597529	87.92042	1.29366
774	27.82086	599076	87.97727	1.29199
775	27.83882	600625	88.03408	1.29032
776	27.85678	602176	88.09086	1.28866
777	27.87472	603729	88.14760	1.28700
778	27.89265	605284	88.20431	1.28535
779	27.91057	606841	88.26098	1.28370
780	27.92848	608400	88.31761	1.28205
781	27.94638	609961	88.37420	1.28041
782	27.96426	611524	88.43076	1.27877
783	27.98214	613089	88.48729	1.27714
784	28.00000	614656	88.54377	1.27551
785	28.01785	616225	88.60023	1.27389
786	28.03569	617796	88.65664	1.27226
787	28.05352	619369	88.71302	1.27065
788	28.07134	620944	88.76936	1.26904
789	28.08914	622521	88.82567	1.26743
790	28.10694	624100	88.88194	1.26582
791	28.12472	625681	88.93818	1.26422
792	28.14249	627264	88.99438	1.26263
793	28.16026	628849	89.05055	1.26103
794	28.17801	630436	89.10668	1.25945
795	28.19574	632025	89.16277	1.25786
796	28.21347	633616	89.21883	1.25628
797	28.23119	635209	89.27486	1.25471
798	28.24889	636804	89.33085	1.25313
799	28.26659	638401	89.38680	1.25156
800	28.28427	640000	89.44272	1.25000

TABLE A *(continued)*

N	$\sqrt{N}$	N^2	$\sqrt{10N}$	$1000/N$	N	$\sqrt{N}$	N^2	$\sqrt{10N}$	$1000/N$
800	28.28427	640000	89.44272	1.25000	850	29.15476	722500	92.19544	1.17647
801	28.30194	641601	89.49860	1.24844	851	29.17190	724201	92.24966	1.17509
802	28.31960	643204	89.55445	1.24688	852	29.18904	725904	92.30385	1.17371
803	28.33725	644809	89.61027	1.24533	853	29.20616	727609	92.35800	1.17233
804	28.35489	646416	89.66605	1.24378	854	29.22328	729316	92.41212	1.17096
805	28.37252	648025	89.72179	1.24224	855	29.24038	731025	92.46621	1.16959
806	28.39014	649636	89.77750	1.24069	856	29.25748	732736	92.52027	1.16822
807	28.40775	651249	89.83318	1.23916	857	29.27456	734449	92.57429	1.16686
808	28.42534	652864	89.88882	1.23762	858	29.29164	736164	92.62829	1.16550
809	28.44293	654481	89.94443	1.23609	859	29.30870	737881	92.68225	1.16414
810	28.46050	656100	90.00000	1.23457	860	29.32576	739600	92.73618	1.16279
811	28.47806	657721	90.05554	1.23305	861	29.34280	741321	92.79009	1.16144
812	28.49561	659344	90.11104	1.23153	862	29.35984	743044	92.84396	1.16009
813	28.51315	660969	90.16651	1.23001	863	29.37686	744769	92.89779	1.15875
814	28.53069	662596	90.22195	1.22850	864	29.39388	746496	92.95160	1.15741
815	28.54820	664225	90.27735	1.22699	865	29.41088	748225	93.00538	1.15607
816	28.56571	665856	90.33272	1.22549	866	29.42788	749956	93.05912	1.15473
817	28.58321	667489	90.38805	1.22399	867	29.44486	751689	93.11283	1.15340
818	28.60070	669124	90.44335	1.22249	868	29.46184	753424	93.16652	1.15207
819	28.61818	670761	90.49862	1.22100	869	29.47881	755161	93.22017	1.15075
820	28.63564	672400	90.55385	1.21951	870	29.49576	756900	93.27379	1.14943
821	28.65310	674041	90.60905	1.21803	871	29.51271	758641	93.32738	1.14811
822	28.67054	675684	90.66422	1.21655	872	29.52965	760384	93.38094	1.14679
823	28.68798	677329	90.71935	1.21507	873	29.54657	762129	93.43447	1.14548
824	28.70540	678976	90.77445	1.21359	874	29.56349	763876	93.48797	1.14416
825	28.72281	680625	90.82951	1.21212	875	29.58040	765625	93.54143	1.14286
826	28.74022	682276	90.88454	1.21065	876	29.59730	767376	93.59487	1.14155
827	28.75761	683929	90.93954	1.20919	877	29.61419	769129	93.64828	1.14025
828	28.77499	685584	90.99451	1.20773	878	29.63106	770884	93.70165	1.13895
829	28.79236	687241	91.04944	1.20627	879	29.64793	772641	93.75500	1.13766
830	28.80972	688900	91.10434	1.20482	880	29.66479	774400	93.80832	1.13636
831	28.82707	690561	91.15920	1.20337	881	29.68164	776161	93.86160	1.13507
832	28.84441	692224	91.21403	1.20192	882	29.69848	777924	93.91486	1.13379
833	28.86174	693889	91.26883	1.20048	883	29.71532	779689	93.96808	1.13250
834	28.87906	695556	91.32360	1.19904	884	29.73214	781456	94.02127	1.13122
835	28.89637	697225	91.37833	1.19760	885	29.74895	783225	94.07444	1.12994
836	28.91366	698896	91.43304	1.19617	886	29.76575	784996	94.12757	1.12867
837	28.93095	700569	91.48770	1.19474	887	29.78255	786769	94.18068	1.12740
838	28.94823	702244	91.54234	1.19332	888	29.79933	788544	94.23375	1.12613
839	28.96550	703921	91.59694	1.19190	889	29.81610	790321	94.28680	1.12486
840	28.98275	705600	91.65151	1.19048	890	29.83287	792100	94.33981	1.12360
841	29.00000	707281	91.70605	1.18906	891	29.84962	793881	94.39280	1.12233
842	29.01724	708964	91.76056	1.18765	892	29.86637	795664	94.44575	1.12108
843	29.03446	710649	91.81503	1.18624	893	29.88311	797449	94.49868	1.11982
844	29.05168	712336	91.86947	1.18483	894	29.89983	799236	94.55157	1.11857
845	29.06888	714025	91.92388	1.18343	895	29.91655	801025	94.60444	1.11732
846	29.08608	715716	91.97826	1.18203	896	29.93326	802816	94.65728	1.11607
847	29.10326	717409	92.03260	1.18064	897	29.94996	804609	94.71008	1.11483
848	29.12044	719104	92.08692	1.17925	898	29.96665	806404	94.76286	1.11359
849	29.13760	720801	92.14120	1.17786	899	29.98333	808201	94.81561	1.11235
850	29.15476	722500	92.19544	1.17647	900	30.00000	810000	94.86833	1.11111

TABLE A *(continued)*

N	$\sqrt{N}$	N^2	$\sqrt{10N}$	$1000/N$	N	$\sqrt{N}$	N^2	$\sqrt{10N}$	$1000/N$
900	30.00000	810000	94.86833	1.11111	950	30.82207	902500	97.46794	1.05263
901	30.01666	811801	94.92102	1.10988	951	30.83829	904401	97.51923	1.05152
902	30.03331	813604	94.97368	1.10865	952	30.85450	906304	97.57049	1.05042
903	30.04996	815409	95.02631	1.10742	953	30.87070	908209	97.62172	1.04932
904	30.06659	817216	95.07891	1.10619	954	30.88689	910116	97.67292	1.04822
905	30.08322	819025	95.13149	1.10497	955	30.90307	912025	97.72410	1.04712
906	30.09983	820836	95.18403	1.10375	956	30.91925	913936	97.77525	1.04603
907	30.11644	822649	95.23655	1.10254	957	30.93542	915849	97.82638	1.04493
908	30.13304	824464	95.28903	1.10132	958	30.95158	917764	97.87747	1.04384
909	30.14963	826281	95.34149	1.10011	959	30.96773	919681	97.92855	1.04275
910	30.16621	828100	95.39392	1.09890	960	30.98387	921600	97.97959	1.04167
911	30.18278	829921	95.44632	1.09769	961	31.00000	923521	98.03061	1.04058
912	30.19934	831744	95.49869	1.09649	962	31.01612	925444	98.08160	1.03950
913	30.21589	833569	95.55103	1.09529	963	31.03224	927369	98.13256	1.03842
914	30.23243	835396	95.60335	1.09409	964	31.04835	929296	98.18350	1.03734
915	30.24897	837225	95.65563	1.09290	965	31.06445	931225	98.23441	1.03627
916	30.26549	839056	95.70789	1.09170	966	31.08054	933156	98.28530	1.03520
917	30.28201	840889	95.76012	1.09051	967	31.09662	935089	98.33616	1.03413
918	30.29851	842724	95.81232	1.08932	968	31.11270	937024	98.38699	1.03306
919	30.31501	844561	95.86449	1.08814	969	31.12876	938961	98.43780	1.03199
920	30.33150	846400	95.91663	1.08696	970	31.14482	940900	98.48858	1.03093
921	30.34798	848241	95.96874	1.08578	971	31.16087	942841	98.53933	1.02987
922	30.36445	850084	96.02083	1.08460	972	31.17691	944784	98.59006	1.02881
923	30.38092	851929	96.07289	1.08342	973	31.19295	946729	98.64076	1.02775
924	30.39737	853776	96.12492	1.08225	974	31.20897	948676	98.69144	1.02669
925	30.41381	855625	96.17692	1.08108	975	31.22499	950625	98.74209	1.02564
926	30.43025	857476	96.22889	1.07991	976	31.24100	952576	98.79271	1.02459
927	30.44667	859329	96.28084	1.07875	977	31.25700	954529	98.84331	1.02354
928	30.46309	861184	96.33276	1.07759	978	31.27299	956484	98.89388	1.02249
929	30.47950	863041	96.38465	1.07643	979	31.28898	958441	98.94443	1.02145
930	30.49590	864900	96.43651	1.07527	980	31.30495	960400	98.99495	1.02041
931	30.51229	866761	96.48834	1.07411	981	31.32092	962361	99.04544	1.01937
932	30.52868	868624	96.54015	1.07296	982	31.33688	964324	99.09591	1.01833
933	30.54505	870489	96.59193	1.07181	983	31.35283	966289	99.14636	1.01729
934	30.56141	872356	96.64368	1.07066	984	31.36877	968256	99.19677	1.01626
935	30.57777	874225	96.69540	1.06952	985	31.38471	970225	99.24717	1.01523
936	30.59412	876096	96.74709	1.06838	986	31.40064	972196	99.29753	1.01420
937	30.61046	877969	96.79876	1.06724	987	31.41656	974169	99.34787	1.01317
938	30.62679	879844	96.85040	1.06610	988	31.43247	976144	99.39819	1.01215
939	30.64311	881721	96.90201	1.06496	989	31.44837	978121	99.44848	1.01112
940	30.65942	883600	96.95360	1.06383	990	31.46427	980100	99.49874	1.01010
941	30.67572	885481	97.00515	1.06270	991	31.48015	982081	99.54898	1.00908
942	30.69202	887364	97.05668	1.06157	992	31.49603	984064	99.59920	1.00806
943	30.70831	889249	97.10819	1.06045	993	31.51190	986049	99.64939	1.00705
944	30.72458	891136	97.15966	1.05932	994	31.52777	988036	99.69955	1.00604
945	30.74085	893025	97.21111	1.05820	995	31.54362	990025	99.74969	1.00503
946	30.75711	894916	97.26253	1.05708	996	31.55947	992016	99.79980	1.00402
947	30.77337	896809	97.31393	1.05597	997	31.57531	994009	99.84989	1.00301
948	30.78961	898704	97.36529	1.05485	998	31.59114	996004	99.89995	1.00200
949	30.80584	900601	97.41663	1.05374	999	31.60696	998001	99.94999	1.00100
950	30.82207	902500	97.46794	1.05263	1000	31.62278	1000000	100.00000	1.00000

TABLE B Random Numbers

12651	61646	11769	75109	86996	97669	25757	32535	07122	76763
81769	74436	02630	72310	45049	18029	07469	42341	98173	79260
36737	98863	77240	76251	00654	64688	09343	70278	67331	98729
82861	54371	76610	94934	72748	44124	05610	53750	95938	01485
21325	15732	24127	37431	09723	63529	73977	95218	96074	42138
74146	47887	62463	23045	41490	07954	22597	60012	98866	90959
90759	64410	54179	66075	61051	75385	51378	08360	95946	95547
55683	98078	02238	91540	21219	17720	87817	41705	95785	12563
79686	17969	76061	83748	55920	83612	41540	86492	06447	60568
70333	00201	86201	69716	78185	62154	77930	67663	29529	75116
14042	53536	07779	04157	41172	36473	42123	43929	50533	33437
59911	08256	06596	48416	69770	68797	56080	14223	59199	30162
62368	62623	62742	14891	39247	52242	98832	69533	91174	57979
57529	97751	54976	48957	74599	08759	78494	52785	68526	64618
15469	90574	78033	66885	13936	42117	71831	22961	94225	31816
18625	23674	53850	32827	81647	80820	00420	63555	74489	80141
74626	68394	88562	70745	23701	45630	65891	58220	35442	60414
11119	16519	27384	90199	79210	76965	99546	30323	31664	22845
41101	17336	48951	53674	17880	45260	08575	49321	36191	17095
32123	91576	84221	78902	82010	30847	62329	63898	23268	74283
26091	68409	69704	82267	14751	13151	93115	01437	56945	89661
67680	79790	48462	59278	44185	29616	76531	19589	83139	28454
15184	19260	14073	07026	25264	08388	27182	22557	61501	67481
58010	45039	57181	10238	36874	28546	37444	80824	63981	39942
56425	53996	86245	32623	78858	08143	60377	42925	42815	11159
82630	84066	13592	60642	17904	99718	63432	88642	37858	25431
14927	40909	23900	48761	44860	92467	31742	87142	03607	32059
23740	22505	07489	85986	74420	21744	97711	36648	35620	97949
32990	97446	03711	63824	07953	85965	87089	11687	92414	67257
05310	24058	91946	78437	34365	82469	12430	84754	19354	72745
21839	39937	27534	88913	49055	19218	47712	67677	51889	70926
08833	42549	93981	94051	28382	83725	72643	64233	97252	17133
58336	11139	47479	00931	91560	95372	97642	33856	54825	55680
62032	91144	75478	47431	52726	30289	42411	91886	51818	78292
45171	30557	53116	04118	58301	24375	65609	85810	18620	49198
91611	62656	60128	35609	63698	78356	50682	22505	01692	36291
55472	63819	86314	49174	93582	73604	78614	78849	23096	72825
18573	09729	74091	53994	10970	86557	65661	41854	26037	53296
60866	02955	90288	82136	83644	94455	06560	78029	98768	71296
45043	55608	82767	60890	74646	79485	13619	98868	40857	19415
17831	09737	79473	75945	28394	79334	70577	38048	03607	06932
40137	03981	07585	18128	11178	32601	27994	05641	22600	86064
77776	31343	14576	97706	16039	47517	43300	59080	80392	63189
69605	44104	40103	95635	05635	81673	68657	09559	23510	95875
19916	52934	26499	09821	87331	80993	61299	36979	73599	35055
02606	58552	07678	56619	65325	30705	99582	53390	46357	13244
65183	73160	87131	35530	47946	09854	18080	02321	05809	04898
10740	98914	44916	11322	89717	88189	30143	52687	19420	60061
98642	89822	71691	51573	83666	61642	46683	33761	47542	23551
60139	25601	93663	25547	02654	94829	48672	28736	84994	13071

SOURCE: The Rand Corporation, *A Million Random Digits with 100,000 Normal Deviates.* New York: The Free Press, 1955. Reproduced with permission of The Rand Corporation.

TABLE C Cumulative Values for the Binomial Probability Distribution

$$P[R \leqslant r] = P[P \leqslant r/n]$$

$n = 1$

π / r	.01	.05	.10	.20	.30	.40	.50
0	0.9900	0.9500	0.9000	0.8000	0.7000	0.6000	0.5000
1	1.0000	1.0000	1.0000	1.0000	1.0000	1.0000	1.0000

$n = 2$

π / r	.01	.05	.10	.20	.30	.40	.50
0	0.9801	0.9025	0.8100	0.6400	0.4900	0.3600	0.2500
1	0.9999	0.9975	0.9900	0.9600	0.9100	0.8400	0.7500
2	1.0000	1.0000	1.0000	1.0000	1.0000	1.0000	1.0000

$n = 3$

π / r	.01	.05	.10	.20	.30	.40	.50
0	0.9703	0.8574	0.7290	0.5120	0.3430	0.2160	0.1250
1	0.9997	0.9927	0.9720	0.8960	0.7840	0.6480	0.5000
2	1.0000	0.9999	0.9990	0.9920	0.9730	0.9360	0.8750
3	1.0000	1.0000	1.0000	1.0000	1.0000	1.0000	1.0000

$n = 4$

π / r	.01	.05	.10	.20	.30	.40	.50
0	0.9606	0.8145	0.6561	0.4096	0.2401	0.1296	0.0625
1	0.9994	0.9860	0.9477	0.8192	0.6517	0.4752	0.3125
2	1.0000	0.9995	0.9963	0.9728	0.9163	0.8208	0.6875
3	1.0000	1.0000	0.9999	0.9984	0.9919	0.9744	0.9375
4	1.0000	1.0000	1.0000	1.0000	1.0000	1.0000	1.0000

$n = 5$

π / r	.01	.05	.10	.20	.30	.40	.50
0	0.9510	0.7738	0.5905	0.3277	0.1681	0.0778	0.0313
1	0.9990	0.9774	0.9185	0.7373	0.5282	0.3370	0.1875
2	1.0000	0.9988	0.9914	0.9421	0.8369	0.6826	0.5000
3	1.0000	1.0000	0.9995	0.9933	0.9692	0.9130	0.8125
4	1.0000	1.0000	1.0000	0.9997	0.9976	0.9898	0.9688
5				1.0000	1.0000	1.0000	1.0000

TABLE C *(continued)*

$n = 6$

π / r	.01	.05	.10	.20	.30	.40	.50
0	0.9415	0.7351	0.5314	0.2621	0.1176	0.0467	0.0156
1	0.9985	0.9672	0.8857	0.6554	0.4202	0.2333	0.1094
2	1.0000	0.9978	0.9841	0.9011	0.7443	0.5443	0.3438
3	1.0000	0.9999	0.9987	0.9830	0.9295	0.8208	0.6563
4	1.0000	1.0000	0.9999	0.9984	0.9891	0.9590	0.8906
5	1.0000	1.0000	1.0000	0.9999	0.9993	0.9959	0.9844
6				1.0000	1.0000	1.0000	1.0000

$n = 7$

π / r	.01	.05	.10	.20	.30	.40	.50
0	0.9321	0.6983	0.4783	0.2097	0.0824	0.0280	0.0078
1	0.9980	0.9556	0.8503	0.5767	0.3294	0.1586	0.0625
2	1.0000	0.9962	0.9743	0.8520	0.6471	0.4199	0.2266
3	1.0000	0.9998	0.9973	0.9667	0.8740	0.7102	0.5000
4	1.0000	1.0000	0.9998	0.9953	0.9712	0.9037	0.7734
5	1.0000	1.0000	1.0000	0.9996	0.9962	0.9812	0.9375
6				1.0000	0.9998	0.9984	0.9922
7					1.0000	1.0000	1.0000

$n = 8$

π / r	.01	.05	.10	.20	.30	.40	.50
0	0.9227	0.6634	0.4305	0.1678	0.0576	0.0168	0.0039
1	0.9973	0.9428	0.8131	0.5033	0.2553	0.1064	0.0352
2	0.9999	0.9942	0.9619	0.7969	0.5518	0.3154	0.1445
3	1.0000	0.9996	0.9950	0.9437	0.8059	0.5941	0.3633
4	1.0000	1.0000	0.9996	0.9896	0.9420	0.8263	0.6367
5	1.0000	1.0000	1.0000	0.9988	0.9887	0.9502	0.8555
6				0.9999	0.9987	0.9915	0.9648
7				1.0000	0.9999	0.9993	0.9961
8					1.0000	1.0000	1.0000

$n = 9$

π / r	.01	.05	.10	.20	.30	.40	.50
0	0.9135	0.6302	0.3874	0.1342	0.0404	0.0101	0.0020
1	0.9966	0.9288	0.7748	0.4362	0.1960	0.0705	0.0195
2	0.9999	0.9916	0.9470	0.7382	0.4628	0.2318	0.0898
3	1.0000	0.9994	0.9917	0.9144	0.7297	0.4826	0.2539
4	1.0000	1.0000	0.9991	0.9804	0.9012	0.7334	0.5000
5	1.0000	1.0000	0.9999	0.9969	0.9747	0.9006	0.7461

$n = 9$

π / r	.01	.05	.10	.20	.30	.40	.50
6	1.0000	1.0000	1.0000	0.9997	0.9957	0.9750	0.9102
7				1.0000	0.9996	0.9962	0.9805
8					1.0000	0.9997	0.9980
9						1.0000	1.0000

$n = 10$

π / r	.01	.05	.10	.20	.30	.40	.50
0	0.9044	0.5987	0.3487	0.1074	0.0282	0.0060	0.0010
1	0.9957	0.9139	0.7361	0.3758	0.1493	0.0464	0.0107
2	0.9999	0.9885	0.9298	0.6778	0.3828	0.1673	0.0547
3	1.0000	0.9990	0.9872	0.8791	0.6496	0.3823	0.1719
4	1.0000	0.9999	0.9984	0.9672	0.8497	0.6331	0.3770
5	1.0000	1.0000	0.9999	0.9936	0.9526	0.8338	0.6230
6	1.0000	1.0000	1.0000	0.9991	0.9894	0.9452	0.8281
7				0.9999	0.9999	0.9877	0.9453
8				1.0000	1.0000	0.9983	0.9893
9						0.9999	0.9990
10						1.0000	1.0000

$n = 11$

π / r	.01	.05	.10	.20	.30	.40	.50
0	0.8953	0.5688	0.3138	0.0859	0.0198	0.0036	0.0005
1	0.9948	0.8981	0.6974	0.3221	0.1130	0.0302	0.0059
2	0.9998	0.9848	0.9104	0.6174	0.3127	0.1189	0.0327
3	1.0000	0.9984	0.9815	0.8369	0.5696	0.2963	0.1133
4	1.0000	0.9999	0.9972	0.9496	0.7897	0.5328	0.2744
5	1.0000	1.0000	0.9997	0.9883	0.9218	0.7535	0.5000
6	1.0000	1.0000	1.0000	0.9980	0.9784	0.9006	0.7256
7				0.9998	0.9957	0.9707	0.8867
8				1.0000	0.9994	0.9941	0.9673
9					1.0000	0.9993	0.9941
10						1.0000	0.9995
11							1.0000

$n = 12$

π / r	.01	.05	.10	.20	.30	.40	.50
0	0.8864	0.5404	0.2824	0.0687	0.0138	0.0022	0.0002
1	0.9938	0.8816	0.6590	0.2749	0.0850	0.0196	0.0032
2	0.9998	0.9804	0.8891	0.5583	0.2528	0.0834	0.0193
3	1.0000	0.9978	0.9744	0.7946	0.4925	0.2253	0.0730
4	1.0000	0.9998	0.9957	0.9274	0.7237	0.4382	0.1938
5	1.0000	1.0000	0.9995	0.9806	0.8821	0.6652	0.3872
6	1.0000	1.0000	0.9999	0.9961	0.9614	0.8418	0.6128
7	1.0000	1.0000	1.0000	0.9994	0.9905	0.9427	0.8062
8				0.9999	0.9983	0.9847	0.9270
9				1.0000	0.9998	0.9972	0.9807
10					1.0000	0.9997	0.9968
11						1.0000	0.9998
12							1.0000

$n = 13$

π / r	.01	.05	.10	.20	.30	.40	.50
0	0.8775	0.5133	0.2542	0.0550	0.0097	0.0013	0.0001
1	0.9928	0.8646	0.6213	0.2336	0.0637	0.0126	0.0017
2	0.9997	0.9755	0.8661	0.5017	0.2025	0.0579	0.0112
3	1.0000	0.9969	0.9658	0.7473	0.4206	0.1686	0.0461
4	1.0000	0.9997	0.9935	0.9009	0.6543	0.3530	0.1334
5	1.0000	1.0000	0.9991	0.9700	0.8346	0.5744	0.2905
6	1.0000	1.0000	0.9999	0.9930	0.9376	0.7712	0.5000
7	1.0000	1.0000	1.0000	0.9988	0.9818	0.9023	0.7095
8				0.9998	0.9960	0.9679	0.8666
9				1.0000	0.9993	0.9922	0.9539
10					0.9999	0.9987	0.9888
11					1.0000	0.9999	0.9983
12						1.0000	0.9999
13							1.0000

$n = 14$

π / r	.01	.05	.10	.20	.30	.40	.50
0	0.8687	0.4877	0.2288	0.0440	0.0068	0.0008	0.0001
1	0.9916	0.8470	0.5846	0.1979	0.0475	0.0081	0.0009
2	0.9997	0.9699	0.8416	0.4481	0.1608	0.0398	0.0065
3	1.0000	0.9958	0.9559	0.6982	0.3552	0.1243	0.0287
4	1.0000	0.9996	0.9908	0.8702	0.5842	0.2793	0.0898
5	1.0000	1.0000	0.9985	0.9561	0.7805	0.4859	0.2120
6	1.0000	1.0000	0.9998	0.9884	0.9067	0.6925	0.3953
7	1.0000	1.0000	1.0000	0.9976	0.9685	0.8499	0.6047
8				0.9996	0.9917	0.9417	0.7880
9				1.0000	0.9983	0.9825	0.9102
10					0.9998	0.9961	0.9713
11					1.0000	0.9994	0.9935
12						0.9999	0.9991
13						1.0000	0.9999
14							1.0000

$n = 15$

π / r	.01	.05	.10	.20	.30	.40	.50
0	0.8601	0.4633	0.2059	0.0352	0.0047	0.0005	0.0000
1	0.9904	0.8290	0.5490	0.1671	0.0353	0.0052	0.0005
2	0.9996	0.9638	0.8159	0.3980	0.1268	0.0271	0.0037
3	1.0000	0.9945	0.9444	0.6482	0.2969	0.0905	0.0176
4	1.0000	0.9994	0.9873	0.8358	0.5155	0.2173	0.0592
5	1.0000	0.9999	0.9978	0.9389	0.7216	0.4032	0.1509
6	1.0000	1.0000	0.9997	0.9819	0.8689	0.6098	0.3036
7	1.0000	1.0000	1.0000	0.9958	0.9500	0.7869	0.5000
8				0.9992	0.9848	0.9050	0.6964
9				0.9999	0.9963	0.9662	0.8491
10				1.0000	0.9993	0.9907	0.9408
11					0.9999	0.9981	0.9824
12					1.0000	0.9997	0.9963
13						1.0000	0.9995
14							1.0000

TABLE C (*continued*)

$n = 16$

π / r	.01	.05	.10	.20	.30	.40	.50
0	0.8515	0.4401	0.1853	0.0281	0.0033	0.0003	0.0000
1	0.9891	0.8108	0.5147	0.1407	0.0261	0.0033	0.0003
2	0.9995	0.9571	0.7892	0.3518	0.0994	0.0183	0.0021
3	1.0000	0.9930	0.9316	0.5981	0.2459	0.0651	0.0106
4	1.0000	0.9991	0.9830	0.7982	0.4499	0.1666	0.0384
5	1.0000	0.9999	0.9967	0.9183	0.6598	0.3288	0.1051
6	1.0000	1.0000	0.9995	0.9733	0.8247	0.5272	0.2272
7	1.0000	1.0000	0.9999	0.9930	0.9256	0.7161	0.4018
8	1.0000	1.0000	1.0000	0.9985	0.9743	0.8577	0.5982
9				0.9998	0.9929	0.9417	0.7728
10				1.0000	0.9984	0.9809	0.8949
11					0.9997	0.9951	0.9616
12					1.0000	0.9991	0.9894
13						0.9999	0.9979
14						1.0000	0.9997
15							1.0000

$n = 17$

π / r	.01	.05	.10	.20	.30	.40	.50
0	0.8429	0.4181	0.1668	0.0225	0.0023	0.0002	0.0000
1	0.9877	0.7922	0.4818	0.1182	0.0193	0.0021	0.0001
2	0.9994	0.9497	0.7618	0.3096	0.0774	0.0123	0.0012
3	1.0000	0.9912	0.9174	0.5489	0.2019	0.0464	0.0064
4	1.0000	0.9988	0.9779	0.7582	0.3887	0.1260	0.0245
5	1.0000	0.9999	0.9953	0.8943	0.5968	0.2639	0.0717
6	1.0000	1.0000	0.9992	0.9623	0.7752	0.4478	0.1662
7	1.0000	1.0000	0.9999	0.9891	0.8954	0.6405	0.3145
8	1.0000	1.0000	1.0000	0.9974	0.9597	0.8011	0.5000
9				0.9995	0.9873	0.9081	0.6855
10				0.9999	0.9968	0.9652	0.8338
11				1.0000	0.9993	0.9894	0.9283
12					0.9999	0.9975	0.9755
13					1.0000	0.9995	0.9936
14						0.9999	0.9988
15						1.0000	0.9999
16							1.0000

TABLE C (*continued*)

n = 18

π / r	.01	.05	.10	.20	.30	.40	.50
0	0.8345	0.3972	0.1501	0.0180	0.0016	0.0001	0.0000
1	0.9862	0.7735	0.4503	0.0991	0.0142	0.0013	0.0001
2	0.9993	0.9419	0.7338	0.2713	0.0600	0.0082	0.0007
3	1.0000	0.9891	0.9018	0.5010	0.1646	0.0328	0.0038
4	1.0000	0.9985	0.9718	0.7164	0.3327	0.0942	0.0154
5	1.0000	0.9998	0.9936	0.8671	0.5344	0.2088	0.0481
6	1.0000	1.0000	0.9988	0.9487	0.7217	0.3743	0.1189
7	1.0000	1.0000	0.9998	0.9837	0.8593	0.5634	0.2403
8	1.0000	1.0000	1.0000	0.9957	0.9404	0.7368	0.4073
9				0.9991	0.9790	0.8653	0.5927
10				0.9998	0.9939	0.9424	0.7597
11				1.0000	0.9986	0.9797	0.8811
12					0.9997	0.9942	0.9519
13					1.0000	0.9987	0.9846
14						0.9998	0.9962
15						1.0000	0.9993
16							0.9999
17							1.0000

n = 19

π / r	.01	.05	.10	.20	.30	.40	.50
0	0.8262	0.3774	0.1351	0.0144	0.0011	0.0001	0.0000
1	0.9847	0.7547	0.4203	0.0829	0.0104	0.0008	0.0000
2	0.9991	0.9335	0.7054	0.2369	0.0462	0.0055	0.0004
3	1.0000	0.9868	0.8850	0.4551	0.1332	0.0230	0.0022
4	1.0000	0.9980	0.9648	0.6733	0.2822	0.0696	0.0096
5	1.0000	0.9998	0.9914	0.8369	0.4739	0.1629	0.0318
6	1.0000	1.0000	0.9983	0.9324	0.6655	0.3081	0.0835
7	1.0000	1.0000	0.9997	0.9767	0.8180	0.4878	0.1796
8	1.0000	1.0000	1.0000	0.9933	0.9161	0.6675	0.3238
9				0.9984	0.9674	0.8139	0.5000
10				0.9997	0.9895	0.9115	0.6762

TABLE C (*continued*)

$n = 19$

π / r	.01	.05	.10	.20	.30	.40	.50
11				0.9999	0.9972	0.9648	0.8204
12				1.0000	0.9994	0.9884	0.9165
13					0.9999	0.9969	0.9682
14					1.0000	0.9994	0.9904
15						0.9999	0.9978
16						1.0000	0.9996
17							1.0000

$n = 20$

π / r	.01	.05	.10	.20	.30	.40	.50
0	0.8179	0.3585	0.1216	0.0115	0.0008	0.0000	0.0000
1	0.9831	0.7358	0.3917	0.0692	0.0076	0.0005	0.0000
2	0.9990	0.9245	0.6769	0.2061	0.0355	0.0036	0.0002
3	1.0000	0.9841	0.8670	0.4114	0.1071	0.0160	0.0013
4	1.0000	0.9974	0.9568	0.6296	0.2375	0.0510	0.0059
5	1.0000	0.9997	0.9887	0.8042	0.4164	0.1256	0.0207
6	1.0000	1.0000	0.9976	0.9133	0.6080	0.2500	0.0577
7	1.0000	1.0000	0.9996	0.9679	0.7723	0.4159	0.1316
8	1.0000	1.0000	0.9999	0.9900	0.8867	0.5956	0.2517
9	1.0000	1.0000	1.0000	0.9974	0.9520	0.7553	0.4119
10				0.9994	0.9829	0.8725	0.5881
11				0.9999	0.9949	0.9435	0.7483
12				1.0000	0.9987	0.9790	0.8684
13					0.9997	0.9935	0.9423
14					1.0000	0.9984	0.9793
15						0.9997	0.9941
16						1.0000	0.9987
17							0.9998
18							1.0000

$n = 50$

r \ π	.01	.05	.10	.20	.30	.40	.50
0	0.6050	0.0769	0.0052	0.0000	0.0000	0.0000	0.0000
1	0.9106	0.2794	0.0338	0.0002	0.0000	0.0000	0.0000
2	0.9862	0.5405	0.1117	0.0013	0.0000	0.0000	0.0000
3	0.9984	0.7604	0.2503	0.0057	0.0000	0.0000	0.0000
4	0.9999	0.8964	0.4312	0.0185	0.0002	0.0000	0.0000
5	1.0000	0.9622	0.6161	0.0480	0.0007	0.0000	0.0000
6	1.0000	0.9882	0.7702	0.1034	0.0025	0.0000	0.0000
7	1.0000	0.9968	0.8779	0.1904	0.0073	0.0001	0.0000
8	1.0000	0.9992	0.9421	0.3073	0.0183	0.0002	0.0000
9	1.0000	0.9998	0.9755	0.4437	0.0402	0.0008	0.0000
10	1.0000	1.0000	0.9906	0.5836	0.0789	0.0022	0.0000
11	1.0000	1.0000	0.9968	0.7107	0.1390	0.0057	0.0000
12	1.0000	1.0000	0.9990	0.8139	0.2229	0.0133	0.0002
13	1.0000	1.0000	0.9997	0.8894	0.3279	0.0280	0.0005
14	1.0000	1.0000	0.9999	0.9393	0.4468	0.0540	0.0013
15	1.0000	1.0000	1.0000	0.9692	0.5692	0.0955	0.0033
16				0.9856	0.6839	0.1561	0.0077
17				0.9937	0.7822	0.2369	0.0164
18				0.9975	0.8594	0.3356	0.0325
19				0.9991	0.9152	0.4465	0.0595
20				0.9997	0.9522	0.5610	0.1013
21				0.9999	0.9749	0.6701	0.1611
22				1.0000	0.9877	0.7660	0.2399
23					0.9944	0.8438	0.3359
24					0.9976	0.9022	0.4439
25					0.9991	0.9427	0.5561
26					0.9997	0.9686	0.6641
27					0.9999	0.9840	0.7601
28					1.0000	0.9924	0.8389
29						0.9966	0.8987
30						0.9986	0.9405
31						0.9995	0.9675
32						0.9998	0.9836
33						0.9999	0.9923
34						1.0000	0.9967
35							0.9987
36							0.9995
37							0.9998
38							1.0000

$n = 100$

π / r	.01	.05	.10	.20	.30	.40	.50
0	0.3660	0.0059	0.0000	0.0000	0.0000	0.0000	0.0000
1	0.7358	0.0371	0.0003	0.0000	0.0000	0.0000	0.0000
2	0.9206	0.1183	0.0019	0.0000	0.0000	0.0000	0.0000
3	0.9816	0.2578	0.0078	0.0000	0.0000	0.0000	0.0000
4	0.9966	0.4360	0.0237	0.0000	0.0000	0.0000	0.0000
5	0.9995	0.6160	0.0576	0.0000	0.0000	0.0000	0.0000
6	0.9999	0.7660	0.1172	0.0001	0.0000	0.0000	0.0000
7	1.0000	0.8720	0.2061	0.0003	0.0000	0.0000	0.0000
8	1.0000	0.9369	0.3209	0.0009	0.0000	0.0000	0.0000
9	1.0000	0.9718	0.4513	0.0023	0.0000	0.0000	0.0000
10	1.0000	0.9885	0.5832	0.0057	0.0000	0.0000	0.0000
11	1.0000	0.9957	0.7030	0.0126	0.0000	0.0000	0.0000
12	1.0000	0.9985	0.8018	0.0253	0.0000	0.0000	0.0000
13	1.0000	0.9995	0.8761	0.0469	0.0001	0.0000	0.0000
14	1.0000	0.9999	0.9274	0.0804	0.0002	0.0000	0.0000
15	1.0000	1.0000	0.9601	0.1285	0.0004	0.0000	0.0000
16	1.0000	1.0000	0.9794	0.1923	0.0010	0.0000	0.0000
17	1.0000	1.0000	0.9900	0.2712	0.0022	0.0000	0.0000
18	1.0000	1.0000	0.9954	0.3621	0.0045	0.0000	0.0000
19	1.0000	1.0000	0.9980	0.4602	0.0089	0.0000	0.0000
20	1.0000	1.0000	0.9992	0.5595	0.0165	0.0000	0.0000
21	1.0000	1.0000	0.9997	0.6540	0.0288	0.0000	0.0000
22	1.0000	1.0000	0.9999	0.7389	0.0479	0.0001	0.0000
23	1.0000	1.0000	1.0000	0.8109	0.0755	0.0003	0.0000
24				0.8686	0.1136	0.0006	0.0000
25				0.9125	0.1631	0.0012	0.0000
26				0.9442	0.2244	0.0024	0.0000
27				0.9658	0.2964	0.0046	0.0000
28				0.9800	0.3768	0.0084	0.0000
29				0.9888	0.4623	0.0148	0.0000
30				0.9939	0.5491	0.0248	0.0000
31				0.9969	0.6331	0.0398	0.0001
32				0.9984	0.7107	0.0615	0.0002
33				0.9993	0.7793	0.0913	0.0004
34				0.9997	0.8371	0.1303	0.0009
35				0.9999	0.8839	0.1795	0.0018

$n = 100$

π / r	.01	.05	.10	.20	.30	.40	.50
36				0.9999	0.9201	0.2386	0.0033
37				1.0000	0.9470	0.3068	0.0060
38					0.9660	0.3822	0.0105
39					0.9790	0.4621	0.0176
40					0.9875	0.5433	0.0284
41					0.9928	0.6225	0.0443
42					0.9960	0.6967	0.0666
43					0.9979	0.7635	0.0967
44					0.9989	0.8211	0.1356
45					0.9995	0.8689	0.1841
46					0.9997	0.9070	0.2421
47					0.9999	0.9362	0.3086
48					0.9999	0.9577	0.3822
49					1.0000	0.9729	0.4602
50						0.9832	0.5398
51						0.9900	0.6178
52						0.9942	0.6914
53						0.9968	0.7579
54						0.9983	0.8159
55						0.9991	0.8644
56						0.9996	0.9033
57						0.9998	0.9334
58						0.9999	0.9557
59						1.0000	0.9716
60							0.9824
61							0.9895
62							0.9940
63							0.9967
64							0.9982
65							0.9991
66							0.9996
67							0.9998
68							0.9999
69							1.0000

TABLE D Areas Under the Standard Normal Curve

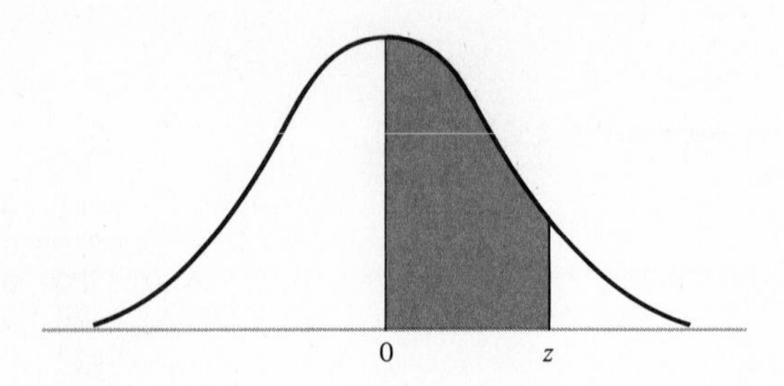

The following table provides the area between the mean and normal deviate value z.

Normal Deviate z	.00	.01	.02	.03	.04	.05	.06	.07	.08	.09
0.0	.0000	.0040	.0080	.0120	.0160	.0199	.0239	.0279	.0319	.0359
0.1	.0398	.0438	.0478	.0517	.0557	.0596	.0636	.0675	.0714	.0753
0.2	.0793	.0832	.0871	.0910	.0948	.0987	.1026	.1064	.1103	.1141
0.3	.1179	.1217	.1255	.1293	.1331	.1368	.1406	.1443	.1480	.1517
0.4	.1554	.1591	.1628	.1664	.1700	.1736	.1772	.1808	.1844	.1879
0.5	.1915	.1950	.1985	.2019	.2054	.2088	.2123	.2157	.2190	.2224
0.6	.2257	.2291	.2324	.2357	.2389	.2422	.2454	.2486	.2518	.2549
0.7	.2580	.2612	.2642	.2673	.2704	.2734	.2764	.2794	.2823	.2852
0.8	.2881	.2910	.2939	.2967	.2995	.3023	.3051	.3078	.3106	.3133
0.9	.3159	.3186	.3212	.3238	.3264	.3289	.3315	.3340	.3365	.3389
1.0	.3413	.3438	.3461	.3485	.3508	.3531	.3554	.3577	.3599	.3621
1.1	.3643	.3665	.3686	.3708	.3729	.3749	.3770	.3790	.3810	.3830
1.2	.3849	.3869	.3888	.3907	.3925	.3944	.3962	.3980	.3997	.4015
1.3	.4032	.4049	.4066	.4082	.4099	.4115	.4131	.4147	.4162	.4177
1.4	.4192	.4207	.4222	.4236	.4251	.4265	.4279	.4292	.4306	.4319
1.5	.4332	.4345	.4357	.4370	.4382	.4394	.4406	.4418	.4429	.4441
1.6	.4452	.4463	.4474	.4484	.4495	.4505	.4515	.4525	.4535	.4545
1.7	.4554	.4564	.4573	.4582	.4591	.4599	.4608	.4616	.4625	.4633
1.8	.4641	.4649	.4656	.4664	.4671	.4678	.4686	.4693	.4699	.4706
1.9	.4713	.4719	.4726	.4732	.4738	.4744	.4750	.4756	.4761	.4767
2.0	.4772	.4778	.4783	.4788	.4793	.4798	.4803	.4808	.4812	.4817
2.1	.4821	.4826	.4830	.4834	.4838	.4842	.4846	.4850	.4854	.4857
2.2	.4861	.4864	.4868	.4871	.4875	.4878	.4881	.4884	.4887	.4890
2.3	.4893	.4896	.4898	.4901	.4904	.4906	.4909	.4911	.4913	.4916
2.4	.4918	.4920	.4922	.4925	.4927	.4929	.4931	.4932	.4934	.4936
2.5	.4938	.4940	.4941	.4943	.4945	.4946	.4948	.4949	.4951	.4952
2.6	.4953	.4955	.4956	.4957	.4959	.4960	.4961	.4962	.4963	.4964
2.7	.4965	.4966	.4967	.4968	.4969	.4970	.4971	.4972	.4973	.4974
2.8	.4974	.4975	.4976	.4977	.4977	.4978	.4979	.4979	.4980	.4981
2.9	.4981	.4982	.4982	.4983	.4984	.4984	.4985	.4985	.4986	.4986
3.0	.49865	.4987	.4987	.4988	.4988	.4989	.4989	.4989	.4990	.4990
4.0	.49997									

TABLE E Student t Distribution

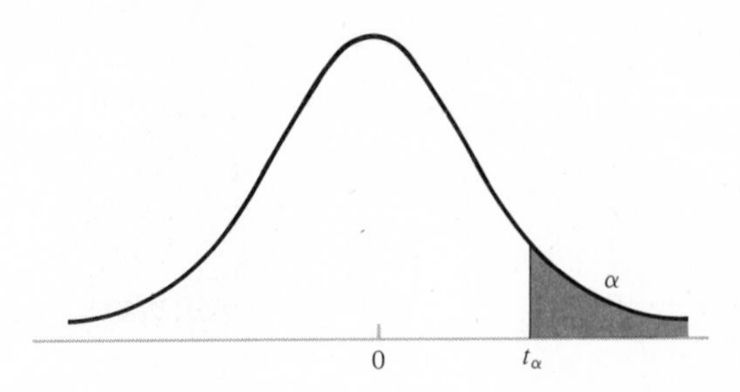

The following table provides the values of t_α that correspond to a given upper-tail area α and a specified number of degrees of freedom.

Degrees of Freedom	Upper-Tail Area α									
	.4	.25	.1	.05	.025	.01	.005	.0025	.001	.0005
1	0.325	1.000	3.078	6.314	12.706	31.821	63.657	127.32	318.31	636.62
2	.289	0.816	1.886	2.920	4.303	6.965	9.925	14.089	22.327	31.598
3	.277	.765	1.638	2.353	3.182	4.541	5.841	7.453	10.214	12.924
4	.271	.741	1.533	2.132	2.776	3.747	4.604	5.598	7.173	8.610
5	0.267	0.727	1.476	2.015	2.571	3.365	4.032	4.773	5.893	6.869
6	.265	.718	1.440	1.943	2.447	3.143	3.707	4.317	5.208	5.959
7	.263	.711	1.415	1.895	2.365	2.998	3.499	4.029	4.785	5.408
8	.262	.706	1.397	1.860	2.306	2.896	3.355	3.833	4.501	5.041
9	.261	.703	1.383	1.833	2.262	2.821	3.250	3.690	4.297	4.781
10	0.260	0.700	1.372	1.812	2.228	2.764	3.169	3.581	4.144	4.587
11	.260	.697	1.363	1.796	2.201	2.718	3.106	3.497	4.025	4.437
12	.259	.695	1.356	1.782	2.179	2.681	3.055	3.428	3.930	4.318
13	.259	.694	1.350	1.771	2.160	2.650	3.012	3.372	3.852	4.221
14	.258	.692	1.345	1.761	2.145	2.624	2.977	3.326	3.787	4.140
15	0.258	0.691	1.341	1.753	2.131	2.602	2.947	3.286	3.733	4.073
16	.258	.690	1.337	1.746	2.120	2.583	2.921	3.252	3.686	4.015
17	2.57	.689	1.333	1.740	2.110	2.567	2.898	3.222	3.646	3.965
18	.257	.688	1.330	1.734	2.101	2.552	2.878	3.197	3.610	3.922
19	.257	.688	1.328	1.729	2.093	2.539	2.861	3.174	3.579	3.883
20	0.257	0.687	1.325	1.725	2.086	2.528	2.845	3.153	3.552	3.850
21	.257	.686	1.323	1.721	2.080	2.518	2.831	3.135	3.527	3.819
22	.256	.686	1.321	1.717	2.074	2.508	2.819	3.119	3.505	3.792
23	.256	.685	1.319	1.714	2.069	2.500	2.807	3.104	3.485	3.767
24	.256	.685	1.318	1.711	2.064	2.492	2.797	3.091	3.467	3.745
25	0.256	0.684	1.316	1.708	2.060	2.485	2.787	3.078	3.450	3.725
26	.256	.684	1.315	1.706	2.056	2.479	2.779	3.067	3.435	3.707
27	.256	.684	1.314	1.703	2.052	2.473	2.771	3.057	3.421	3.690
28	.256	.683	1.313	1.701	2.048	2.467	2.763	3.047	3.408	3.674
29	.256	.683	1.311	1.699	2.045	2.462	2.756	3.038	3.396	3.659
30	0.256	0.683	1.310	1.697	2.042	2.457	2.750	3.030	3.385	3.646
40	.255	.681	1.303	1.684	2.021	2.423	2.704	2.971	3.307	3.551
60	.254	.679	1.296	1.671	2.000	2.390	2.660	2.915	3.232	3.460
120	.254	.677	1.289	1.658	1.980	2.358	2.617	2.860	3.160	3.373
∞	.253	.674	1.282	1.645	1.960	2.326	2.576	2.807	3.090	3.291

SOURCE: E.S. Pearson and H.O. Hartley, *Biometrika Tables for Statisticians*, Vol. I. London: Cambridge University Press, 1966. Partly derived from Table III of Fisher and Yates, *Statistical Tables for Biological, Agricultural and Medical Research*, published by Longman Group Ltd., London (previously published by Oliver & Boyd, Edinburgh, 1963). Reproduced with permission of the authors and publishers.

TABLE F Chi-Square Distribution

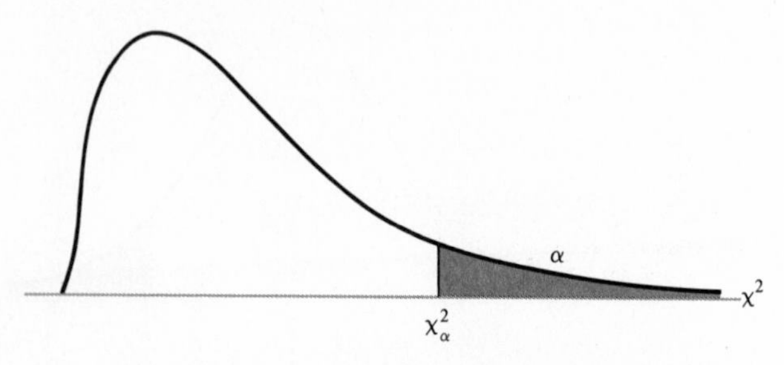

The following table provides the values of χ^2_α that correspond to a given upper-tail area α and a specified number of degrees of freedom.

Degrees of Freedom	Upper-Tail Area α						
	.99	.98	.95	.90	.80	.70	.50
1	$.0^3157$	$.0^3628$	.00393	.0158	.0642	.148	.455
2	.0201	.0404	.103	.211	.446	.713	1.386
3	.115	.185	.352	.584	1.005	1.424	2.366
4	.297	.429	.711	1.064	1.649	2.195	3.357
5	.554	.752	1.145	1.610	2.343	3.000	4.351
6	.872	1.134	1.635	2.204	3.070	3.828	5.348
7	1.239	1.564	2.167	2.833	3.822	4.671	6.346
8	1.646	2.032	2.733	3.490	4.594	5.527	7.344
9	2.088	2.532	3.325	4.168	5.380	6.393	8.343
10	2.558	3.059	3.940	4.865	6.179	7.267	9.342
11	3.053	3.609	4.575	5.578	6.989	8.148	10.341
12	3.571	4.178	5.226	6.304	7.807	9.034	11.340
13	4.107	4.765	5.892	7.042	8.634	9.926	12.340
14	4.660	5.368	6.571	7.790	9.467	10.821	13.339
15	5.229	5.985	7.261	8.547	10.307	11.721	14.339
16	5.812	6.614	7.962	9.312	11.152	12.624	15.338
17	6.408	7.255	8.672	10.085	12.002	13.531	16.338
18	7.015	7.906	9.390	10.865	12.857	14.440	17.338
19	7.633	8.567	10.117	11.651	13.716	15.352	18.338
20	8.260	9.237	10.851	12.443	14.578	16.266	19.337
21	8.897	9.915	11.591	13.240	15.445	17.182	20.337
22	9.542	10.600	12.338	14.041	16.314	18.101	21.337
23	10.196	11.293	13.091	14.848	17.187	19.021	22.337
24	10.856	11.992	13.848	15.659	18.062	19.943	23.337
25	11.524	12.697	14.611	16.473	18.940	20.867	24.337
26	12.198	13.409	15.379	17.292	19.820	21.792	25.336
27	12.879	14.125	16.151	18.114	20.703	22.719	26.336
28	13.565	14.847	16.928	18.939	21.588	23.647	27.336
29	14.256	15.574	17.708	19.768	22.475	24.577	28.336
30	14.953	16.306	18.493	20.599	23.364	25.508	29.336

Degrees of Freedom	Upper-Tail Area α						
	.30	.20	.10	.05	.02	.01	.001
1	1.074	1.642	2.706	3.841	5.412	6.635	10.827
2	2.408	3.219	4.605	5.991	7.824	9.210	13.815
3	3.665	4.642	6.251	7.815	9.837	11.345	16.268
4	4.878	5.989	7.779	9.488	11.668	13.277	18.465
5	6.064	7.289	9.236	11.070	13.388	15.086	20.517
6	7.231	8.558	10.645	12.592	15.033	16.812	22.457
7	8.383	9.803	12.017	14.067	16.622	18.475	24.322
8	9.524	11.030	13.362	15.507	18.168	20.090	26.125
9	10.656	12.242	14.684	16.919	19.679	21.666	27.877
10	11.781	13.442	15.987	18.307	21.161	23.209	29.588
11	12.899	14.631	17.275	19.675	22.618	24.725	31.264
12	14.011	15.812	18.549	21.026	24.054	26.217	32.909
13	15.119	16.985	19.812	22.362	25.472	27.688	34.528
14	16.222	18.151	21.064	23.685	26.873	29.141	36.123
15	17.322	19.311	22.307	24.996	28.259	30.578	37.697
16	18.418	20.465	23.542	26.296	29.633	32.000	39.252
17	19.511	21.615	24.769	27.587	30.995	33.409	40.790
18	20.601	22.760	25.989	28.869	32.346	34.805	42.312
19	21.689	23.900	27.204	30.144	33.687	36.191	43.820
20	22.775	25.038	28.412	31.410	35.020	37.566	45.315
21	23.858	26.171	29.615	32.671	36.343	38.932	46.797
22	24.939	27.301	30.813	33.924	37.659	40.289	48.268
23	26.018	28.429	32.007	35.172	38.968	41.638	49.728
24	27.096	29.553	33.196	36.415	40.270	42.980	51.179
25	28.172	30.675	34.382	37.652	41.566	44.314	52.620
26	29.246	31.795	35.563	38.885	42.856	45.642	54.052
27	30.319	32.912	36.741	40.113	44.140	46.963	55.476
28	31.391	34.027	37.916	41.337	45.419	48.278	56.893
29	32.461	35.139	39.087	42.557	46.693	49.588	58.302
30	33.530	36.250	40.256	43.773	47.962	50.892	59.703

SOURCE: From Table IV of Fisher and Yates, *Statistical Tables for Biological, Agricultural and Medical Research*, published by Longman Group Ltd., London (previously published by Oliver & Boyd, Edinburgh, 1963). Reproduced with permission of the authors and publishers.

TABLE G F Distribution

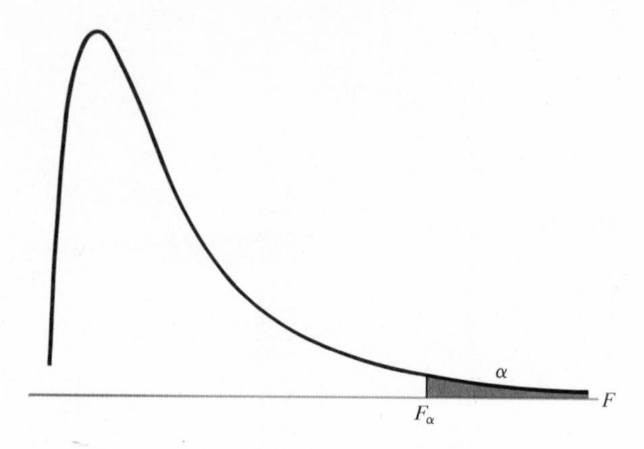

The following table provides the values of F_α that correspond to a given upper-tail area α and a specified degrees of freedom pair. The values of $F_{.05}$ are in lightface type, while those for $F_{.01}$ are given in boldface type. The number of degrees of freedom for the *numerator* mean square is indicated at the head of each *column*, while the number of degrees of freedom for the *denominator* mean square determines which *row* is applicable.

Degrees of Freedom in Denominator	Degrees of Freedom in Numerator											
	1	2	3	4	5	6	7	8	9	10	11	12
1	161	200	216	225	230	234	237	239	241	242	243	244
	4,052	**4,999**	**5,403**	**5,625**	**5,764**	**5,859**	**5,928**	**5,981**	**6,022**	**6,056**	**6,082**	**6,106**
2	18.51	19.00	19.16	19.25	19.30	19.33	19.36	19.37	19.38	19.39	19.40	19.41
	98.49	**99.00**	**99.17**	**99.25**	**99.30**	**99.33**	**99.36**	**99.37**	**99.39**	**99.40**	**99.41**	**99.42**
3	10.13	9.55	9.28	9.12	9.01	8.94	8.88	8.84	8.81	8.78	8.76	8.74
	34.12	**30.82**	**29.46**	**28.71**	**28.24**	**27.91**	**27.67**	**27.49**	**27.34**	**27.23**	**27.13**	**27.05**
4	7.71	6.94	6.59	6.39	6.26	6.16	6.09	6.04	6.00	5.96	5.93	5.91
	21.20	**18.00**	**16.69**	**15.98**	**15.52**	**15.21**	**14.98**	**14.80**	**14.66**	**14.54**	**14.45**	**14.37**
5	6.61	5.79	5.41	5.19	5.05	4.95	4.88	4.82	4.78	4.74	4.70	4.68
	16.26	**13.27**	**12.06**	**11.39**	**10.97**	**10.67**	**10.45**	**10.29**	**10.15**	**10.05**	**9.96**	**9.89**
6	5.99	5.14	4.76	4.53	4.39	4.28	4.21	4.15	4.10	4.06	4.03	4.00
	13.74	**10.92**	**9.78**	**9.15**	**8.75**	**8.47**	**8.26**	**8.10**	**7.98**	**7.87**	**7.79**	**7.72**
7	5.59	4.74	4.35	4.12	3.97	3.87	3.79	3.73	3.68	3.63	3.60	3.57
	12.25	**9.55**	**8.45**	**7.85**	**7.46**	**7.19**	**7.00**	**6.84**	**6.71**	**6.62**	**6.54**	**6.47**
8	5.32	4.46	4.07	3.84	3.69	3.58	3.50	3.44	3.39	3.34	3.31	3.28
	11.26	**8.65**	**7.59**	**7.01**	**6.63**	**6.37**	**6.19**	**6.03**	**5.91**	**5.82**	**5.74**	**5.67**
9	5.12	4.26	3.86	3.63	3.48	3.37	3.29	3.23	3.18	3.13	3.10	3.07
	10.56	**8.02**	**6.99**	**6.42**	**6.06**	**5.80**	**5.62**	**5.47**	**5.35**	**5.26**	**5.18**	**5.11**
10	4.96	4.10	3.71	3.48	3.33	3.22	3.14	3.07	3.02	2.97	2.94	2.91
	10.04	**7.56**	**6.55**	**5.99**	**5.64**	**5.39**	**5.21**	**5.06**	**4.95**	**4.85**	**4.78**	**4.71**
11	4.84	3.98	3.59	3.36	3.20	3.09	3.01	2.95	2.90	2.86	2.82	2.79
	9.65	**7.20**	**6.22**	**5.67**	**5.32**	**5.07**	**4.88**	**4.74**	**4.63**	**4.54**	**4.46**	**4.40**
12	4.75	3.88	3.49	3.26	3.11	3.00	2.92	2.85	2.80	2.76	2.72	2.69
	9.33	**6.93**	**5.95**	**5.41**	**5.06**	**4.82**	**4.65**	**4.50**	**4.39**	**4.30**	**4.22**	**4.16**
13	4.67	3.80	3.41	3.18	3.02	2.92	2.84	2.77	2.72	2.67	2.63	2.60
	9.07	**6.70**	**5.74**	**5.20**	**4.86**	**4.62**	**4.44**	**4.30**	**4.19**	**4.10**	**4.02**	**3.96**
14	4.60	3.74	3.34	3.11	2.96	2.85	2.77	2.70	2.65	2.60	2.56	2.53
	8.86	**6.51**	**5.56**	**5.03**	**4.69**	**4.46**	**4.28**	**4.14**	**4.03**	**3.94**	**3.86**	**3.80**

TABLE G *(continued)*

Degrees of Freedom in Denominator	Degrees of Freedom in Numerator											
	1	2	3	4	5	6	7	8	9	10	11	12
15	4.54 **8.68**	3.68 **6.36**	3.29 **5.42**	3.06 **4.89**	2.90 **4.56**	2.79 **4.32**	2.70 **4.14**	2.64 **4.00**	2.59 **3.89**	2.55 **3.80**	2.51 **3.73**	2.48 **3.67**
16	4.49 **8.53**	3.63 **6.23**	3.24 **5.29**	3.01 **4.77**	2.85 **4.44**	2.74 **4.20**	2.66 **4.03**	2.59 **3.89**	2.54 **3.78**	2.49 **3.69**	2.45 **3.61**	2.42 **3.55**
17	4.45 **8.40**	3.59 **6.11**	3.20 **5.18**	2.96 **4.67**	2.81 **4.34**	2.70 **4.10**	2.62 **3.93**	2.55 **3.79**	2.50 **3.68**	2.45 **3.59**	2.41 **3.52**	2.38 **3.45**
18	4.41 **8.28**	3.55 **6.01**	3.16 **5.09**	2.93 **4.58**	2.77 **4.25**	2.66 **4.01**	2.58 **3.85**	2.51 **3.71**	2.46 **3.60**	2.41 **3.51**	2.37 **3.44**	2.34 **3.37**
19	4.38 **8.18**	3.52 **5.93**	3.13 **5.01**	2.90 **4.50**	2.74 **4.17**	2.63 **3.94**	2.55 **3.77**	2.48 **3.63**	2.43 **3.52**	2.38 **3.43**	2.34 **3.36**	2.31 **3.30**
20	4.35 **8.10**	3.49 **5.85**	3.10 **4.94**	2.87 **4.43**	2.71 **4.10**	2.60 **3.87**	2.52 **3.71**	2.45 **3.56**	2.40 **3.45**	2.35 **3.37**	2.31 **3.30**	2.28 **3.23**
21	4.32 **8.02**	3.47 **5.78**	3.07 **4.87**	2.84 **4.37**	2.68 **4.04**	2.57 **3.81**	2.49 **3.65**	2.42 **3.51**	2.37 **3.40**	2.32 **3.31**	2.28 **3.24**	2.25 **3.17**
22	4.30 **7.94**	3.44 **5.72**	3.05 **4.82**	2.82 **4.31**	2.66 **3.99**	2.55 **3.76**	2.47 **3.59**	2.40 **3.45**	2.35 **3.35**	2.30 **3.26**	2.26 **3.18**	2.23 **3.12**
23	4.28 **7.88**	3.42 **5.66**	3.03 **4.76**	2.80 **4.26**	2.64 **3.94**	2.53 **3.71**	2.45 **3.54**	2.38 **3.41**	2.32 **3.30**	2.28 **3.21**	2.24 **3.14**	2.20 **3.07**
24	4.26 **7.82**	3.40 **5.61**	3.01 **4.72**	2.78 **4.22**	2.62 **3.90**	2.51 **3.67**	2.43 **3.50**	2.36 **3.36**	2.30 **3.25**	2.26 **3.17**	2.22 **3.09**	2.18 **3.03**
25	4.24 **7.77**	3.38 **5.57**	2.99 **4.68**	2.76 **4.18**	2.60 **3.86**	2.49 **3.63**	2.41 **3.46**	2.34 **3.32**	2.28 **3.21**	2.24 **3.13**	2.20 **3.05**	2.16 **2.99**
26	4.22 **7.72**	3.37 **5.53**	2.98 **4.64**	2.74 **4.14**	2.59 **3.82**	2.47 **3.59**	2.39 **3.42**	2.32 **3.29**	2.27 **3.17**	2.22 **3.09**	2.18 **3.02**	2.15 **2.96**

Photo Credits: p.2, NASA; p.8, p.66, Charles Gatewood; p.92, Bob Vose from Black Star; p.152, Carew from Monkmeyer; p. 202, Magnum; p. 244, FIAT; p. 284, NBC; p. 328, Harbrace; p. 380, Lederle Laboratories Division, American Cyanamid Company; p. 402, Harvey Stein; p. 434, Whitestone Photo for Boston Symphony Orchestra; p. 460, Standard Oil Company, (N.J.), USDA, Cities Service Company, "West Texas Today"; p. 500, Wide World.

Abbreviated Answers to Even-Numbered Exercises

2-2 Answers may vary.

2-4 (a)

Class Interval	Frequency
$3,000–under 7,000	10
7,000–under 11,000	26
11,000–under 15,000	8
15,000–under 19,000	3
19,000–under 23,000	1
23,000–under 27,000	2
	50

2-6 (b)

Class Interval	Combined Frequency
200–under 300	150
300–under 400	350
400–under 500	250
500–under 600	400
600–under 700	150
700–under 800	50

(d) Separate graphs are better.

2-8 Intervals of width $.05 would be too small, while $1 and $2 intervals would be too wide.

2-10 (a)–(c) Class intervals must not overlap and must encompass all possible values.

2-12

Class Interval	(a) Relative Frequency	(b) Cumulative Relative Frequency
0–less than 5	.620	.620
5–less than 10	.162	.782
10–less than 15	.102	.884
15–less than 20	.064	.948
20–less than 25	.024	.972
25–less than 30	.006	.978
30–less than 60	.022	1.000

2-14

Class Interval	(a) Relative Frequency
0–under 5,000	.36
5,000–under 10,000	.27
10,000–under 15,000	.18
15,000–under 20,000	.09
20,000–under 25,000	.07
25,000–under 30,000	.03
	1.00

(b) Frequency	(c) Cumulative Frequency
360	360
270	630
180	810
90	900
70	970
30	1,000
1,000	

2-16 Answers may vary.

2-18 $\bar{X} = 93.8$ for CompuQuick
$\bar{X} = 192.125$ for Dial-a-Pute

2-20 26.0 feet

2-22 $\bar{X} = 2.75$ $m = 2$ mode $= 2$

2-24 $m = \$2{,}129$

2-26 Answers may vary.

2-28 $s^2 = 184.30$ $s = 13.58$

2-30 $\bar{X} = 26.15$ $s = 21.11$

2-32 $\bar{X} = 2.75$ $s = .63$

2-34

Firm	A	B	C	D	E
Proportion	.0624	.178	.0629	.113	.125

2-36

	(1)	(2)	(3)	(4)
(a) P	.07	.05	.017	.039
(b)	reject	accept	accept	reject

2-38

Class Interval	Relative Frequency	Cumulative Frequency
20–under 25	.18	18
25–under 30	.23	41
30–under 35	.15	69
35–under 40	.13	76
40–under 45	.07	76
45–under 50	06	82
50–under 55	.05	87
55–under 60	.05	92
60–under 65	.08	100
	1.00	

2-40 (a) 4 (b) 2.0 (c) 1.33 (d) 1.15

2-42 (a) .10 (b) .04 (c) .03 (d) .07 (e) .04 (f) .05

2-44 (a)

Sex	*Frequency*
Male	220
Female	130
	350

(b)

Marital Status	*Frequency*
Married	171
Single	179
	350

(c)

Occupation	*Frequency*
Blue collar	155
White collar	173
Professional	22
	350

2-46 $\bar{X} = 126.2 \quad s^2 = 385.41$

2-48 (a) $\bar{X} = 77.615 \quad s^2 = 208.55$
(b) $\bar{X} = 78.077 \quad s^2 = 176.14$
(c) error $= -0.462$

3-2 Answers may vary.

3-4 Answers may vary.

3-6 Answers may vary.

3-8 Answers may vary.

3-10 (a) convenience (b) judgment
(c) random
(d) judgment or convenience

3-12 Answers may vary.

3-14 A judgment-convenience sample applies.

3-16 Solti, Maag, Abbado, Rozhdestvensky, Bloomfield, Krips, Prêtre, Frühbeck de Burgos, Newman, Schippers

3-18 Beecham, Dragon, Golschmann, Karajan, Krips, Pedrotti, Rignold, Scherchen, Stein, Svetlanov

4-2 (a) $\{(H,H,T),(H,T,H),(T,H,H)\}$
(b) $\{(H,T,H),(T,H,T)\}$
(c) $\{(H,H,H),(H,T,H),(T,H,H),(T,T,H)\}$
(d) $\{(H,H,T),(H,T,H),(T,H,H),(T,T,T)\}$

4-4 (a) (1, 4), (4, 1), (2, 3), (3, 2), (3, 3), (1, 5), (5, 1), (2, 4), (4, 2), (1, 6), (6, 1), (2, 5), (5, 2), (3, 4), (4, 3)
15/36
(b) (2, 6), (6, 2), (3, 5), (5, 3), (4, 4), (3, 6), (6, 3), (4, 5), (5, 4), (5, 5), (4, 6), (6, 4), (5, 6), (6, 5), (6, 6)
15/36
(c) (1, 1), (1, 2), (2, 1), (4, 6), (6, 4), (5, 5), (5, 6), (6, 5), (6, 6)
9/36 = 1/4
(d) (1, 1), (2, 2), (3, 3), (4, 4), (5, 5), (6, 6)
6/36 = 1/6
(e) (2, 1), (3, 1), (4, 1), (5, 1), (6, 1), (3, 2), (4, 2), (5, 2), (6, 2), (4, 3), (5, 3), (6, 3), (5, 4), (6, 4), (6, 5)
15/36
(f) (6, 3), (5, 4), (6, 4), (6, 5)
4/36 = 1/9

4-6 Sample space $= \{0, 1, 2, 3, 4, 5, 6, 7, 8, 9, 10\}$
(a) $\{3, 4, 5, 6, 7\}$ (b) $\{8, 9, 10\}$
(c) $\{0, 1, 2, 3, 4\}$
(d) $\{0, 1, 2, 3, 4, 9, 10\}$
(e) $\{0, 1, 2, 3, 4, 5, 6\}$

4-8 (a) and (b) are not collectively exhaustive, while events in (c) are.

4-10 (a) (1) $\{1, 3\}$ (2) $\{3, 4\}$ (3) $\{2, 3\}$ (4) $\{0, 1, 2, 3\}$ (5) $\{3\}$
(b) (1) $\{0, 2, 4\}$ (2) $\{0, 1, 2\}$ (3) $\{0, 1, 4\}$ (4) $\{4\}$ (5) $\{0, 1, 2, 4\}$

4-12 (a) .7 (b) .5 (c) .7 (d) .3 (e) .6

4-14 Only the Silver Ghost probabilities are correct.

4-16 (a) 1/13 (b) 1/26
(c) 1/2 (d) 3/13

4-18 (a) .7 (b) 0 (c) 0 (d) 1

4-20 (a) 1/2 (b) 1/2
(c) (1) 1/4 (2) 1/4

4-22 (a) 3/8 (b) 5/8

4-24 (a)

Quality of Shipment	Actions		Marginal Probability
	Reject	Accept	
Good	.055	.845	.900
Bad	.095	.005	.100
Marginal Probability	.150	.850	1.000

(b) .75 (c) not independent

4-26 (a) .729 (b) .001 (c) .243

4-28 (a) 720 (b) 364 (c) 40,320
(d) 138,600 (e) 66,045

4-30 (a) 6 (b) 3 (c) 6 (d) 1,944

4-32 C_{13}^{52}; P_{13}^{52}

4-34 (a) 20! (b) $2(10!)^2$
(c) $10!(2^{10})$

4-36 (a) (1) {K-H, Q-H, J-H, 10-H, 9-H, 8-H, 7-H, 6-H, 5-H, 4-H, 3-H, 2-H, A-H}
(2) {A-S, A-H, A-D, A-C}
(3) {3-S, 3-H, 3-D, 3-C, 2-S, 2-H, 2-D, 2-C, A-S, A-H, A-D, A-C}
(4) {8-S, 8-H, 8-D, 8-C, 9-S, 9-H, 9-D, 9-C, 10-S, 10-H, 10-D, 10-C}
(b) (1) {A-H}
(2) {8-S, 8-H, 8-D, 8-C, 9-S, 9-H, 9-D, 9-C, 10-S, 10-H, 10-D, 10-C, A-S, A-H, A-D, A-C}
(3) { }
(4) {A-S, A-H, A-D, A-C}
(5) {3-S, 3-H, 3-D, 3-C, 2-S, 2-H, 2-D, 2-C, A-S, A-H, A-D, A-C}

4-38 1,327,104,000

4-40 (a) neither (b) mutually exclusive
(c) both
(d) collectively exhaustive

4-42 (a) $P[R_1] = .6$ $P[R_2 | R_1] = .4$
$P[W_1] = .4$ $P[R_2 | W_1] = .7$
$P[W_2 | R_1] = .6$
$P[W_2 | W_1] = .3$
(b) $P[R_1 \text{ and } R_2] = .24$
$P[W_1 \text{ and } R_2] = .28$
$P[R_1 \text{ and } W_1] = .36$
$P[W_1 \text{ and } W_2] = .12$
(c) (1) .64 (2) .36

4-44 (a) .000005 (b) 5

4-46 (a) (1) 1/15 (2) 1/12 (3) 1/24
(4) 1/4 (5) 4/15 (6) 1/6
(7) 1/8 (8) 17/24
(9) 31/60 (10) 2/3
(b) (1) No (2) No
(3) *A* and *B*

4-48 .427

4-50 (a) .809 (b) .005

4-52 (a) .8 (b) .2

4-54 (a) .50 (b) 1/3 (c) 1/4

5-2

w	$P[W = w]$
−$1	1/2
+1	1/2
	1

5-4 (a)

w	$P[W = w]$
−$1	20/36
+1	14/36
+2	1/36
+3	1/36
	1

(b)

w	$P[W = w]$
−$2	20/36
+2	14/36
+4	1/36
+6	1/36
	1

5-6

w	$P[W = w]$
−$1	20/38
+1	18/38
	1

$E(W) = -\$2/38 = -\$0.053$

5-8 $E(X) = 2.1$ $\sigma^2(X) = 1.29$

5-10 (a) \$155.25 for High-Volatility Engineering
\$103.50 for Stability Power

5-12 (a) (1) .1641 (2) .2734
(3) .0078 (4) .2734
(b) They must be the same.

5-14 (a) .1390 (b) .1224 (c) .7822
(d) .0848 (e) .0009 (f) .3105

5-16 $E(P) = .10$

5-18 (a) .0000 (b) .5905 (c) .4095
(d) .0815 (e) .0815

5-20 0.0000 (rounded)

5-22 (a) .1347 (b) .1285 (c) .9672
(d) .3743 (e) .0942 (f) .9096

5-24 (a)

$\bar{x}$	$P[\bar{X} = \bar{x}]$
90	1/25
95	4/25
100	4/25
105	4/25
110	8/25
120	4/25
	1

(b) $E(\bar{X}) = 106$ $\sigma^2(\bar{X}) = 72$
$\sigma(\bar{X}) = 8.49$

5-26

x	$P[X = x]$
5	36/108
10	48/108
20	16/108
50	4/108
100	4/108
	1

5-28 $E(X) = 14.63$ $\sigma^2(X) = 360.90$
$\sigma(X) = 19.0$

5-30 (a) .04 (b) .10 (c) .03

5-32 (a)

$\bar{x}$	$P[\bar{X} = \bar{x}]$
19.5	1/10
20.0	1/10
20.5	1/10
21.0	1/10
21.5	2/10
22.0	2/10
22.5	1/10
23.5	1/10
	1

$\sigma^2(\bar{X}) = 1.29$
$\sigma_{\bar{X}} = 1.14$

(b) $\sigma = 1.85$ $\sigma_{\bar{X}} = 1.31$

6-2 (a) .4332 (b) .1915 (c) .2420
(d) .0062 (e) .0968 (f) .9861
(g) .97585 (h) .0606

6-4 (a) 5′4.175″ (b) 5′5.9″
(c) 5′8.7″ (d) 5′10.625″
(e) 6′2.1″ (f) 6′3.825″

6-6 (a) .0456 (b) .0228 (c) .50003

6-8 (a) .01 (b) .10 (c) .50
(d) 2.00 (e) 10.00

6-10 (a) 4 (b) .4938 (c) .99865
(d) .0401 (e) .0994 (f) .3085

6-12 .9544

6-14 (a) .9876 (b) .7888
(c) Greater variability reduces reliability.
(d) 1 (approximately)
(e) Larger sample sizes increase reliability.

6-16 (a) .3164 (b) 1.3791 (c) .1000
(d) .3000 (e) .4359

6-18 (1) .5000 (2) .9429
(3) 1 (approximately)
Choose rule (3).

6-20 .8904 for $n = 64$; .9544 for $n = 100$; 1 (approximately) for $n = 400$. The probabilities increase with n because of greater reliability.

6-22 (a) .0116 for $N = 500$;
.0150 for $N = 1{,}000$
(b) .0681 for $N = 500$;
.0708 for $N = 1{,}000$
(c) No

6-24 (a) .0228 (b) .9544
(c) .99865 (d) .1587

6-26 (a) .6828 (b) .9876

6-28 .00135

6-30 (a) .0207 (b) .0036

6-32 (a) .4525 (b) .0475
(c) 0 (approximately)
(d) 0 (approximately) (e) .9734

6-34 (a) .0228 (b) .0062

7-2 (a) interval (b) either
(c) point (d) interval

7-4 (a) 5 (b) 1.94 (c) No

7-6 (a) 100.53 ± 4.96 (b) $69.2 \pm .15$
(c) \$12.00 ± .73

7-8 105.6 ± 2.88

7-10 (a) 52,346 ± 1,058 (c) 77.38%

7-12 (a) .25 ± .050 (b) .1 ± .0244
(c) .5 ± .074

7-14 (a) .2 ± .024 (b) .4 ± .017
(c) .01 ± .0087

7-16 (a) .37 ± .21 (b) $1,200
(c) .3821

7-18 (a) $8.00 ± 2.62 (b) 15.03 ± .29
(c) 27.30 ± 1.06

7-20 31.2 ± 2.23

7-22 90 ± 18.0

7-24 $73,249 ± 14,817

7-26 (a) 661 (b) 385
(c) 166; one-fourth as large
(d) 2,642; increases 4 times

7-28 (a) 370 for $N = 10{,}000$;
357 for $N = 5{,}000$
(b) 0 (approximately)

7-30 (a) 385 (b) 350; too many by 35

7-32 1,246 ± 22.5

7-34 (a) 8,068 (b) 9,220
(c) 9,604 (d) 3,458

7-36 100 ± 12.7

7-38 (a) 1,537 (b) 385
(c) 97 (d) 16

8-2 $\alpha = .1788$; $\beta = .1587$

8-4 (a) (1) correct (2) Type II error
(3) correct (4) Type I error
(b) (1) Type II error (2) correct
(3) Type I error (4) correct
(c) (1) Type I error (2) correct
(3) correct (4) Type II error

8-6 (a) H_0 : unsuccessful; H_1 : successful
(b) H_0 : successful; H_1 : unsuccessful

8-8 (a) $\alpha = .1379$; $\beta = .1539$
(b) α will decrease to .0968; β increased to .2061

8-10 (a) lower-tailed (b) $\bar{X}^* = 28.38$
Accept H_0 if $\bar{X} \geqslant 28.36$
Reject H_0 if $\bar{X} < 28.36$
(c) accepted
(d) $\bar{X}^* = 28.53$; accepted

8-12 (a) upper-tailed; $z_{.01} = 2.33$
(b) $z = 6.25$; rejected

8-14 (a) $z_{.01} = -2.33$; $\bar{X}^* = 512.9$
(b) accepted

8-16 (a) $\bar{X}^* = 103.28$
Accept H_0 if $\bar{X} \leqslant 103.28$
Reject H_0 if $\bar{X} > 103.28$
(b) $\bar{X}^* = 102.62$
(c) $\beta = .1949$ for $n = 64$;
$\beta = .0681$ for $n = 100$
Use $n = 100$.

8-18 (a) $\bar{X}_1^* = -1.96$; $\bar{X}_2^* = 1.96$
Accept H_0 if $-1.96 \leqslant \bar{X} \leqslant 1.96$
Reject H_0 either if $\bar{X} < -1.96$ or if $\bar{X} > 1.96$
(b) .8300 (c) accepted

8-20 *Reject* H_0 and conclude that marijuana *lengthens* dream time.

8-22 (a) $t = -3$; lower-tailed;
$t_{.01} = -2.896$; rejected
(b) $t = 1.667$; upper-tailed;
$t_{.05} = 1.711$; accepted
(c) $t = .333$; two-sided; $t_{.05} = 1.753$; accepted
(d) $t = -7.0$; two-sided;
$t_{.005} = 2.797$; rejected

8-24 Yes

8-26 (a) .0475 (b) .1335
(c) .0668 (d) .0548

8-28 (a) $H_0 : \mu = .06$; $H_1 \mu \neq .06$; two-sided
(b) rejecting (c) accepting
(d) $\bar{X}_1^* = .0594$; $\bar{X}_2^* = .0606$
(e) (1)$\beta = .0918$ (2)$\beta = .83995$

8-30 (a) $H_0 : \mu \geqslant 2$; $H_1 : \mu < 2$; lower-tailed
(b) Use t (c) *Reject* H_0; terminate

8-32 Kept

9-2 Answers may vary; $Y = 20$; $b = .2727$;
$\bar{Y}_X = 20 + .2727X$

9-4 (a) GPA dependent, IQ independent
(b) IQ dependent, GPA dependent
(c) Same as (a)

9-6 (a) $\bar{Y}_X = 10.0 + 1.0X$
(b) $\bar{Y}_X = 30.0 - 1.0X$
(c) $\bar{Y}_X = -15.0 + .5X$

9-8 (a) $\bar{Y}_X = -3.197 + .3218X$
(c) $s_{Y \cdot X} = .52$; $s_Y = 1.53$; Yes

9-10 $\mu_{Y \cdot 150} = 1{,}250 \pm 1.96$;
$Y_I = 1{,}250 \pm 19.6$

9-12 (a) $2.5 \pm .01$ (b) $2.5 \pm .196$

9-14 (a) 27 ± 7.25 (b) rejected
(c) rejected

9-16 (a) .987 (b) $-.80$
(c) .975 (d) $-.90$

9-18 (a) $\bar{Y}_X = 276.55 + .4121X$
(b) positive (d) .78

9-20 (a) 78% (b) .88 (c) .88

9-22 (a) $r^2 = .4$; $r = .6325$
(b) $r^2 = .5$; $r = -.7071$

9-24 (a) $\bar{Y}_X = -.40 + .1X$

9-26 (a) $12{,}500 \pm 1{,}209.3$
(b) $10{,}000 \pm 1{,}263.0$
(c) $15{,}000 \pm 1{,}263.0$

9-28 (a) $80,000 ± 78.40
(b) $80,000 ± 784

10-2 (a) 3.189 (b) 3.254
(c) 2.975 (d) 3.708

10-4 (a) (1) 21.847 (2) 26.741
(3) 21.294 (4) 126.612
(b) 4.808
(c) (1) $21.847 \pm .943$
(2) $26.741 \pm .943$
(3) $21.294 \pm .943$
(4) $126.612 \pm .943$
(d) (1) 21.847 ± 9.47
(2) 26.741 ± 9.47
(3) 21.294 ± 9.47
(4) 126.612 ± 9.47

10-6 (a) $b = .6333$; $a = 1.667$;
$s_{Y \cdot X_1} = 701.2$
(b) $a = -59$; $b_1 = .6765$; $b_2 = 323.5$
$S_{Y \cdot 12} = 434.8$
(c) Yes

10-8 (a) $S_{Y \cdot 12} = 1.759$; $R^2_{Y \cdot 12} = .786$;
78.6% (b) 96.4% (c) Yes
(d) .214 for A; .036 for B; a proportional reduction of .832

11-2 $.19 \pm .412$

11-4 $2 \pm .392$

11-6 *Reject* H_0 and use ladybugs.

11-8 Conclude that there is a greater index for the older siblings.

11-10 *Reject* H_0 and conclude that radial tires are better.

11-12 He should conclude that the candidate has equal strength in both areas.

11-14 1.5 ± 1.41

11-16 (a) $2{,}000 \pm 1{,}430.6$ (b) rejected

11-18 accepted

11-20 accepted

11-22 (a) Conclude that batting averages are higher for home games.
(b) $.028 \pm .023$

11-24 (a) $\bar{d} = 5.4$; $s_{d\text{-paired}} = 6.096$
(b) rejected

11-26 (a) (1) $H_0 : \mu_A = \mu_B$
(2) $d_1^* = -3.30$; $d_2^* = 3.30$
(3) $d = 3.60$; rejected
(b) (1) $H_0 : \mu_A \geqslant \mu_B$
(2) $d^* = -2.76$
(3) $d = 3.60$; accepted
(c) (1) $H_0 : \mu_A \leqslant \mu_B$
(2) $d^* = 2.76$
(3) $d = 3.60$; rejected

11-28 rejected

12-2 (a) accepted (b) rejected
(c) rejected (d) accepted

12-4 (a)

	(1) *Single*	(2) *Married*	Total
(1) *A*	12	18	30
(2) *B*	28	42	70
Total	40	60	100

(b) 12.699
(c) 1; $\chi^2_{.01} = 6.635$
(d) rejected

12-6 (a)

	(1) *Male*	(2) *Female*	Total
(1)	27.65	22.35	50
(2)	24.34	19.66	44
(3)	32.08	25.92	58
(4)	18.81	15.19	34
(5)	22.12	17.88	40
Total	125	101	226

(b) 10.683
(c) 4
(d) 13.277; accepted
(e) 9.488; rejected

12-8 Conclude that quality is affected by temperature.

12-10 (a) $13.38 \leq \sigma^2 \leq 35.18$
(b) $56.80 \leq \sigma^2 \leq 244.28$
(c) $.313 \leq \sigma^2 \leq 1.129$
(d) $4.90 \leq \sigma^2 \leq 14.61$

12-12 (a) $2.23 \leq \sigma^2 \leq 5.86$
(b) $1.49 \leq \sigma \leq 2.42$
(c) Yes

12-14 (a) $56.20 \leq \sigma^2 \leq 83.68$
(b) *Accept* H_0

12-16 Yes

12-18 (a) $2{,}929{,}800 \leq \sigma^2 \leq 6{,}302{,}000$
(b) No

12-20 Yes

13-2 (a) 6.93; accepted
(b) 3.88; rejected
(c) 3.88; rejected

13-4 (a) $\bar{X}_1 = 87.8$; $\bar{X}_2 = 64.2$; $\bar{X}_3 = 76.0$; $\bar{\bar{X}} = 76.0$
(b) $SST = 1{,}392.4$; $SSE = 2{,}001.6$
(c) $MST = 696.2$; $MSE = 166.8$; $F = 4.17$
(d) rejected

13-6 (a) No (b) Answers may vary.

13-8 (a) 87.8 ± 12.59
(b) 23.6 ± 17.80; yes

13-10 (a) $SSE = 72$; $MST = 21.0$; $MSE = 6.0$; $F = 3.5$
(b) accepted

13-12 (a) $MST = 200{,}000$; $SSE = 1{,}200{,}000$; $MSE = 80{,}000$; $F = 2.5$
No
(b) $MST = 300{,}000$; $SSE = 1{,}100{,}000$; $MSE = 68{,}750$; $F = 4.36$
No
(c) $MSA = 200{,}000$; $MSB = 300{,}000$; $SSE = 300{,}000$; $MSE = 25{,}000$; $F = 8.0$ for A; $F = 12.0$ for B
Reject both.
(d) Two-factor analysis is more discriminating.

13-14 (a) Yes (b) No (c) No
(d) Yes (e) Yes

13-16 (a) $SSE = 75$; $MST = 39.25$; $MSCOL = 19.25$; $MSROW = 35.75$; $F = 6.28$ for X; $F = 3.08$ for Y; $F = 5.72$ for Z
(b) The population means are significantly different only for X and Z.

13-18 (a) rejected (b) rejected
(c) accepted (d) accepted

13-20 No

13-22 (a) $SSE = 40$; $MSA = 15.60$; $MSB = 16.25$; $MSE = 2.00$; $F = 7.80$ for A; $F = 8.13$ for B
(b) *Reject* the null hypotheses for both A and B.

13-24 (a) $\bar{X}_{1\cdot} = 326.0$; $\bar{X}_{2\cdot} = 381.0$; $\bar{X}_{3\cdot} = 432.5$; $\bar{X}_{4\cdot} = 536.5$; $\bar{X}_{\cdot 1} = 385.50$; $\bar{X}_{\cdot 2} = 387.25$; $\bar{X}_{\cdot 3} = 440.25$; $\bar{X}_{\cdot 4} = 463.00$; $\bar{X}_A = 473.50$; $\bar{X}_B = 437.25$; $\bar{X}_C = 396.25$; $\bar{X}_D = 369.00$; $\bar{\bar{X}} = 419$; $SST = 25{,}271.5$; $SSCOL = 18{,}071.5$; $SSROW = 96{,}326$; Total $SS = 139{,}776$; $SSE = 107$
(b) $MST = 8{,}423.83$; $MSE = 17.83$; $F = 472.45$
(c) Yes

14-2 Yes

14-4 (a) lower-tailed
(b) $W = 25$
(c) accepted

14-6 (a) (1) lower-tailed
(2) 6 (4) rejected
(b) (1) upper-tailed
(2) 62 (4) accepted
(c) (1) lower-tailed
(2) 19 (4) rejected
(d) (1) upper-tailed
(2) 13 (4) accepted

14-8 leaded

14-10 (b) .16; upper-tailed (c) 2.68
(d) *Reject* H_0 and conclude that the new supplement is more effective.

14-12 (a) (1) upper-tailed
(2) $z_{.05} = 1.64$
(3) $z = -1.98$; rejected
(b) (1) lower-tailed
(2) $z_{.01} = -2.33$
(3) $z = -1.58$; accepted
(c) (1) upper-tailed
(2) $z_{.01} = 2.33$
(3) $z = 2.44$; rejected
(d) (1) lower-tailed
(2) $z_{.05} = -1.64$
(3) $z = -2.05$; rejected

14-14 $V = 5.5$; rejected

14-16 (b) (1) $R_a = 7$; accepted
(2) $R_a = 8$; rejected
(3) $R_a = 2$; rejected
(4) $R_a = 8$; rejected

14-18 (a) 95 (b) .68 (c) .4964

14-20 *Accept* H_0 and conclude the appearance of randomness.

14-22 .61

14-24 $-.50$

14-26 $K = 5.18$; *accept* H_0 and conclude that identical treatment means exist.

14-28 accepted

14-30 rejected

14-32 .88

Index

Index

A

D

E

F

G

H

I

J

K

L

M

N

O

P

Q

R

S

T

U

V

W

Y

5 6 7 8 9 0 1 2 3 4

Symbol	Description
$\pi_1, \pi_2, \pi_3, \ldots$	Values of population proportions in testing for equality of several population proportions p.448
π_A, π_B	Proportions of populations *A* and *B* which are compared in two-sample hypothesis tests p.420
R	(1) A random variable denoting the number of successes obtained from several trials of a Bernoulli process p.171 (2) The test statistic for the matched-pairs sign test for comparing two populations; *R* is the number of positive sign differences for the paired difference between two sample observations p.513
R_a	Number of runs of type *a* obtained in a sample; used as the test statistic for the number-of-runs test p.524
$R_{Y\cdot 12}$, $R_{Y\cdot 123}$	Sample multiple correlation coefficient; the square root of the coefficient of multiple determination p.397
$R^2_{Y\cdot 12}$, $R^2_{Y\cdot 123}$	Sample coefficient of multiple determination; represents the proportion of variation in the dependent variable *Y* that can be explained by the multiple regression equation of *Y* on independent variables X_1, X_2 (and X_3) p.397 (10-11)
r	(1) One of several possible values of the random variable *R* p.171 (2) The sample correlation coefficient; expresses the strength and direction of the relationship between variables *X* and *Y* p.369 (9-28) (3) Number of rows in an analysis-of-variance layout p.465
r^2	Sample coefficient of determination; represents the proportion of the variation in the dependent variable *Y* that can be explained by linear regression on the independent variable *X* p.365 (9-23)
r_i	The marginal total in the *i*th row of a contingency table p.440
r_s	Spearman rank correlation coefficient p.532 (14-7)
r^2_{Y1}	Sample coefficient of determination for regression of the dependent variable *Y* on the independent variable X_1; used in multiple correlation analysis p.387 (9-23)
$r^2_{Y\cdot 12}$, $r^2_{Y3\cdot 12}$	Coefficients of partial determination for multiple regression of the dependent variable *Y* on the independent variables X_1, X_2 (and X_3); measures the proportional reduction in previously unexplained variation in *Y* by adding X_2 or X_3 to the multiple regression analysis while incorporating the effects of the other independent variables p.398 (10-12)
ρ (rho)	Population correlation coefficient; estimated by *r* p.369
ρ^2	Population coefficient of determination; estimated by r^2 p.366 (9-22)

Symbol	Description
$S_{Y\cdot 12}$, $S_{Y\cdot 123}$	Standard error of the estimate ab the multiple regression plane p.388 (10-5)
SSA	Sum of squares (between colum for a two-factor analysis of varia *A* represents one of the factors ab which inferences are to be m p.485
SSB	Sum of squares (between rows) f two-factor analysis of variance wh (1) *B* represents the blocking v able p.481 (2) *B* represents one of the fac about which inferences are to made p.485
$SSCOL$	Sum of squares (between colum for a latin-square design in a th factor analysis of variance p.491 (13-22)
SSE	Error sum of squares: (1) (within columns) for a one-fa analysis of variance p.466 (13-4) (2) (residual) for a two-factor three-factor analysis of varia p.492 (13-24)
$SSROW$	Sum of squares (between rows) f latin-square design in a three-fa analysis of variance p.491 (13
SST	Sum of squares: (1) (between columns) for the tr ments in a one-factor analysi variance p.466 (13-3) (2) (between letters) in a three-fa analysis of variance using a la square design p.491 (13-2
s	Sample standard deviation p.55 (2-8)
s^2	Sample variance p.54 (2-8)
s^2_A, s^2_B	Variances of samples from pop tions *A* and *B*; used in two-sam hypothesis tests p.405
s_X	Sample standard deviation of independent variable *X*; used in relation analysis p.372
s_Y	Sample standard deviation of dependent variable *Y*; used in relation analysis p.347 (9-13)
$s_{Y\cdot X}$	Standard error of the estimate a the regression line; $s^2_{Y\cdot X}$ is an unbi estimator of $\sigma^2_{Y\cdot X}$ p.347 (9-12
s_d	The estimator of the standard e σ_d for the difference between: (1) two sample means p.405 (1 (2) two sample proportions p (11-22)
$s_{d\text{-paired}}$	Sample standard deviation matched-pairs differences p.409 (11-9)
$s_{d\text{-small}}$	Sample standard deviation of difference in means of indepen samples p.424 (11-27)
Σ	Summation sign p.40
σ (sigma)	Population standard deviation
σ^2	Population variance p.54 (2-5
$\sigma(X)$	Standard deviation of random able *X* p.165 (5-6)
$\sigma^2(X)$	Variance of random variable p.163 (5-4)
σ^2_0	Value of the population vari assumed under the null hypot p.454